SHALLOW-WATER ACOUSTICS

SHALLOW-WATER ACOUSTICS

Proceedings of the Second International Shallow-Water Acoustics Conference (SWAC'09)

Shanghai, China 16-20 September 2009

EDITORS

Jeffrey Simmen
University of Washington
Seattle, Washington, USA

Ellen S. Livingston
Office of Naval Research
Arlington, Virginia, USA

Ji-Xun Zhou
Georgia Institute of Technology
Atlanta, Georgia, USA

Feng-Hua Li
Institute of Acoustics, CAS
Beijing, China

SPONSORING ORGANIZATIONS
Office of Naval Research (USA)
Chinese Academy of Sciences
Acoustical Society of China
Acoustical Society of America
National Natural Science Foundation of China

American Institute of Physics

Melville, New York, 2010
AIP CONFERENCE PROCEEDINGS ■ 1272

Editors

Jeffrey Simmen
Applied Physics Laboratory
University of Washington
Seattle, WA 98105
E-mail: simmen@apl.washington.edu

Ji-Xun Zhou
School of Mechanical Engineering
Georgia Institute of Technology
Atlanta, GA 30332-0405
E-mail: jixun.zhou@me.gatech.edu

Ellen S. Livingston
Ocean Acoustics Program
Office of Naval Research
Arlington, VA 22203
E-mail: ellen.livingston@navy.mil

Feng-Hua Li
State Key Laboratory of Acoustics
Institute of Acoustics, CAS
Chinese Academy of Sciences
Beijing, China 100190
E-mail: lfh@mail.ioa.ac.cn

Cover image: Dark and light color bands showing impressions of a propagating internal wave packet on the sea surface, captured in August 2006 on the New Jersey continental shelf during a shallow-water acoustics experiment. Photo taken by Mohsen Badiey aboard R/V Sharp.

L.C. Catalog Card No. 2010910324

ISBN 978-0-7354-0822-7

ISSN 0094-243X
Printed in the United States of America

CONTENTS

REVERBERATION AND SCATTERING

SIGNAL PROCESSING, ENGINEERING AND APPLICATIONS

PREFACE

For an eight-year period starting in 1995, the Office of Naval Research and the Chinese Academy of Sciences developed and sustained a program of cooperation and collaboration in ocean acoustics between the United States and China, focused primarily on fundamental research in shallow-water acoustics. At that time the United States had a growing emphasis in shallow-water acoustics, a transition that began following the end of the Cold War, and China had four decades of experience and continued interest in this research field because of their large continental shelf. Shallow-water acoustics is an area of ocean acoustics that is challenging and ripe with many research opportunities, owing to the range of complex properties resulting from boundary and volume interaction in highly dynamic and inhomogeneous environments. This cooperative program involved many researchers and institutions from both countries and consisted of many interactions in the US and in China, as well as in other Pacific Rim countries. These interactions evolved into coherent initiatives that produced notable scientific achievements, in the form of unique multinational field experiments, analyses of one-of-a-kind datasets, novel theory and model developments, and numerous scientific journal publications.

More importantly, this period of cooperation was a period for building not only scientific ties, but also cultural and personal bonds, between individuals of these countries. It provided an opportunity to get to know and appreciate each others foods, history, traditions, and ways of life. Many of us have built long-lasting friendships and have many wonderful memories from these interactions that in their own small way contribute to a more stable and friendly world.

For a period of about six years following this cooperative era, the research ties in ocean acoustics between the United States and China grew into smaller, more independent efforts. International economic and political landscapes changed, making visits, interactions, and joint field experimentation more difficult. However, science is endowed with a perspective and approach that has no political or cultural boundaries, and sharing ideas and tools is the best way to achieve the common goal of discovery and understanding. And so, our mission with the Second International Conference on Shallow-Water Acoustics in Shanghai was to help invigorate the cooperation in ocean-acoustics research between the US and China that was once so energetic. Just prior to this conference, we had already taken some steps to begin this revival: We held a small workshop in Seattle in January 2009 to discuss research areas of common interest and to explore experiments that might form the basis of a future US-China collaboration. And several months later we had a small joint acoustics experiment that took place on Lake Washington, near Seattle, involving both Chinese and American acousticians.

The conference in Shanghai, we hoped, would be a strong catalyst for this renewed cooperation, and our primary goal for this meeting, in addition to exchanging scientific information, was to seed new friendships and ties between participating scientists from the different countries and institutes. In particular, we hoped that the next generation of leaders in the field of ocean acoustics would begin to build connections following their introductions at the meeting. To help encourage these new ties, many of the senior and key individuals responsible for building the past program of US-China ocean-acoustics collaboration attended the conference, including Art Baggeroer, Peter Dahl, Ellen Livingston, Jim Lynch, Yuanliang Ma, Jeff Simmen, Bob Spindel, Renhe Zhang, and Jixun Zhou. Unfortunately, several other individuals critical to the success of the past cooperation were absent from the meeting, including Warren Denner and Dinghua Guan, who are no longer with us, and Ching-Sang Chiu, Xianyi Gong, Bill Kuperman, Qihu Li, Steve Ramp, and Shi'e Yang, who were unable to attend for other reasons.

We chose this conference venue to contrast strongly with that of the First International Conference on Shallow-Water Acoustics, held in 1997 in Beijing, and to parallel how the new China contrasts with China then. Beijing was the appropriate conference site in 1997 because it is the center government, the capital of China; Shanghai, the center of business, was appropriate in 2009, because of the dominance of economics to developing China. China had seen enormous change in the twelve-year period since the first conference in Beijing in 1997. Then, Beijing had relatively few cars and many bicycles on the roads, and the styles of dress and architecture were uniform and traditional. Chinese scientific facilities were inadequately supported and were lacking of young scientists. Since then, in only a dozen years, China has prospered and changed dramatically. Today, the cities of China are filled with all makes of cars, with high-rise hotels, fancy shopping malls, and scientific facilities, funding, and research community are significantly improved. Our 2009 meeting site in the modern Jinmao Tower, located in Shanghai's expanding business district of Pudong, and until 2008 having the world's highest hotel, was symbolic of this tremendous change in China. It provided strong contrast against the venue in 1997 of the old-style and historic Grand Hotel in Beijing.

At the Second International Conference on Shallow-Water Acoustics we heard close to seventy scientific presentations, focused on shallow-water acoustics but spread across different topical areas, including propagation, inversion, signal processing, reverberation & target scattering, and miscellaneous topics. Each of the eight sessions had two chairpersons, one from China and one from another country, symbolizing linkage to China. The scientific presentations were preceded by opening remarks from Jeffrey Simmen, Jing Tian, Ellen Livingston, and Renhe Zhang. This conference volume is a collection of most of the scientific papers that were presented at this Shanghai meeting, and it covers many of the important areas of research in the field of shallow-water acoustics today.

We would like to express our gratitude to the Chinese acousticians and colleagues of the Chinese Academy of Sciences who in recent years patiently supported and encouraged this conference and volume, including co-chair Jing Tian, Renhe Zhang, Shi'e Yang, Yuanliang Ma, Xiaoming Wang, Fenghua Li, Lianghao Guo and Qi Li. We'd also like to express appreciation to the Chinese Academy of Sciences Institute of Acoustics and State Key Laboratory of Acoustics for assisting in the organization, to Jian Xu, Jinfeng Mao, Ying Li, Jie Yang and Fengxiang Ge for providing logistical support, to Eric Thorsos for reviewing and editing abstracts for the meeting, and to Jeffrey Donnell and James Martin for editing the English in some of the Chinese papers in this proceedings. And our gratitude goes to Frank Herr and the Office of Naval Research and Charles Schmid and the Acoustical Society of America for providing financial and moral support for this meeting and volume. Finally, and most importantly, we'd like to thank all of the conference attendees, who had to make significant adjustments and arrangements to come to the meeting and present papers at the Second International Conference on Shallow-Water Acoustics.

Jeffrey Simmen, Applied Physics Laboratory, University of Washington
Ellen S. Livingston, Ocean Acoustics Program, Office of Naval Research
Ji-Xun Zhou, School of Mechanical Engineering, Georgia Institute of Technology
Feng-Hua Li, State Key Laboratory of Acoustics, Institute of Acoustics, Chinese Academy of Sciences

Fifteen Years of U.S.–China Cooperative Research in Underwater Acoustics: 1995–2010

Ellen Livingston[a] and Jeffrey Simmen[b]

[a] *Ocean Acoustics Program, Office of Naval Research, Arlington, VA 22203, USA*
[b] *Applied Physics Laboratory, University of Washington, Seattle, WA 98105, USA*

Abstract. The Office of Naval Research and the Chinese Academy of Sciences established and continue to encourage an unprecedented collaboration between the United States and China in the basic research field of ocean acoustics. This paper describes the progression of events in the scientific relationship. This cooperative research effort, primarily focused on shallow-water acoustics, has varied over fifteen years. It has involved numerous researchers and institutions from the U.S., China, and other countries, and it has resulted in productive workshops, conferences, field experiments, research collaborations, and scientific publications.

Keywords: Shallow Water, Acoustics, Propagation, Scattering, China
PACS: 43

Starting as early as 1995, the shallow-water acoustics research communities of the United States and China have worked together to build a cooperative international relationship for scientific exchange. The Second International Conference on Shallow-Water Acoustics in September 2009 is another milestone in a fifteen-year period of productive work, extensive travel, and continued successful exchange of ideas between these two communities.

Shallow-water acoustics is a complex research area. Shallow-water propagation involves the interaction of acoustic energy with a rough and bubbly air–sea interface, with complicated temperature and salinity structures (and biologics) in the water column, and with an irregular and inhomogeneous seabed. Predictive models require an understanding of all of these effects on the acoustic field, and models must often be stochastic in nature due to the non-deterministic spatial and temporal variability of the medium. Full understanding of the influence of the ocean environment on acoustic propagation and scattering requires an interdisciplinary approach encompassing acoustics, geology and geophysics, oceanography, air-sea interaction, and marine biology, as well as the underlying physics and mathematics of the integrated theory, and ocean engineering to ensure successful data collection in this observational science. The U.S. and China have successfully worked together to investigate many of the complex phenomena in this demanding research area.

YELLOW SEA EXPERIMENT, THE FIRST U.S.–CHINA FIELD COLLABORATION IN ACOUSTICS

The Yellow Sea Experiment was a pilot study designed primarily to study propagation loss due to shallow-water internal waves and solitons in an area with a flat, homogeneous seabed and a well-developed thermocline. The experiment was preceded by, and essentially an outcome of, a workshop in Monterey, California in

CP1272, *Shallow-Water Acoustics, Second International Shallow-Water Acoustics Conference*
edited by J. Simmen, E. S. Livingston, J. X. Zhou, and F. H. Li

1995. This meeting established a dialog between about twenty Chinese and U.S. acousticians who discussed their research interests, scientific approaches, available resources, and possible experiments. The Monterey meeting resulted in the development of an experiment plan and proposals of contributed equipment and funding to carry out the first joint U.S.-China experiment in the Yellow Sea.

The Yellow Sea Experiment took place in the summer of 1996. The participating scientists came from the National (State Key) Acoustics Laboratory in Beijing, Institute of Acoustics of the Chinese Academy of Sciences in Beijing, South China Sea Institute of Oceanology in Guangzhou, Ocean University in Qingdao, Applied Physics Laboratory at the University of Washington in Seattle, and Georgia Institute of Technology in Atlanta. They simultaneously measured internal waves and acoustic arrivals on vertical hydrophone arrays to study the hypothesized modal effects of internal waves on acoustic transmission loss. Data were also collected to address research topics in reverberation, geoacoustic inversion, and ambient noise. The Chinese research vessels *Shi-Yan 2* and the *Shi-Yan 3* were used to deploy sources and receivers and make complementary oceanographic measurements (Fig. 1). Guangzhao Zhou, then President of the Chinese Academy of Sciences, visited the scientific team on board R/V *Shi-Yan 3* at the end of the experiment.

FIGURE 1. Retrieval of instrumentation during the joint U.S.–China experiment in the Yellow Sea in August 1996. R/V *Shi-Yan 3* is in the background.

FIGURE 2. Poster for the First International Conference on Shallow-Water Acoustics, held in Beijing in April 1997.

THE FIRST INTERNATIONAL SHALLOW-WATER ACOUSTICS CONFERENCE IN CHINA

The first international Shallow-Water Acoustics Conference (SWAC) took place in April 1997 in Beijing, China (Fig. 2). The organizers from the U.S. and China envisioned a conference that would further the development of a cooperative shallow-water research community between their countries but also encourage participation from additional international investigators on both sides of the Pacific. The result was a week-long meeting with over 120 international participants from 17 countries. In addition to a full schedule of presentations, the conference included ample opportunities for detailed technical discussions between smaller groups, as well as social and personal interactions. The first results from the Yellow Sea Experiment '96 were presented here by the U.S. and Chinese experiment participants.

Topics for the technical sessions included theoretical and computational acoustics, sound propagation and seabed properties, reverberation and bottom scattering, effects of water column variability, near-surface acoustic behavior, signal processing, ambient noise, high frequency acoustics in shallow water, and tomography and inverse techniques. SWAC'97 Proccedings[1] were published within the year. These proceedings include introductory remarks and over 100 technical articles from the presentations given at the conference. All of the conference session topics were well represented, and the proceedings volume includes a list of conference organizers, sponsors, and participants.

POST-CONFERENCE VISITS AND WORKSHOPS

Following the SWAC'97 conference, groups of U.S. government officials and scientists visited and engaged research institutions and universities around China. Oceanographic research institutions visited were the Institute of Oceanology in Qingdao, South China Sea Institute of Oceanology in Guangzhou, Ocean University of Qingdao, and First Institute of Oceanography in Qingdao (Fig. 3). Acoustic research institutions visited were the Institute of Acoustics in Beijing, National (State Key) Laboratory of Acoustics in Shanghai, Hangzhou Applied Acoustics Research Institute, Harbin Engineering University, Jiaotong University in Xi'an, Northwest Polytechnical Institute in Xi'an, and Dalian Scientific Test and Control Technology Institute. At these institutions, the U.S. visiting scientists gave lectures about their experimental work and current research interests. The U.S. guests met Chinese research investigators, faculty and students, and were shown the institutional research facilities, including acoustic laboratories and tanks.

U.S.–China scientific collaboration was further developed and expanded through one-on-one and small group interactions and at international conferences. Individuals and small groups of Chinese scientists made occasional visits to the United States. China jointly hosted several meetings and workshops that involved American acousticians. These included the Sino–U.S. Marine Science and Technology workshop in Qingdao, 1998, the High-Frequency Acoustics Workshop in Hangzhou, 1999, (Fig.

4) and the International Workshop on Acoustical Engineering and Technology in Harbin in 2000.

After the successful conclusion of SWAC'97, planning workshops and small meetings were organized to investigate the possibility of conducting a major U.S.–China shallow-water acoustics experiment. From 1997 to 2000, a total of seven workshops were held to establish experiment objectives, to decide on participating institutions, to establish experiment resources, and to design a data collection plan. These workshops were held on the mainland U.S., in Alaska (Fig. 5) and Hawaii, in China, and in Singapore and had multi-national participation. There was eventual agreement on experiment objectives, on having dual experiment sites in the East China Sea (ECS) and the South China Sea (SCS), on measurement and data collection plans, and on committed equipment and research vessels for acoustic, oceanographic, seafloor, and sea surface measurements.

FIGURE 3. Visit of American scientists with Chinese scientists in Qingdao, China in April 1997.

FIGURE 4. Some American and Chinese attendees at the High-Frequency Acoustics Workshop at the Hangzhou Applied Acoustics Research Institute in January 1999.

FIGURE 5. Participants in an ASIAEX experiment planning workshop held in Anchorage, Alaska in July 1999.

THE ASIAN SEAS INTERNATIONAL ACOUSTICS EXPERIMENT (ASIAEX)

ASIAEX had two components, an experiment in the East China Sea (ECS) and an experiment in the South China Sea (SCS; Fig. 6). The objectives and participants differed for each site, but the total experiment addressed the most important problems in low-frequency shallow-water acoustics research that had been identified at SWAC'97. In 2000, in pilot experiment cruises involving the U.S. R/V *Revelle* and the Taiwanese R/V *Ocean Researcher I*, extensive seafloor and oceanographic surveys were conducted at ECS and SCS ASIAEX sites to understand the ocean environment around them for planning the experiment and for post-experiment analysis. The full shallow-water acoustic ASIAEX experiments in the ECS and the SCS began in 2001.

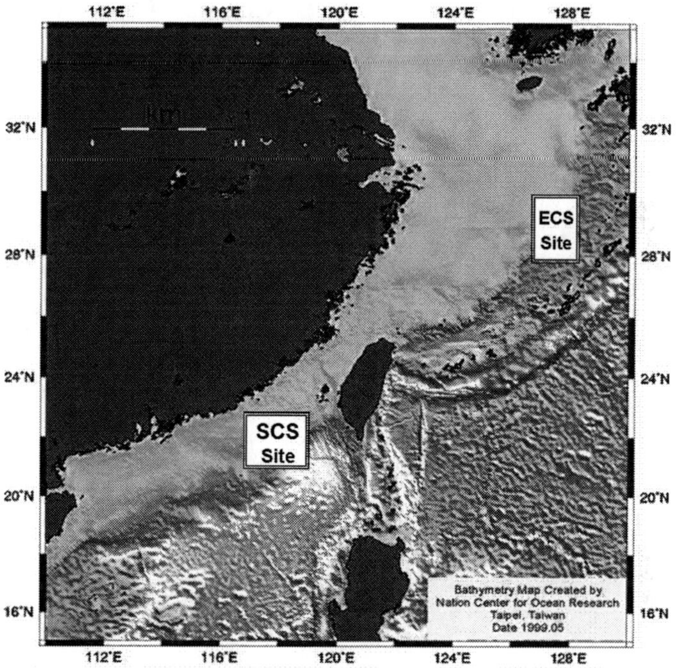

FIGURE 6. The location of the East China Sea (ECS) and South China Sea (SCS) sites for the 2001 experiments of ASIAEX.

The ECS component of ASIAEX focused on boundary interactions and geoacoustic inversion. That is, the experiment was designed to investigate the effects of surface roughness, bottom roughness and sediment inhomogeneity on shallow-water acoustic propagation and reverberation. In addition, acoustic data were collected for several complementary acoustic inversion methods that were used to establish a geoacoustic description of the ECS seabed. The site was chosen on the continental shelf with gentle bathymetric variability and a weak thermocline. The bottom was measured for sediment thickness, subbottom layer thickness, and sediment roughness, and cores were taken for sediment compressional speed and other sediment parameters. The sea surface root-mean-square wave height and wind speed were measured, and frequent CTD casts measured water column variability. The acoustic sources and receivers included towed and moored sources, explosive sources, and several types of vertical line receiving arrays.

Participating scientists came from the U.S., China, Korea, and Taiwan. Participating research institutions in the U.S. included the Applied Physics Laboratory of the University of Washington, Georgia Institute of Technology, Marine Physical Laboratory of Scripps Institution of Oceanography, and University of Rhode Island. Research institutions in China included the Dalian Scientific Test and Control Technology Institute, Hangzhou Applied Acoustics Research Institute, Harbin Engineering University, Institute of Acoustics in Beijing, Northwestern Polytechnical University in Xi'an, Ocean University of Qingdao, and South China Sea Institute of Oceanology in Guangzhou. The Korean Ocean Research and Development Institute and the National Taiwan University also participated. The Chinese research vessels R/V *Shi-Yan 2* and R/V *Shi-Yan 3* (Fig. 7) and the U.S. R/V *Melville* provided the data collection platforms for ASIAEX-ECS. The R/V *Melville* paid a port call at Shanghai after the data collection was completed.

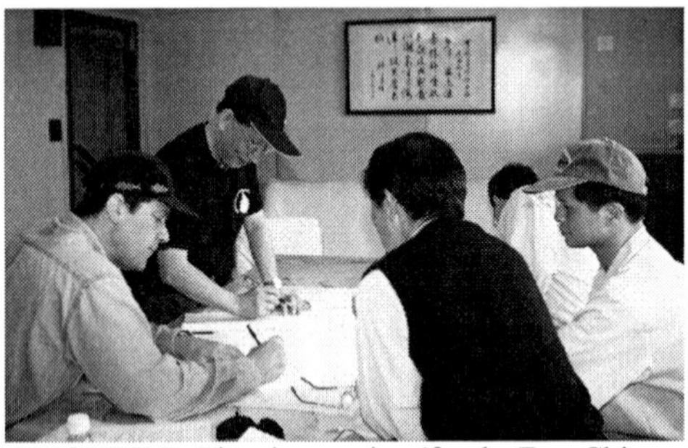

FIGURE 7. Planning session (for the East China Sea component of ASIAEX) in the ward room of the R/V *Shi-Yan 3*.

FIGURE 8. Recovery of a Low-Cost (LOCO) mooring in the South China Sea during the 2001 field effort.

The SCS component of ASIAEX addressed water column variability and its impact on acoustic propagation loss. In particular, the experiment was designed to measure the effects of strong variability in the ocean volume on acoustic propagation. This site was chosen to include both the continental shelf and shelfbreak and was a site known for the regular presence of large nonlinear internal waves and interesting mesoscale and fine scale oceanography. Moored and ship-towed sources were used to provide acoustic signals. These were received on a bottom-mounted combined vertical and horizontal (L-shaped) line array and on a single phone moored receiver. A bottom survey and chirp sonar imagery of the sediment layering provided estimates of seabed properties. Extensive measurements (Fig. 8) of the water column were made on an array of temperature, conductivity, current, and pressure sensor moorings deployed up the slope and across the shelf in the experiment area. Additional temperature, pressure and salinity measurements were taken at the acoustic receiving array, on other moorings located nearby, and by CTD measurements. The variability of the acoustic arrivals was analyzed over time in relation to the measured oceanographic variability.

Principal investigating scientists came from the U.S., Taiwan, and Singapore. U.S. research institutions included Florida Atlantic University, NASA Goddard Space Flight Center, Naval Postgraduate School, Naval Research Laboratory, University of North Carolina, and Woods Hole Oceanographic Institution. Taiwanese research institutions included the Chinese Naval Academy of Taiwan, National Sun Yat-Sen University, National Taiwan University, National Taiwan Ocean University, and Taiwan Fisheries Research Institute. The National University of Singapore also participated. The Taiwanese R/V *Ocean Researcher I,* R/V *Ocean Researcher II,* R/V *Ocean Researcher III,* and the R/V *Fisheries Researcher I* provided data collection capability, as well as mooring deployment and recovery platforms, for ASIAEX-SCS, and altogether these research vessels participated in fifteen research cruises in 2000 and 2001.

ASIAEX POST-EXPERIMENT WORKSHOPS AND SPECIAL ISSUE OF THE *IEEE JOURNAL OF OCEANIC ENGINEERING*

Acoustic and oceanographic measurements for the ASIAEX experiments were completed by the summer of 2001. Three workshops were held to summarize the collected data, to arrange for data exchange, to present early results, and to provide for a full discussion and coordination of data analysis to include all participants. The first was held in Maui in October 2001 to cover both experiments, the second in Chengdu, China in October 2002 (Fig. 9) primarily for the East China Sea component, and the last in Taiwan in March 2003 primarily for the South China Sea component. The workshops included participating scientists and students, program managers, and others from the ASIAEX research community.

FIGURE 9. Group photo of participants in the ASIAEX post-experiment workshop held in Chengdu, China in October 2002.

FIGURE 10. Chinese scientists and engineers from the Hangzhou Applied Acoustics Research Institute visiting the Applied Physics Laboratory of the University of Washington in September 2009, just prior to executing a joint acoustics experiment on Lake Washington.

The idea for a special journal issue of ASIAEX experimental results was conceived during these workshops. The participating scientists generated a set of papers that were published together in a special issue of the *IEEE Journal of Oceanic Engineering* in October 2004.[2] The volume contains overviews of the ECS and SCS components of the experiment and extensive analyses of the data collected. There are over 30 papers covering various aspects of the ocean and seabed variability measured in ASIAEX, the impact of this variability on the measured acoustic signals, and the implications for acoustic applications. Additional papers based on the ASIAEX data have been published in the *Journal of the Acoustical Society of America*, and elsewhere.

POST-ASIAEX INSTITUTION VISITS AND JOINT U.S.–CHINA MEETINGS

Following the ASIAEX experiments, U.S.–China scientific collaboration was also carried out through individual and small group interactions and at international conferences. China hosted the International Conferences on Theoretical and Computational Acoustics in Beijing, 2001, and in Hangzhou, 2005, and the International Workshop on Acoustical Engineering and Technology in Harbin, 2002. The Western Pacific Acoustics Conferences in Seoul, 2006, and in Beijing, 2009, and the Pacific Rim Underwater Acoustics Conferences in Vancouver, 2007, and Xi'an, 2009, also brought together many U.S. and Chinese investigators. Some of these collaborations have resulted in student and post-doctoral appointments for Chinese acousticians at U.S. universities.

Finally, individual invited visits to universities and institutions have produced some very productive collaboration. Several years of research and planning through exchange visits (Fig. 10) between scientists at the Applied Physics Laboratory of the University of Washington (APL-UW) and the Hangzhou Applied Acoustics Research Institute (HAARI) resulted in a small joint experiment in the waters of Lake Washington, Seattle, in September 2009. The Cooperative Array Performance Experiment (CAPEx) compared the performances of two vertically suspended acoustic receiving arrays, one a conventional pressure sensor array and the other a vector sensor array, using a towed source at many ranges. APL-UW scientists had visited the HAARI to plan this experiment and scientists from HAARI came to APL-UW to carry out the experiment.

THE SECOND INTERNATIONAL CONFERENCE ON SHALLOW-WATER ACOUSTICS

Plans for the Second International Conference on Shallow-Water Acoustics began in earnest early in 2007. Representatives from the U.S. acoustics research community visited the Institute of Acoustics in Beijing (Fig. 11) and the Hangzhou Applied Acoustic Research Institute in June 2007 to confirm Chinese support for a proposed U.S.-China conference. There had been significant growth and development of the Chinese acoustics research community since 1997 and the U.S. had advanced its understanding of many important shallow-water issues for acoustics. There would be ample material for presentation and discussion at such a conference. Shanghai was chosen as the preferred site. Additional meetings were needed to choose the specific conference topics, hotel and dates, and to develop conference support. A U.S.–China Shallow-Water Workshop was held at APL-UW in January 2009. The workshop hosted six Chinese and ten U.S. scientists for two days of technical and planning discussions. This was the final planning workshop for the conference in Shanghai. Funding and approvals for the conference were completed several months later.

FIGURE 11. Meeting held at the Institute of Acoustics in Beijing in June 2007 to plan for the Second International Conference on Shallow-Water Acoustics.

FIGURE 12. Jeff Simmen, Ellen Livingston, and Jixun Zhou (left to right) attending the Second International Conference on Shallow-Water Acoustics at the Grand Hyatt in Shanghai, China in September 2009.

The Second International Conference on Shallow-Water Acoustics was held in Shanghai during 16–20 September 2009 (Fig. 12). The U.S. Office of Naval Research and the Chinese Academy of Sciences were the principal co-sponsors of the conference. It had almost 100 attendees, representing 32 institutions from 8 countries. The conference consisted of 66 technical presentations, each lasting about 20 minutes, followed by a short question and answer period. Introductory and concluding remarks were given by representatives from the Office of Naval Research, Institute of Acoustics of the Chinese Academy of Sciences, and the Applied Physics Laboratory of the University of Washington. The technical presentations were organized into sessions on propagation, acoustic inversion, signal processing, reverberation and target scattering, and miscellaneous topics. The conference proceedings includes a list of conference organizers, sponsors, and conference participants.

INSTITUTION VISITS AND CONFERENCES FOLLOWING THE SHANGHAI GATHERING

Groups of ten to twenty U.S. scientists and government officials attending the Second International Conference on Shallow-Water Acoustics were able to visit

Chinese research institutions during their stay in China from 13–28 September 2009. They were hosted at the HAARI, Naval University of Engineering in Wuhan, Institute of Acoustics in Beijing, Northwestern Polytechnical University in Xi'an, and Harbin Engineering University (Fig. 13). In addition, they attended the Western Pacific Acoustics Conference in Beijing and the Pacific Rim Underwater Acoustics Conference in Xi'an. Several U.S. scientists will attend the International Symposium on Acoustic Engineering and Technology to be held in Harbin China in July 2010.

FIGURE 13. American team visiting Harbin Engineering University in September 2009.

THE FUTURE

We have seen that since 1995, the shallow-water acoustics research communities of the United States and China have cultivated an ongoing relationship for productive scientific exchange. The publication of this proceedings volume of the Second International Conference on Shallow-Water Acoustics marks another step in this development. Scientists, both young and old, from both nations must continue to strive to nurture this valuable U.S.–China cooperative research in underwater acoustics.

REFERENCES

1. R. Zhang and J. Zhou, *Shallow-Water Acoustics*, Beijing: China Ocean Press, 1997.
2. *IEEE J. Ocean. Eng.* **29** (October 2004).

Finite-element modeling in ocean acoustics: Where are we heading?

Finn B. Jensen* and Mario Zampolli†

**NATO Undersea Research Centre, 19126 La Spezia, Italy*
†TNO, 2509 JG The Hague, The Netherlands

Abstract. This paper summarizes 10 years of NURC experience with the application of a state-of-the-art FE code [Zampolli *et al.*, JASA **122**, 1472–85 (2007)] to propagation and scattering problems in ocean acoustics. We show benchmark results for low-frequency propagation in a range-dependent waveguide with an elastic bottom, and object scattering results for spheres and cylinders placed near the seafloor. Despite the computational burden associated with the FE method, its generality in treating propagation in layered fluid-elastic media of complex geometry without theoretical approximations makes it very attractive for acoustic benchmarking. And, as experience shows, what is computationally impractical today, will be easily done on a desktop computer 10 years from now.

INTRODUCTION

The technique of choice for solving structurally and geometrically complex vibration problems is the finite element (FE) method. In ocean acoustics the use of FE methods remains a computational challenge and spatial 2D domain sizes are typically limited to 10x1000 λ. Localized scattering problems involving objects of a few tens of acoustic wavelengths in dimension have recently received much attention, and this may be a class of problems better suited for the FE modeling approach.

Developing the capability to accurately solve range-dependent propagation problems in ocean acoustics has been going on for many years [1, 2]. New numerical codes have become available since then for solving general range-dependent elastic problems, based either on marching schemes that ignore backscattering (or adopt a single-scatter approximation) or techniques that solve the full boundary-value problem. In an attempt to establish a set of benchmark solutions for seismo-acoustic problems, we solve several low-frequency elastic propagation problems involving upslope bathymetries, with a commercial FE code being used to generate reference solutions.

High-fidelity computer models of acoustic scattering from 3-D elastic objects included inside layered acoustic media are of high importance in underwater acoustics. The models allow quick studies of many insonification scenarios for a variety of objects within different background media configurations, with the scope of determining optimal detection and classification strategies for different frequencies. In underwater acoustic applications, low-frequency sonars (typically operating at frequencies < 50 kHz for meter-sized objects) penetrate the outer shell of the target and hence provide information about its internal structure and material properties. We again use a commercial FE code to generate broadband scattering responses for problems involving a cylinder on the seafloor and spherical shells that are proud, half-buried or fully buried in the sediment.

CP1272, *Shallow-Water Acoustics, Second International Shallow-Water Acoustics Conference*
edited by J. Simmen, E. S. Livingston, J. X. Zhou, and F. H. Li

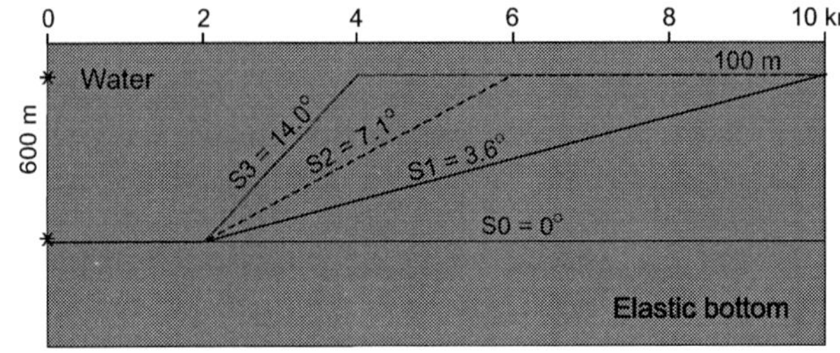

Bottom	#1	#2	#3
c_p [m/s]	2000	2400	5000
c_s [m/s]	800	1200	2500
α_p [dB/λ_p]	0.2	0.2	0.2
α_s [dB/λ_s]	0.2	0.2	0.2
ρ [kg/m^3]	2000	2000	2500

FIGURE 1. Test environments: Slope 0 is the flat bottom calibration case with a constant water depth of 600 m. The three upslope cases all have constant water depth out to 2 km, followed by bottom slopes of 3.6, 7.1 and 14.0 ° until the water depth reaches 100 m. The slope changes occur exactly between 2 and 10 km for S1, between 2 and 6 km for S2, and between 2 and 4 km for S3. Three elastic bottoms are considered with the properties given in the table.

SEISMO-ACOUSTIC TEST PROBLEMS

We consider upslope propagation in a homogeneous water column overlying a homogeneous elastic bottom, see Fig. 1. The water sound speed and density are taken to be 1500 m/s and 1000 kg/m^3, respectively. The bottom properties are given in the table of Fig. 1, which lists bottom types ranging from moraine (c_s =800 m/s), over chalk (c_s =1200 m/s) to basalt (c_s =2500 m/s), hence covering the full range of elastic sea floors with significant shear effects. To be able to ignore the effects of thin sediment covers with low shear speeds, we consider low enough frequencies (5 and 15 Hz) so that the acoustic penetration into the sediment is several hundred meters.

As shown in Fig. 1, we consider three different bottom slopes, 3.6°, 7.1° and 14.0°, which are quite steep slopes that may result in significant mode coupling and hence could be considered a severe test of the capability of a range-dependent seismo-acoustic code to handle sloping bathymetry. We also solve the flat-bottom calibration case, since a reference solutions here can be obtained from one of the legacy propagation codes such as OASES [3]. All problems were done for two source depths, 100 and 590 m, whereas the receiver was placed at 30-m depth. The deep source just 10 m off the bottom will excite an interface wave (Scholte-Stoneley wave) in cases where the shear speed is lower than the water speed, i.e. for bottoms 1 and 2. Hence, the 590-m source case is designed to test how well the models handle propagation of interface waves along a seabed with abruptly changing slopes.

THE NUMERICAL MODELS

Most of the models employed in the previous seismo-acoustic benchmarking effort from 1997 [2] were research codes, which are not generally available to the acoustics community. Only the FEPES model [4] developed at NRL has been widely distributed, but this model, though computationally efficient, has shown its limitations in handling shear effects in strongly range-dependent environments.

An improved version of the elastic PE (parabolic equation) was subsequently developed using a coordinate transformation technique to handle variable slopes instead of the stair-step approximations employed in FEPES. This new code, named ROTVARS [5], is expected to perform well for elastic bottoms with variable slopes, as long as backscattering is negligible (the usual PE one-way approximation).

The second model used in this study is a commercial finite-element code FEM developed by COMSOL AB [6], which allows problem formulation in a 2D axisymmetric geometry. This is a general-purpose code which solves the full elastic boundary-value problem and hence should provide accurate results even when there is significant backscatter.

Both ROTVARS and FEM were carefully set up with appropriately defined numerical radiation condition on the boundary delimiting the physical domain in depth (ROTVARS and FEM) and range (FEM only). ROTVARS works with a sponge layer in the bottom between 2000 and 4000-m depth, where the attenuation increases linearly from $\alpha_{p,s} =$ 0.2 dB/$\lambda_{p,s}$ to 10 dB/$\lambda_{p,s}$. FEM, on the other hand, uses the more efficient approach of PML's (Perfectly Matched Layers) of only one wavelength thickness placed at a depth of 1100 m and at a range of 14 km, see Fig. 2. Then solutions were checked for convergence with respect to the spatial discretizations utilized by each code. Whereas FEM provided converged results to within the line thickness for all test cases, we observed a non-uniform convergence for ROTVARS, where an excessively fine spatial discretization of, say, $\lambda/100$ would cause diverging results. Hence, it was not always possible to push ROTVARS to high solution accuracy.

In terms of computation times, ROTVARS provides results in a few seconds even for the 15 Hz cases, whereas FEM takes longer, around 3 min for a 5 Hz result and 10 min for a 15 Hz result.

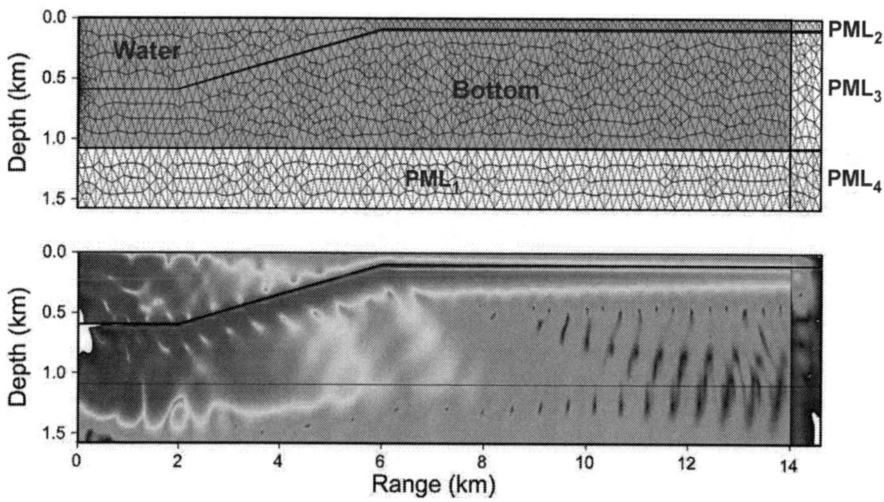

FIGURE 2. Coarse FE mesh for test environment (upper panel) and computed transmission loss (lower panel). The source depth is 590 m and the frequency 5 Hz.

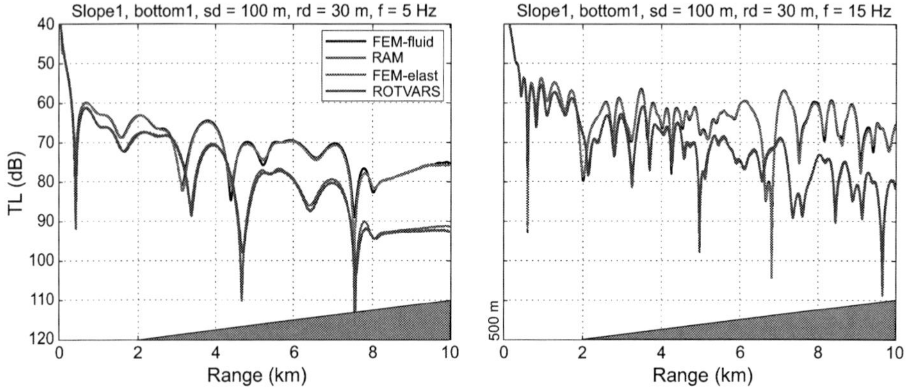

FIGURE 3. TL comparisons for Slope 1 with bottom type 1. SD = 100 m, RD = 30 m, F = 5 Hz (left) and 15 Hz (right).

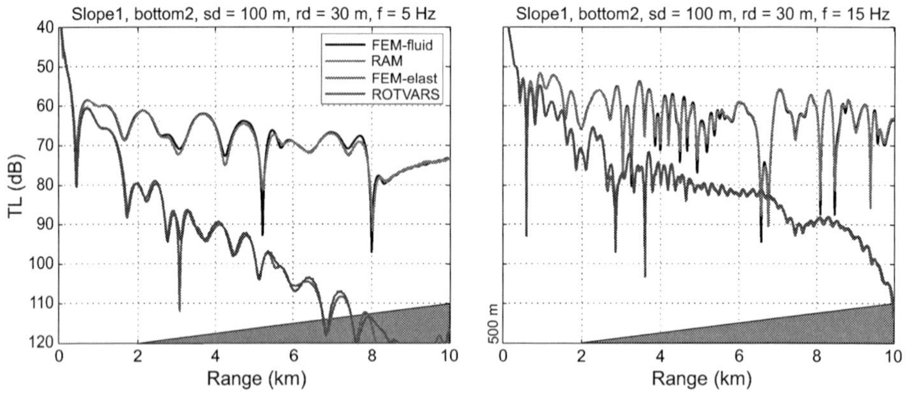

FIGURE 4. TL comparisons for Slope 1 with bottom type 2. SD = 100 m, RD = 30 m, F = 5 Hz (left) and 15 Hz (right).

SLOPE RESULTS

We briefly comment on a set of results illustrating the general agreement between the model predictions for sloping environments. The format is the one shown in Fig. 3 with the 5-Hz results to the left and the 15-Hz results to the right. Each plot also displays two fluid results from FEM and RAM [7] obtained by ignoring shear in the bottom. The importance of shear is immediately evident by comparing TL levels for the fluid and the elastic solutions. Figure 3 is a benign environment (slope 1, bottom 1) where shear effects are 10–15 dB at longer ranges. The agreement between ROTVARS and FEM results is very good, particularly at 15 Hz, where we have benchmark quality results.

Figure 4 shows results for slope 1 with bottom 2, where the shear speed is 1200 m/s. Here shear effects are huge, around 40–50 dB at 10 km, and both models give similar

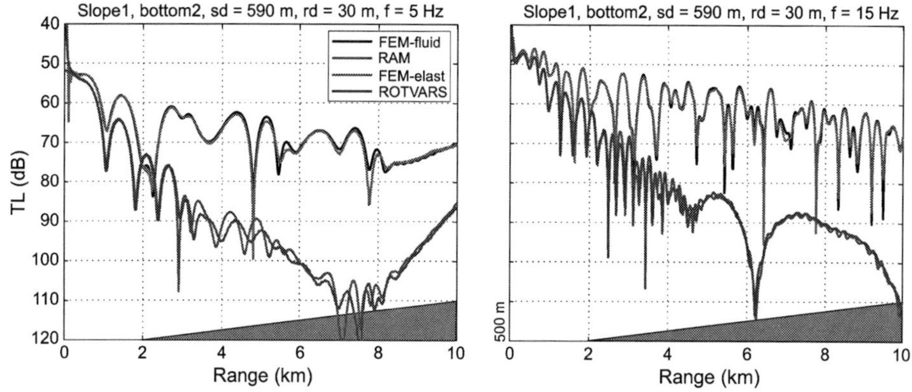

FIGURE 5. TL comparisons for Slope 1 with bottom type 2. SD = 590 m, RD = 30 m, F = 5 Hz (left) and 15 Hz (right).

results, with best agreement at 15 Hz. Taking the FEM solution to be the reference, we conclude that ROTVARS provides the correct answer to within the convergence accuracy of the code. Note that there are small disagreements even for the two fluid results at 5 Hz.

By changing the source depth to 590 m for slope 1 and bottom 2, we obtain the results shown in Fig. 5. With the source being close to the bottom, an interface wave is excited, and its contribution to the acoustic field at the receiver at 30-m depth is clearly seen in the 5-Hz result at ranges beyond 8 km, where the TL level rises by around 20 dB. No interface wave contribution is observed at 15 Hz, due both to a weaker excitation and a stronger attenuation. Whereas mean levels are in good agreement between FEM and ROTVARS for this particular case, there are some differences in the multipath interference structure, particularly at 5 Hz and for ranges between 3 and 8 km. Since the FEM results were confirmed to be accurate by comparison with a newly developed virtual source technique [8], it is clear that ROTVARS here introduces phase errors which, in our opinion, cannot be attributed solely to the aforementioned numerical convergence problems for the code. It appears that ROTVARS phase errors are only significant when an interface wave is excited.

Finally we look at a worst-case scenario with a hard, steeply sloping basalt bottom. As shown in Fig. 6, backscattering is very strong in the FEM-fluid solution and there is poor agreement with the RAM 1-way result. Note that the effect of backscattering on mean levels is as high as 5–10 dB at ranges beyond 4 km. Much better results are obtained by including shear and using ROTVARS because the elastic bottom is less reflective than the equivalent fluid bottom. Even though there is clear evidence of backscatter also in the elastic FEM solutions, particularly at 15 Hz, there is quite good agreement between FEM and ROTVARS for this extreme situation of a very steep slope combined with a hard elastic bottom. In summary, it appears that ROTVARS can be applied with some confidence for elastic bottoms with slopes as steep as 7–14°. Note that the shear results in Fig. 6 have been shifted down by 20 dB to avoid overlap with the fluid solutions. In fact, mean levels are only weakly affected by shear for a type-3 bottom.

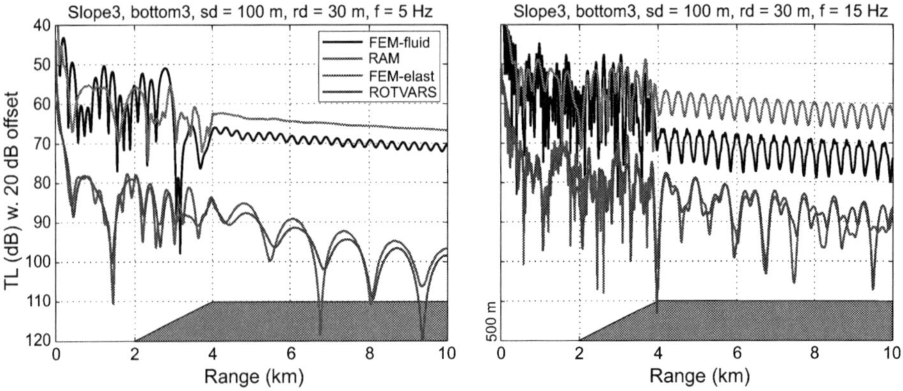

FIGURE 6. TL comparisons for Slope 3 with bottom type 3. SD = 100 m, RD = 30 m, F = 5 Hz (left) and 15 Hz (right).

COMMENTS ON SEISMO-ACOUSTIC TESTS

A set of test problems for benchmarking range-dependent seismo-acoustic models were defined, with the possibility to investigate the effects of bottom type, bottom slope, frequency and source depth (interface wave or not) on the solution accuracy. Overall, the agreement on predicted TL levels for ROTVARS and FEM is very good. Backscattering is found to be weak even for upslope geometries as steep as 14°, indicating that a one-way parabolic-equation model such as ROTVARS is applicable to wide class of seismo-acoustic propagation problems. The FEM model is ideal for generating reference solutions for these types of propagation problems, but the computational cost is about a factor 100 higher than for ROTVARS.

OBJECT SCATTERING TEST PROBLEMS

The finite-element method described in Refs. [9, 10] is capable of computing the scattering from axially-symmetric elastic targets subject to non-axisymmetric incident acoustic fields, such as for example plane waves or spherical waves. The elastic targets can have internal structures and fluid partitions, provided that also these are axially symmetric. The model is obtained by expressing all the unknown field quantities, namely the pressure in the fluid parts of the computational domain and the three-dimensional displacement vector in the elastic parts of the computational domain, as a Fourier series in the azimuthal coordinate ϕ. The unknown coefficients of the Fourier series depend only on the cylindrical coordinates r and z, while the Fourier modes $\exp(im\phi)$ depend on the azimuth only. The modal order is $m = -M, \ldots, M$. Applying this expansion in the finite-element variational equations, allow us to integrate explicitly the azimuthal coordinate, so that the original (large) 3-D scattering problem is transformed into a series of $2M$ independent (smaller) 2-D problems. The solution of the individual small 2-D problems is

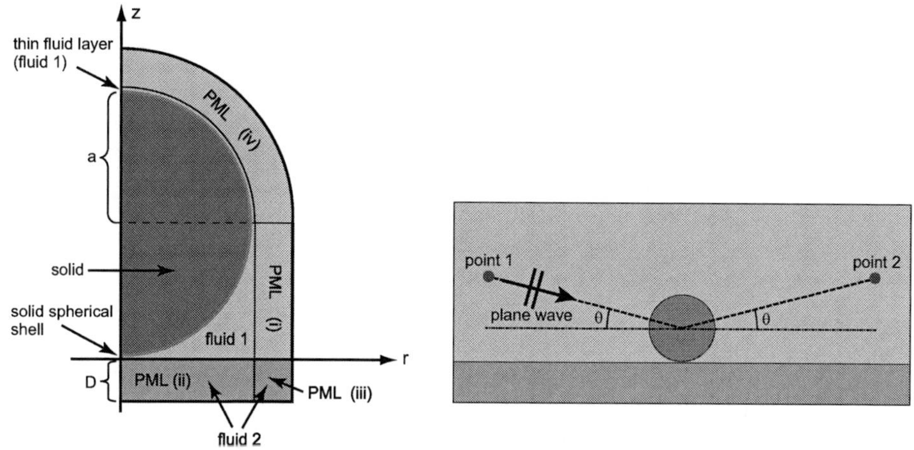

FIGURE 7. FE computational domain for scattering from an elastic layered sphere proud on the interface between two different fluid halfspaces. Right panel shows the problem geometry.

numerically much more efficient than the solution of a single (large) 3-D problem, with improvements in memory efficiency and speedup typically of up to two orders of magnitude. This approach makes it possible to solve 3-D scattering problems in the low- to mid-frequency regime (several tens of kHz for meter-sized objects) on standard desktop workstations, and hence treat a class of scattering problems of interest to the underwater acoustics community.

To illustrate the application of the FE method to such problems, we present here the scattering from a 1-m diameter filled elastic spherical shell, proud, buried and half-buried in a sand sediment (modeled as a fluid). The geometry is represented schematically in Fig. 7. The drawing shows how the absorbing PML can be easily applied also in the presence of an interface between two different fluids, such as the water and the sand considered here. The geometry for the completely buried case and for the half-buried case is shown in Fig. 8. The farfield is computed in each case by a discrete approximation of the Helmholtz-Kirchhoff surface integral, using the pressure and the normal displacement on the target surface obtained from the FE solution in conjunction with the Green's functions for the two-layered medium, see Ref. [11].

Since the overall geometry of the problem is axially symmetric, the FE model described above can be applied to compute the scattered field, including the multiple scattering interactions between the target and the seabed. The incident field in this case consists of the incident and reflected plane waves in the upper medium, and the transmitted wave in the lower medium.

All material properties are given in Table 1. The critical angle of reflection at the bottom is around 20°. The outer radius of the sphere is $a = 0.5$ m, and the shell is 1 cm thick. Two different insonification angles are considered: subcritical insonification at $\theta = 15°$ and supercritical insonification at $\theta = 25°$. In what follows we present the target strength, $\mathrm{TS} = 20\,\log_{10}\left(R|p^{\mathrm{scat}}|/|p^{\mathrm{inc}}|\right)$, computed from the scattered pressure p^{scat} at the farfield location. The farfield receiver is chosen here to be at a distance of $R = 50$ m

TABLE 1. Material properties. A p-wave attenuation of 0.5 dB/λ is assumed for the sediment.

	c_p[m/s]	c_s[m/s]	ρ [kg/m^3]
Water	1500	0	1000
Sediment	1600	0	1800
Fiberglass casing	3500	1400	3000
Solid filling	2500	1200	2000

from the target. The amplitude of the incident plane wave is $|p^{\text{inc}}| = 1$ and the frequency band is 0.1–10 kHz.

The FE solution is compared to a solution obtained with the scattering module of the wavenumber propagation code OASES [12, 13] which uses an analytical model to compute the scattering from the layered elastic sphere and the full wavenumber integral for the two-layered medium Green's function. For the first two examples, converged solutions up to 10 kHz were obtained with OASES using the single-scatter approximation, in which the scattering from the target in an ideal infinite medium is described using spherical harmonics. The scattered field is propagated into the farfield by the wavenumber propagation model. This formulation makes it possible to obtain solutions at higher frequencies at the the cost of neglecting multiple scattering between the scatterer and the fluid-fluid interface beyond the first bounce. Figures 9 and 10 show the close agreement between the FE results and the OASES single-scatter approximation for the proud and buried cases. As shown in Fig. 10(a) for the buried sphere, there are some disagreements on predicted backscatter levels above 4 kHz at subcritical incidence, where the target strength is more than 60 dB below the levels at the lower frequencies.

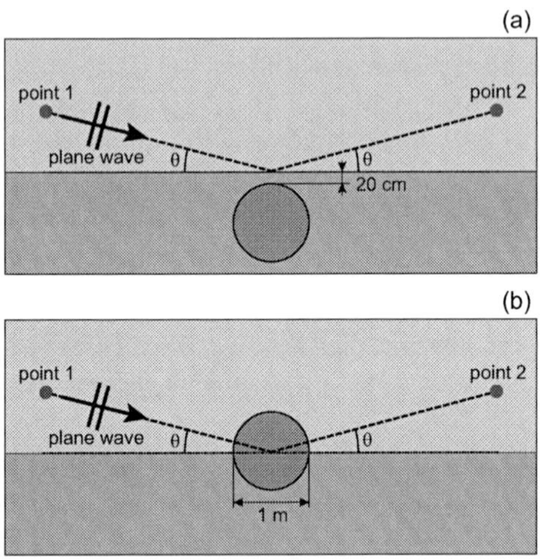

FIGURE 8. Scattering from a buried and a partially buried solid-filled spherical shell. Problem geometry for the buried case (a) and the half-buried case (b).

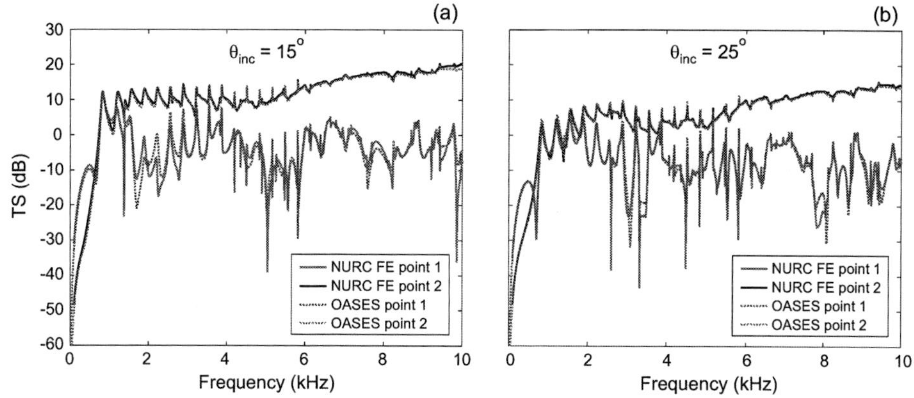

FIGURE 9. Scattering from proud sphere at (a) subcritical incidence and (b) supercritical incidence.

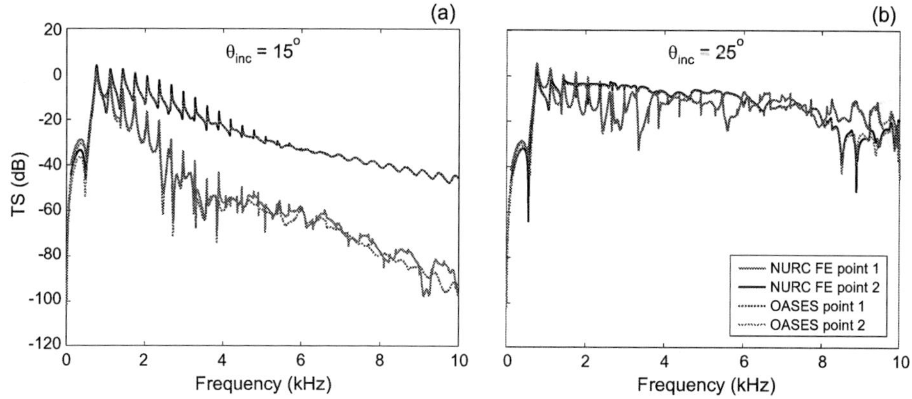

FIGURE 10. Scattering from buried sphere at (a) subcritical incidence and (b) supercritical incidence.

Multiple scattering effects between the water-sand interface and the sphere could not be neglected in the half-buried case, and it was therefore necessary to resort to a functionality of OASES in which the target response is described by its nodal surface stiffness matrix [14]. The target in the waveguide is then replaced by an equivalent impedance surface obtained from the nodal stiffness matrix. This approach is numerically more intensive, and the OASES solutions are converged only to about 5.5 kHz for the half-buried sphere. The FE results shown in Fig. 11 agree closely with the OASES full scattering solution up to this frequency. This, in turn, is an indication that the FE code is capable of computing the full scattering solution, including multiple interactions between the target and the seabed, in cases where the overall geometry comprising the target and the environment is axially symmetric.

When the overall geometry is not axially symmetric, such as the example shown in Fig. 12, the FE model described above can still be applied in the single-scatter approximation. The target is here an air-filled cylindrical shell with one hemispherical

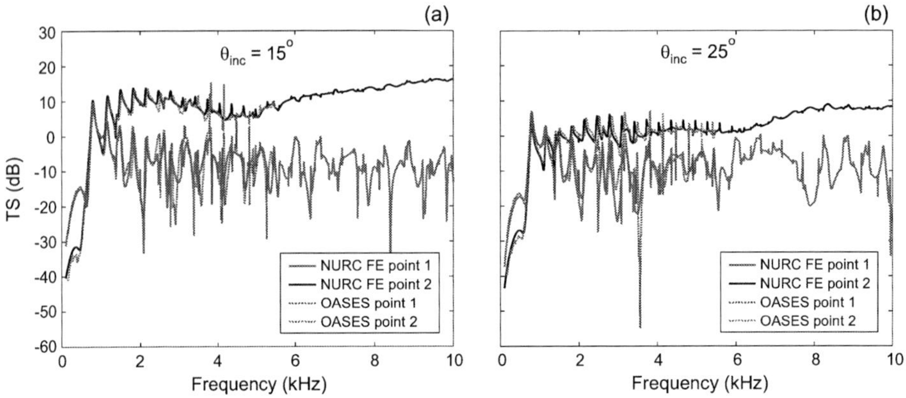

FIGURE 11. Scattering from half-buried sphere at (a) subcritical and (b) supercritical incidence.

endcap and one flat endcap. The overall length is 2 m and the outer diameter is 0.5 m. The shell is 1 cm thick and has the same material properties as the spherical shell of the previous examples, Table 1. The target is considered to be in an ideal infinite space of water, and the incident field is given by the superposition of the incoming and the boundary-reflected plane waves. The FE problem is solved and the pressure and normal displacements on the target surface are used in the Helmholtz-Kirchhoff integral together with the layered-medium Green's functions to compute the farfield pressure. This approach takes into account only the first bounce between the target and the seabed. Nevertheless, Fig. 13 shows that a fully three-dimensional boundary element solution obtained at FOI in Sweden [15] agrees with the FE result except around 3 kHz for supercritical incidence. At subcritical insonification the agreement in the backscattering at low frequencies improves, while discrepancies on the order of 3 dB appear at the higher frequencies, above 4 kHz. The disagreement between the two models is likely due to model approximations such as: (i) neglecting multiple scattering between the seabed and the cylinder and (ii) neglecting the effects of physical contact between the surface of the cylinder and the seabed. The latter may influence the results significantly

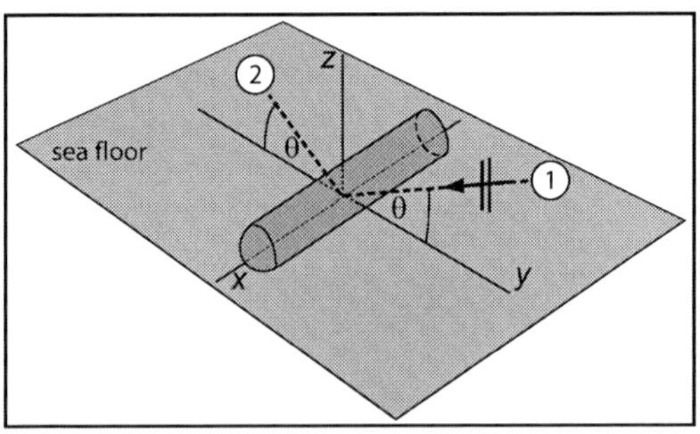

FIGURE 12. Cylindrical target on the planar interface between a water halfspace and a sand halfspace.

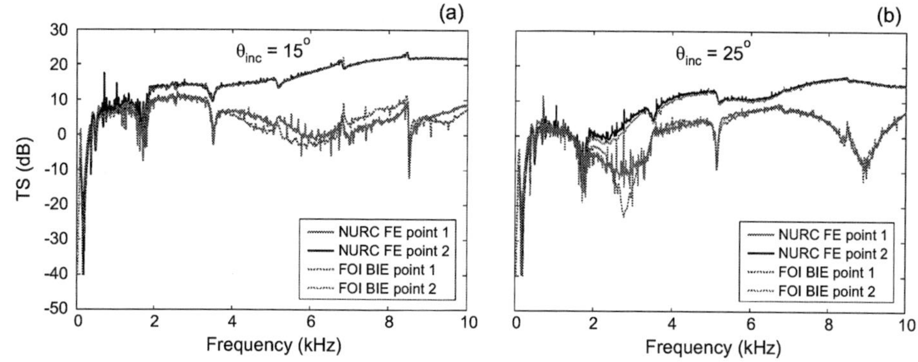

FIGURE 13. Scattering from a 2-m long cylinder on the seabed at (a) subcritical and (b) supercritical incidence.

at frequencies where shell-borne structural waves are prominent in the echo, since such waves are sensitive to changes in the boundary conditions. These aspects are investigated in more detail in Ref. [16], in which the FE model described here and a fully 3-D FE model implemented on a parallel computer are compared to multi-aspect (0–90°), broadband measurements (2–30 kHz) for a proud aluminum cylinder of 60 cm length.

COMMENTS ON OBJECT SCATTERING TESTS

The problem of scattering from objects inside layered media is computationally challenging, but possible to solve on standard desktop workstations using a finite element model. Although the FE model presented here is restricted to treating axially symmetric targets, a large number of scenarios of practical interest can be modeled accurately. In those cases where the overall geometry comprising the target and the environment is axially symmetric, the FE model provides the full scattering solution. In other cases, such as the example of a cylinder on the seabed, the FE model can still be used to obtain the single-scatter solution, in which only one bounce between the target and the seabed is taken into account. The comparison between data and models presented in Refs. [16, 17] shows the practical application of the model. The complexity of the computational problem increases by orders of magnitude for fully 3-D targets and environments, which to this day requires using massively parallel computing approaches.

CONCLUSIONS

We have shown finite-element results for propagation in range-dependent waveguides and for scattering from objects placed near the seafloor. Despite the computational burden associated with the FE method, its generality in treating wave propagation in layered fluid-elastic media of complex geometry makes it very attractive for acoustic benchmarking. With the ever-increasing computer power made available, we expect, over the next decade or so, to see the finite-element method becoming one of the standard modeling tool in ocean acoustics.

REFERENCES

1. Jensen, F.B. and Ferla, C.M., "Numerical solutions of range-dependent benchmark problems in ocean acoustics," *J. Acoust. Soc. Amer.* **87**,1499–1510 (1990).
2. Goh, J.T., Schmidt, H., Gerstoft, P. and Seong, W., "Benchmarks for validating range-dependent seismo-acoustic propagation codes," *IEEE J. Oceanic Engn.* **22**, 226–236 (1997).
3. Schmidt, H., "OASES ver. 3.2: User guide and reference manual," Dept. Mechanical Engn., MIT (2004). Available at http://acoustics.mit.edu/faculty/henrik/oases.html
4. Collins, M.D., "An energy conserving parabolic equation for elastic media," *J. Acoust. Soc. Amer.* **94**, 975–982 (1993).
5. Outing, D.A., Siegmann, W.L., Collins, M.D. and Westwood, E.K., "Generalization of the rotated parabolic equation to variable slopes," *J. Acoust. Soc. Amer.* **120**, 3534–3538 (2006).
6. COMSOL AB, "COMSOL Multiphysics 3.2 – User's Guide," Stockholm (2005), http://www.comsol.com
7. Collins, M.D., "A split-step Padé solution for the parabolic equation method," *J. Acoust. Soc. Amer.* **93**, 1736–1742 (1993).
8. Abawi, A.T. and Porter, M.B., "Propagation in an elastic wedge using the virtual source technique," *J. Acoust. Soc. Amer.* **121**, 1374–1382 (2007).
9. Zampolli, M., Tesei, A., Jensen, F.B., Malm, N. and Blottman, J.B., "A computationally efficient finite element model with perfectly matched layers applied to radiation and scattering from axisymmetric objects," *J. Acoust. Soc. Amer.* **122**, 1472–1485 (2007).
10. Zampolli, M., Jensen, F.B. and Tesei, A., "Benchmark problems for acoustic scattering from elastic objects in the free field and near the seafloor," *J. Acoust. Soc. Amer.* **125**, 89–98 (2009).
11. Zampolli, M., Tesei, A., Canepa, G. and Godin, O.A., "Computing the far field scattered or radiated by objects inside layered fluid media using approximate Green's functions," *J. Acoust. Soc. Amer.* **123**, 4051–4058 (2008).
12. Schmidt, H. and Glattetre, J., "A fast field model for the three-dimensional wave propagation in stratified environments based on the global matrix method," *J. Acoust. Soc. Amer.* **78**, 2105–2114 (1985).
13. Lucifredi, I. and Schmidt H., "Subcritical scattering from buried elastic shells," *J. Acoust. Soc. Amer.* **120**, 3566–3583 (2006).
14. Schmidt, H., "Virtual source approach to scattering from partially buried elastic targets," in *High-Frequency Ocean Acoustics*, edited by M.B. Porter, M. Siderius and W.A. Kuperman (AIP, Melville, New York, 2004), pp. 456–463.
15. Karasalo, I. and Mattsson, J., "Accurate numerical modelling of scattering by 3D bodies and shells in a fluid-solid medium," in *Proc. Fourth Europ. Conf. Underwater Acoustics*, edited by A. Alippi, and G.B. Cannelli (CNR-IDAC, Rome, Italy, 1998), pp. 691–696.
16. Williams, K.L., Kargl, S.G., Thorsos, E.I., Lopes, J.L., Burnett, D.S., Zampolli, M. and Marston, P.L., "Acoustic scattering from an aluminum cylinder in contact with a sand sediment: Measurements, modeling, and interpretation," *J. Acoust. Soc. Amer.*, submitted for publication (2010).
17. Tesei, A., Zampolli, M. and Canepa, G., "At-sea measurements of acoustic-elastic scattering by a 1.5 m long cylinder made of composite materials," in *Proc. Second Int. Conf. Underwater Acoustic Measurements*, edited by J.S. Papadakis and L. Bjorno (Univ. of Crete, Heraklion, Greece, 2007), pp. 505–510.

Nonlinear Internal Wave Interactions with Low Frequency Shallow Water Sound – What is left to Do?

James F. Lynch, Timothy F. Duda, Ying-Tsong Lin and Arthur E. Newhall

Woods Hole Oceanographic Institution, Woods Hole, MA 02543 USA

Abstract. This paper discusses what future research should be done in the area of nonlinear internal wave interactions with low frequency shallow water sound.

Keywords: Low frequency acoustics, shallow water acoustics, nonlinear internal waves.
PACS: 43.30.Bp, 43.30.Cq.

INTRODUCTION

Nonlinear internal waves are quite common in the shallow waters of continental shelves, and have been shown to have strong effects on acoustic propagation and scattering. An early study in the Yellow Sea [1] showed a strong absorption of low-frequency (50-1000 Hz) sound due to scattering by internal wave packets, and this work provided the impetus for many later studies. In the 1995 Shallow Water Acoustics in a Random Medium (SWARM) experiment [2], strong mode coupling occurring with across-internal-wave propagation geometry was emphasized, with the main acoustic tracks being directed across-shelf. However, also as part of SWARM, Badiey et al. [3] performed transmissions in an along-shelf (and thus along the nonlinear internal wavefronts) geometry. They observed a very strong ducting of low frequency (roughly 50-500 Hz) sound between nonlinear internal waves. This effect had been predicted by computer modeling [4] and normal mode theory [5], but not previously observed. A strong and acoustically important ducting effect of order 6-10 dB was observed. The anticipation of this real world effect by theory and models, its appreciable size, and its three-dimensionality are all noteworthy. Interestingly, this pattern of discovery of 3-D, internal wave induced acoustics effects has continued. The ducting and lateral mode scattering due to curved internal waves, previously examined for straight internal waves, was recently described theoretically by Duda et al. [6] and Lynch et al. [7], and it now appears that such curved ducting effects have been observed by Reeder and his collaborators in the South China Sea [8]. Another internal wave induced, 3-D acoustic propagation effect, the so-called "horizontal Lloyd's mirror" was predicted theoretically by Lynch et al. [9], and has recently been clearly observed by Badiey and Katznelson and their collaborators [10]. This effect combines a frequency and mode number reflection/refraction from the front of the

CP1272, *Shallow-Water Acoustics, Second International Shallow-Water Acoustics Conference*
edited by J. Simmen, E. S. Livingston, J. X. Zhou, and F. H. Li

internal wave packet with a direct path from the source, and is clearly seen due to the sharp "leading edge" of the internal wave trains.

The prediction and observation of strong 3-D acoustic propagation effects due to nonlinear internal waves should be cause for ocean acousticians to feel somewhat confident in their skills in describing the acoustic interaction with nonlinear internal waves. We have certainly come a long way since the first observations by Zhou et al. [1] that nonlinear internal waves are important acoustically. So we are perhaps ready to ask the question that every child asks on a long journey: "Are we there yet?" And, the reply, we believe, is the usual one given: "We are getting close, but there is still a little way left to go." In this article, we would like to, in a conversational manner, discuss just how far we have to go before we can say that this area is essentially understood theoretically, has been measured to satisfaction experimentally, and can be well modeled computationally.

DISCUSSION

There are, in fact, many facets of the acoustics/nonlinear internal wave interaction problem that are either currently being pursued or should be pursued in the near future. We will try to be inclusive in listing them, realizing, however, that some omission may be possible. The topics we will discuss include: 1) the frequency directional spectrum of the internal waves, 2) crossing and "diffusive ducting" of energy between internal waves, 3) the effects of wave movement upon acoustics, 4) the intensity statistics one obtains from internal wave scattered acoustic energy, 5) the development of an ocean model for internal waves that can provide good input for acoustics calculations, and 6) the predictability of the internal wave and acoustic fields. We again state that acoustics has come a long way in this area in that all these topics are, in a very real sense, advanced topics which can only be approached after significant initial progress has been made.

The first topic, the frequency-directional spectrum of the internal wave field, is obvious from synthetic aperture radar (SAR) images of the sea surface, which routinely show curved internal waves, multiple internal wave packets, crossing wave packets, and spatially and temporally variable (non-stationary) wave signatures. The coastal nonlinear internal wave field is quite analogous to the coastal surface wave field in that it apparently needs a frequency-directional spectrum that is a function of geographical coordinate and clock time for its proper description. However, even at a spectral level, that is a lot of information to require, and how one approaches adequately approximating these spectra for practical use remains "to be determined."

The second topic springs directly from the first topic. The acoustics of the crossing waves is currently being explored by A. Shmelev et al. [11] of the Woods Hole Oceanographic Institution (WHOI), and includes some interesting mode coupling effects which allow sound energy to transit from one internal wave horizontal duct to another. The finite length of the internal waves along crest also allows energy transfer from one acoustic duct to another, only adiabatically, and this topic is being pursued by K. McMahon from the Rensselaer Polytechnic Institute [12]. These are both fully 3-D acoustics effects, and it will be interesting to see what their quantitative magnitudes and signatures are.

The third topic, motion of the internal waves, is interesting both from a Doppler effect and from a "moving beam" standpoint. The Doppler shifting seen by sound horizontally reflected from the front of a planar wave internal wave packet front has been recently described by McMahon [12]. Her work is now being extended to horizontal reflection of a beam from curved internal wave fronts, which may produce some exciting results, as the reflection of a beam off a moving curved surface can produce "swept beams" with extremely high phase velocities, mimicking a high velocity source in the ocean. The effects of moving internal waves on a vertical beam of mid-frequency sound (a few kHz) was recently discussed by Tang et al. [13], who showed that the internal waves can redirect the sound energy into the bottom, and thus create a localized patch of bottom reverberation that moves with the speed of the internal wave, an annoying "false target" or "clutter" for sonar operators. Prior to these newer works, it was shown that moving internal waves imparted temporal variability to coupled mode propagation [14]. Evolution of the nonlinear wave packets was excluded from that study. Including the variability of the internal waves would improve analysis.

The acoustic intensity statistics one obtains in the vicinity of a field of nonlinear internal waves are also of great interest. Due to the directional spectrum of the internal waves typically extending over a 30-60 degree sector, the "strong focusing" region that one encounters (as graphically described in Katznelson's popular azimuthal sector diagram [10]) is not just zero to five degrees - one can expect strong focusing effects over a much larger angular regime. This means that the scintillation index diagram is a complicated range and azimuth dependent entity, and that the statistics are not just that of a saturated, phase random multipath medium. Just how much the acoustic fluctuations deviate from saturated, simple statistics depends on the wave field and the frequency. To date, there is no simple formulation for this, and work needs to be done. However, to get these statistics correctly, one needs to have a model that includes all the fully 3-D spatial and time dependent scattering and focusing mechanisms at once, which is non-trivial, so hopes for a "simple" model may be unfounded.

The development of a useful fully regional scale physical oceanographic model that includes the oceanography through the internal wave field has yet to be accomplished. There have been supercomputer models like the Stanford Unstructured Non-hydrostatic Terrain-following Adaptive Navier-Stokes simulator (SUNTANS) model that have actually worked at nonlinear internal wave scales [15]. However, these models take a long time for one calculation, and it is doubtful that they will ever be at the stage where one can make ocean realizations with them and do Monte Carlo calculations for statistics (as was mentioned previously). There is a need for such a model, and recently Duda et al. have proposed a "nested" model that can be used along with regional models and supply at least a good approximation to the full nonlinear internal wave field [16].

We finally come to the prediction of the nonlinear coastal internal wave field, which when coupled with fully 3-D acoustic propagation models, provides predictions of the acoustic field. The coastal nonlinear internal wave field is an interesting mixture of rather deterministic looking wave trains, but with enough randomness in them (e.g. in amplitude, shape, arrival time, etc.) that one would have much trepidation in using any fully deterministic approach. Indeed, the basic fact that the internal waves are nonlinear means that what one sees at a given place and time depends sensitively on the initial conditions of the wave generation and propagation; it is well known that there are a plethora of non-linear internal wave theory solutions, and that a given region can produce a variety of wave train shapes and sizes. How one approaches constructing a hybrid "partially random/partially deterministic" description of the nonlinear internal wave field also needs to be ascertained. Until that is done, prediction in any optimal sense will not be achievable.

We come to the end of this brief discussion with the sense that there is still much to be done, particularly on the ocean/acoustic model interface problem. But much has been done, and there are good ideas floating about in all the current areas that we have discussed, so that we can perhaps be confident in stating that if the SWAC conference series is continued in another decade or so, we will be hearing the solutions to most, if not all of the problems discussed above.

ACKNOWLEDGMENTS

This work was supported by the Office of Naval Research.

REFERENCES

1. Zhou, J., X. Zhang and P.H. Rogers, "Resonant interaction of sound wave with internal solitons in the coastal zone," *J. Acoust. Soc. Am.*, vol. 90, pp. 2042-2054, 1991.
2. Apel, J.R., M. Badiey, C.-S. Chiu, S. Finette, R. Headrick, J. Kemp, J.F. Lynch, A.E. Newhall, M.H. Orr, B.H. Pasewark, D. Tielbuerger, A. Turgut, K. von der Heydt, and S.N. Wolf, "An overview of the 1995 SWARM shallow-water internal wave acoustic scattering experiment," *IEEE J. Oceanic Eng.*, vol. 22, pp. 465-500, 1997.
3. Badiey, M., Y. Mu, J.F. Lynch, J.R. Apel, and S.N. Wolf, "Temporal and azimuthal dependence of sound propagation in shallow water with internal waves," *IEEE J. Ocean. Eng.*, vol. 27, pp. 117-129, 2002.
4. Finette, S., and R. Oba, "Horizontal array beamforming in an azimuthally anisotropic internal wave field, " *J. Acoust. Soc. Am.*, vol. 114, pp. 131–144, 2003.
5. Katznelson, B.G., and S.A. Pereselkov, "Low-frequency horizontal acoustic refraction caused by internal wave solitons in a shallow sea," *Acoust. Phys.*, vol. 46, pp. 684–691, 2000.
6. Duda, T.F., *Initial results from a Cartesian three-dimensional parabolic equation acoustical propagation code*, WHOI Tech. Rept., WHOI-2006-041, 2006.
7. Lynch, J.F., Y.-T. Lin, T.F. Duda, and A.E. Newhall, "Acoustic ducting, reflection, refraction, and dispersion by curved nonlinear internal waves in shallow water," *IEEE J. Oceanic Eng.*, vol. 35, pp. 12-27, 2010.
8. Reeder, D.B., T.F. Duda and B. Ma, "Short-range acoustic propagation variability on a shelf area with strong nonlinear internal waves," *Oceans '08 Kobe Conference Proceedings*, IEEE, 2008.
9. Lynch, J.F., J. Colosi, G.G. Gawarkiewicz, M. Badiey, B. Katznelson, J.H. Miller, W. Siegmann, C.-S. Chiu, and A.E. Newhall. "Consideration of fine-scale coastal oceanography and 3-D acoustics effects for the ESME sound exposure model," *IEEE J. Oceanic Eng.*, vol. 31, pp. 33-48, 2006.
10. Badiey, M., J.F. Lynch, Y.-T. Lin, and B. Katznelson, "Horizontal reflection of a low frequency sound signal from a moving nonlinear internal wave front," (A) *J. Acoust. Soc. Am.*, vol 127, No. 3, Pt. 2, page 1974, 2010.

11. Lynch, J.F., A.A. Shmelev, Y.-T. Lin, and A.E. Newhall, "Acoustic ducting and refraction by sea bottom relief in shallow water," (A) *J. Acoust. Soc. Am.*, vol 127, No. 3, Pt. 2, page 1786, 2010.
12. McMahon, K.G., L.K. Reilly-Raska, J.F. Lynch, T.F. Duda, and W.L. Siegmann, "Horizontal Lloyd's mirrors arising from propagation through straight and curved nonlinear wave fronts," (A) *J. Acoust. Soc. Am.*, vol 127, No. 3, Pt. 2, page 1963, 2010.
13. Tang, D.J. and F. Henyey, "Reverberation clutter from combined internal wave refraction and bottom backscatter." (A) *J. Acoust. Soc. Am.*, vol 127, No. 3, Pt. 2, page 1974, 2010.
14. Duda, T.F. and J.C. Preisig, "A modeling study of acoustic propagation through moving shallow-water solitary wave packets," *IEEE J.Ocean. Eng.*, vol 24, pp. 16-32, 1999.
15. Fringer, O.B., M. Gerritsen, and R.L. Street. "An unstructured-grid, finite-volume, nonhydrostatic, parallel coastal-ocean simulator," *Ocean Modeling*, **14** (3-4), 139-278, 2006.
16. Lynch, J.F., T.F. Duda, Y.-T. Lin, A.E. Newhall, and P.F.J. Lermusiaux, "A model for four dimensional coastal internal waves and applications to acoustics," (A) *J. Acoust. Soc. Am.*, vol 126, No. 4, Pt. 2, page 1974, 2009.

Mid-Frequency Acoustics in SW06

Dajun Tang

*Applied Physics Laboratory, University of Washington, 1013 NE 40th Street
Seattle, WA 98105, USA.*

Abstract. This overview of the mid-frequency (1–10 kHz) acoustics programs that were part of the Shallow Water 2006 Experiment includes descriptions of design goals, the experimental environment, and instrumentation deployed. Recent progress on the data analysis from the acoustics experiments is summarized.

Keywords: Mid-frequency, shallow water, propagation.
PACS: 43.30.Ft, 43.30 Re, 43.30 Bp

INTRODUCTION

To improve basic understanding of shallow water acoustics and its dependence on oceanographic and geoacoustic environments, the Office of Naval Research (ONR) sponsored a multi-disciplinary, multi-institution experiment off the coast of New Jersey in the summer of 2006, designated "Shallow Water '06" (SW06). A general description of the scope and execution of the fieldwork is given in Tang et al. [1]. This paper summarizes the mid-frequency (1–10 kHz) component of the experiment, including the various field measurements and some of the research results.

The mid-frequency band in underwater acoustics is unique in that with moderate attenuation, sound can travel to 10s of kilometers and provide an effective means for remote sensing. Sound in this frequency range is of wavelengths that offer finer resolution for detection purposes. Therefore, many practical sonar systems operate in the mid-frequency band. Few experiments emphasizing basic mid-frequency acoustics research have been fielded. A major challenge in understanding and modeling mid-frequency acoustics in shallow water is the environmental variability typical of shallow waters; e.g., internal waves and sediment heterogeneity have strong impacts on mid-frequency sound propagation. Modeling mid-frequency sound with either a modal or ray approach is complicated by a large number of modes or multiple boundary interactions, respectively.

The goal of SW06 was to study in detail the impact of various environmental processes on acoustics. The experiment was centered on the 80-m isobath of the U.S. continental shelf off Atlantic City, New Jersey. In such shallow water regions, the performance of mid-frequency systems is strongly influenced by the bottom, sea surface, and the water column. This influence can be in the form of sound scattering, refraction and attenuation. Uncertainties in any or all of these categories are common, and thus knowing the various mechanisms governing acoustic interaction with the environment is essential to improve sonar performance in shallow water. To gain quantitative understanding of how sound interacts with these environmental uncertainties, it is also essential to measure all relevant environmental parameters at sufficient spatial and temporal resolutions to facilitate model/data comparisons.

CP1272, *Shallow-Water Acoustics, Second International Shallow-Water Acoustics Conference*
edited by J. Simmen, E. S. Livingston, J. X. Zhou, and F. H. Li

ENVIRONMENTAL MEASUREMENTS

The mid-frequency efforts of SW06 were concentrated on receiving arrays moored in the central area at 39° 1.5 N and 73° 2.5 W, where the water depth is about 80 m. The bottom has sandy sediments. During the experiment, a typical strong summer thermocline existed, and in addition, warm, salty slope water underlay the colder mid-depth water. Thus, there was a sound speed minimum at ~32 m depth, forming a sound channel. In this central area, extensive environmental measurements were made to support the mid-frequency modeling as well as to assess model uncertainties. Three sets of different environmental measurements are described here.

The New Jersey shelf during the summer has an active internal wave field. One part of SW06 was designed to provide sufficient oceanographic measurements of non-linear internal waves (NLIWs) so that the two-dimensional sound speed fields (range and depth) in the acoustic path at the time of sound transmissions could accommodate deterministic modeling of the effects of NLIWs on sound propagation. NLIWs were measured by a towed conductivity-temperature-depth (CTD) chain, which provided a two-dimensional sound speed field over time (Figure 1). The measurements dedicated to acoustics lasted for 2.5 hours, during which the CTD chain was towed from the R/V *Endeavor* in seven circuits around the propagation path between the moored receiving array and the stationary source deployed from the R/V *Knorr*. One and one-half circuits in the middle of this time period were selected for analysis, as the large waves at the front of the packets passed the acoustic path during this time; detailed data analysis is given in Henyey et al. [2].

Acoustic interaction with the sea bottom is, in many cases, an essential component of sound propagation in a shallow water waveguide. *In situ* geoacoustic properties of the seabed, however, are difficult to obtain (e.g., see [3–5]). As part of the experimental effort in SW06, a unique Sediment Acoustic-speed Measurement System (SAMS) was developed to measure the sediment sound speed directly to a depth of 1.7 m (Figure 2). This is accomplished by using a vibro-core fitted with a hydrophone at its tip to penetrate into the bottom in controlled depth-steps while 10 acoustic sources located above the bottom transmit sound in the band of 2–35 kHz. This source and receive arrangement offers many ray paths crisscrossing the sediment volume, so sediment sound speed as a function of depth is measured directly. Data analyses demonstrate that SAMS can provide reliable sediment sound speeds. Calibration data taken in the water column were first investigated to establish the system uncertainty — about 3 m/s. In-sediment data were taken at three positions. Sediment sound speeds and uncertainties measured in SW06 at positions 1, 2, and 3 were 1618 ± 11, 1598 ± 10, and 1600 ± 20 m/s, respectively [6] (Figure 3).

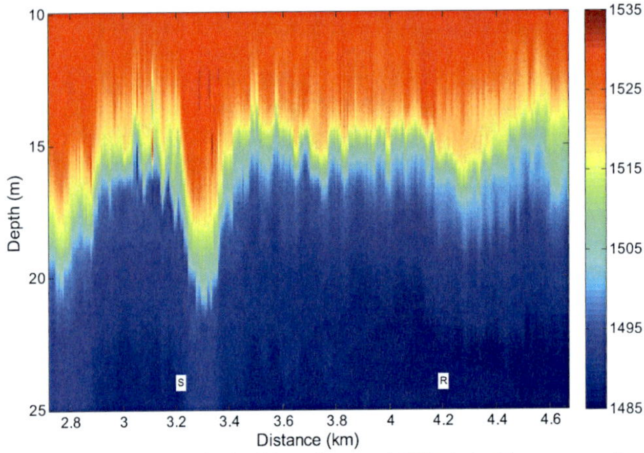

FIGURE 1. Sound speed field obtained from the towed CTD chain. The source and receiver locations relative to the waves are shown.. The horizontal axis is distance projected to 300° [2].

FIGURE 2. Sediment Acoustic-speed Measurement System (SAMS) during deployment. Seven sources (ITC1032's) are mounted on the horizontal beam and three on the triangular frame. A single hydrophone is fitted inside the vertical shaft that is driven into the sediment in a controlled manner by a vibro-core at the top of the system. The SAMS was operated from on board of the R/V Knorr, which was under dynamic position.

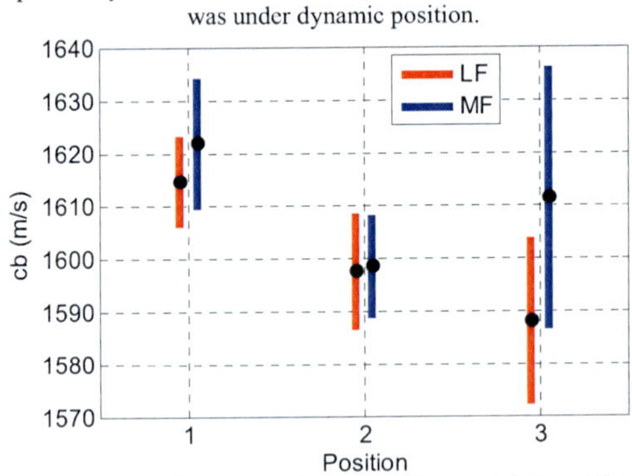

FIGURE 3. Measured sediment sound speeds with uncertainties at three positions [6].

To support the modeling of bottom backscatter, an existing sediment conductivity probe (In situ Measurement of Porosity, second generation—IMP2 [7]) and a new laser liner scanner were deployed jointly for the first time to measure bottom roughness spectra over the same location [8]. This offered an opportunity to compare the measured spectra from both instruments. Bottom roughness has been measured extensively in coastal waters where diver support is available. The measurements are gathered by analogue and digital stereo photography [9–11], laser line scanning [12–13] and sediment conductivity probes [7]. The IMP2 has a single electrode probe mounted on a 4-m-long track. It samples vertical profiles with 1-mm vertical resolution and at 1-cm horizontal increments along the track. Bottom roughness is inferred by a detection of sudden changes in conductivity in each vertical profile [7]. To improve spatial resolution and coverage of bottom roughness measurements, a self-contained laser line scanning system (Seafloor Laser Scanner—SLS), was mounted on the IMP2 carriage[8]. During SW06 the combined system was deployed to 80 m depth to measure the seafloor roughness. Measurements began with the IMP2 run over the 4-m-long track followed by laser scanning over the same piece of seafloor. The swath of the laser line projected on the seafloor, constrained by the height of the IMP2 carriage from the seafloor, was 30 cm. The overall area sampled was 30 cm × 350 cm with a sub-mm resolution in both vertical and horizontal directions [8]. The joint deployment of the instruments improved overall measurements in three areas: (1) extension of the measurements to higher wavenumbers and to two dimensions, thus improving the signal-to-noise ratio of the measured power spectra; (2) cross-comparison of the two independent measurement methods confirmed the validity of these techniques; (3) the increased SLS spatial resolution detected shells and shell fragments widely distributed on the seafloor, which may play an important role in acoustic backscatter at certain frequencies [8]. Sample results from the roughness measurements are given in Fig. 4

ACOUSTICS MEASUREMENTS

The acoustic propagation measurements were made on the moored receiving array (MORAY1) [14], which has two four-element vertical sub arrays with the top four elements at depths 25.0, 25.2, 25.5, and 26.4 m, and the bottom four at 50.0, 50.2, 50.5, and 51.4 m. The mid-frequency sound source, an ITC-2015 transducer, was deployed off the stern of the R/V *Knorr* at 30 m depth, close to the sound axis to emphasize the propagation of energy trapped in the channel. The waveform used in analysis is a 20-ms linear frequency-modulated chirp, which spanned the 1.5 to 10.5-kHz frequency band with a Hann window and 10 percent taper. Transmissions were repeated about every 20 s. The propagation experiment was divided into two segments. The first was at either 0.5-km or 1-km range over an extended period of time, lasting over 6 hours and including 1500 ping repetitions. This provides ample data for statistical analysis of field fluctuations. The second segment was at 10 stations along the 80-m isobath at ranges 1, 2,..., 10 km. At each station, 100 pings were transmitted.

The major goal for the long-range propagation experiment is to understand transmission loss (TL). This includes the mean intensity and the fluctuation of intensity. During SW06, the combination of the sound channel and ambient internal waves had a large impact on mid-frequency sound propagation. It was found that most of the sound is trapped in the sound channel where water column variability dominated the acoustics field fluctuations. Because of the presence of closely packed caustics at convergence regions, ray theory cannot be used to calculate sound intensity levels accurately. The intensity was estimated from data and compared to an incoherent mode sum model (Figure 5). The comparison indicates that the mean intensity can be modeled to within 2 dB for most of the ranges and frequencies studied, but there are exceptions where the difference between model and data exceed 4 dB. Although the differences between the model and data are small, they are statistically significant [15].

Sound intensity fluctuations at 1–9-km range stations were investigated through the single frequency scintillation index. The intensity fluctuations are due to water column sound speed variability from ambient internal waves; the data were collected on a day when there were no strongly non-linear internal waves present. Two findings were not anticipated before the experiment: intensity is only correlated over a narrow frequency band (50–200 Hz) and the bandwidth is independent of center frequency and range; the intensity probability distribution peaks at zero for all frequencies, and follows an exponential distribution, indicating possible saturation. Saturation is a result of contributions from multiple, statistically independent sources. However, it is not yet understood why at a range as short as 1 km, the intensity could be saturated as measured. Deep fades in intensity were observed which manifest as uncertainty in predicting mid-frequency sound propagation in shallow water [16].

The short-range propagation experiment was conducted in concurrence with the sound field measurements using the towed CTD chain. The goal was to model sound propagation in the presence of non-linear internal waves in the acoustic path in a deterministic manner. The acoustic range as nominally 1000 m, the source was positioned at 30 m depth, and the receiver used in this analysis was 25 m deep. The acoustic source was deployed from the stern of the R/V *Knorr*, which was positioned dynamically at a fixed point. The acoustic path was at a bearing of 300 degrees. The transmitted acoustic signal spanned the frequencies 1.5 to 10.5 kHz. A set of direct-path propagation data was collected at the 50-m range as calibration, so the transmitted waveforms were known.

Acoustic transmission started at 14:14 and ended 18:49 UTC August 13 with a nominal 14-s ping repetition rate, resulting in 1100 pings recorded. During this period, a NLIW wave train passed the acoustic path. Extensive modeling of the internal waves was made [2] to predict the impact of the NLIW on acoustic propagation. The uncertainty of the source position excluded the possibility of coherent ping-to-ping processing and motivated use of the highest arrival of each ping as reference time zero to align all the pings. The highest arrival almost always occurred during the arrival of sound propagating near the minimum sound speed depth. Internal waves have little effect on the sound speed near its minimum, making this a good reference. We define the reduced travel time as the actual travel time shifted by the amount corresponding to the reference time.

The arrivals and modeling results show that the arrivals have a changing structure versus geo-time (Figure 6). They tend to get closer to one another starting from about 16:30. Notably, three of the arrival times have oscillatory patterns. Different arrivals were identified through ray tracing and details are in given in Henyey et al. [2]. The overall spread in arrival times is reduced by about 1 ms after 16:30, with the reduction occurring between arrivals 4 and 5. Using the measured NLIW waveform and traveling speed, we were able to replicate the various oscillations of all the arrivals observed in the acoustic data (Figure 6).

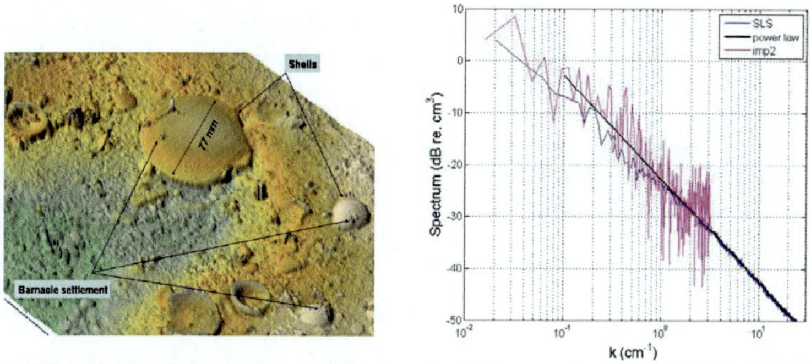

FIGURE 4. Shaded relief map of the seafloor reconstructed from a subset of the SLS data (left), where fine features can be identified such as barnacle settlements on shell surfaces. Comparison of one-dimensional power spectra from IMP2 data (pink), one-dimensional average spectra from SLS along-track data (blue), and power-law fit (black) for the first deployment.

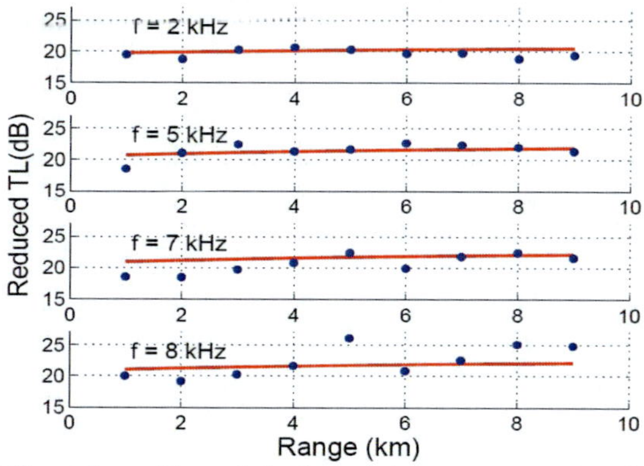

FIGURE 5. Comparisons of transmission losses between measurements and a model at four frequencies, each with a 100-Hz bandwidth. The data are from the mean of 100 pings and the model is the incoherent sums of all trapped modes.

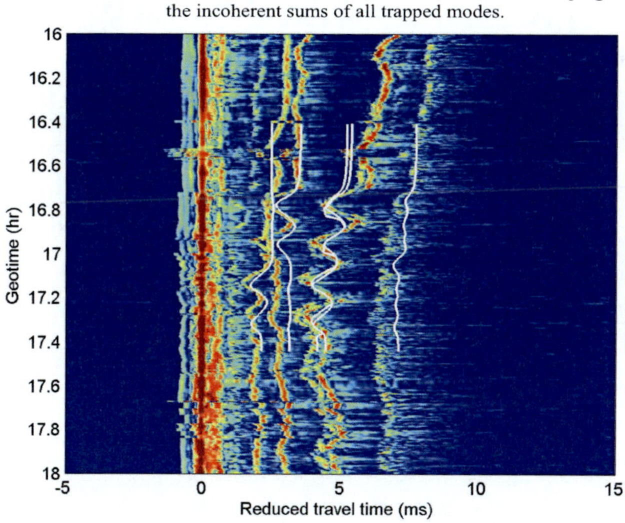

FIGURE 6. Comparison of acoustic data with ray tracing results (white curves) for arrivals (from left to right as indicated by the white curves) between 16:24 and 17:26. The times, temporal extent, and magnitude of travel time oscillations were calculated deterministically, and are close to the measured values.

ACKNOWLEDGMENTS

The following people contributed to the article: Frank S. Henyey, Kevin L. Williams, Daniel Rouseff, Peter H. Dahl, and Jie Yang of the Applied Physics Laboratory, University of Washington; Jorge Quijano of Portland State University; Zhongkang Wang of the Hangzhou Applied Acoustics Research Institute (China); and Jee Woong Choi of Hanyang Univerisity (Korea). The work is funded by the Office of Naval Research.

REFERENCES

1. D. Tang, J. Moum, J. Lynch, P. Abbott, R. Chapman, P. Dahl, T. Duda, G. Gawarkiewicz, S. Glenn, J. Goff, H. Graber, J. Kemp, A. Maffei, J. Nash, and A. Newhall, *Oceanography* **20** (2007).
2. F. S. Henyey, K. L. Williams, J. Yang, and D. Tang, *IEEE J. Ocean. Eng.* (in press).
3. A. Turgut and T. Yamamoto, *J. Acoust. Soc. Am.* **87**, 2376-2383 (1990).
4. A. Barbegelata, M. D. Richardson, B. Miaschi, E. Muzi, P. Guerrini, L. Troiano and T. Akal, "ISSAMS: An in situ sediment acoustic measurement system," in *Shear Waves in Marine Sediments*, edited by J. M. Hoven, M. D. Richardson, and R. D. Stoll, Kluwer Academic Publishers, 1991.
5. S. S. Fu, R. H. Wilkens, and L. N. Frazer, *J. Acoust. Soc. Am.* **99**, 234-242 (1996).
6. J. Yang, D. Tang, and K. L. Williams, *J. Acoust. Soc. Am.* **124**, EL116 (2008).
7. D. Tang, *IEEE J. Ocean. Eng.* **29**, 929-939 (2004).
8. C.-C. Wang, and D. Tang, *IEEE J. Ocean. Eng.* **34**, 459-465 (2009).
9. K. B. Briggs, *IEEE J. Ocean. Eng.* **14**, 360–367 (1989).
10. A. P. Lyons, T. Akal, and E. Pouliquen, "Measurement of seafloor roughness with close range digital photogrammetry," *in IEEE OCEANS '98 Conference Proceedings*, New York, 1998.
11. A. P. Lyons, W. L. J. Fox, T. Hasiotis, and E. Pouliquen, *IEEE J. Ocean. Eng.* **27**, 515–524 (2002).
12. A. M. Crawford and A. E. Hay, *IEEE J. Ocean. Eng.* **23**, 12–19 (1998).
13. K. D. Moore and J. S. Jaffe, *IEEE J. Ocean. Eng.* **29**, 525-545 (2004).
14. P. H. Dahl, J. W. Choi, N. J. Williams, and H. C. Graber, *J. Acoust. Soc. Am.* **124**, EL163 (2008).
15. D. Tang, F. S. Henyey, Z. Wang, K. L. Williams, D. Rouseff, P. H. Dahl , J. Quijano, and J.W. Choi, *J. Acoust. Soc. Am.* **124**, EL85 (2008).
16. D. Tang, F. S. Henyey, Z. Wang, K. L. Williams, D. Rouseff, P. H. Dahl, J. Quijano, and J.W. Choi, *J. Acoust. Soc. Am.* **124**, EL91 (2008).

Prufer Transformations for the Normal Modes in Ocean Acoustics

Arthur B. Baggeroer

Massachusetts Institute of Technology, Cambridge, MA USA 02139

Abstract. In 1926 Prufer introduced a method of transforming the second order Sturm-Liouville (SL) equation into two nonlinear first order differential equations for the phase ϕ and "magnitude", $|\phi^2 + \dot{\phi}^2|$ for a Poincare phase space representation, $(\phi, \dot{\phi})$. The useful property is the phase equation decouples from the magnitude one which leads to a nonlinear, two point boundary value problem for the eigenvalues, or SL numbers. The transformation has been used both theoretically, *e.g.* Atkinson, [1960] to prove certain properties of SL equations as well as numerically *e.g* Bailey [1978]. This paper examines the utility of the Prufer transformation in the context of numerical solutions for modes of the ocean acoustic wave equation. (Its use is certainly not well known in the ocean acoustics community.) Equations for the phase, ϕ, and natural logarithm of the "magnitude", $\ln(|\phi^2 + \dot{\phi}^2|)$ lead to same decoupling and a fast and efficient numerical solution with the SL eigenvalues mapping to the horizontal wavenubers. The Prufer transformation has stabilty problems for low order modes at high frequecies, so a numerically stable method of integrating the phase equation is derived. This seems to be the first time the these stability issues have been highlighted to provide a robust algorthim for the modes.

Keywords: wave equation, Prufer transformation, stability
PACS: 43.20B

INTRODUCTION

The solution for the normal modes of the acoustic wave equation can be formulated in terms of a Sturm-Liouville equation. The Prufer transformation method for solving Sturm-Liouville equations was introduced by Prufer in 1926 [16]. The method is straightforward: i) the second order Sturm-Liouville differential equation is represented in a state, or phase, space; ii) a transformation is made to polar coordinates; iii) two differential equations for the phase and magnitude of the polar reperesentation are derived. The important results are i) the equation for the phase is decoupled from that for the magnitude, so one obtains a single nonlinear differential equation with a two point boundary condition which can be solved iteratively for the Sturm-Liouville eigenvalues and ii) the equation for the natural logarithm of the magnitude is a direct integration of a function of the phase and eigenvalue. In addition to the Prufer transformation Milne [12] introduced a method where one aslo uses a magnitude and phase representation for the solution; however, there seems to be less note of it in the literature. Finally, while the WKBJ method uses magnitude and phase representations, it is not the same as the Prufer transform; in particular, the Prufer method is an exact one.

The Prufer method has become well known. Several textbooks now include a discussion of the Prufer method and its several variations [2, 17]. Most discussions emphasize an-

CP1272, *Shallow-Water Acoustics, Second International Shallow-Water Acoustics Conference*
edited by J. Simmen, E. S. Livingston, J. X. Zhou, and F. H. Li

alytic solutions which can be obtained using the method. Two often cited articles are by Birkhoff and Rota, [8] and Bailey, *et al.* [6]. Birkhoff and Rota used the Prufer transformation to derive several important properties of Sturm-Liouville equations while Bailey *et al.* introduced a numerical algorithm based upon the Prufer transformation. Several variations of the basic Prufer transformation have also been introduced. One is the "modified" Prufer transformation which has been used to derive asymptotic solutions to the Sturm-Liouville equation; however, this "modified" method can not be applied to most acoustic normal mode programs [8, 9].

The Prufer transformation has also been introduced in the context of invariant imbedding [7]. With the invariant imbedding approach one also derives uncoupled nonlinear Ricatti differential equations and iterates their solution for the eigenvalues. These equations satisfy an initial value problem instead of a two point boundary value problem so one can solve them up to the interval endpoint. In an acoustics context this implies solving for the eigenvalues as a function of depth up to the desired depth. The approach also involves switching between two Ricatti equations to keep stability aspects under control. The arguments made for invariant imbedding are i) it can be generalized to higher order systems and ii) the differential equations are simpler and easier to evaluate for numerical integration purposes. The first argument is in general correct; however, elastic propagation is probably the most important higher order extension for ocean acoustics and one can extend the Prufer transformation to include elastic coupling at interfaces. In answer to the second argument we introduce Prufer transformations which are just as simple as those for invariant imbedding and which highlight the numerical issues relevant to obtaining stable solutions. In addition, one needs to iterate for the eigenvalue so the advantage of solving an initial value problem, which is the main advantage of the invariant imbedding approach, *versus* a two point boundary problem is not an issue.

In this article we consider numerical algorithms using the Prufer transformation for acoustic normal modes in an oceanic waveguide. Bailey *et al.* were the first to develop such algorithms for the general Sturm-Liouville equation and they produced a code SLEIGEN which is publicly available. Here, we examine some of the numerical issues in more detail. We also introduce an alternative phase, or state, space which leads to simpler equations for the phase and magnitude differential equations and this results in a modest reduction in the computation required. Second, we consider the stability issues in integrating the phase equation. In contrast to the apparently arbitrary means of chosing directions of integration to shoot from the end points to a midpoint of the interval in order to satisfy the two point boundary problem as suggested by Bailey *et al.*[6], we demonstrate that the directions of integration are crucial in obtaining correct solutions for truly ducted problems where the top and/or bottom boundary conditions are numerically decoupled from the solution in the duct. We also consider applying the Prufer method to a double duct problem, one of the more challenging acoustic wave guide problems [14], to highlight the direction of integration issue.

The algorithms based upon Sturm sequences and Richardson extrapolation developed by Porter and Reiss have been extensively tested and are the basis for the popular codes

SuperSNAP and KRAKEN [14?] [1]. While we have not done exhaustive comparisions for either speed or accuracy, our numerical results are in excellent agreement with those obtained by SuperSNAP and KRAKEN. For example, group speeds, one of the more sensitive calculations for a normal mode program, agree to seven figures.
Finally, the importance of fast and accurate numerical methods for normal mode solutions does not need to be emphasized in ocean acoustics. There are many methods now available. It is not our intent here to contrast the Prufer transformation approach to many normal mode programs now available. We suggest the bibliography of Chapter 5 of Jensen *et al.*, [11], for a excellent summary of many of these programs. We simply offer it as an interesting method which some may find useful for numerical problems.

NOTATION AND WAVE EQUATION FORMULATION

We consider the velocity potential, $\phi_n(z)$, for a laterally homogeneous medium where the sound speed may vary continuously in depth and the density is piecewise constant. The wave equation for the normal modes is given by [2]

$$\frac{\partial^2\phi_n(z)}{\partial z^2} - \left(k_n^2 - (\frac{\omega}{c(z)})^2\right)\phi_n(z) = 0, \tag{1}$$

where $\phi_n(z)$ is the n th mode for the velocity potential;
k_n is the horizontal wave number for the n th mode;
ω is the temporal radian frequency;
$c(z)$ is the sound speed profile.

At a density interface at depth Z the continuity equations for a velocity potential are

$$\begin{aligned} \rho_-\phi_n(Z_-) &= \rho_+\phi_n(Z_+) \\ \frac{\partial\phi_n(z)}{\partial z}|_{Z_-} &= \frac{\partial\phi_n(z)}{\partial z}|_{Z_+} \end{aligned} \tag{2}$$

where Z_+, Z_-, ρ_+ and ρ_- indicate the limiting values approaching the top and bottom of the interface.
In ocean acoustics the surface boundary condition at $Z = 0$ is given by

$$\phi_n(0) = 0; \tag{3}$$

[1] SuperSNAP is a modification of SNAP where H. Schmidt has replaced the numerical kernel of SNAP (SACLANTCTR Normal mode Acoustic Propagation model) with a implementation based upon the Porter and Riess approach [10?].
KRAKEN: a fabulous Scandinavian sea monster (Norwegian),[15]

[2] The notation for the dependence of k_n and $\phi_n(z)$ upon the frequency, ω, is suppressed throughout the paper.

The bottom boundary condition at $z = H$ usually assumes one of two forms: a rigid bottom or a half space. For a rigid bottom, the condition is

$$\frac{\partial \phi_n(z)}{\partial z}|_H = 0. \tag{4}$$

This boundary condition leads to a discrete spectrum for the horizontal wavenumbers, k_n and a complete, orthonormal set of modes, $\phi_n(z)$; however, only a finite of these have real wavenumbers and correspond to propagating modes while there are an infinite number which have purely imaginary wavenumbers and correspond to modes evanescent in range. For a half space boundary the sound speed and density are constant for $z > H$ and equal to $c(H)$ and $\rho(H)$ respectively. The condition is that the propagating modes be bounded within the half space, or

$$|\phi_n(z)| \leq M_\phi < \infty, \; z > H. \tag{5}$$

This condition leads to a spectrum with both discrete and continuous components. In this paper we are concerned only with the discrete, propagating modes. These have wavenumbers, k_n, in the range [11]

$$\frac{\omega}{c(H)} \leq k_n \leq \frac{\omega}{c_{min}} \tag{6}$$

where c_{min} and c_{max} are the minimum and maximum of the sound speed profile. [3] For the velocity potential the modes are orthonormal with respect to aweighting by the density, or

$$\int_0^{Z_{max}} \rho(z)\phi_m(z)\phi_n(z)dz, \tag{7}$$

where Z_{max} is equal to H for a rigid bottom or equal to ∞ for a half space bottom.
We now transform this wave equation to a normalized Sturm-Liouville form. First, we define λ_{min} to be the wavelength given by

$$\lambda_{min} = \frac{c_{min}}{f} \tag{8}$$

where $f = \omega/2\pi$. We next define the normalized depth dimension, ζ, to be

$$\zeta = z/\lambda_{min}, \tag{9}$$

so typically the depth is scaled in terms of the wavelength at the sound channel axis. It is also convenient to convert to slowness quantities with $s(z) = 1/c(z)$, $s_n = k_n/\omega$ and $s_{max} = 1/c_{min}$ where $s(z)$ is the column slowness and s_n is the modal phase slowness.

[3] The half space boundary condition does not lead to a "proper" Sturm-Liouville equation. The consequences of this are discussed in [11]. The continuous components can be found with a trivial extension of the method.

We now define two quantities which lead to the Sturm-Liouville form; we have

$$q^2(\zeta) = 1 - \left(\frac{s(\zeta\lambda_{min})}{s_{max}}\right)^2, \qquad (10)$$

$$\sigma_n^2 = 1 - \left(\frac{s_n}{s_{max}}\right)^2. \qquad (11)$$

We note that

$$0 \leq q^2(\zeta) \leq 1$$

since s_{max} is the maximum slowness in the profile. Similarly, we can show from Eq. 6 that $s_n < s_{max}$ and for all the propagating modes

$$0 < \sigma_n^2 < 1.$$

It now is straightforward to show that with this scaling and these substitutions that the wave equation has the following normalized, Sturm-Liouville form

$$\frac{d^2\phi_n(\zeta)}{d\zeta^2} + (2\pi)^2(\sigma_n^2 - q^2(\zeta))\phi_n(\zeta) = 0. \qquad (12)$$

The continuity conditions at a density interface (Eq. 2) and the boundary conditions (Eqs. 3 - 5) remain the same. Eq. 12 the standard Schrodinger equation for the modes of a single degree of freedom, harmonic oscillator where $(2\pi q(\zeta))^2$ scales to potential energy and $(2\pi\sigma_n)^2$ to the modal energy.[4]

THE "STANDARD" PRUFER TRANSFORMATION

The Prufer transformation starts with using a state space, or dynamical system, representation for the Sturm-Liouville equation as commonly done in problems of optimal control and filtering. We convert the second order equation into two coupled first order equations and representing the solution in a two-dimensional state space. In the standard Prufer transformation one starts with $\phi_n(\zeta)$ and its first derivative, $\psi_n(\zeta)$. [5] We define the notation

$$\psi_n(\zeta) = \frac{d\phi_n(\zeta)}{d\zeta}; \qquad (13)$$

this leads to the state equations for Eqn. 12:

$$\frac{d}{d\zeta}\begin{bmatrix} \phi_n(\zeta) \\ -- \\ \psi_n(\zeta) \end{bmatrix} = \begin{bmatrix} 0 & | & 1 \\ ---------- & | & -- \\ -(2\pi)^2(\sigma_n^2 - q^2(\zeta)) & | & 0 \end{bmatrix} \begin{bmatrix} \phi_n(\zeta) \\ -- \\ \psi_n(\zeta) \end{bmatrix}. \qquad (14)$$

---

[4] The $(2\pi)^2$ factoring is introduced for convenience in preserving the bounding of $q^2(\zeta)$ and σ_n^2 within the interval $(0,1)$.

[5] In the literature the Prufer transformation is usually discussed in terms of the general Sturm-Liouville equation.[8, 6] This leads to a slightly different definition of the second state variable. There is also a modified Prufer method which involves derivative of $q^2(\zeta)$.[8, 9]

For the standard Prufer transformation we consider the state vector to specify a point in the two dimensional phase space and make a transformation to polar coordinates, or actually to the phase and the natural log of the magnitude. We define [6]

$$\Theta(\zeta,\sigma_n) = \arctan(\psi_n(\zeta),\phi_n(\zeta)) \tag{15}$$

and

$$M(\zeta,\sigma_n) = \ln\left[\sqrt{\phi_n(\zeta)^2+\psi_n(\zeta)^2}\right], \tag{16}$$

where $\arctan(y,x)$ indicates the four quadrant arctangent function. Direct differentiation of these definitions yield the following equations for the phase and ln of the magnitude

$$\begin{aligned}\frac{d\Theta(\zeta,\sigma_n)}{d\zeta} &= -(2\pi)^2\left[\sigma_n^2-q^2(\zeta)\right]\cos^2(\Theta(\zeta,\sigma_n))-\sin^2(\Theta(\zeta,\sigma_n)) \qquad (17)\\ &= -\frac{1}{2}\left[1+(2\pi)^2\left(\sigma_n^2-q^2(\zeta)\right)\right]+\frac{1}{2}\left[1-(2\pi)^2\left(\sigma_n^2-q^2(\zeta)\right)\right]\cos(2\Theta(\zeta,\sigma_n))\\ \frac{dM(\zeta,\sigma_n)}{d\zeta} &= \frac{1}{2}\left[1-(2\pi)^2(\sigma_n^2-q^2(\zeta))\right]\sin(2\Theta(\zeta,\sigma_n)) \qquad (18)\end{aligned}$$

The continuity condition at a density interface at $\zeta_i = Z_i/\lambda_{min}$ is

$$\tan(\Theta(\zeta,\sigma_n))|_{Z_+/\lambda_{min}} = \frac{1}{\gamma}\tan(\Theta(\zeta,\sigma_n))|_{Z_-/\lambda_{min}}, \tag{19}$$

$$\gamma \triangleq \frac{\rho_+}{\rho_-}. \tag{20}$$

Note that the phase does not jump across a principal value at a density interface. The surface boundary condition is

$$\Theta(0,\sigma_n) = \left(n-\frac{1}{2}\right)\pi,\ n=1,2,\cdots \tag{21}$$

and the bottom boundary condition is

$$\Theta(H/\lambda_{min},\sigma_n) = \left[\begin{array}{ll} 0 & \text{rigid} \\ \arctan(2\pi\sqrt{1-(\frac{s(H)}{s_{max}})^2-\sigma_n^2},1) & \text{half space}\end{array}\right. \tag{22}$$

Each value of n leads to one of the normal modes; consequently, each mode has distinct boundary conditions so one does not need to count zero crossings to determine to which mode an eigenvalue corresponds. The bottom boundary condition can also be satisfied at

[6] Most discussions use the magnitude and not its logarithm; however, the large dynamic range of the magnitude makes it very useful numerically to use the logarithm. The importance of this was recognized in Ref. [6] in the discussion of an algorithm for implementing the standard Prufer transformation.

$m\pi$ offsets, but it is only the difference in the phase between the two boundary conditions which is significant.

These are the standard Prufer equations most commonly found in the literature, *e.g.* [8] The equation for the phase, $\Theta(\zeta, \sigma_n)$, is a one dimensional, nonlinear equation and is not coupled to the magnitude equation and the equation for $M(\zeta, \sigma_n)$, the ln of the magnitude involves a direct integration of a nonlinear function of the phase, $\Theta(\zeta, \sigma_n)$; consequently, the fundamental issue of the Prufer transform method is to find the eigenvalues, σ_n^2, which leads to a trajectory that satisfies the boundary conditions upon $\Theta(0, \sigma_n)$ and $\Theta(H/\lambda_{min}, \sigma_n)$. Hence, the numerical issues primarily concern the ease of integrating Eq. 17.

There are several numerical issues in integrating Eq. 17 for problems in ocean acoustics. The first concerns the stability of of the solutions. In most problems there are two regions for the modes. When $\sigma_n^2 > q^2(\zeta)$, the modes decay towards the surface and bottom boundaries; while for $\sigma_n^2 < q^2(\zeta)$, the modes have an oscillatory form. These conditions translate into conditions for the stability loci of $\Theta(\zeta, \sigma_n)$ *versus* depth and in turn upon the direction for integrating Eq. 17. This is a manifestation of the growing and decaying exponential solutions in the original formulation of the problem; the stability points and the directions of integration relate to obtaining the appropriate exponential solutions.

The computation time is dominated by the ease in which the right side of Eq. 17 can be evaluated since the phase trajectory must be integrated many times for the trial solutions *versus* σ_n^2; consequently, it is important to keep the expression for $\dot{\Theta}(\zeta, \sigma_n)$ as simple as possible. For example, the Eq. 17 requires the evaluation of one trigonometric function plus three additions and two multiplications for each integration step. We have found an alternate Prufer transformation which reduces this computation somewhat[4]. Finally, in many ocean acoustics problems there are discontinuities, or very rapid changes, in the sound speed at interfaces, most importantly at and within the seabed. This implies that the derivative in Eq. 17 above has at the discontinuities of $q^2(\zeta)$ and the need to treat these interfaces carefully.

INTEGRATING THE PHASE EQUATION

We consider the integration of the phase equation, Eq. 17, for the standard Prufer. We first examine a stability analysis at a fixed depth, ζ_o, of Eq.17.

In Figure 1 we plot the derivatives, Eqs. 17 and 18, for the phase and magnitude when $\sigma_n > q(\zeta_o)$. Here, the phase derivative is always negative so the phase decreases continuously in any region satisfying this condition. This corresponds to the oscillatory region for the modes. The phase is decreasing and hence changing the sign of the modeshape. Note that when $\sigma_n >> q(\zeta_o)$, or deep within a duct, the phase is decreasing approximately linearly.

In Figure 2 we plot the same phase and magnitude expressions when $\sigma_n < q(\zeta_o)$. Here the phase derivative has stability points when $\dot{\Theta} = 0$, or

$$\Theta(\zeta, \sigma_n) = \arctan\left(2\pi\sqrt{[\sigma_n^2 - q^2(\zeta_o)]}\right).$$

The circular points are stable while the square ones are unstable, *i.e.* a solution near a stable point tends to converge to the stability point while it diverges from an unstable one. Note that this analysis assumes integration in the positive direction, or down from the surface; the stability points are reversed if integrating in the negative direction, or up from the bottom. The derivative for the magnitude, $\dot{M}$, is negative at an unstable point and positive at a stable one. This region corresponds to evanescent region for the modes. Here the solution "locks" near a stable point and is "repelled" from an unstable one, so it is important to have the integration algorithm for the phase equation which exploit this behavior. It is numerically very difficult to keep a solution near an unstable equilibrium point over an extended region. These stability points of the phase equation for the Prufer method dictate the direction of integration in each of the evanescent regions.

We consider the integration scenario for a sound speed profile with a sound channel axis. For this $q^2(\zeta)$ has a minimum near the channel axis and increases towards both the surface and bottom. Integrating down from the surface boundary condition the phase "locks" onto the stability point which is greatest to the initial condition and this leads to solutions in the evanescent region with increasing magnitude. As the depth approaches the axial region, the separation between the stable and unstable stability points narrows and the stability point eventually vanishes when $q^2(\zeta) = \sigma_n^2$; here the solution enters the oscillatory, or mode formation, region. The phase continues to decrease until the bottom evanescent region is re-established. One wants to have a solution with decreasing magnitude to satisfy the bottom boundary condition; however, numerically the solution locks onto a stable point leading to an increasing magnitude. It becomes impossible to keep the solution near the unstable point with the decreasing magnitude especially if for an extended evanescent region.

The way around this is to integrate a second solution up from the bottom boundary condition. The reversal of the stability points leads to the desired increase in the magnitude as the depth decreases. The characteristic equation, or condition for eigenvalue, is imposed by requiring that the two solutions match, therefore their derivatives as well, at an interior point in the oscillating, or mode formation, region. Usually, we have chosen the sound channel axis for this point.

More generally, the stability analysis highlights one of the important numerical issues in determining the normal modes. In the evanescent region the local solution has both exponentially growing and decaying components. In many applications, especially for low order modes at high frequencies, the equations are quite "stiff" and integrating them directly has numerical difficulties because of the wide dynamic range between the two exponentials. One method for circumventing this is by a change of variables referred to as a Ricatti transformation [1]. This transformation can highlight the desired solution by a sign choice; nevertheless, one still needs to be attentive to the direction of integration since Ricatti equations become unstable if integrated in the wrong direction over long extents. In fact, much the same issue appears with an invariant imbedding approach.[7] Here, the suggestion is made to alternate between the two Ricatti equations to circumvent the stability issue. The important aspect of the Prufer transform is that an analysis of its stability points indicates the preferred direction of integration so the desired solution is numerically stable.

Fig. 3 illustrates the solution for the phase trajectory obtained by integrating from each boundary and requiring equality for the solution at the minimum of the sound speed channel. This is done by iterating the eigenvalue until convergence is obtained at σ_2^2. Fig. 4. illustrates the solution for the magnitude obtained when the phase trajectory solution is used. The mode itself is given by

$$\phi(\zeta) = exp^{M(\zeta|\sigma_2^2)} \cos(\theta(\zeta|\sigma_2^2))$$

CONCLUSION

We have outlined the application of the Prufer method for solving the wave equation in ocean acoustics. We have also highlighted a significant numerical analysis problem with the method and a solution for it based upon be judicious of where the boundary conditions are applied. The method has been used extensively by students at MIT where it has proven to be relibable and acccurate. We have not seen it used in the ocean acoustics literature. A subsequent article extends this to a second and faster set of Prufer equations as well as addresses the more complicated problem of a double duct which again highlights the need for choosing the point where to apply the boundary conditions carefully.

REFERENCES

1. F. S. Acton, *Numerical Methods That Work* Harper and Row, New York, 1970
2. U. M. Ascher, R. M. M. Mattheij and R. D. Russell, *Numerical Solution of Boundary Value Problems for Ordinary Differential Equation,* Prentice Hall, Englewood Cliffs, New Jersey, 1988
3. Atkinson, F.V *Discrete and Continuous Boundary Value Problems, Academic Press, New York, 1964*
4. Baggeroer, A.B., *Prufer Transformation for Acoustic Normal Modea*, to appear
5. P. B. Bailey, "Sturm-Liouville Eigenvalues via a Phase Function," *SIAM J. Applied Math* 14, pp. 242-249, (1966)
6. P. B. Bailey, M. K. Gordon and L. F. Shampine, "Automatic solution of the Sturm-Liouville problem," *ACM Transactions on Mathematical Software*, 4(3), 193-208, (September 1978)
7. R. Bellman and G. M. Wing, An Introduction to Invariant Imbedding, pp. 140 - 142, John Wiley & Sons, New York, (1975)
8. G. Birkhoff and G. -C. Rota, *Elements of Ordinary Differential Equations*, John Wiley & and Sons, New York, 1978 (Chapter 10),
9. B. A. Hargrave, "Numerical approximation of eigenvalues of Sturm-Liouville systems," *J. of Computational Physics,* 20, 381-396, (1976)
10. F. B. Jensen and M. C. Ferla, "SNAP: The SACLANTCEN normal mode acoustic propagation model," SACLANTCEN Memorandum SM-121 (1979)
11. F. B. Jensen, W. A. Kuperman, M. Porter and H. Schmidt, *Computational Ocean Acoustics*, American Institute of Physcis Press, New York, 1994
12. W. E. Milne, "The numerical determination of characteristic numbers," *Physical Review* 35, pp. 863-867, (1930)
13. W. H. Munk, "Sound speed channel in an exponentially stratified ocean with applications to SOFAR," *J. of the Acoustical Soc. of America,* 66(1), pp. 250 - 255, (1979)
14. Porter, M. B. and Reiss, E. L., "A numerical method for ocean acoustic normal modes," *J. of the Acoustical Soc. of America*, 76(1), pp. 244-252, (July 1984)
15. Porter, M.B., "The KRAKEN Normal Mode Program," Naval Research Laboratory Report NRL/MR/5120-92-6920, (May 22, 1992)
16. "*Neue Herteitung der Sturm-Liouvilleschen Reihenkentwicklung steiger Funktionen,*" *Math. Ann*, pp. 499 - 518, (1926)
17. D. Zwillinger, *Handbook for Differential Equations*, Academic Press, Inc, Boston, (1989)

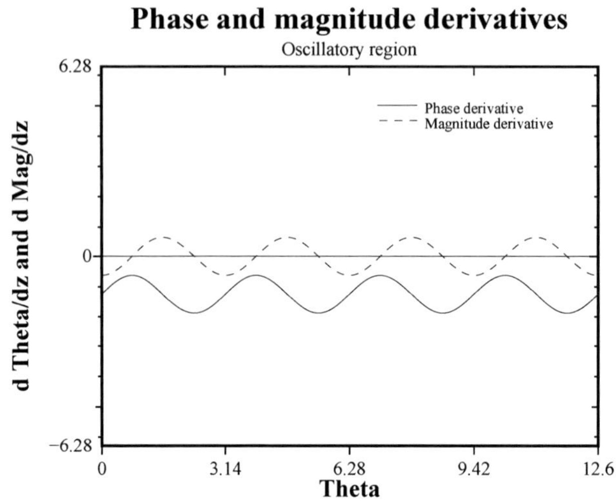

FIGURE 1. Solution for the Prufer phase and magnitude within the mode formation, or duct region, of the wave guide. There are no stable points where the phase derivative vanishes and the magnitude continues to increase when integrated from bottom to top.

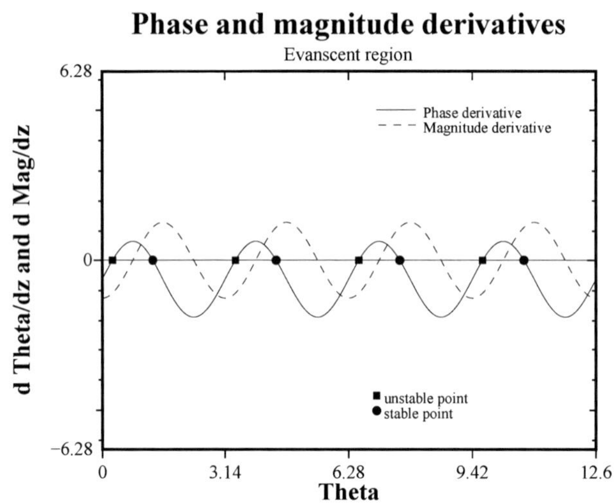

FIGURE 2. Solution for the Prufer phase and magnitude within the evansescent, or edges, of the waveguide. There are stable points to which the solution converges. By integrating in the correct direction, *i.e.* from the edges one obtains the correct stability point.

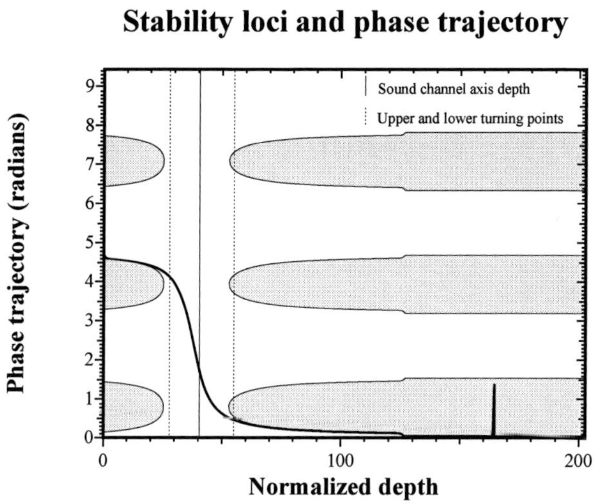

FIGURE 3. Solution for the Prufer (Poincare) phase of the 2nd normal mode w/ Munk SVP Profile

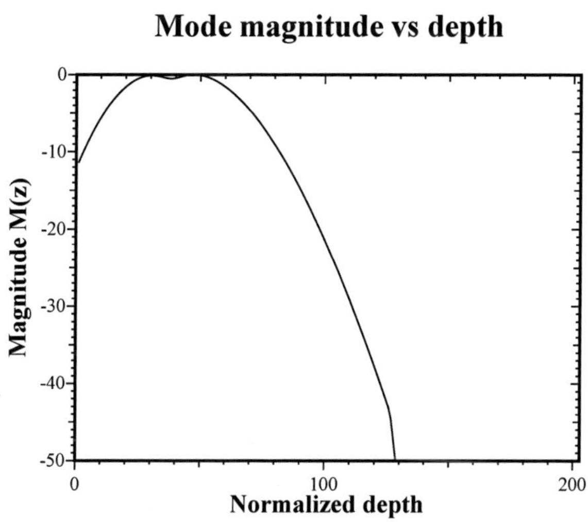

FIGURE 4. Solution for the Prufer (Poincare) ln magnitude of the 2nd normal mode w/ Munk SVP Profile

Accounting for water-column variability in shallow-water waveguide characterizations based on modal eigenvalues

Kyle M. Becker* and Megan S. Ballard†

*The Pennsylvania State University, Applied Research Laboratory, P.O. Box 30, State College, PA 16804-0030 USA
†Applied Research Laboratories, University of Texas at Austin, Austin, TX 78713-8029 USA

Abstract. The influence of water-column variability on the characterization of shallow-water waveguides using modal eigenvalue information is considered. This work is based on the relationship between the acoustic pressure field in shallow water and the depth-dependent Green's function through the Hankel transform. In many practical situations, the Hankel transform can be approximated by a Fourier transform, in which case the Green's function is approximated by a horizontal wave number spectrum with discrete peaks corresponding with individual modal eigenvalues. In turn, the wave number data can be used in inverse algorithms to determine geoacoustic properties of the waveguide. Wave number spectra are estimated from measurements of a point-source acoustic field on a horizontal aperture array in the water column. For range-dependent waveguides, techniques analogous to using a short-time Fourier transform are employed to estimate range-dependent wave number spectra. In this work, water-column variability due to linear internal waves and mesoscale features are considered. It will be shown that these two types of variability impact the estimation of range-dependent modal eigenvalues in different ways. Approaches for accounting for these different types of variability will be discussed as they apply to waveguide characterization.

Keywords: Shallow Water, Modes, Variability
PACS: 43.30.Bp, 43.20.Mv, 43.60.Pt

INTRODUCTION

Historically, the seabed has been assigned the dominant role in influencing acoustic propagation in shallow-water ocean waveguides [1]. However, more recently, the role of water-column variability has been explored for its influence on low-frequency, shallow-water acoustics[2, 3, 4, 5]. Interest in the combined interaction of spatial and temporal variability of the shallow water waveguide environment provided motivation for the Shallow Water Experiment 2006 (SW06)[6, 7]. A subset of SW06 experiments were devoted to range-dependent geoacoustic inversion for both water column and sediment properties from acoustic field data. Of interest to this work, are inversion approaches that use modal eigenvalue information as input data. Detailed descriptions of the approach and recent results can be found in the literature[8, 9]. In this paper, the input eigenvalue data are considered for a range-dependent shallow-water waveguide. The objective is to provide an intuitive interpretation of how different forms of water column variability effect modal eigenvalues and the consequent impact on inversion.

The remainder of the paper is organized as follows. Characterization of the shallow-water waveguide using the Hankel transform is reviewed, including the case of range-

CP1272, *Shallow-Water Acoustics, Second International Shallow-Water Acoustics Conference*
edited by J. Simmen, E. S. Livingston, J. X. Zhou, and F. H. Li

dependent waveguides. In this formulation, the kernel of the integral transform is the depth-dependent Green's function. Next, singularities of the Green's function, that occur along the real horizontal wave-number axis, are related to properties of the waveguide. Assuming a stationary seabed, the relationship between water-column variability and modal-eigenvalue variability is explored. In particular, two cases of water-column variability are examined. In case 1, the impact of small fluctuations about a range-independent mean sound speed profile are considered. In case 2, the impact of a mean sound speed profile that changes with range is examined. The impact on horizontal wave number estimation for cases 1 and 2 are shown to be distinct and observable in measured data. The results are summarized in the conclusion.

SHALLOW-WATER WAVEGUIDE CHARACTERIZATION

This work is based on the Hankel transform pair relationship between the acoustic pressure $p(r;z,z_s)$ due to a time harmonic point source ($e^{-i\omega t}$) and the depth-dependent Green's function $g(k_r;z,z_s)$[10]

$$\begin{aligned} p(r;z,z_s) &= \int_0^\infty g(k_r;z,z_s)J_0(k_r r)k_r dk_r \\ g(k_r;z,z_s) &= \int_0^\infty p(r,z,z_s)J_0(k_r r)r dr. \end{aligned} \tag{1}$$

In (1), J_0 is the zero-order Bessel function, k_r is horizontal wave number, and r is the range between the source (at depth z_s) and the receiver (at depth z). Converting to a two-sided integral and assuming $k_r r >> 1$, the depth-dependent Green's function is approximated by the horizontal wave number spectrum

$$g(k_r;z,z_s) \approx \frac{e^{i\pi/4}}{\sqrt{2\pi k_r}} \int_{-\infty}^{\infty} w(L)p(r;z,z_s)\sqrt{r}e^{-ik_r r}dr. \tag{2}$$

In Eq. (2), $w(L)$ is a window function of length L, applied to the pressure field data. In practice, the length of the window function is less than the total data aperture. By sliding the window in range, range-dependent wave number spectra, indicating local changes in the waveguide environment are obtained[11]. Employing high-resolution spectral estimation methods, improved range and wave number resolution can be achieved using relatively short window apertures[12].

In Ref. [13] it is shown that for shallow-water waveguides the depth-dependent Green's function is characterized by a finite valued continuum along with a set of resonances that occur at discrete values of k_r along the real horizontal wave number axis. These resonances occur at exactly the horizontal wave numbers corresponding with propagating modes supported by the waveguide environment. The modes are solutions to a depth-dependent Helmholtz equation that depend on the propagation medium and satisfy the boundary conditions. The normal mode solution is in fact equivalent to the

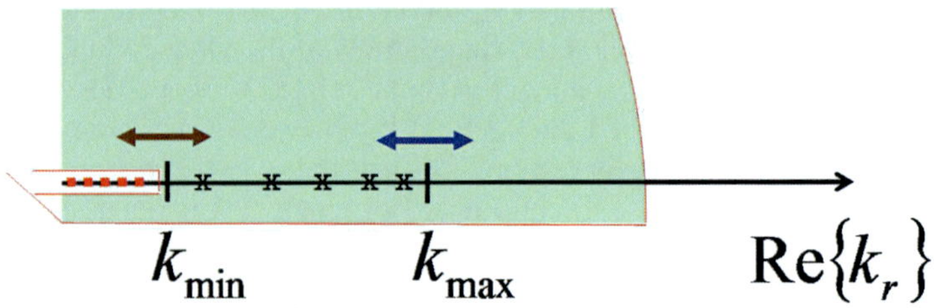

FIGURE 1. Range of modal eigenvalues along the real horizontal wave number axis with influence of water column (blue arrow) and sediment (brown arrow) sound speeds indicated.

sum of residues which results from closing the contour of integration around the singularities of the Green's function when carrying out the integration in (1) over the entire k_r axis[10, 14]. Further, in closing the contour, it can be shown that the discrete modal wave numbers occur along the horizontal wave number axis between values related to the maximum and minimum sound speeds defining the waveguide. Schematically, this is shown in Fig. 1 where $k_{max} = \omega / c_{min}$ and c_{min} is the minimum sound speed in the waveguide, and $k_{min} = \omega / c_{max}$ with c_{max} the maximum sound speed in the waveguide. For the perfectly trapped modes in the waveguide, c_{max} occurs in the sediment and therefore k_{min} is controlled by the sediment properties. Likewise, for many ocean acoustic environments, the minimum sound speed in the waveguide occurs in the water column. In this case, a critical angle exists at the water-bottom interface and k_{max} is controlled by water-column properties. This was the case observed in the SW06 experimental work and also considered in previous studies. The influence on the trapped mode wave number domain caused by changes in either of the limiting water column or sediment sound speeds is indicated by the blue and brown arrows, respectively. Based on these properties, by application of (2) with a sliding window, intuition can be gained about the physical nature of waveguide variability through examination of the resulting values of discrete horizontal wave numbers as a function of range. In particular, it will be shown that the nature of changes in the limiting waveguide sound speed values can be inferred.

NUMERICAL EXAMPLES

In this section, wave-number estimation for the acoustic field in range-dependent shallow-water waveguides is considered. Although, range dependence can occur in both the seabed and the water column, here the focus is on range dependence caused by spatial and temporal variability of the water column sound speed profiles. In particular, two distinct classes of water column variability are examined. Case 1 is illustrated in Fig. 2, where the depth-dependent sound speed profiles as a function of range are defined in Ref. [5]. As indicated, maximum variability occurs at a depth of 25 m corresponding with the bottom of the mixed layer. The sound speed at the surface and bottom are constant with range. The minimum and maximum sound speed profile excursions relative to

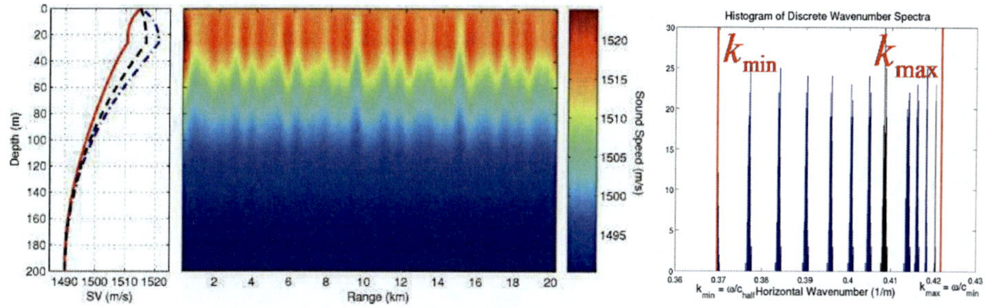

FIGURE 2. Sound speed field used in [5] with minimum and maximum sound speed profile excursions shown in left-most panel. Resulting histogram of individual wave number estimates obtained at 100 Hz is shown in right panel

the mean profile are indicated in the left panel. For this case, the minimum sound speed in the waveguide occurs at the water/bottom interface and is range-independent. The mean sound speed profile in depth is thus stationary in range. For range-independent sediment properties, this type of range dependence in the water column results in perturbations in the modal wave numbers as a function of range. Estimates of individual wave numbers as a function of range can be represented by a distribution about a mean value. Histograms of individual wave numbers for a frequency of 100 Hz are shown in the right panel of Fig. 2. In this case, there is no change observed in the limiting values of k_{min} or k_{max}, indicating stationarity of the mean sound speed profile as a function of range. In addition, as concluded in [5], the mean wave number values are consistent with values for the range-averaged sound speed profile and range-independent boundary conditions.

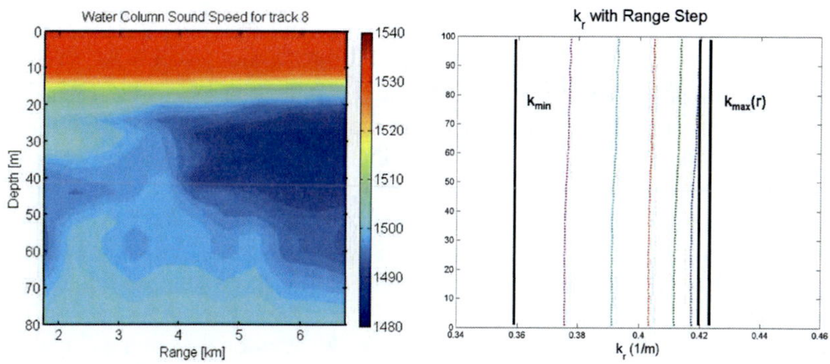

FIGURE 3. Right Panel: Water column sound speed field indicating decreasing sound speed minimum as a function of range. Left Panel: Shift in Horizontal wave numbers with range resulting from change in sound speed minimum. The two solid lines in the right of the figure indicate the total shift in k_{max} for the range of c_{min} values observed in the waveguide.

The second case considered was derived from from water column sound speed measurements made as a function of range and depth during the SW06 experiments. As indicated in the left panel of Fig. 3, it was observed that the minimum sound speed in the water column decreased as a function of range along the track. The mean sound speed profile as a function of depth is thus non-stationary with range. Using this derived sound speed field, the expected shifts in modal eigenvalues as a function of range are shown in the right panel. The shifts for low order modes are directly correlated with the decrease in sound speed minimum observed with range. The shifts corresponds with a move to the right of the upper limit in the wave number domain indicated in Fig. 1, with the maximum shift indicated by the dark lines representing $k_{max}(r)$.

DATA EXAMPLES

Here we examine whether the effects of case 1 or case 2 water-column variability, as described in the preceding section, are observables in measured data. In the preceding section, wave numbers as a function of range were estimated for the different water column sound speed fields using a normal mode code. Here they are estimated by applying a sliding-window auto-regressive spectral estimator to 125 Hz acoustic pressure data measured as a function of range. The acoustic data were recorded on a moored vertical line array of receivers for a signal emitted from a towed source. Water column properties along the range track were measured using a towed conductivity-temperature-depth chain. Details of the experiment and the analysis scheme can be found in [9] along with geoacoustic inversion results for the sediment.

During the experiment, acoustic measurements were made repeatedly over the same track throughout the course of the day. As a result, observed variations in the acoustic field as a function of range could be attributed to variations in the water column as they occurred over the day. In Figs. 4 and 5, the observed water column sound speed fields are shown for measurements separated by 5 hours and denoted run 2 and run 8. Throughout the course of the day, a persistent cooling in the water column occurred at far ranges along the track resulting in a decrease in the sound speed. The impact of this cooling on the resulting wave number estimates is indicated in the right panels of the figures. In Fig. 4, although there is variability along the track with range, no persistent cooling is observed and the resulting wave number estimates for the low-order modes are predominantly range-independent. This is highlighted by the dashed line plotted on top of the mode 1 eigenvalue estimate and is consistent with case 1 described in the previous section. In Fig. 5, the mode 1 eigenvalue is observed to increase in value as a function of range and is consistent with the persistent cooling of the water column as in case 2. Finally, as indicated in the left panel of both figures, the higher-order modes (mode 5) exhibit similar range-dependent behavior that can be attributed to range-dependence in the underlying sediments[9].

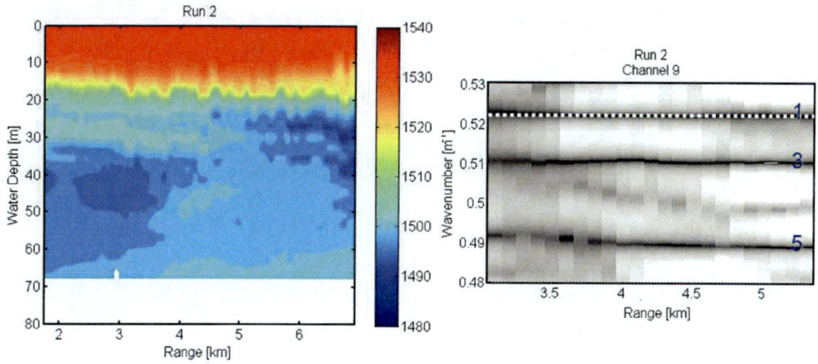

FIGURE 4. Left Panel: Water column sound speed observed during run 2 indicating small change in minimum sound speed with range. Right Panel: Resulting wave number estimates with range. Respective mode numbers are labeled. Dotted line indicates constant value.

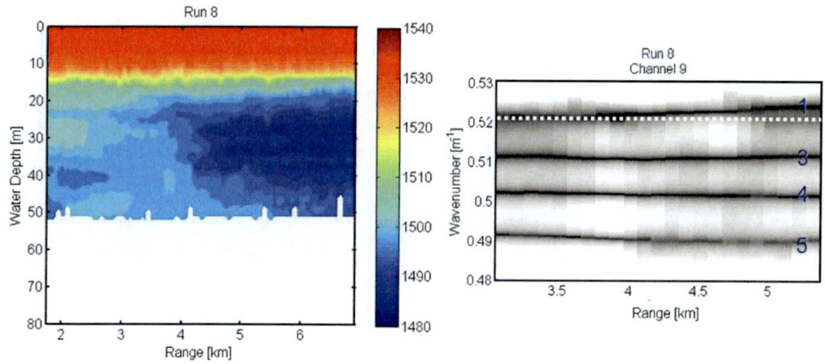

FIGURE 5. Left Panel: Water column sound speed observed during run 8 indicating systematic change in minimum sound speed with range. Right Panel: Resulting wave number estimates with range. Respective mode numbers are labeled. Dotted line indicates constant value. Mode 1 eigenvalue increases with range corresponding to change in minimum sound speed.

CONCLUSIONS

The impact of-water column variability on modal eigenvalue estimates was presented for two cases of interest. By examining range/wave number estimates, it is possible to gain intuition regarding the type of variability affecting the waveguide over the course of a measurement. Small variations about a mean wave number value for the low-order modes indicate that the mean sound speed properties of the channel are stationary with range, reflecting case 1. During run 8, the observed wave number shifts with range indicated a persistent cooling of water column temperature consistent with a decrease in the sound speed minimum, as in case 2. In addition, it was observed that while low-order modal eigenvalues were most influenced by water-column variability, the estimated eigenvalue for mode 5 was consistent between both runs. This reflects that higher order modes are most influenced by the bottom properties. This analysis and the resulting observations provide an intuitive way to infer limiting sound speed values in a waveguide and to estimate their range variability.

ACKNOWLEDGMENTS

We would like to thank the organizing committee of SWAC09 for inviting us to participate in the meeting. We would also like to thank the US Office of Naval Research Ocean Acoustics Program and Global for supporting our research efforts and travel. Finally, we thank our collaborators at the Woods Hole Oceanographic Institution for providing the VLA and other resources during SW06.

REFERENCES

1. G. V. Frisk, Report on the office of naval research shallow water acoustics workshop april 24-26,1991, Tech. Rep. WHOI-92-06, Woods Hole Oceanographic Institution (1992).
2. J. C. Preisig, and T. F. Duda, *IEEE Journal of Oceanic Engineering* **22**, 256–269 (1997).
3. C. Higham, and C. Tindle, *Journal of the Acoustical Society of America* **113**, 2515–2522 (2003).
4. T. F. Duda, *IEEE Journal of Oceanic Engineering* **29**, 118–125 (2004).
5. K. M. Becker, and G. V. Frisk, *Journal of the Acoustical Society of America* **123**, 658–666 (2008).
6. D. Tang, J. N. Moum, J. F. Lynch, P. A. amd Ross Chapman, P. H. Dahl, T. F. Duda, G. Gawarkiewicz, S. Glenn, J. A. Goff, H. Graber, J. Kemp, A. Maffei, J. D. Nash, and A. Newhall, *Oceanography* **20**, 156–167 (2007).
7. N. R. Chapman, and J. F. Lynch, *IEEE Journal of Ocean Engineering* **35**, 1 (2010).
8. S. D. Rajan, J. F. Lynch, and G. V. Frisk, *Journal of the Acoustical Society of America* **82**, 998–1017 (1987).
9. M. S. Ballard, K. M. Becker, and J. A. Goff, *IEEE Journal of Ocean Engineering* **35**, 28–42 (2010).
10. G. V. Frisk, *Ocean and Seabed Acoustics*, Prentice Hall, 1994.
11. K. Ohta, and G. V. Frisk, *IEEE Journal of Ocean Engineering* **22**, 501–521 (1997).
12. K. M. Becker, and G. V. Frisk, *Journal of the Acoustical Society of America* **120**, 1423–1434 (2006).
13. G. V. Frisk, and J. F. Lynch, *Journal of the Acoustical Society of America* **76**, 205–216 (1984).
14. F. B. Jensen, W. A. Kuperman, M. B. Porter, and H. Schmidt, *Computational Ocean Acoustics*, American Institute of Physics, 1994.

Theory and Modeling the Horizontal Refraction in Shallow Water

Boris Katsnelson[a] and Mohsen Badiey[b]

[a]*Voronezh State University, 1 Universitetskaya sq, Voronezh 394006 Russia*
[b]*University of Delaware, Newark, DE, 19716 USA*

Abstract. Overview of the so horizontal refraction of sound waves in shallow water waveguide is presented, as one of the main mechanism of 3D acoustical effects.

Keywords: shallow water, horizontal refraction, focusing/defocusing
PACS: 43.30.Bp, 43.30.Dr, 43.30.Es

INTRODUCTION

Term "horizontal refraction" was introduced apparently by D.Weston [1] for description of sound propagation in waveguide with slowly varying depth and now is commonly accepted expression. This concept especially after work [2] turned out very useful for modeling of the sound field in shallow water medium in presence of inhomogeneous structures, having remarkably anisotropic character such as coastal edge, area of the shelf break, temperature front, nonlinear internal waves (NIW) etc. In the last time it was shown that horizontal refraction can lead to significant acoustic effects for sound propagation [3-7] provided by both bottom relief and volume inhomogeneities. In shallow water horizontal refraction has specific features such as frequency and modal dependence that can be manifested in structure of received signals.

THEORETICAL DESCRIPTOPN OF HORIZONTAL REFRACTION

For theoretical description of horizontal refraction in shallow water we start with Helmholtz equation for the sound field from the point source at $\vec{R}_0 = (\vec{r}_0 = 0, z_0)$ of constant frequency ω (Green's function) $\Psi(\vec{r}, z)$:

$$[\nabla_{\perp}^2 + \frac{\partial^2}{\partial z^2} + k^2(\vec{r}, z)]\Psi(\vec{r}, z) = 0 \quad , \tag{1}$$

where $\nabla_{\perp} = (\frac{\partial}{\partial x}, \frac{\partial}{\partial y})$, $\vec{R} = (\vec{r}, z)$, $k = \frac{\omega}{c}$ and $\Psi(\vec{r}, z)_{\vec{R} \to \vec{R}_0} \to \frac{\exp(ik_0|\vec{R} - \vec{R}_0|}{4\pi|\vec{R} - \vec{R}_0|}$.

CP1272, *Shallow-Water Acoustics, Second International Shallow-Water Acoustics Conference*
edited by J. Simmen, E. S. Livingston, J. X. Zhou, and F. H. Li

Sound field in the waveguide is presented in the form of decomposition over waveguide modes (local modes):

$$\Psi(\vec{r},z)=\sum_{l} a_l(\vec{r})\psi_l(\vec{r};z) \quad , \tag{2}$$

where functions $\psi_l(\vec{r};z)$ and $q_l(\vec{r})$ are eigen functions and eigen values of the corresponding Sturm-Liouville problem:

$$\begin{aligned}
&\frac{\partial^2\psi_l(\vec{r};z)}{\partial z^2}+\left[k^2(\vec{r};z)-q_l^2(\vec{r})\right]\psi_l(\vec{r};z)=0,\\
&\psi_l(\vec{r};z)\big|_{z=0}=0,\\
&\left[\psi_l(\vec{r};z)+g(q_l(\vec{r}))\frac{\partial\psi_l(\vec{r};z)}{\partial z}\right]_{z=H(\vec{r})}=0.
\end{aligned} \tag{3}$$

where

$$\Delta_{\_}a_l(\vec{r})+q_l^2(\vec{r})a_l(\vec{r})=\sum_m\left[\vec{B}_{lm}\nabla_{\_}a_m(\vec{r})+D_{lm}a_m(\vec{r})\right] \tag{4}$$

here $\Delta_{\_}=\dfrac{\partial^2}{\partial x^2}+\dfrac{\partial^2}{\partial y^2}$.

The right side of equation (4) is responsible for mode coupling, in the left side we have redistribution of the sound field in horizontal plane. If we neglect mode coupling then

$$\Delta_{\_}a_l(\vec{r})+q_l^2(\vec{r})a_l(\vec{r})=0 \tag{5}$$

Equation (5) is 2-dimensional version of Helmholtz equation. It describes redistribution of the amplitude and other parameters of the mode number l in horizontal plane. In other words we have 2D waves, for which it is possible to construct different approximations. First of all it is ray approximation, which in accordance with 2D character can be called as theory of horizontal rays [2]. Amplitude of the 1th mode should be found in the form

$$a_l(\vec{r})=A_l(\vec{r})\exp[i\theta_l(\vec{r})] \quad , \tag{6}$$

where $A_l(\vec{r})$ and $\theta_l(\vec{r})$ are amplitude and eikonal correspondingly. After substitution of (6) into (5) and the corresponding evaluations we can get well known eikonal equation:

$$(\nabla_r\theta_l)^2=(q_l^0)^2 n_l^2(\vec{r}) \quad , \tag{7}$$

where refraction index in horizontal plane is introduced: $n_l(\vec{r}) = q_l(\vec{r})/q_l^0$

On the base of this equation it is possible to construct trajectories of horizontal rays, $\vec{r}_l = \vec{r}_l(t)$ corresponding to definite mode as a function of parameter t:

$$\frac{d\vec{r}_l}{dt} = c_l^g \vec{\tau}_l, \qquad \frac{d\vec{q}_l}{dt} = c_l^g \nabla_{\perp} q_l \quad . \tag{8}$$

The following approximation is parabolic equation (PE) in horizontal plane. It is also can be obtained in a standard way from equation (5). Amplitude of mode l is presented in the form

$$a_l(\vec{r}) = F_l(\vec{r}) e^{iq_l^0 x} \quad , \tag{9}$$

where $F_l(\vec{r})$ is amplitude of the sound field, smoothly changing along predominating direction of propagation (in our approximation axis X). Equation for the amplitude $F_l(\vec{r})$ is

$$\frac{\partial F_l}{\partial x} = \frac{i}{2q_l^0}\frac{\partial^2 F_l}{\partial y^2} + \frac{iq_l^0}{2}\left(n_l^2(x,y) - 1\right)F_l \quad , \tag{10}$$

where $n_l(x,y) = q_l(x,y)/q_l^0$ refraction index in horizontal plane. Below we consider some acoustical effects due to two phenomena having anisotropic character. Remark, that as usually refraction index in horizontal plane differs from 1 not significantly

$$n_l^2 = 1 + \mu_l(\vec{r}), \tag{11}$$

where correction $\mu_l(\vec{r})$, provided by perturbation of different nature in water layer has the order $\mu_l < 10^{-2} - 10^{-3}$ however its influence can be rather remarkable due to accumulating character of impact of internal waves on sound signals propagating approximately along wave front of NIW.

Temperature front

Temperature front is rather widespread phenomena in coastal area, for example in Barents sea [7]. Due to variation of temperature profile across temperature front we have variation of refraction index in horizontal plane. It can be presented as small perturbation of refraction index (11). Perturbation of refraction index is rather small, however due to long distance of propagation along wave front, influence of this correction can be significant.

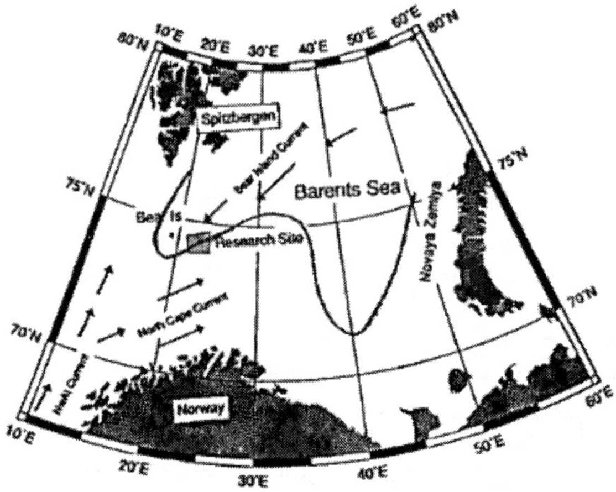

FIGURE 1. Polar front in Barents sea [8]

Horizontal refraction in area of temperature front has the form of multipath (Fig.2) area which can be formed due to refraction of rays

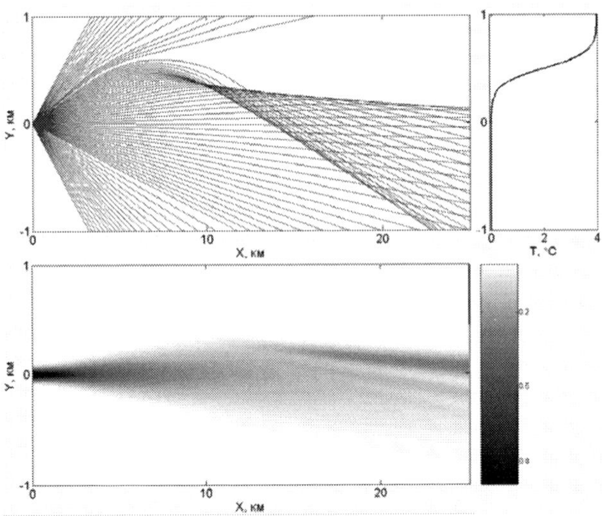

FIGURE 2. Distribution of the sound field in horizontal plane in area of temperature front. Temperature profile across front is shown in the right side. Upper figure shows rays pattern, lower figure is result of PE modeling [4].

Remark, that intensity distribution in horizontal plane depends on frequency, so if we will measure spectrum of received signal then it will have different shapes in different places.

Nonlinear Internal Waves

Nonlinear internal waves (NIW) or soliton-like internal waves are rather widespread phenomena in shallow water. As a rule they are being generated in area of the shelf break when water layer (more exactly thermocline layer) rises up or down in period of tide, approximately twice per day. They propagate towards the coast in the form of train consisting of a few separate waves or solitons with amplitude up to 10-20 m at the velocity below 1 m/s. Wave front of this train is rather long (up to 20-25 km) and almost plane. Influence of trains of NIW on sound propagation as a factor of horizontal refraction can be considered using calculation of correction μ_l due to NIW. If we suppose that envelope of train NIW is function of coordinates in horizontal plane and time (more exactly geotime T) and has the form $\zeta_s(\vec{r},T)$ then mentioned correction can be found using perturbation theory [3]

$$\mu_l = -\nu_l \zeta_s(\vec{r},T) \ , \tag{12}$$

$$\nu_l = \frac{2Qk^2}{(q_l^0)^2} \int_0^H \left[\psi_l^0(z)\right]^2 N^2(z)\Phi(z)dz \ . \tag{13}$$

As we can see this correction, and in turn ray pattern and redistribution of the sound field in horizontal plane depends on mode number and frequency. This dependence for conditions of the SWRM'95 and SW06 area has the form, presented in the FIGURE 3.

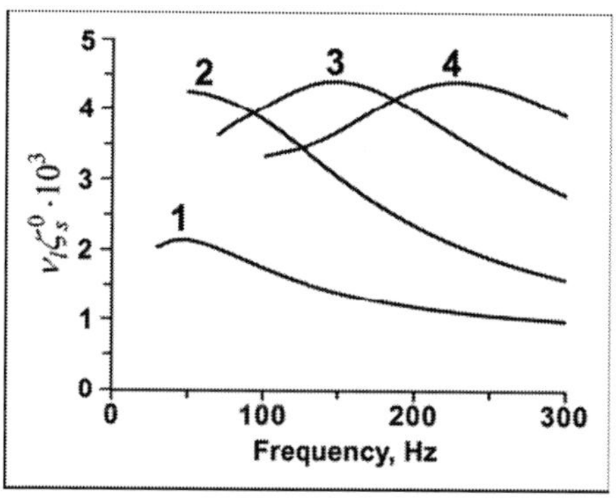

FIGURE 3. Frequency dependence of correction to refraction index in horizontal plane for amplitude of NIW 10 m for lowest modes. Number of modes is shown in the figure [7].

Behavior of the sound field in area of this train is determined by position of the source. If source is placed at the top of NIW (remark that Z axis is directed down) we have defocusing situation, where rays radiated from the source merge out from direction of an acoustic track. If the source is placed between two maximum then we have focusing of rays, and correspondingly of the sound intensity between two adjacent solitons. The calculated behavior of horizontal rays and the corresponding PE solutions are shown in the FIGURE 4 for conditions of the SWARM'95 experiment, where this effect of focusing/defocusing of sound waves in horizontal plane was registered [5].

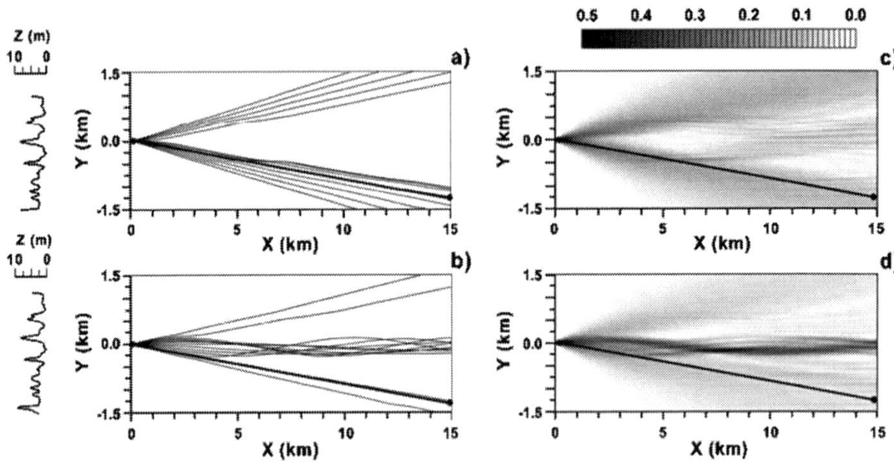

FIGURE 4. Horizontal ray (a,b) and PE (c,d) modeling for SWARM"95 conditions. The corresponding envelope of NIW train is shown in the left side of the figure. Upper part (a,c) is defocusing situation, lower part (b,d) is focusing in horizontal plane [6].

Remark, that in spite of manifestation of this effect in horizontal plane, is possible to observe effect of focusing/defocusing using vertical line array (VLA). The point is that focusing and defocusing of horizontal rays should take place for all modes simultaneously (in different degrre0, it means that we should have synchronicity in vertical plane, more exactly, if, during motion of NIW train the source falls in the trough between solitons then acoustical energy should be concentrated on VLA, in opposite case energy merges aside from VLA. The corresponding experimental observation were done in the SWARM'95 experiment [4] and figure 5 illustrates focusing/defocusing effects as a function of geotime where results of modeling and experiment for temporal fluctuations of depth distribution of intensity is shown. We see that taking into account horizontal refraction in modeling (middle panel) demonstrates the same behavior like experimental data (upper panel), at the same time calculations without horizontal refraction (adiabatic approximation) presented at lower panel gives principally another character. Typical time period between maximums at the middle and upper panels is about 13-15 minutes that corresponds to Vaisala period for given thermocline. More neat details of these fluctuations can be obtained using frequency and modal filtering because we have frequency and modal dependence in refraction index. As a metric for intensity fluctuations we will take scintillation index $SI_{l\omega}$ defined by (14).

$$SI_{l\omega}^2 = (\langle I_{l\omega}^2 \rangle - \langle I_{l\omega} \rangle^2) / \langle I_{l\omega} \rangle^2 \tag{14}$$

where $I_{l\,\omega}$ is modal intensity in absence of internal waves, $\langle \; \rangle$ denotes averaging over geotime. In the figure 6 we can see comparison of experimental value of scintillation index (right panel) and theoretical values (left panel). Taking into account of horizontal refraction provides good agreement between results of experiment and modeling.

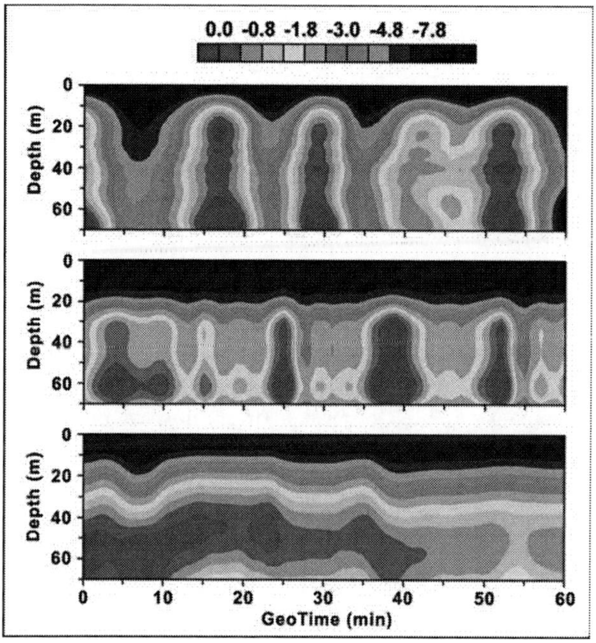

FIGURE 5. Temporal dependence of depth distribution of sound intensity for the SWARM'95 experiment. Upper panel is experimental data, middle panel is modeling taking into account horizontal refraction, lower panel is modeling without horizontal refraction [6].

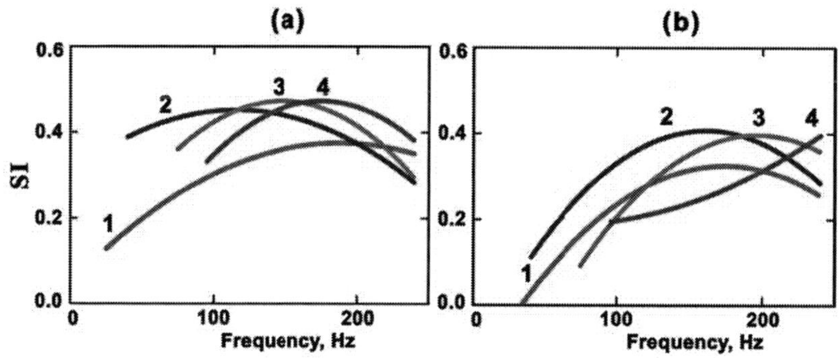

FIGURE 6. Frequency dependence of scintillation index for modal intensities. Left panel is theory, right panel is experiment. Number of modes are shown [7].

CONCLUSION

Horizontal refraction is remarkable mechanism in shallow water which should be taken into account in acoustical experiments in shallow water. Both approaches - vertical modes and horizontal rays or vertical modes and PE in horizontal plane give a good way for description of behavior of the sound field in horizontal plane.
Fluctuations of amplitude (intensity) can reach 3-4 dB and even up to 10 dB and depend on mode number and frequency as well as on position of the source and receiver. This dependence can be rather complex.
Frequency and mode filtering is necessary for theoretical interpretation of fluctuations, or in more general form - for correct interpretation of 3D effects it is necessary more and more sophisticated experiments (long-time observation, complex, broadband signals, moving sources, vertical and L-shaped arrays etc) and data processing.

.

ACKNOWLEDGMENTS

Authors would like to thank Dr. E. Livingston and Dr. J. Simmen for invitation and support of our participation in the Conference on shallow water acoustics.

This work (during a few years) was carried out in collaboration with US colleagues from WHOI (Dr. Lynch and his team) and supported by ONR.

REFERENCES

1. D.E.Weston. Guided propagation in a slowly varying medium. Proc. Phys.Soc. 1959, v.73, p.365-384 Horizontal refraction in a three-dimensional medium of variable stratification Proc. Phys.Soc. 1961, v.78, p.46-52
2. Barridge R., Weinberg G. Horizontal rays and vertical modes In: Wave propagation and underwater acoustics, ed J.B. Keller and J.S. Papadakis , 1977, Springer-Verlag, pp. 86–150
3. Katsnelson, B. G., and S. A. Pereselkov, "Low-frequency horizontal acoustic refraction caused by internal wave solitons in a shallow sea," *Acoust. Phys.* **46**, 684-691(2000).
4. B. G. Katsnel'son J. Lynch A. V. Tshoidze Space–Frequency Distribution of Sound Field Intensity in the Vicinity of the Temperature Front in Shallow Water Acoustical Physics, Vol. 53, No. 5, pp. 611–617 (2007),
5. Badiey, M., Y. Mu, J. F. Lynch, J. R. Apel, and S. N. Wolf, "Temporal and azimuthal dependence of sound propagation in shallow water with internal waves," *IEEE J. Ocean. Eng.* **27**, 117-129 (2002).
6. Badiey, M., Katsnelson, B., Lynch, J., Pereselkov S., Siegmann, W., "Measurement and modeling of 3-D sound intensity variations due to shallow water internal waves," J. Acoust. Soc. Am., Vol. 117, No. 2, pp. 613-625, 2005.
7. Badiey, M., Katsnelson, B., Lynch, J., Pereselkov S.Frequency dependence and intensity fluctuations due to shallow water internal waves," J. Acoust. Am., 122(2), pp. 747-760, 2007.
8. J.F.Lynch, G.Jin, R.Pawlowich, D. Ray, A.PlueddermanR.H.Bourke, A.R.Parsons, R.Muenich. Acoustic travel time perturbation due to shallow water internal waves and internal tides in the Barents Sea Polar front: Theory and experiment. Journal of the Acoust. Soc. Am., 1996, v.99(2), p. 803-821

The Effects of Sloping Bottoms on the Vertical Correlation

Zhaohui Peng[a], Lujun Wang[b], and Zhenglin Li[a]

[a]*State Key Laboratory of Acoustics, Institute of Acoustics, Chinese Academy of Sciences, Beijing 100190, P.R. China.*
[b]*Naval Academy of Armament, Beijing 100161, P.R. China.*

Abstract. Using the parabolic equation code RAM, the sound field vertical correlations are studied as functions of vertical spans and ranges in sloping bottom environments. The wedge mode theory is used to explain the numerical results. It is shown that there is significant difference of the sound field vertical correlation between the upslope propagation and the downslope propagation. In comparison to the vertical correlation in flat bottom environments, downslope propagation increases the vertical coherence and upslope propagation decreases the vertical coherence. An oscillating phenomenon of the vertical correlations caused by the interference between effective modes is observed. For upslope propagation and downslope propagation, the radius of the oscillation decrease or increase, respectively.

Keywords: Vertical correlation, slope bottom.
PACS: 43.30

I. INTRODUCTION

It is well established that the correlation of sound pressures is an important factor for the performance of acoustic arrays. The special correlations of the sound field, including horizontal longitudinal correlation, horizontal transversal correlation, and vertical correlation, have received much attention [1~5]. Compared with the flat bottom, the sloping penetrable bottom (for example: continental shelves) has been shown to have strong effects on sound propagation. There has been much interest in underwater sound propagation over a sloping bottom [6~15]. However, few studies have examined the sound field's vertical correlation in a shallow water environment and with a sloping bottom.

The vertical correlations of the sound field over a downsloping bottom and an upsloping bottom are studied in the present paper. The vertical correlation is defined as

$$\rho = \frac{\left|\int_{\omega_1}^{\omega_2} p(z_s, z_r, r, \omega) p^*(z_s, z_r + dz, r, \omega) d\omega\right|}{\sqrt{\int_{\omega_1}^{\omega_2} \left|p(z_s, z_r, r, \omega)\right|^2 d\omega \int_{\omega_1}^{\omega_2} \left|p(z_s, z_r + dz, r, \omega)\right|^2 d\omega}}, \tag{1}$$

CP1272, *Shallow-Water Acoustics, Second International Shallow-Water Acoustics Conference*
edited by J. Simmen, E. S. Livingston, J. X. Zhou, and F. H. Li

where p is the sound pressure, ω is the circular frequency, z_s is the source depth, z_r and $z_r + dz$ are receiver depths, and r is the range between the source and the receiver.

The sound pressure field generated by a time harmonic point source in a range-dependent waveguide has a normal mode representation, [6]

$$p(r,z) = \sum_n \Phi_n(r)\Psi_n(z;r), \tag{2}$$

where $\Psi_n(z;r)$ is the eigenfunction, and $\Phi_n(r)$ is the mode amplitude, which satisfies

$$\frac{1}{r}\frac{d}{dr}\left(r\frac{d\Phi_n}{dr}\right) + k_n^2\Phi_n = -\frac{\delta(r)\psi_n(z_s)}{2\pi r} - \sum_m A_{mn}\Phi_m - 2\sum_m B_{mn}\frac{d\Phi_m}{dr}. \tag{3}$$

Here, k_n is the eigenvalue, A_{mn} and B_{mn} are mode coupling coefficients, which are defined as

$$A_{mn}(r) = \int \frac{1}{r}\frac{\partial}{\partial r}\left(r\frac{\partial \psi_m}{\partial r}\right)\frac{\psi_n}{\rho}dz, \tag{4}$$

$$B_{mn} = \int \frac{\psi_n}{\rho}\frac{\partial \psi_m}{\partial r}dz. \tag{5}$$

By Substituting Eq.(2) to Eq.(1), the vertical correlation can be expressed as

$$\rho(z,z+dz) = \frac{\left|\int_{\omega_1}^{\omega_2}\left(\sum_m |\Phi_m|^2 \psi_m(z)\psi_m^*(z+dz) + \sum_{m\neq l}\Phi_m\Phi_l^*\psi_m(z)\psi_l^*(z+dz)\right)d\omega\right|}{\sqrt{\int_{\omega_1}^{\omega_2}\left|\sum_m \Phi_m\psi_m(z)\right|^2 d\omega \int_{\omega_1}^{\omega_2}\left|\sum_l \Phi_l\psi_l(z+dz)\right|^2 d\omega}}. \tag{6}$$

The first term of the numerator on the right side of Eq.(6) varies slowly with circular frequency ω, and the second term varies much faster. Averaged in the frequency band, Eq. (6) can be simplified as

$$\rho(z,z+dz) = \frac{\left|\sum_m |\Phi_m(\omega_0)|^2 \psi_m(z,\omega_0)\psi_m^*(z+dz,\omega_0)\right|}{\sqrt{\left|\sum_m \Phi_m(\omega_0)\psi_m(z,\omega_0)\right|^2 \left|\sum_m \Phi_l(\omega_0)\psi_l(z,\omega_0)\right|^2}}. \tag{7}$$

II. NUMERICAL SIMULATIONS

Shallow water environments with a homogenous water column over either a downsloping bottom or an upsloping bottom are used for numerical simulations. The bottom sloping angle varies from 0 to 1 degree. For downslope propagation, sea depth at the range of 40 km is 100 m, and for upslope propagation, sea depth at the source range is 100 m. The source depth varies at 5 m, 30 m, and 80 m, respectively. The receiver depth is 50 m. The frequency band is 100-300 Hz with the step of 2 Hz. A popular parabolic equation model, RAM [16, 17], is used to predict the sound field in a range dependent waveguide. The input parameters are shown in Table 1.

TABLE 1. Acoustic parameters for numerical simulations

	Water Column (iso-speed layer)	Bottom (half space)
Sound speed (m/s)	1510	1587
Density (g/cm^3)	1.0	1.8
Attenuation (dB/λ)	0	0.3

Figure 1 compares the vertical correlations as a function of receiver separation at different sloping angles, where the range is 20 km and the receiver depth is 50 m. A total of four curves are shown in each subplot. The solid curve is the vertical

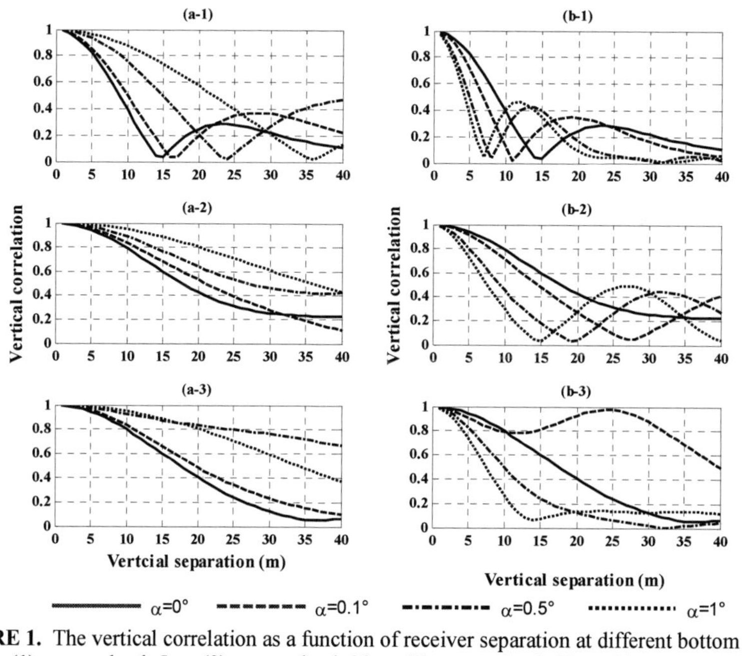

FIGURE 1. The vertical correlation as a function of receiver separation at different bottom sloping angles. (1) source depth 5 m; (2) source depth 30 m; (3) source depth 80 m. (a) represents upslope propagation; (b) represents downslope propagation.

correlation with sloping angle of 0 degrees (i.e. flat bottom). The dashed curve, the dashdotted curve, and the dotted curve are vertical correlations with sloping angle of 0.1, 0.5, and 1 degree, respectively. The upper two subplots (denoted as -1) show the results for source depth of 5 m, the middle two subplots (denoted as -2) for source depth of 30 m, and the lower two subplots (denoted as -3) for source depth of 80 m. The vertical correlations for both upsloping bottoms and downsloping bottoms are taken into consideration. The left three subplots (denoted as a-) show results for upslope propagation, and the right three subplots (denoted as b-) for downslope propagation.

In the same way, Fig. 2 compares the vertical correlations as a function of range at different bottom sloping angles. The receiver separation is 6 m. The receiver depth is 50 m.

In this paper, the vertical correlation radius is defined as the separation between two receivers, at which the vertical correlation falls below 0.5. From Fig.1 and Fig.2, it can be seen that the vertical correlation is dependent on the bottom sloping angle. For downslope propagation, the larger the bottom sloping angle, the higher the vertical correlation, and the larger the vertical correlation radius. For upslope propagation, the larger bottom sloping angle, the lower the vertical correlation, and the shorter the vertical correlation radius. For example, for the source at 30 m in depth, the vertical correlation coefficients corresponding to a flat bottom, downsloping bottom with angle of 1 degree, and upsloping bottom with angle of 1 degree (as shown in Fig.1) are 0.92, 0.98, and 0.65, respectively.

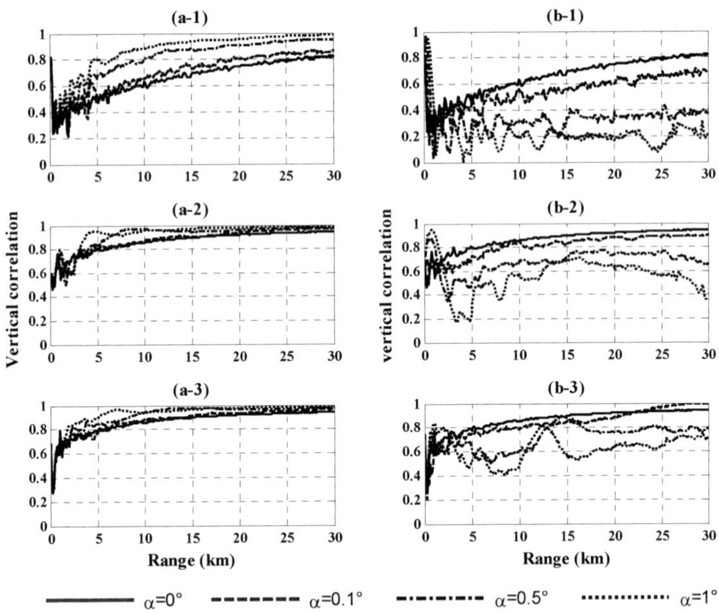

FIGURE 2. The vertical correlation as a function of range at different bottom sloping angles. (1) source depth 5 m; (2) source depth 30 m; (3) source depth 80 m. (a) represents upslope propagation; (b) represents downslope propagation.

The vertical correlation usually decreases with the receiver separation. But sometimes it increases with the receiver separation, especially for receiver separation, as shown in Fig. 1. We call that phenomenon “oscillation.” The receiver separation, at which the vertical correlation coefficient turns to increase, is defined as radius of the oscillation. For downslope propagation, the larger the bottom sloping angle, the larger the radius of the oscillation. However, for upslope propagation, the larger the bottom sloping angle, the smaller the radius of the oscillation. For example, when the source depth is 5 m, the radiuses of the oscillation corresponding to a flat bottom, bottom downsloping angle 0.1 degree, bottom downsloping angle 1 degree, bottom upsloping angle 0.1 degree, and bottom upsloping angle 1 degree are 15 m, 17 m, 36 m, 11 m, and 7 m, respectively.

From Fig.2, one can also see that the vertical correlation usually increases with range for both the downsloping case and the upsloping case. The larger the sloping angle, the more quickly the vertical correlation increases near to 1.0. For the upslope propagation case, the tendency of the vertical correlation to increase with range may break for a sloping bottom with large angles.

III. PHYSICAL EXPLANATION

Before the discussion of the vertical correlation in a range-dependent environment, the vertical correlation in a range-independent environment (bottom sloping angle being equal to 0, solid curves shown in Fig.1 and Fig.2) is taken into consideration.

Due to mode stripping [18] (higher modes with larger grazing angles suffer greater loss), the number of effective modes decreases with range. That is, the larger the distance, the larger the vertical correlation coefficient, as shown in Fig.2. When the number of effective modes is small enough, the influence of mode interference (the second term on the right side in Eq.(12) cannot be neglected. The mode interference is the cause of the vertical correlation oscillation phenomenon. When the bottom sloping angle is small enough, and the coupling energy between modes is negligible, the aforementioned discussion is still valid.

When the bottom sloping angle is large enough, the distribution of energy among normal modes is very complex due to both mode coupling and mode stripping. The “wedge mode” theory [10] is shown to be a fast and accurate way to handle sound propagation over sloping bottoms. The wedge mode theory solves the wave function in “wedge” coordinates (R,θ) rather than the usual Cartesian coordinates. As shown in Fig.3, the wedge coordinate system is a rotated cylindrical coordinate system with a vertical $R-\theta$ plane and a horizontal axis (not shown). The $r-z$ coordinate system in the figure is an ordinary cylindrical system with a vertical z axis. The relationship between the wedge coordinate system (R,θ) and the ordinary cylindrical system (r,z) is:

$$R=\sqrt{(r-r_{apex})^2+z^2}\ ,\quad \theta=\tan^{-1}\left(\frac{z}{r-r_{apex}}\right). \tag{8}$$

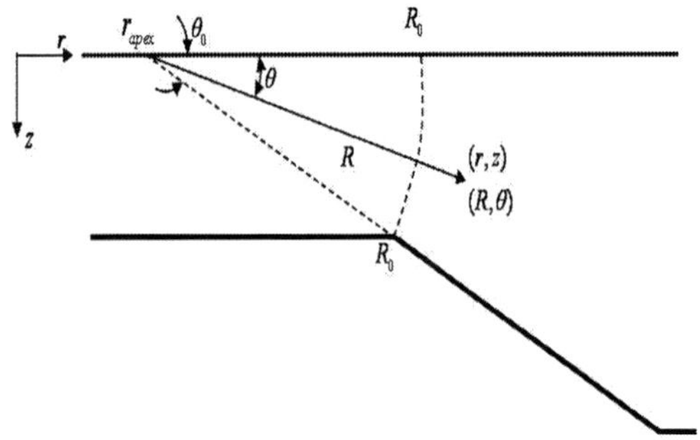

FIGURE 3. "Wedge" coordinate system used for downslope propagation

In wedge coordinates (R,θ) , the solution of the sound field can be expressed by a series of wedge modes. These wedge modes have wave fronts that are curved into arcs of circles centered on the wedge apex [13]:

$$u(R,\theta)=\sum_n A_n(R)W_n(R,\theta)e^{i\int_{R_{ref}}^{R}\alpha_n(R')dR'}, \tag{9}$$

where $A_n(R)$ are the wedge mode amplitude. Eigenvalues α_n and eigenfunctions W_n satisfy

$$\left(\frac{1}{R^2}\frac{\partial^2}{\partial\theta^2}+k^2\right)W_n=\alpha_n^2 W_n, \tag{10}$$

and corresponding boundary conditions.

For a constant sound speed profile in the water column and perfectly reflecting (rigid) bottom, the wedge modes do not depend on R, i.e. $\partial W_n/\partial R=0$. In this case the wedge modes are theoretically uncoupled. For the sloping penetrable bottom, the coupling of energy between wedge modes is also negligible [12]. Therefore, the wedge mode theory is convenient to explain sound propagation in a shallow water environment with a sloping bottom.

The differential of Eq. (8) gives

$$dR=\frac{\cos\theta dr+\sin\theta dz}{r-H/\tan\theta_0},\quad d\theta=\frac{-\sin\theta dr+\cos\theta dz}{r-H/\tan\theta_0}, \tag{11}$$

where H is the sea depth in shallower water and θ_0 is the sloping angle.

Only vertical correlation is of concern in this paper, that is, $dr=0$. Equation (11) can be rewritten as

$$dR = \frac{\sin\theta dz}{r - H/\tan\theta_0}, \quad d\theta = \frac{\cos\theta dz}{r - H/\tan\theta_0}. \tag{12}$$

From Eq. (12), one can deduce that

(1) For downslope propagation, while the sloping angle increases, the dimensions of the range step dR and the angle step $d\theta$ at a constant range r decrease, and so the corresponding vertical correlation increases. As shown in subplots (a-1), (a-2), and (a-3) of Fig.1, the larger the sloping angle, the larger the vertical correlation, and the larger the oscillation radius.

(2) In contrast, for upslope propagation, while the sloping angle increases, the dimensions of the range step dR and the angle step $d\theta$ at a constant range r increase, and so the corresponding vertical correlation decreases. As shown in subplots (b-1), (b-2), and (b-3) of Fig.1, the larger the sloping angle, the lower the vertical correlation, and the smaller the oscillation radius.

Due to mode stripping, the number of effective wedge modes increases with range, and the vertical correlation increases with range for both upslope propagation and downslope propagation, as shown in Fig.2. But for upslope propagation, when the sloping angle is large enough, the transfer of energy into the bottom and the coupling of energy between adjacent modes, as the modes reach cutoff, will increase the number of effective modes, and then break the tendency of vertical correlation to increase with range.

IV. SUMMARY

Sound field vertical correlations are studied as functions of vertical spans and ranges in sloping bottom environments by using numerical simulations. The wedge mode theory is used to explain the results. It is shown that the vertical correlation is dependent on the bottom sloping angle.

When the sloping angle is small, the vertical correlation of the sound field over a sloping bottom has the same characteristics as that over a flat bottom.

There is significant difference of the vertical correlation between the upslope propagation and the downslope propagation. For downslope propagation, the larger the bottom sloping angle, the higher the vertical correlation, and the larger the vertical correlation radius. For upslope propagation, the larger the bottom sloping angle, the lower the vertical correlation, and the shorter the vertical correlation radius.

The vertical correlation increases with range for both upslope propagation and downslope propagation. But for upslope propagation, the tendency will break due to mode cutoff.

The oscillating phenomenon of the vertical correlations caused by the interference between normal modes is observed, and the upsloping bottom and downsloping bottom will decrease or increase the radius of the oscillation.

ACKNOWLEDGMENTS

The work was supported by the National Natural Science Foundation of China under grant number 10734100.

REFERENCES

1. Cox H., *J.Acoust.Soc. Am.*, **54**, 1743-1746 (1973).
2. Morgan D R, Smith T M., *J.Acoust.Soc. Am.*, **87**, 737-747 (1990)
3. Zhu R C, Guan D H., *J.Acoust.Soc. Am.*, **92,** 956-961 (1992)
4. Wang Q, Zhang R., *J. Acoust. Soc. Am.*, **92**, 932-938 (1992)
5. Li Z L, Zhang R H, Yan J et al., *IEEE J. Ocean. Eng.*, **29**,973-979 (2004)
6. McDaniel S T., *J. Acoust. Soc. Am.*, **72**, 916-923 (1982)
7. Knobles D P, Stotts S A, Koch R A., *J. Acoust. Soc. Am.,* **113**, 781-787 (2003)
8. Fawcett J A., *J. Acoust. Soc. Am.*, **93**,: 2627-2632 (1993)
9. Westwood F K., *J. Acoust. Soc. Am.*, **87**, 1539-1545 (1990)
10. Arnold I M, Felsen L B. *J. Acoust. Soc. Am.*, **76**, 850-860 (1984)
11. Jensen F B, Tindle C T., *J. Acoust. Soc. Am.*, **82**, 211-216 (1987)
12. Tindle C T, Hobaek H, Muir T G., *J. Acoust. Soc. Am.*, **81**, 287-294 (1987)
13. Primack H, Gilbert E., *J. Acoust. Soc. Am.*, **90**, 3254-3262 (1991)
14. Buckingham M J, Tolstoy A., *J. Acoust. Soc. Am.*, **87**, 1511-1513 (1990)
15. Z. Li, R. Zhang, J. Yan, *Acta Acustica*, **28**, 425-428 (2003)
16. Collins M D., *J.Acoust.Soc. Am.*, **93**, 1736-1742 (1993)
17. Collins M D. "User's Guide for RAM Version 1.0 and 1.0p". Naval Research Laboratory, Washington DC, 1995
18. Wan L, Zhou J X, Rogers P H, et al., Acoustical Physics, **55**, 383-392 (2009)

Investigation of Interference Phenomena of Broadband Acoustic Vector Signals in Shallow Water

Shengchun PIAO and Qunyan REN

College of Underwater Acoustic Engineering, Harbin Engineering University Harbin, 15001, China

Abstract. Although the ocean environment in shallow water is very complex, there still exists stable interference pattern for broadband low frequency sound propagation. The waveguide invariant concept is introduced to describe the broadband interference structure of the acoustic pressure field in a waveguide and now it is widely used in underwater acoustic signal processing. Acoustic vector sensor can measure the particle velocity in the ocean and provides more information for the underwater sound field. In this paper, the interference phenomena of broadband vector acoustic signals in shallow water are investigated by numerical simulation. Energy spatial-frequency distributions are shown for energy flux density vector obtained by combination of pressure and particle velocity signals and they are analyzed according to normal mode theory. Comparisons of the interference structure between the scale acoustic field and vector acoustic field also have been made. The waveguide invariant concept is extended to describe the interference structure of vector acoustic field in shallow water. A method for extraction of the waveguide invariant from interference patterns in vector acoustic field spectrograms is presented, which can be used in matched-field processing and geoacoustic inversion. It is shown that this method may have more advantages than the traditional methods which calculate the waveguide invariant using measured sound pressure in the ocean.

Keywords: Waveguide Invariant; Acoustic Vector Field; Shallow Water; Normal Mode Theory.
PACS: 43.30.Bp, 43.30.Wi, 43.30.Zk, 43.60.Lq

1. INTRODUCTION

When acoustic signals propagate in shallow water, they are greatly affected by the ocean environment. However, the random changes of the waveguide have little effect on low frequency sound transmission. There exists stable interference structure for broadband low frequency signals due to multi-path arrivals or dispersive characteristics of the waveguide. The waveguide invariant was first introduced by Russian researchers to summarize and analyze the interference phenomenon[1,2]. The invariant is determined by the sound speed profile, the positions of source and receiver, and the characterization of the sediment, and it has been widely used in matched field processing[3], LOFAR[4], geoacoustical parameters inversion[5,6] etc. All the mentioned applications are restricted in pressure field, although the vector components of acoustic field provide more information than pressure components. The theory and technique of underwater vector hydrophone is developing very fast. If we could

CP1272, *Shallow-Water Acoustics, Second International Shallow-Water Acoustics Conference*
edited by J. Simmen, E. S. Livingston, J. X. Zhou, and F. H. Li

introduce the invariant to vector field, it will extend the scope of vector hydrophone application and be greatly useful for sonar systems. This paper presents some research results of the acoustic vector field characterization in shallow water.

2. BASIC THEORY OF THE INVARIANT

The acoustic pressure field caused by a single point omni-directional source in underwater acoustic waveguide can be expressed as a finite sum of normal modes[1]:

$$p(r,z) = \pi i \sum_l \varphi_l(z_0)\varphi_l(z) H_0^{(1)}(\xi_l r) \approx \sum_l \frac{1}{\sqrt{r}} B_l e^{i\xi_l r} \tag{1}$$

where $B_l = i\sqrt{2\pi}\varphi_l(z_0)\varphi_l(z)e^{-\frac{\pi}{4}i}\Big/\sqrt{\xi_l}$, r is the range from source, ξ_l and φ_l are the eigenvalue and eigen-function of normal mode l, respectively. We have used the asymptotic form of Hankel function in this equation. The received sound intensity or pressure spectrum I can be written as:

$$I = |p|^2 = pp^* = \frac{1}{r}\sum_{l(l=m)} B_l^2 + \frac{2}{r}\sum_{l,m(l\neq m)} B_l B_m^* \cos(\Delta\xi_{l,m} r) \tag{2}$$

As expressed by Eq. 2, the intensity consists of two terms: the first one varies slowly with range and frequency; the second term is the causation of the interference pattern in shallow water, which oscillates with range and frequency due to mode interference.

In the range-frequency plane, we consider the condition of constant spectral level, $dI=0$, that is:

$$dI = \frac{\partial I}{\partial r}dr + \frac{\partial I}{\partial \omega}d\omega = 0 \tag{3}$$

From Eq. 2, we can get the following two partial derivatives:

$$\frac{\partial I}{\partial r} = -\omega \sum_{l,m} B_l B_m^* \left(\frac{1}{v_l} - \frac{1}{v_m}\right) \sin \Delta\xi_{l,m} r \tag{4-1}$$

$$\frac{\partial I}{\partial \omega} = -r \sum_{l,m} B_l B_m^* \left(\frac{1}{u_l} - \frac{1}{u_m}\right) \sin \Delta\xi_{l,m} r \tag{4-2}$$

where u_l and v_l are the group and phase velocity of normal mode l respectively. Substituting Eq.4-1 and Eq.4-2 into Eq.3, the waveguide invariant can be denoted as[1]:

$$\beta \equiv \frac{r}{\omega}\frac{d\omega}{dr} = -\frac{d(1/v)}{d(1/u)} = -\left(\frac{u}{v}\right)^2 \frac{dv}{du} \tag{5}$$

where u and v are the average values of group and phase velocity, respectively.

Fig.1 is the interference pattern of intensity of the Pekeris waveguide calculated using Kraken[10] for the environment composed of a 100m homogeneous fluid layer overlying a homogeneous fluid half-space. The fluid has a sound speed of 1500m/s, and the bottom has a sound speed of 1550m/s. The ratio of bottom density to fluid density is 1.6. The source depth and the receiver depth are 20m and 30m, respectively. The source frequency changes from 50 to 200Hz, and the maximum range of the receiver is 5000m. Without statement, the following simulations are all calculated using these environmental parameters by Kraken code. The slopes of the visible striations in Fig.1 are related to the waveguide invariant. It should be noticed that the interference structure does not vary continuously with frequency, which is mainly caused by abrupt change of the number of propagation normal mode, which has been illustrated by the left side of Fig. 1.

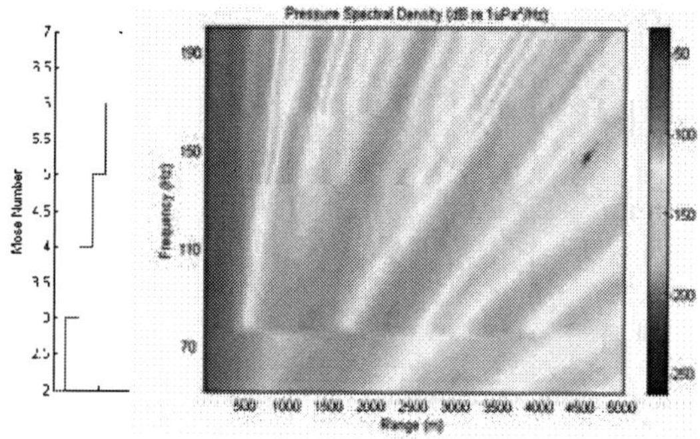

FIGURE 1. The Simulated Range/Frequency Spectrum of Pressure Intensity.

3. THE INVARIANT CHARACTERISTIC OF VECTOR FIELD

In acoustic vector field, we can gain the acoustic signals with higher SNR than that in pressure field, so it is necessary to introduce the waveguide invariant to acoustic vector theory. Apply Euler operator to Eq. 1, the horizontal particle velocity and vertical particle velocity can be written as follows:

$$V_r(r,z)=\frac{-1}{i\omega\rho}\frac{\partial}{\partial r}\sum_l\frac{1}{\sqrt{r}}B_l e^{i\xi_l r}\sum_l\frac{i}{r^{\frac{3}{2}}\omega\rho}B_l e^{i\xi_l r}-\sum_l\frac{\xi_l}{\sqrt{r}\omega\rho}B_l e^{i\xi_l r} \tag{6}$$

$$V_z(r,z)=\frac{-1}{i\omega\rho}\frac{\partial}{\partial z}\sum_l\frac{1}{\sqrt{r}}B_l e^{i\xi_l r}=\sum_l\frac{ik_{lz}}{\sqrt{r}\omega\rho}B_l^{'}e^{i\xi_l r} \tag{7}$$

where $k_{lz}=\sqrt{k^2-\xi_l^2}$, $B_l^{'}$ represents partial differential respect to z .

If the receiver is several depths away from the source, the contribution of the first term in Eq. 6 can be ignored, then the horizontal particle velocity and vertical particle velocity can be rewritten as:

$$V_r(r,z)\approx\sum_l\frac{C_l}{\sqrt{r}}e^{i\xi_l r} \tag{8}$$

$$V_z(r,z)=\sum_l\frac{D_l}{\sqrt{r}}e^{i\xi_l r} \tag{9}$$

where $C_l=\frac{\xi_l}{\omega\rho}B_l$, $D_l=\frac{ik_{lz}}{\omega\rho}B_l^{'}$. It should be noticed that these expressions have the same form as Eq. 1, and following similar steps which have been applied to the intensity in the second section, we can get the same expression for the invariant of received particle velocity intensity or particle velocity spectrum as Eq. 5. The calculated range/frequency spectrums of horizontal and vertical particle velocity are showed in Fig. 2.

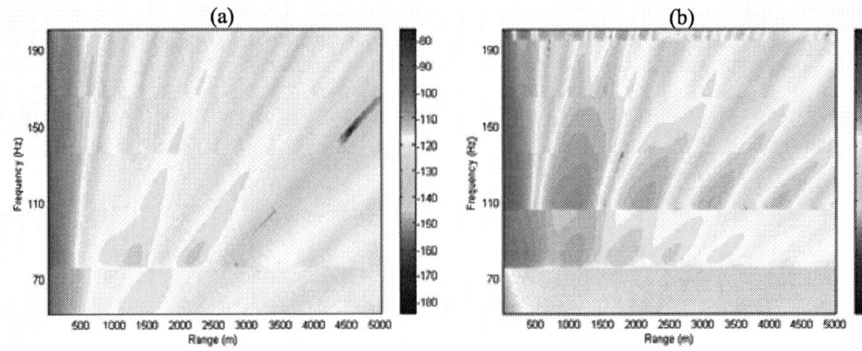

FIGURE 2. The Simulated Range/Frequency Spectrum.
(a) The Horizontal Particle Velocity Intensity (b) the Vertical Particle Velocity Intensity.

There also exist visible striations in the above figure like in Fig. 1, and the slopes of the striations are related to the waveguide invariant too. The acoustic vector hydrophone has many advantages than traditional pressure hydrophone in underwater acoustic applications. So it is useful to introduce the concept of waveguide invariant to acoustic vector field, even the expressions are similar to that for acoustic pressure field.

4. THE COMBINATION PROCESS PRESSURE AND VECTOR COMPONENTS

Let's recall to Eq. 2, which includes two parts. The second part (coherent terms) is the causation of the interference pattern appeared in the spectrum of intensity. The first part (incoherent terms) only varies slowly with range and frequency, has no positive contribution to the interference structure, on the contrary, it sometimes has negative effect or even destroy the interference structure that formed by the coherent terms. This remind us that, if we could get an interference structure that only consists of the coherent terms, it will be great helpful for extracting the waveguide invariant by processing the interference pattern in underwater acoustic applications. If we only consider two propagation normal modes in the waveguide, Eq. 2 can be re-written as:

$$I = \frac{1}{r}\left(B_1^2 + B_2^2\right) + \frac{2}{r} B_1 B_2^* \cos\left((\xi_1 - \xi_2)r\right) \tag{10}$$

According to the above equation, the received intensity should vary (co)sinusoidally with a spatial frequency of $(\xi_1 - \xi_2)$. Indeed, the spatial period not always equal to $2\pi/(\xi_1 - \xi_2)$ as expected (The spatial period will be $\pi/(\xi_1 - \xi_2)$ if we apply absolute operator to Eq. 10.

The propagation of a 70Hz sound signal in shallow water waveguide is considered. The sound transmission loss curves of the summation of the two parts, the first part and the second part of Eq. 10 are shown in (a), (b) and (c) of Fig. 3, respectively. In this waveguide, there exist two propagation normal modes for 70Hz signal, and the wave numbers are approximately 0.2920603 and 0.2884514, respectively. The spatial period for this given frequency is 1741m, and the distance between the successive peaks of TL in (a) and (c) are 1800m and 869m respectively. The spatial period calculated from (c) is approximately equal to $\pi/(\xi_1 - \xi_2)$, determined by the difference of the wave numbers, this could help us in geoacoustic inversion when one use the eigenvalues to obtain the environmental parameters. Whereas the spatial period of (a) is much larger than that calculated by analytical Eq. 10.

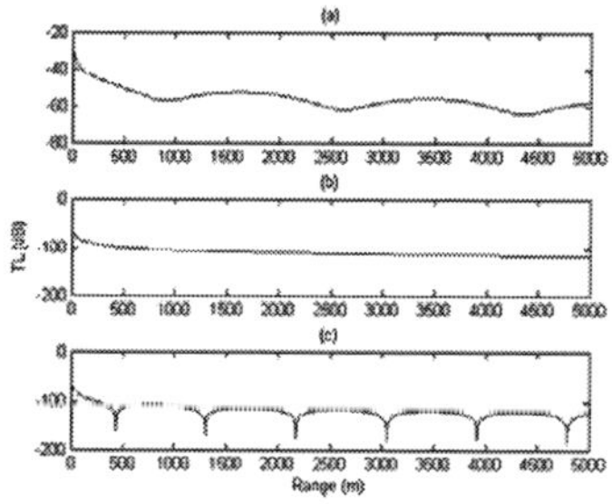

FIGURE 3. Transmission Loss Curves of Different Terms of Eq. 10 versus Range for 70Hz.

We can draw conclusion that the oscillations caused by the coherent terms are destroyed by the incoherent terms, and the information of normal mode eigenvalues that consisted in the spectrum of pressure intensity has been covered. The difference between the coherent terms and the summation of both terms is shown more clearly in Fig. 4. The curves for increasing frequencies are progressively offset in 1-dB steps.

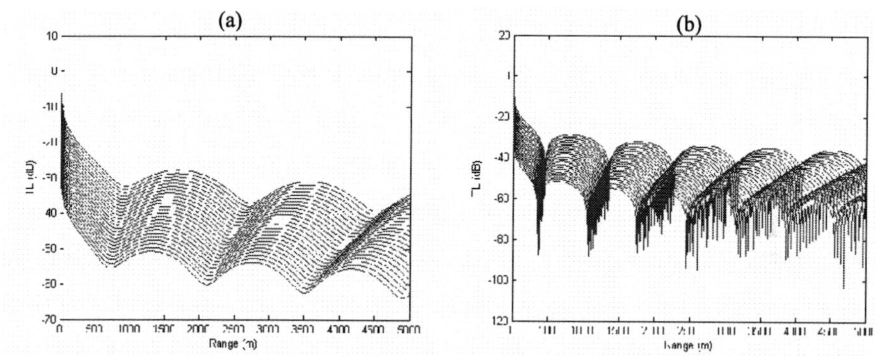

FIGURE 4. Transmission Loss Curves versus Range with frequency change from 50 to 75 Hz (a) The Intensity Transmission Loss Curves (b) The Coherent Terms of Intensity.

Obviously, the coherent terms of intensity transmission loss curves vary more rapidly than the intensity transmission loss curves. This indicates that we could get more uniform interference pattern of the broadband low frequency sound propagation in shallow water, if we could weaken or even eliminate the contribution of the incoherent terms.

Now, we will process the intensity flux of broadband low frequency in shallow water, which is the combination of the pressure and vector components. From Eqs.1, 6 and 7, we can get the horizontal and vertical intensity flux as follows:

$$Ir = pV_r^* = \sum_l \frac{1}{\sqrt{r}} B_l e^{i\xi_l r} \left(\sum_m \frac{-i}{r^{\frac{3}{2}}\omega\rho} B_m^* e^{-i\xi_m r} - \sum_m \frac{\xi_m}{\sqrt{r}\omega\rho} B_m^* e^{-i\xi_m r} \right) \tag{11}$$

$$= \frac{-i}{r^2\omega\rho} \left(\sum_{l,m(l=m)} B_l B_l^* + \sum_{l,m(l\neq m)} B_l B_m^* e^{ir(\xi_l - \xi_m)} \right) - \frac{1}{r\omega\rho} \sum_{l(l=m)} \xi_l B_l B_l^* - \frac{1}{r\omega\rho} \sum_{l,m(l\neq m)} \xi_m B_l B_m^* e^{ir(\xi_l - \xi_m)}$$

$$Iz = pV_z^* = \sum_l \frac{1}{\sqrt{r}} B_l e^{i\xi_l r} \left(\sum_m \frac{-ik_{mz}}{\sqrt{r}\omega\rho} (B_m')^* e^{-i\xi_m r} \right) = \sum_{l,m} \frac{1}{\sqrt{r}} B_l e^{i\xi_l r} \frac{-ik_{mz}}{\sqrt{r}\omega\rho} (B_m')^* e^{-i\xi_m r} \tag{12}$$

$$= \frac{-i}{r\omega\rho} \left(\sum_{l=m} k_{lz} B_l (B_l')^* + \sum_{l,m(l\neq m)} k_{mz} B_l (B_m')^* e^{ir(\xi_l - \xi_m)} \right)$$

When the receiver is several depths away from the source, the contribution of first term in Eq. 11 can be ignored, here we get:

$$real(Ir) = -\frac{1}{r\omega\rho} \sum_{l(l=m)} \xi_l B_l B_l^* - \frac{1}{2r\omega\rho} \sum_{l,m(l\neq m)} \xi_m B_l B_m^* \cos[r(\xi_l - \xi_m)] \tag{13}$$

$$imag(Ir) = -\frac{1}{2r\omega\rho} \sum_{l,m(l\neq m)} \xi_m B_l B_m^* \sin[r(\xi_l - \xi_m)] \tag{14}$$

$$real(Iz) = \frac{-1}{2r\omega\rho} \sum_{l,m(l\neq m)} k_{mz} B_l (B_m')^* \sin[r(\xi_l - \xi_m)] \tag{15}$$

$$imag(Iz) = \frac{-1}{r\omega\rho} \left(\sum_{l=m} k_{lz} B_l (B_l')^* + \frac{1}{2} \sum_{l,m(l\neq m)} k_{mz} B_l (B_m')^* \cos[r(\xi_l - \xi_m)] \right) \tag{16}$$

Eqs. 13, 16 have the same form as Eq. 2 except their coefficients, and the Eqs. 14, 15 only have oscillatory terms which is the causation of the interference structure in shallow water waveguide. It should be figured out that the image part of the horizontal intensity flux and the real part of the vertical intensity flux have more uniform interference structure than that of pressure intensity and particle velocity intensity.

Figure (a) and (b) in Figure 5 are the transmission loss curves of the image part of horizontal intensity flux and the real part of vertical intensity flux versus range with frequency from 50 to 75 Hz, respectively. The curves increased with frequencies are progressively offset in 1-dB steps. The transmission loss curves oscillate almost as same as the curves of (b) in Fig.4, and vary more rapidly than the intensity transmission loss curves.

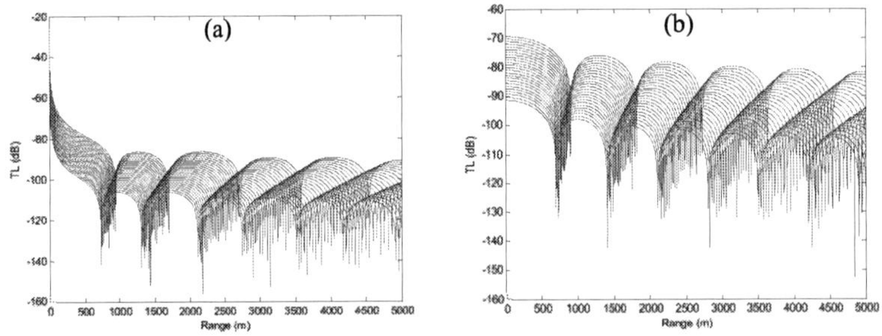

FIGURE 5. Transmission Loss Curves versus Range with Frequency change from 50Hz to 75 Hz. (a) The Image Part of the Horizontal Intensity Flux (b) The Real Part of Vertical Intensity Flux.

It is shown that more uniform interference structure of broadband low frequency propagation can be obtained than that of pressure intensity in shallow water. The difference is shown more clearly in Fig. 6, which is the simulated range/frequency spectrum of the image part of horizontal intensity flux and the real part of vertical intensity flux.

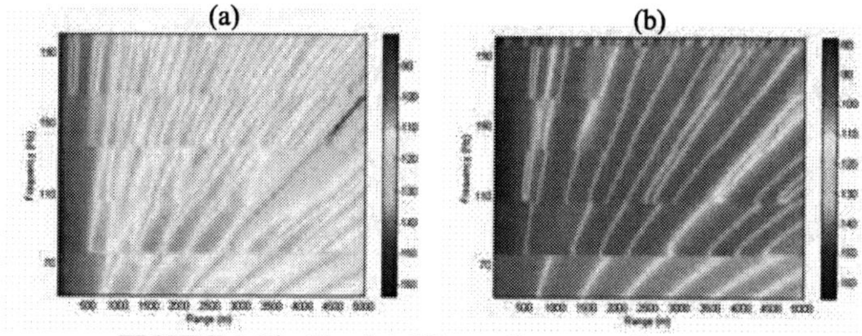

FIGURE 6. The Simulated Range-Frequency Spectrum.
(a) The Image Part of Horizontal Intensity Flux (b) the Real Part of Vertical Intensity Flux.

5. THE EFFECT OF THE ENVIRONMENTAL PARAMETERS ON THE INTERFERENCE PATTERNS

The shallow water sound propagation is great affected by the environmental parameters, the effect of typical environmental parameters on the waveguide invariant distribution of scalar field has been discussed in Ref 7. In this section, we will focus on the environmental influences on the interference pattern for acoustic scalar field and acoustic vector field.

i. The simulations of the shallow water with typical sound profile

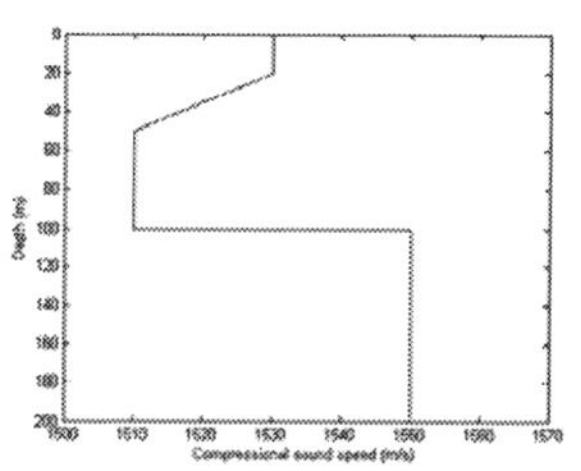

FIGURE 7. The Sound Profile of the Shallow Water Environment.

In this section, we just give the simulated range-frequency spectrum results that calculated by Kraken code using the sound profile illustrated in Fig. 7. In Fig. 8, (a), (c) and (e) are the results of the pressure intensity, the image part of the horizontal intensity flux and the real part of the vertical intensity flux of the receiver at depth 30m, and the (b), (d) and (f) are the results of the receiver at depth 50m. The other environmental parameters of the fluid are same with those in section two.

The spectrums with a typical shallow water sound profile are great different from the results of the homogeneous sound profile. The interference patterns are also different when the receiver depth varies, especially for the results of the image part of the horizontal intensity flux and the pressure intensity. On the other hand, the interference pattern of the real part of the vertical intensity flux is not so sensitive to receiver depth.

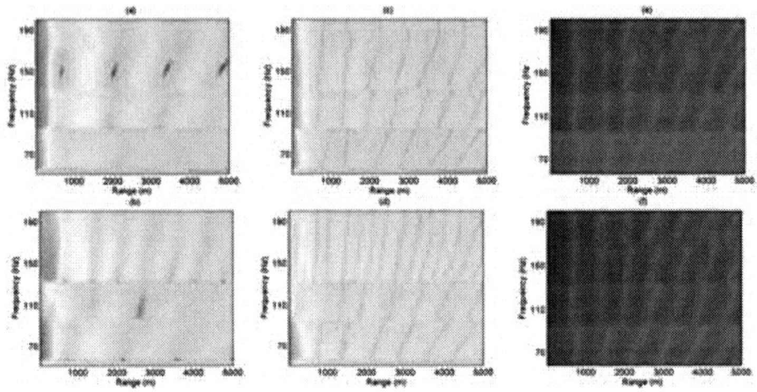

FIGURE 8. Simulated Range/Frequency Spectrums (a), (c) and (e) are the Results of Pressure Intensity, the Image Part of the Horizontal Intensity Flux and the Real Part of the Vertical Intensity Flux for Receiver at Depth 30m. (b), (d) and (f) are the Results for Receiver at Depth 50m.

ii. The simulations with different bottom types

Case 1: The sound speed of the bottom is 1700m/s. The ratio of the bottom density to the fluid is 1.65. Case 2: The sound speed of the bottom is 1800m/s. The ratio of the bottom density to the fluid is 1.7. The other environmental parameters are the same with those in section two and the results are shown below.

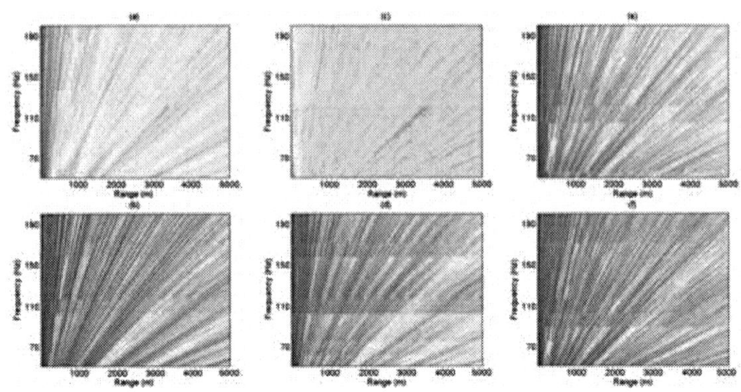

FIGURE 9. Simulated Range/Frequency Spectrums. (a), (c) and (e) are the Spectrums of the Pressure Intensity, the Image Part of the Horizontal Intensity Flux and the Real Part of the Vertical Intensity Flux for Case 1 Respectively, (b), (d) and (f) are the results for Case 2.

Obviously, there are more striations in the two cases due to more propagation normal modes in these waveguide. The slopes of the striations for case 1 and case 2 are slightly different. We also should notice that, when the number of the propagation normal modes is large, the difference between the interference structures can't be clearly visible just from the images, the figures of case 2 is such an example.

6. STRIATION PROCESSING OF THE SPECTROGRAM TO ESTIMATE THE WAVEGUIDE INVARIANT VALUE

The stable interference phenomenon of broadband in shallow water is characterized by the environment. The extraction of the acoustic data from interference pattern can be used to determine the properties of the environment. K. D. Heaney[5,6] inversed geoacoustic parameters by analyzing range-frequency spectrogram, and the method he used has been discussed by Laura A Brooks1, M R F Kidner[8]. In this section, we will apply the approach that used by K. D. Heaney to both of the pressure intensity and the intensity flux to estimate the waveguide invariant.

D'Spain and Kuperman[9] have showed that the equation governing the striations in a range independent environment is:

$$\omega = \omega_0 \left(r/r_0 \right)^{\beta} = C \cdot r^{\beta} \tag{17}$$

Where C, ω_0 and r_0 are specific constants. Heaney states that the waveguide invariant is approximately equal to the normalized slope[6]:

$$\beta \equiv \frac{\Delta f}{f} \bigg/ \frac{\Delta r}{r} \tag{18}$$

The geometry for the determination of the waveguide invariant is illustrated in Fig. 10., and it is the same with Ref. (6). Assume that (r_2, ω_2) is the middle point in range and frequency, between two adjacent striations governed by $\omega = C_2 \cdot r^{\beta}$ and $\omega = C_3 \cdot r^{\beta}$, respectively.

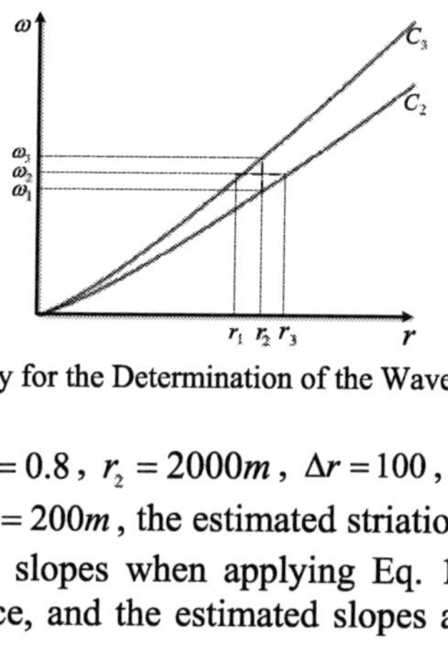

FIGURE 10. Geometry for the Determination of the Waveguide Invariant

If we set $\omega_0 = 2\pi * 50$, $\beta = 0.8$, $r_2 = 2000m$, $\Delta r = 100$, the normalized striation slope by Eq. 18 is 0.8001, if $r_2 = 200m$, the estimated striation slope is 0.8264. That is to say, we could get different slopes when applying Eq. 18 to the spectrogram at different ranges from the source, and the estimated slopes are different for different ranges and frequencies.

In order to estimate the waveguide invariant using analytic expression Eq. 5 [4], we first calculate the group and phase velocities for each frequency, then a cubic spline is fitted for the slowness of the group velocity as a function of the slowness of the phase velocity, then we use the differential of this spline to represent the value of waveguide invariant. The results for different frequencies of the Pekeris waveguide of section 2 are shown in table 1.

TABLE 1. Estimated Waveguide Invariant Values for Different Frequencies.

Frequency (Hz)	90	100	110	120	130
Estimated Waveguide Invariant	1.3232	1.2490	1.3914	1.2318	1.1843

The estimated waveguide values are not constant for different frequencies. As discussed by Daniel Rouseff and Roubert[7], the waveguide invariant should rather be treated as a distribution. Now we will apply the numerical method of Eq. 18 to the spectrogram of Fig. 1, Fig. 9 and Fig. 10 to estimate the waveguide invariant values at different frequencies and different ranges. The selected frequencies are 90, 100, 110,120 and 130 Hz, and the ranges for each frequency are, 1500, 2000, 2500, 3000 and 3500m, respectively. The results applying the approach to Fig. 1, 9 and 10 are shown in Table 2, 3, 4, respectively:

TABLE 2. The Estimated Waveguide Invariant Values from Fig.1

	1500m	2000m	2500m	3000m	3500m
90 Hz	0.80053	0.88318	0.9608	0.9608	1.1979
100 Hz	0.63745	0.82618	0.67281	0.78731	0.96988
110 Hz	0.61318	1.2455	1.0405	1.1392	1.4275
120 Hz	0.75524	1.2884	1.9909	0.83646	2.1878
130 Hz	0.80208	0.86532	1.1038	1.0901	1.245

TABLE 3. The Estimated Waveguide Invariant Values from Fig.6 (a)

	1500	2000	2500	3000	3500
90 Hz	0.86308	1.2249	1.2459	1.1437	1.0691
100 Hz	1.9358	1.0229	0.98019	0.98932	1.6581
110 Hz	1.0346	1.1956	1.2421	1.3099	1.1496
120 Hz	1.2375	1.377	0.60784	1.4402	1.166
130 Hz	0.81409	1.1846	1.1233	1.1959	1.2497

TABLE 4. The Estimated Waveguide Invariant Values from Fig.6 (b)

	1500	2000	2500	3000	3500
90 Hz	1.0365	1.2384	1.2937	1.2327	1.0036
100 Hz	0.82047	2.0959	0.93225	1.1415	1.3156
110 Hz	1.1924	1.2108	1.1524	1.2598	1.2847
120 Hz	0.82605	0.69489	1.2853	1.4283	1.2721
130 Hz	1.1677	1.3358	1.1176	1.2747	1.0331

From the above three tables, we can obtain that the estimated waveguide invariant values for these three kinds of spectrograms will change with frequency and ranges and the estimated waveguide invariant values for different kind of spectrograms will be different. We got different invariant for different range and frequencies. Then it is better to treat the waveguide invariant as a distribution and we could get different distributions from different spectrograms, which may be more useful for geoacoustic parameters inversion and sonar applications.

Table 5 is the comparison of the mean waveguide invariant values with the theoretical values for each frequency of different spectrograms, we denoted the result from Fig. 1(The intensity spectrogram) as β_I, the result from Fig. 9 (The spectrogram of image part of horizontal intensity flux) as β_R, the result from Fig. 10 as (The spectrogram of real part of vertical intensity flux) β_Z and the theoretic result as β_T.

The results showed that. Although the waveguide invariant could be treated as a distribution, the mean values of the β_R and β_Z are more close to the true values than that of β_I.

TABLE 5. Comparison of the Mean Waveguide with the Theoretical Values for each Frequency of Different Spectrograms.

	β_I	β_R	β_Z	β_T
90Hz	0.80053	1.161	1.1093	1.3232
100Hz	0.77873	1.2611	1.3173	1.2490
110Hz	1.0932	1.22	1.1864	1.3914
120Hz	1.4118	1.1013	1.1657	1.2318
130Hz	1.0213	1.1858	1.1135	1.1843

7. CONCLUSIONS

A numerical research on the interference phenomena of broadband acoustic vector field in shallow water has been conducted according to normal mode theory. The waveguide invariant was introduced into the acoustic vector field in underwater acoustics. After analyzing the causation of the interference structure, we get more uniform interference structure by combination of pressure and particle velocity components.

The effect of the environmental parameters such as the sound profile, receiver depth and bottom properties on the interference structure was also discussed. The basic method to extract the waveguide invariant from the spectrogram was applied to different kind of spectrograms. All the simulations indicates that more uniform interference structure and information about the waveguide environment can be obtained from the combination process of the acoustic pressure and particle velocity components. It will be useful for application of vector hydrophone and research on broadband low frequency sound propagation.

ACKNOWLEDGMENTS

This work was supported by NSFC. Fund number: 50979019.

REFERENCES

1. L. Brekhovskikh and Y. Lysanov, ''Fundamentals of Ocean Acoustics, '' 2nd edn, Springer-Verlag, 1991.
2. S. D. Chuprov, ''Interference structure of a sound field in a layered ocean,'' in *Ocean Acoustics, Current State*, edited by L. M. Brekhovskikh and I. B. Andreevoi Nauka, Moscow, 1982, pp. 71–91.
3. Evan K. Westwood, Broadband matched-field source localization, *J. Acoust. Soc. Am.*, 91(5), 1992, pp. 2777–2789.
4. Liang An; Zhi-qiang Wang; Ji-ren Lu, ''Calculating the waveguide invariant by passive sonar lofargram image,'' in *Mechatronics and Machine Vision in Practice*, 14th International Conference on Signal and Image Processing. 2007. M2VIP 2007.

5. K. D. Heaney, ''Rapid geoacoustic characterization using a surface ship of opportunity, '' *IEEE J. Ocean. Eng*, 29(1), 2004, pp. 88–99.
6. K. D. Heaney, ''Rapid geoacoustic characterization: Applied to range-dependent environments,'' *IEEE J. Ocean. Eng*, 29(1), 2004, pp. 43–50.
7. Daniel Rouseff and Robert C.Spindel, ''Modeling the waveguide invariant as a distribution,'' in *Ocean acoustic interference phenomena and signal processing*, San Francisco, Canlifornia, 2002, pp. 137-148.
8. Laura A. Brooks, Michael R. F. Kidner, Anthony C. Zander, Colin H. Hansen and Z. Yong Zhang, "Striation processing of spectrogram data," Proceedings of the 13th International Congress of Sound and Vibration, Vienna, Austria, 2-6 July,2006, pp.1-8.
9. G. L. D'Spain and W. A. Kuperman, "Application of waveguide invariants to analysis of spectrograms from shallow water environments that vary in range and azimuth," *J. Acoust. Soc. Am.*, 106(5), 1999, pp. 2454–2468.
10. Michael B. Porter. The KRAKEN Normal Mode Program [M], Saclant Undersea Research Centre, 2001.

Gaussian beam tracing for ocean acoustics

Michael B. Porter and Paul Hursky

3366 N. Torrey Pines Ct., Suite 310, La Jolla, CA, U.S.A.

Abstract. Gaussian beam tracing methods have emerged as a standard approach for modeling sound propagation in the ocean. The first implementations were developed in the 1970's by Bucker and evolved significantly. Today there are actually some four different types of Gaussian beam algorithms. They are quite different in terms of both the beam characteristics and their performance.This paper will review the development of the methods and their application to typical ocean acoustic problems.

Keywords: ocean, acoustics, beam, propagation
PACS: 43.30.m

INTRODUCTION

Ray tracing methods were widely used in the 1960's for ocean acoustic applications. However, for range-dependent problems, interest shifted to the parabolic equation (PE) method in the early 1970's. Typically the PE offered more accurate results, but ray methods continued to play an important role in broadband applications such as ocean acoustic tomography, and for reverberation modeling.

In our view ray methods have a great deal to offer, as we shall attempt to demonstrate in this short paper. While there are often more accurate alternatives for a specified scenario, the modeling accuracy is often limited by environmental uncertainty, not the model. Let us consider a number of attractions of ray methods. Broadband calculations, e.g. to provide arrival times or a direct time-series, require essentially no more work. Reverberation modeling is easily done by incorporating scattering formulas. The ray models are largely insensitive to frequency. They readily handle backscatter, as well as Doppler effects due to dynamics in the source, receivers, boundaries, or volume. For these and many other reasons we find ray methods to be the most important in our ensemble of modeling tools.

We use the term 'ray method' somewhat loosely here. In fact, our preferred approach is a generalization sometimes called beam tracing that almost always produces a more accurate result. In the next sections we review that development.

CONVENTIONAL RAY TRACING

The conventional ray tracing approach is illustrated in Fig. 2 for a canonical deep water profile shown in Fig. 1. One begins by tracing a fan of rays from the source as shown in Fig. 2a (the source depth is set to 3000 m). The next step is to identify the *eigenrays* which are those rays connecting the source to a particular receiver location. An example of the eigenrays is shown in Fig. 2b for a receiver at 100 km in range and 800 m in depth.

CP1272, *Shallow-Water Acoustics, Second International Shallow-Water Acoustics Conference*
edited by J. Simmen, E. S. Livingston, J. X. Zhou, and F. H. Li

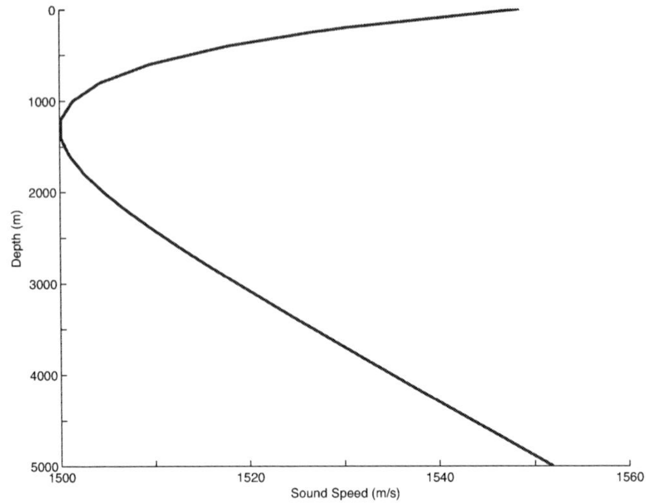

FIGURE 1. A canonical deep water sound speed profile due to Munk.

The sound energy propagates along the eigenrays providing an echo with a particular amplitude and travel-time. After computing the eigenrays for a grid of receiver locations, the sound field is calculated as shown in Fig. 2c.

It is frequently stated that ray methods are independent of frequency. However, only the rays themselves are independent of frequency. The interference field seen in Fig. 2c depends on the number of wavelengths in each eigenray path. The frequency here was chosen to be 50 Hz. Most people would consider this a very low frequency and assume that ray methods are inapplicable. However, as we shall see, the accuracy is excellent.

While the above process sounds simple, it is in fact complicated to program. Thus, all of the early ray models that we are familiar with produced a lot of artifacts beyond those that are really intrinsic to the ray approximation itself. Part of the difficulty is that the identification of eigenrays is really a nonlinear root finding problem in which the take-off angle of the ray must be adjusted to get a ray that passes through a given receiver position. We consider next what we consider to be a better approach based on beam tracing.

BEAM TRACING

Beam tracing was first introduced by Bucker in the 1970's. The first step of the beam tracing algorithm is to construct a fan of rays from the source, just as in a ray model. Next we fatten (term of art) each ray into a beam as shown in Fig. 3a. Finally, we sum up the contributions of each beam to form the sound field as shown in Fig. 3b. This result may be seen to provide excellent agreement with a reference solution shown in Fig. 3c calculated by the SCOOTER wavenumber integration model.

Without getting into details of the algorithm, we can say that we find it has many advantages. There is no need to identify eigenrays. Further, one can see that the caustics (the artificially bright spots where the ray theory predicts infinite intensity) are smoothed out. Energy diffracts naturally into the shadow zones. In summary, the algorithm is both simpler and more accurate.

This brief description necessarily omits important information. In fact, the ray theoretic result shown in Fig. 2c was also produced using the same beam tracing code. We simply changed the type of beam: to recover the ray theoretic result, we use a *geometric beam*[2] with the shape of a triangular hat, as typically used in finite-element modeling. The width of the beams (the width of the hat) is set to precisely follow the geometric spreading of adjacent rays. In effect, this produces a linear interpolation of the sound field between each ray and therefore recovers ray theory, within the convenient framework of a beam tracing code.

To smooth out the caustics we use a Gaussian beam. Again there are several choices. One approach is to use Gaussian beams that satisfy the wave equation. These can further be approximated using a paraxial approximation to yield simple differential equations governing the spreading of the beam as discussed in Ref. [5]. However, such beams have a balloon-like quality in that if we squeeze their width down at the waist, then they expand elsewhere. As such it is usually not possible to keep their width narrower than a few wavelengths. Then they tend to be too large to fit within the waveguide or satisfy the paraxial assumption of their derivation. We have achieved the best results using geometric Gaussian beams, i.e. beams whose width expands in proportion to the geometric spreading of adjacent rays. Such beams can be arbitrarily narrow since they are sized by the user-specified fan of rays. To avoid caustics, we limit their minimum width to a few wavelengths as suggested in Ref. [3]. We can see in Fig. 3b that such beams neck down near the caustics but are never singular.

We consider next a couple of extensions.

RANGE-DEPENDENT SOUND SPEED PROFILES

The beam-tracing framework is equally applicable to problems with range-dependent sound speed profiles. We require the sound-speed to be given by a matrix corresponding to a rectangular grid of range-depth coordinates. The sound speed is interpolated linearly in range between profiles and the ray-tracing is done with steps that are dynamically adjusted so that no step straddles a profile. A beam curvature correction is also needed in crossing the discontinuity (in the first derivative of the sound speed) at each profile.

As an example we consider a North Atlantic scenario in the Gulf Stream. A mesoscale eddy generates a double-duct profile; however, with range the eddy dissipates and the profile returns to a conventional one with a single duct. We can see in Fig. 4 how the energy due to a source in the upper duct (source depth of 300 m) is gradually directed down into the main SOFAR axis.

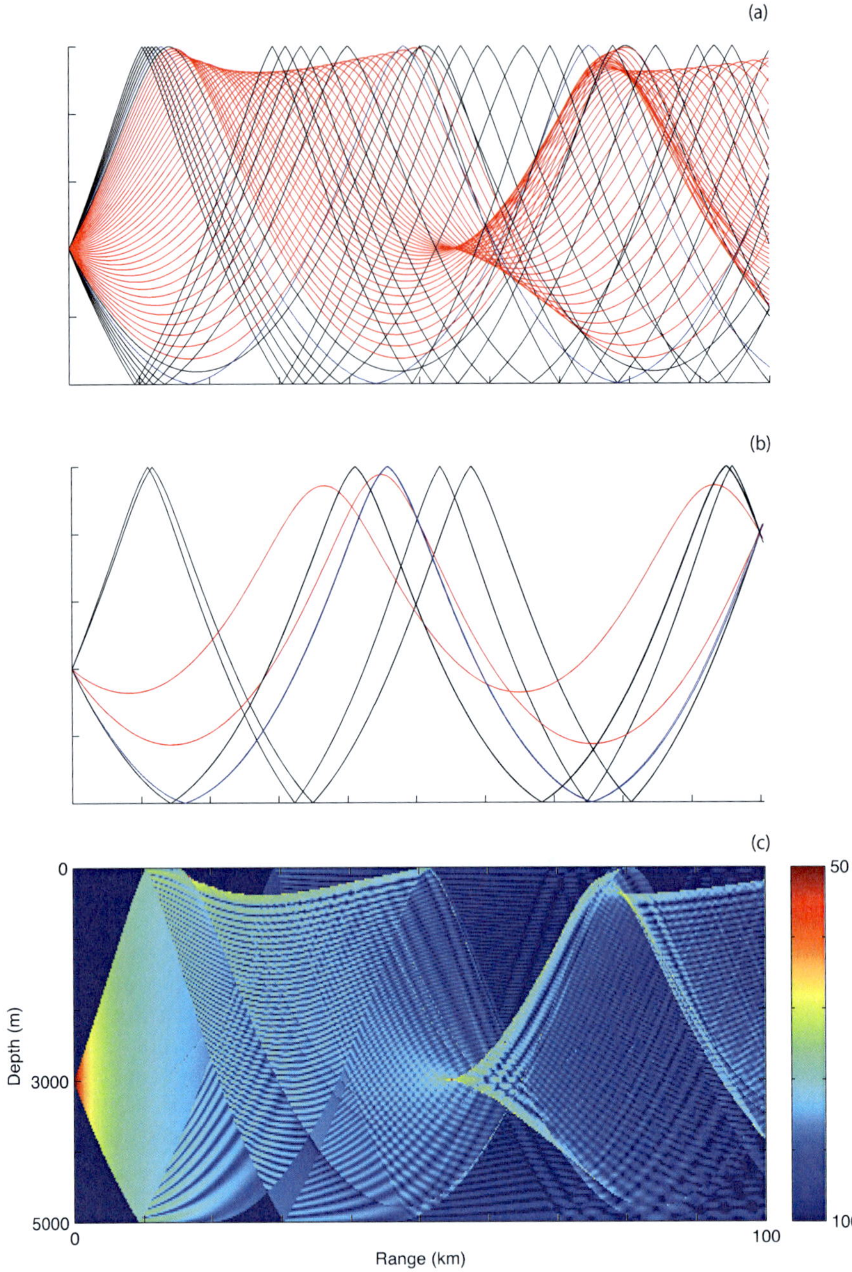

FIGURE 2. Steps in conventional ray tracing: (a) fan of rays from the source at 3000 m depth, (b) eigenrays connecting the source to a receiver at a range of 100 km and a depth of 800 m, and (c) transmission loss at 50 Hz, calculated using eigenray information over a grid of receiver locations.

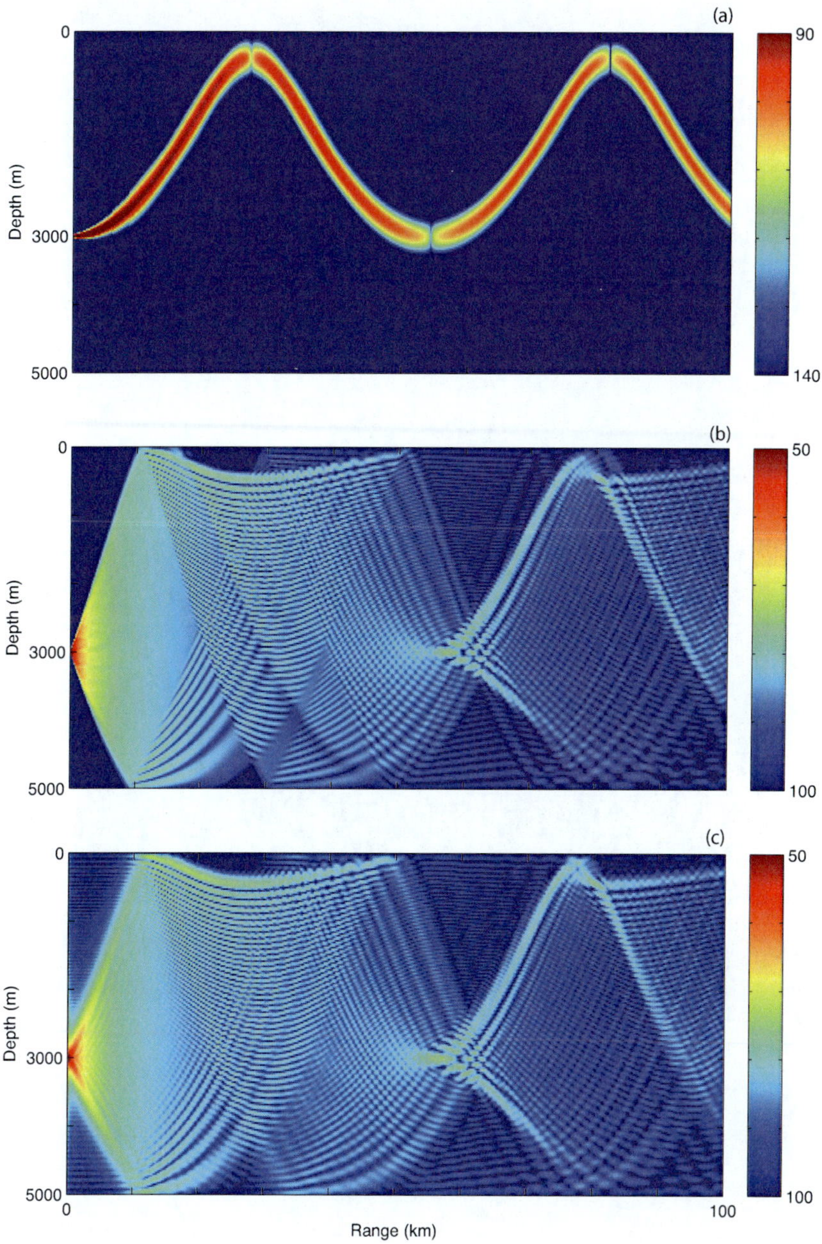

FIGURE 3. Overview of beam tracing: (a) each ray in the ray fan is plumped up into a Gaussian beam, (b) transmission loss is calculated by summing the contributions of all the beam. These may be compared to the (c) reference solution calculated using a wavenumber integration program.

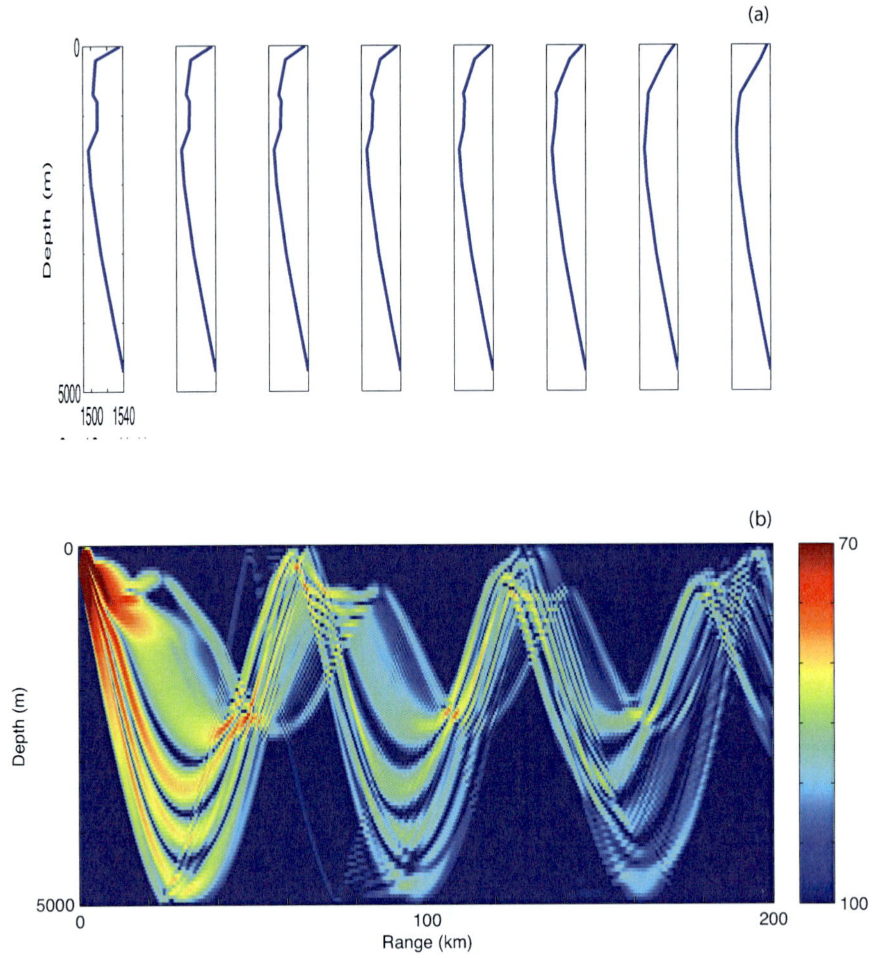

FIGURE 4. Propagation in the Gulf Stream environment: (a) Range-dependent sound speed profiles, and (b) transmission loss for a 50 Hz source at 300 m in depth.

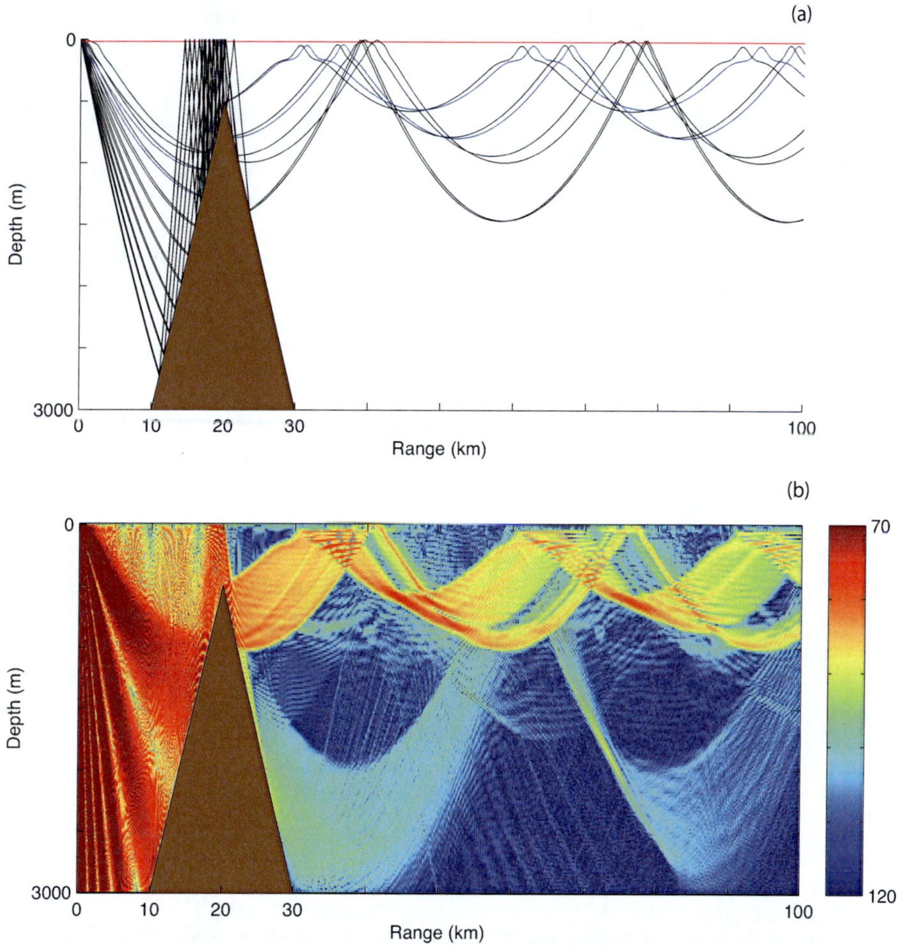

FIGURE 5. Propagation over the Dickins seamount: (a) ray trace, and (b) transmission loss for a 230 Hz source at 18 m in depth.

RANGE-DEPENDENT BATHYMETRY

Range dependence in the water depth is also easily included. The water depth is specified as a vector of range-depth pairs, with range increasing monotonically, but otherwise unconstrained. The depth is interpolated linearly, or in a smoother curvilinear fashion as best describes the actual environment. As is well known, the ray reflects off the bottom with the reflection angle matching the incidence angle. The beam curvature also needs to be corrected if the bottom is curved, or if there is a gradient in the sound speed at the bottom. As an example we show a result for propagation over the Dickins seamount for a source depth of 18 m and a source frequency of 230 m. The ray trace in the upper figure shows the angle changes as the rays interact with the seamount. As expected there are similar results in the sound level, showing steep angle energy being redirected into the main axis of the channel.

CONCLUSION

In the space of a short paper, one can only begin to show the capabilities of beam tracing models. As we have seen, they are easily modified to handle problems with complicated variation in the sound speed and water depth. Complicated geoacoustic bottom models can be incorporated approximately using tabulated reflection coefficients. Three-dimensional models are also being developed and are now very practical.

REFERENCES

1. V. Červený, *Seismic Ray Theory* (Cambridge University Press, Cambridge, UK, 2001).
2. M.B. Porter and Y-C. Liu, "Finite-Element Ray Tracing"', in *Proceedings of the International Conference on Theoretical and Computational Acoustics*, edited by D. Lee and M.H. Schultz (World Scientific, Singapore, 1994), pp. 947–956.
3. H. Weinberg and R.E. Keenan, "Gaussian ray bundles for modeling high-frequency propagation loss under shallow-water conditions," J. Acoust. Soc. Am. **100**, 1421–1431 (1996).
4. H.P. Bucker, "A simple 3-D Gaussian beam sound propagation model for shallow water," J. Acoust. Soc. Am. **95**, 2437–2440 (1994).
5. V. Červený, M.M. Popov, and I. Pšenčík, "Computation of wave fields in inhomogeneous media — Gaussian beam approach," Geophys. J. R. Astr. Soc. **70**, 109–128 (1982).
6. M.B. Porter and H.P. Bucker, "Gaussian beam tracing for computing ocean acoustic fields," J. Acoust. Soc. Am. **82**, 1349–1359 (1987).

COUPLED MODE MODELING OF 3-D PROPAGATION AND SCATTERING AROUND CONICAL SEAMOUNT

Henrik Schmidt[a] and Wenyu Luo[a]

[a]*Department of Mechanical Engineering*
Massachusetts Institute of Technology
77 Massachusetts Avenue, 5-204
Cambridge, MA 02139

Abstract. A three-dimensional propagation and scattering model is developed for an ocean with axisymmetric bathymetry and an offset acoustic source. Based on the same theoretical foundation as the formulation presented by Taroudakis [M. I. Taroudakis, J. Comput. Acoust. 4, 101-121 (1996)], the present approach combines a spectral decomposition in azimuth with a coupled-mode theory for two-way, range-dependent propagation. However, the earlier formulations were severely limited in terms of frequency, size and geometry of the seamount, the seabed composition, and the distance between the source and the seamount, and were therefore severely limited in regard to realistic seamount problems. Without changing the fundamental theoretical foundation, this approach applies a number of modifications to the numerical formulation, leading to orders of magnitude in numerical efficiency for realistic problems. Further, by using a standard normal mode model for determining the fundamental modal solutions and coupling matrices, and by applying a simple superposition principle, the computational requirements are made independent of the distance between the seamount and the source and receivers, and dependent only on the geometry of the seamount and the source frequency. Therefore, realistic propagation and scattering scenarios can be modeled, including effects of seamount roughness and realistic sedimentary structure. Numerical examples show that strong mode coupling may occur at the boundary of a seamount. In addition, coherent artificial backscatter can be diffused in the model by the use of stair steps of random horizontal step sizes, which results in a faster convergence rate in terms of the number of ring-shaped sectors. [Work supported by the Office of Naval Research.]

Keywords: three-dimensional propagation and scattering, seamount, mode coupling.
PACS: 43.30.Bp, 43.30.Gv, 43.20.Fn

1. INTRODUCTION

Over the past several decades, a large number of numerical models have been developed for acoustic propagation in the ocean. Most of these models provide solutions for two-dimensional (2D) (range and depth) problems, and they provide satisfactory solutions for the majority of propagation problems where the environmental dependence on azimuth is insignificant.

However, there are several classes of propagation problems where three-dimensional (3D) effects cannot be ignored. These include, for example, propagation over and around a seamount, or through a strong eddy. For such problems, the 2D models, in which energy does not couple between planes of constant azimuth with respect to the source, often fail to provide accurate solutions, whereas a 3D model is able to provide accurate solutions for the field in range, depth, and azimuth. An important 3D effect, as shown in [1] and this paper, is that the azimuthal coupling increases the width of shadows behind islands and seamounts, which can be explained physically by horizontal refraction [2][3].

The $N \times 2D$ approach introduced first by Perkins and Baer [4] assumes out-of-plane scattering to be insignificant, and is therefore applicable only to problems with weak transverse environmental variability.

CP1272, *Shallow-Water Acoustics, Second International Shallow-Water Acoustics Conference*
edited by J. Simmen, E. S. Livingston, J. X. Zhou, and F. H. Li

In 1996, a coupled-mode formulation for the solution of the Helmholtz equation in the presence of a conical seamount was developed by Taroudakis [5]. In his work, the conical seamount was divided into a number of rings, in each of which a series expansion of the acoustic pressure in terms of normal modes and an azimuthal Fourier series was applied. The expansion coefficients were obtained by solving linear systems of equations resulting from the continuity conditions at the vertical interfaces separating the ring-shaped environmental sectors. The formulation introduced was theoretically exact, but the numerical implementation was inefficient, and associated with numerical stability issues, which severely limited the applicability to realistic seamount problems. Firstly, the applied expansion involved Hankel functions of high orders, for which the numerical evaluation is inherently unstable for high Fourier orders and small arguments. Secondly, when the source is far from the seamount, the number of azimuthal modes required for convergence is too large to make this formulation applicable. Finally, the choice of radial Hankel functions in Taroudakis' formulation yields unstable solutions of the mode coupling equations at high azimuthal orders.

To alleviate some of these issues, Eskenazi modified Taroudakis' model by applying the direct global matrix (DGM) approach [6] to obtain stable coupled-mode systems [7]. However, Eskenazi's model was still inefficient for distant source problems, and the use of the global coupled-mode equations still limited the applicability to relatively simple, canonical seamount problems.

Here a 3D spectral coupled-mode model is developed, based on the same theoretical foundation as used by previous investigators, but applying a number of numerical tools which extend the numerical feasibility to problems characterized by large Fourier orders, such as those involving distant sources and realistic seamounts of significant extent. The seamount propagation problem is reformulated into a scattering problem. As demonstrated in this paper, this reformulation has significant impact in terms of numerical efficiency. The modifications are as follows.

(i) Functions $J_m(\cdot)$ and $H_m^{(1)}(\cdot)$ are used as the two linearly independent range solutions, instead of $H_m^{(1)}(\cdot)$ and $H_m^{(2)}(\cdot)$ in Taroudakis' approach, due to the fact that $J_m(\cdot)$ and $H_m^{(1)}(\cdot)$ remain linearly independent numerically for both large and small arguments [8], whereas the two Hankel functions become numerically indistinguishable at high order m.

(ii) Normalized Bessel and Hankel functions are used to avoid overflow and underflow problems; in addition, the asymptotic forms of the normalized Bessel and Hankel functions for small and large arguments are used.

(iii) The efficiency is improved dramatically by introducing the superposition representation of the external field with respect to the seamount [9][10][11].

(iv) The single-scatter approximation used in this model vastly improves numerical efficiency, and as has been shown in the past, is highly accurate for most ocean acoustic propagation problems.

(v) A noteworthy feature of the present approach, which is also true in previous models, is that the coupling matrices are independent of azimuthal orders, so they need to be calculated only once.

(vi) This model is perfectly scalable, and therefore it can be easily parallelized to run on computer clusters, making it applicable to large-scale 3D problems.

2. THEORY

The geometry considered involves a point source, offset horizontally relative to a conical seamount, as illustrated in Fig. 1. This is a 3D problem in which the acoustic field depends not only on range and depth, but also on azimuth.

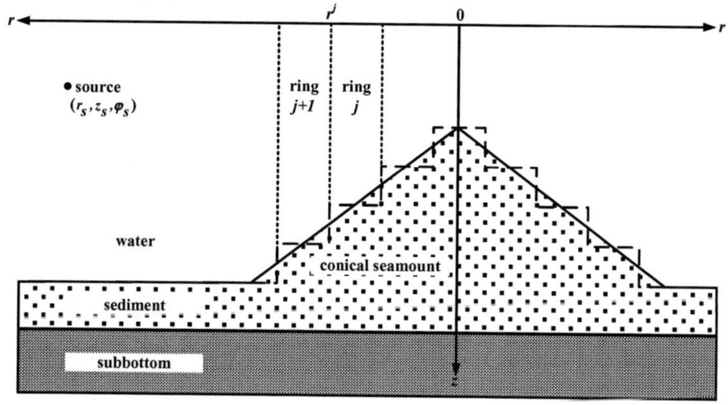

Fig. 1: An ocean waveguide with a conical seamount, which is approximated by a number of range-independent ring-shaped sectors.

In the coupled-mode approach, a number of range-independent ring-shaped sectors are used to approximate such a conical seamount. The notation r^j is used to denote the range at the interface between ring j and ring $j+1$, for $j = 1,2,\ldots,K-1$, with K denoting the total number of ring-shaped sectors.

A cylindrical coordinate system is introduced, centered at the axis of the seamount. The location of the point source is denoted by $\mathbf{r}_s = (r_s, z_s, \varphi_s)$, and the location of a field point is denoted by $\mathbf{r} = (r, z, \varphi)$. The 3D Helmholtz equation in the outer medium containing the point source is [5][10][11]

$$\frac{1}{r}\frac{\partial}{\partial r}\left(r\frac{\partial p}{\partial r}\right)+\frac{1}{r^2}\frac{\partial^2 p}{\partial \varphi^2}+\rho(z)\frac{\partial}{\partial z}\left(\frac{1}{\rho(z)}\frac{\partial p}{\partial z}\right)+\frac{\omega^2}{c^2(z)}p \\ = -\frac{\delta(r-r_s)}{r}\delta(z-z_s)\delta(\varphi-\varphi_s), \tag{1}$$

where $p = p(r,z,\varphi)$ is the acoustic pressure (factoring out the harmonic time dependence $e^{-i\omega t}$, ω being the angular frequency), and $\rho(z)$ and $c(z)$ are density and sound speed profiles in the outer source ring, respectively.

In ring j, which is a range-independent environment, the 3D Helmholtz equation is the homogeneous version of Eq. (1), with $\rho(z)$, $c(z)$, and $p(r,z,\varphi)$ being replaced by $\rho^j(z)$, $c^j(z)$, and $p^j(r,z,\varphi)$, respectively.

The environmental representations used by Taroudakis [5] and Eskenazi [7] are very similar. Both divided the seamount environment into three regions, one being the innermost, cylindrical sector containing the summit of the seamount, one containing all the ring-shaped sectors between the central cylinder and the source range, and the third being a semi-infinite ring outside the source range. The difference between the two lies in the numerical implementation of the spectral coupled-mode representation of the field. Taroudakis [5] used unnormalized Hankel functions of the first and second kind in all sectors for representing the range-dependence of the field. However, as demonstrated by Ricks *et al.* [8], this choice is numerically unstable at high Fourier orders. Recognizing this, Eskenazi [7] used the stable, normalized Bessel function of the first kind and Hankel function of the first kind.

The present representation of the field differs from Taroudakis' [5] and Eskenazi's [7] in that it eliminates the virtual sector boundary at the source range, and instead extends the outermost sector to the base of the seamount. Therefore, in this representation, region I is the innermost cylindrical sector, and region II contains all the ring-shaped sectors between the central cylinder and the base of the seamount, and region III is a semi-infinite ring outside the seamount. The acoustic field in region III is then represented as a superposition of the unperturbed field produced by the source in the absence of the seamount, and a scattered field produced by the seamount. This basically reformulates the seamount propagation problem into a scattering problem. Although rather trivial, this reformulation has significant impact in terms of numerical efficiency, as demonstrated in the following.

By using K to denote the total number of sectors and r_I to denote the range of the base of a seamount, the pressure field is expressed as follows.

(I) Inner cylinder $r \le r^1$,

$$p^1(r,z,\varphi) = \sum_{m=0}^{\infty}\sum_{n=1}^{\infty} b^1_{mn} J^1_{mn}(r)\Psi^1_n(z)\Phi_m(\varphi). \tag{3}$$

(II) Intermediate sectors $r^1 < r \le r_I$, where r_I is the radius of the base of the seamount. In ring j, i.e., $r^{j-1} < r \le r^j$,

$$p^j(r,z,\varphi) = \sum_{m=0}^{\infty}\sum_{n=1}^{\infty}\left[a^j_{mn} H^j_{mn}(r) + b^j_{mn} J^j_{mn}(r)\right]\Psi^j_n(z)\Phi_m(\varphi). \tag{4}$$

(III) Outer region containing source $r > r_I$,

$$p(r,z,\varphi) = p_i(r',z) + \sum_{m=0}^{\infty}\sum_{n=1}^{\infty} a^K_{mn} H^K_{mn}(r)\Psi^K_n(z)\Phi_m(\varphi), \tag{5}$$

where $p_i(r',z)$ is the 2D normal mode solution in the absence of the seamount [6], and r' is the range of a field point from the source.

Here, a_{mn}^j and b_{mn}^j are coupling coefficients which are determined by applying source conditions and boundary conditions at each vertical interface. Functions $\Psi_n^j(z)$ are local, depth-dependent eigenfunctions and $\Phi_m(\varphi)$ are azimuthal eigenfunctions [10][11].

3. NUMERICAL IMPLEMENTATION

Compared to the earlier formulations, there are three principal reasons for the improved numerical efficiency of the present formulation. The first is the use of the superposition principle for handling the source field. The second is the use of asymptotic representations of the normalized range functions. The third is the use of a marching solution of the coupled-mode equations, based on the single-scatter approximation.

In Eskenazi's and Taroudakis' formulations, convergence for a field point outside the seamount, i.e. $r > r_I$, requires the number of azimuthal orders of at least $k_0 r_s$. In contrast, the use of the superposition principle reduces the minimum number of azimuthal orders to $k_0 r_I$, where r_I is the radius of the base of the conical seamount. Therefore, for a fixed source frequency, the number of azimuthal orders required depends only on the size of the seamount, not the source distance. For realistic seamount problems this can lead to orders of magnitude in numerical efficiency.

Both of Eskenazi's and Taroudakis' formulations suffered from potential overflow and underflow problems in the numerical evaluation of the radial Bessel and Hankel functions. Although providing numerical stability of the solution for the modal coefficients, the use of the normalized Bessel and Hankel functions does not in itself solve this numerical problem for small and large arguments with respect to Fourier order. However, using the asymptotic representations for the Bessel and Hankel functions leads to closed form asymptotics for the normalized functions as well, therefore eliminating the overflow and underflow problems [10][11].

4. NUMERICAL RESULTS

The schematic of the problem is shown in Fig. 2. The waveguide consists of a 5000 m inhomogeneous water column limited above by a pressure-release flat sea surface and below by a homogeneous fluid half space with a compressional speed of 2000 m/s, a density of 1 g/cm^3, and an attenuation of 0.1 dB/λ. A false bottom at depth 7000 m is introduced to take into account the contribution from the continuous spectrum. The seamount is 100 km from the source, with a radius at the base of 20 km, and the same acoustic properties as the bottom. Two different heights of the seamount are considered, H = 1000 m and H = 3800 m, respectively. The source depth is 100 m and the source frequency is 10 Hz.

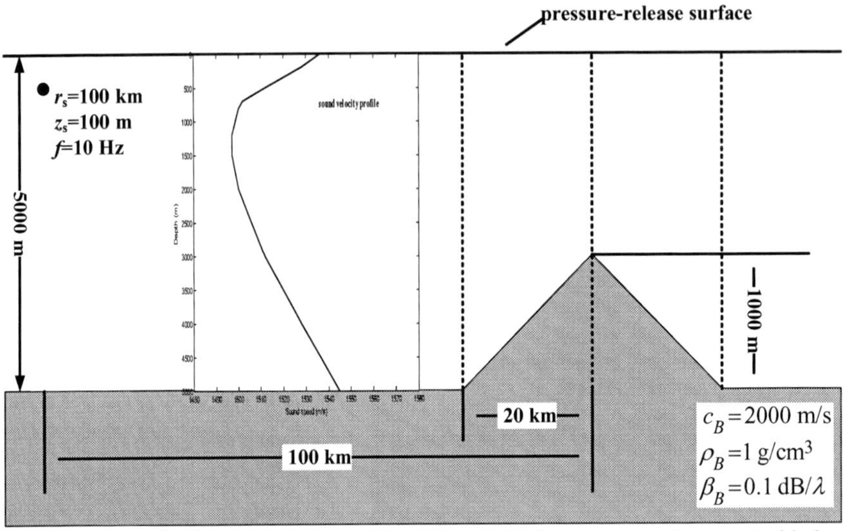

Fig. 2: Schematic of a deep water waveguide with a conical seamount and a penetrable bottom.

In this problem, the modal starting field is limited to contain only the waterborne modes, which are the first 12 modes; otherwise a maximum of 74 modes are used. In addition, for the purpose of azimuthal convergence, the number of azimuthal modes is taken to be 898.

To compare the results of the $N \times 2\text{D}$ model and those of the 3D model introduced in this paper, the height of the seamount is set to vary from 1000 m to 3800 m. Results of transmission loss (TL) in the horizontal plane at depth 300 m are shown in Fig. 3.

From Fig. 3 (a) and Fig. 3 (b) it can be seen that when the height of the seamount is 1000 m, which is relatively small, the 3D effects are insignificant and the $N \times 2\text{D}$ model is a good approximation of the 3D model. However, as the height of the seamount rises to 3800 m, from the results shown in Fig. 3 (c) and Fig. 3 (d), it can be seen that the approximation of the $N \times 2\text{D}$ model to the 3D model deteriorates, and the span of the shadow zone in the 3D result is more pronounced than in the $N \times 2\text{D}$ result. This example indicates that the azimuthal coupling can be important for seamount problems, and the broadening of shadows behind seamounts by horizontal refraction can be captured by implementing a true 3D model.

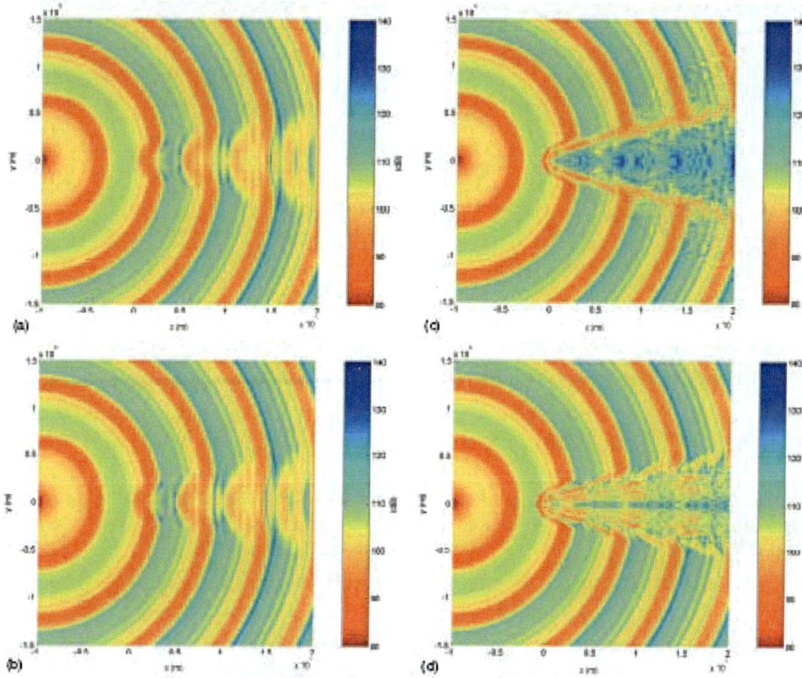

Fig. 3: TL in the horizontal plane at depth 300 m for two different seamount heights, H=1000 m and H=3800 m, respectively. The source frequency is 10 Hz. (a) H=1000 m, 3D result, (b) H=1000 m, $N \times 2\mathrm{D}$ *result; (c) H=3800 m, 3D result, (d) H=3800 m,* $N \times 2\mathrm{D}$ *result.*

A major advantage of the normal mode to ocean acoustic propagation is the inherent decomposition of the field in physically significant components, the normal modes for the waveguide. To investigate the mode coupling occurring around a conical seamount, a single incident mode is considered. For an incident mode to excite a nonzero scattered field, this incident mode must interact with the seamount. Otherwise, the propagation of this incident mode is not affected by the seamount and the scattered field is zero. For an incident mode to interact with the seamount, the oscillating region of the mode shape of this mode must extent to a depth below the summit of the seamount.

To illustrate this, two different source frequencies are considered, 10 and 20 Hz, respectively. It is assumed that mode 10 is the sole incident mode incident on the seamount. Fig. 4 shows the mode shapes of several modes in the ocean surrounding the seamount. At 10 Hz, the oscillating region of mode 10 extends to a depth below the summit of the seamount; however, as the source frequency is 20 Hz, the amplitude of mode 10 is evanescent below summit of the seamount. As a result, mode 10 will not interact with this seamount of height 1000m at 20 Hz. At 10 Hz, however, mode 10 has amplitude below the summit, and scattering will occur.

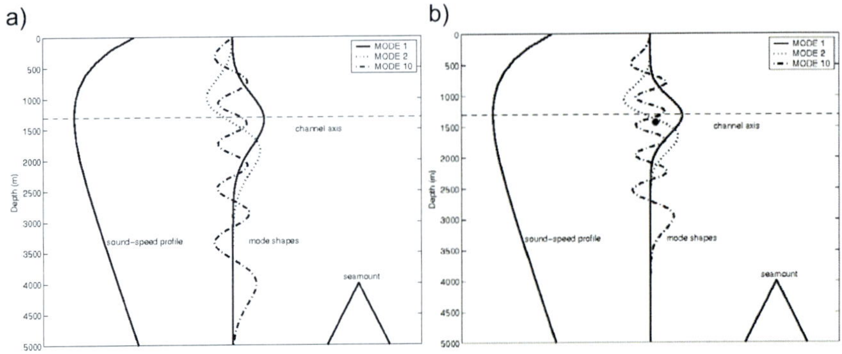

Fig. 4: Mode shapes of selected normal modes away from the seamount. (a) Source frequency 10 Hz: Mode 10 extends to a depth below thesummit of the seamount. (b) Source frequency 20 Hz: Mode shape 10 is evanescent at the summit of the seamount.

Figure 5 shows contours of the scattered mode amplitudes in dB versus azimuthal angle and mode number at 10 Hz. The incident mode, mode 10, interacts with the seamount, and as observed, strong mode coupling occurs into a spectrum of lower and, in particular, higher order modes. It is also observed that the forward scatter is much stronger than the backscatter, with most of the modal spreading occurring for propagation straight across the seamount.

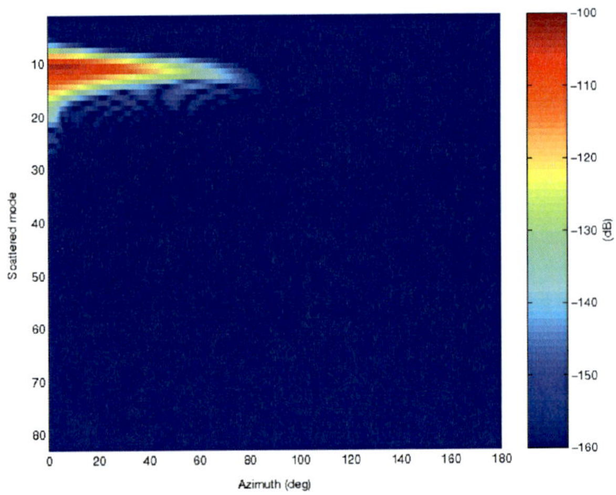

Fig. 5: Contours of scattered mode amplitudes in dB vs azimuthal angle and mode number for mode 10 incident on the seamount. The source frequency is 10 Hz.

5. CONCLUSIONS

The coupled normal mode method is a very efficient, simple and accurate method to solve range-dependent problems. First, since the normal mode method solves the full-wave Helmholtz equation, it is applicable to cases where the backscattering is not negligible. Second, it is free of angular limitations for propagation angles less than critical, and therefore provides accurate solutions at long ranges. This is achieved by including evanescent modes by introducing a false bottom or by calculating the branch-line integral. Third, as an important feature of the spectral coupled-mode approach, the coupling matrices are independent of azimuthal orders, which makes this approach applicable to problems of scattering from azimuthal symmetric features.

Here a spectral coupled-mode model has been developed and applied to analyze the propagation and scattering around conical seamounts. Although based on a theoretical foundation identical to that used in earlier related work, this approach provides more efficient and stable numerical solutions for realistic seamount problems. The keys to its superior performance are the fundamental superposition principle applied to introduce the source, the choice of normalized range solutions which can be evaluated accurately and remain numerically stable even at high azimuthal orders, and the use of a marching, single scatter approximation for the solution of the coupled-mode equations.

The numerical examples show that azimuthal coupling can be important for seamount problems. When the azimuthal variation is weak, the out-of-plane scattering is negligible and therefore, as shown by the results, the $N \times 2D$ approach is a good approximation of the true 3D approach. However, when the azimuthal variation is strong, azimuthal coupling can not be ignored. In this case, the shadows behind seamounts are broadened by horizontal refraction. A true 3D approach, instead of the $N \times 2D$ approach, can capture this feature. The dependence of the 3D effects on the environment is also illustrated through the numerical example, which shows that for seamounts with steep slopes and reaching far into the sound channel in terms of wavelengths, the 3D effects are significant. In a situation such as this, a true 3D model is required, rather than an $N \times 2D$ model.

6. ACKNOWLEDGEMENTS

This work was supported by the Office of Naval Research, Grant N00014-1-08-1-0013

REFERENCES

[1] **M.D. Collins and B.E. McDonald**, Three-dimensional effects in global acoustics, *J. Acoust. Soc. Am.*, 97 (3), pp. 1567-1575, 1995.

[2] **C.H. Harrison**, Three-dimensional ray paths in basins, troughs, and near seamounts by use of ray invariants, *J. Acoust. Soc. Am.*, 62 (6), pp. 1382-1388, 1977.

[3] **M.J. Buckingham**, Theory of acoustic propagation around a conical seamount, *J. Acoust. Soc. Am.*, 80 (1), pp. 256-277, 1986.

[4] **J.S. Perkins and R.N. Baer**, An approximation to the three-dimensional parabolic-equation method for acoustic propagation, *J. Acoust. Soc. Am.*, 72 (2), pp. 515-522, 1982.

[5] **M.I. Taroudakis**, A coupled-mode formulation for the solution in the presence of a conical sea-mount, *J. Comput. Acoust.*, 4 (1), pp. 101-121, 1996.

[6] **F.B. Jensen, W.A. Kuperman, M.B. Porter, and H. Schmidt**, *Computational Ocean Acoustics*, American Institute of Physics, 1994.

[7] **J. Eskenazi**, A computer model for sound propagation around conical seamounts, Master's thesis, Dept. of Ocean Engineering, Mass. Inst. Tech., 2001.

[8] **D.C. Ricks and H. Schmidt**, A numerically stable global matrix method for cylinderly layered shells excited by ring forces, *J. Acoust. Soc. Am.*, 95 (6), pp. 3339--3349, 1994.

[9] **G.A. Athanassoulis and K.A. Belibassakis**, All-frequency normal-mode solution of the three-dimensional acoustic scattering from a vertical cylinder in a plane-horizontal waveguide, *J. Acoust. Soc. Am.*, 101 (6), pp. 3371--3384, 1997.

[10] **W. Luo**, Three-dimensional propagation and scattering around a conical seamount, Ph.D. thesis, Dept. of Mechanical Engineering, Mass. Inst. Tech., 2007.

[11] **W. Luo and H. Schmidt**, Three-dimensional propagation and scattering around a conical seamount, *J. Acoust. Soc. Am.*, 125 (1), pp. 52--65, 2009.

Transport Theory for Shallow Water Propagation with Rough Boundaries

Eric I. Thorsos, Frank S. Henyey, W. T. Elam, Brian T. Hefner, Stephen A. Reynolds, and Jie Yang

Applied Physics Laboratory, University of Washington, 1013 NE 40th Street
Seattle, WA 98105, USA.

Abstract. At frequencies of about 1 kHz and higher, forward scattering from a rough sea surface (and/or a rough bottom) can strongly affect shallow water propagation and reverberation. The need exists for a fast, yet accurate method for modeling such propagation where multiple forward scattering occurs. A transport theory method based on mode coupling is described that yields the first and second moments of the field. This approach shows promise for accurately treating multiple forward scattering in one-way propagation. The method is presently formulated in two space dimensions, and Monte-Carlo rough surface PE simulations are used for assessing the accuracy of transport theory results.

Keywords: Mid-frequency, shallow water propagation, surface scattering.
PACS: **43.30 Bp, 43.30.Hw**

INTRODUCTION

Modeling of shallow water propagation can be especially challenging at frequencies above about 1 kHz, because in this regime forward scattering can be significant and affect the level and properties of the propagated field. One approach to modeling forward scattering in such propagation scenarios is to use a Monte Carlo method where multiple realizations of the rough surface or bottom are constructed, and propagation simulations are made for each realization using, for example, rough surface PE [1]. Averaging these results over realizations can then yield the desired average properties of the propagated field, and the field statistics can be obtained as well. This approach is reliable, but computationally intensive, and is effectively limited to two-dimensional (2-D) problems.

The need exists for a modeling method that yields directly the average intensity of the propagated field and can accurately account for forward scattering while being computationally efficient. An approach using transport theory is described here that satisfies these requirements. The work described has been restricted to the 2-D propagation problem in order that results can be verified by comparison with 2-D rough surface PE simulations. However, because of the method's computational efficiency, extensions to reverberation and to a 3-D geometry appear practical. Here we focus on the effects of rough sea surface scattering at a frequency of 3 kHz in a waveguide with a constant depth of 50 m.

Transport theory can be viewed as any method that gives evolution equations for the moments of the field. The moments of interest here are the coherent pressure field denoted by $<p>$, the first moment of the field, and the average absolute square of the total pressure denoted by $<|p|^2>$, a second moment of the field. In these expressions, $< >$ denotes an average over an ensemble of rough sea surface realizations. For transport theory, that average is formally done and the average represents the limit for an arbitrarily large number of realizations. For the rough surface PE simulation results used for comparison, the averages have been done over an ensemble of 50 realizations. When the propagating field is confined to a narrow angular range, $<|p|^2>$ is proportional to the average total intensity, and for simplicity it will be referred to by that term; similarly, the terms coherent intensity and incoherent intensity will also be used. In the results that follow, the quantities that will be displayed are the

CP1272, *Shallow-Water Acoustics, Second International Shallow-Water Acoustics Conference*
edited by J. Simmen, E. S. Livingston, J. X. Zhou, and F. H. Li

average total intensity, the coherent intensity given by $|<p>|^2$, and the average incoherent intensity given by subtracting the coherent intensity from the average total intensity: $<|p|^2>-|<p>|^2$.

TRANSPORT THEORY

The starting point in our transport theory approach [2] is to expand the pressure field in unperturbed modes (i.e., modes for a waveguide with a flat sea surface at $z=0$ and a flat water-sediment interface):

$$p(x,z)=\sum_{j=1}^{N} a_j(x)\phi_j(z), \tag{1}$$

with mode amplitudes $a_j(x)$ and mode eigenfunctions $\phi_j(z)$. Because of space limitations, only an outline of the derivation and the final equations will be given here, with a full derivation to be given elsewhere. First-order perturbation theory is used to evaluate the coupling between modes leading to a set of coupled equations for the evolution of the mode amplitudes:

$$\frac{da_j}{dx}=ik_j a_j+ih(x)\sum_{m=1}^{N} M_{jm}a_m, \tag{2}$$

where k_j is the horizontal wave number for mode j (k_j^2 is the corresponding mode eigenvalue), $h(x)$ is a realization of the rough surface height, and $M_{jm}=\dfrac{1}{2k_j}\dfrac{d\phi_j}{dz}\dfrac{d\phi_m}{dz}$ with the derivatives evaluated at $z=0$. In (2) terms of order $N h^2$ have been ignored, where N is the number of modes retained. It is convenient to write (2) in the form

$$\frac{dA}{dx}=i\boldsymbol{K}\boldsymbol{A}+ih(x)\boldsymbol{M}\boldsymbol{A}, \tag{3}$$

where $\boldsymbol{A}$ is a vector of the mode amplitudes a_j, $\boldsymbol{K}$ is a diagonal matrix of horizontal wave numbers k_j, and $\boldsymbol{M}$ is a matrix with elements M_{jm}. Stopping at this stage would yield an alternate Monte Carlo simulation method using rough surface realizations.

Transport theory gives evolution equations for moments of the mode amplitudes at the cost of some approximations. The evolution equations for the first moment of mode amplitudes are obtained by formally averaging (3) to get

$$\frac{d<\boldsymbol{A}>}{dx}=i\boldsymbol{K}<\boldsymbol{A}>+i\boldsymbol{M}<h(x)\boldsymbol{A}>. \tag{4}$$

Following Van Kampen [3] the transport theory approximation for (4) can be written

$$\frac{d<\boldsymbol{A}>}{dx}=i\boldsymbol{K}<\boldsymbol{A}>-\boldsymbol{M}\boldsymbol{N}<\boldsymbol{A}>, \tag{5}$$

where the elements of the matrix $\boldsymbol{N}$ (not to be confused with the number of modes N) are given by

$$N_{ij}=\pi W(k_i-k_j)M_{ij}, \tag{6}$$

with $W(k_x)$ the surface roughness spectrum normalized such that $\int_{-\infty}^{\infty} W(k_x)dk_x = h^2$, the mean square height. In (6) only the real part of $k_i - k_j$ is used, and corrections to these horizontal wave numbers [3] required by causality have been ignored.

To obtain the transport equation for the second moment, we start with

$$\frac{d}{dx}(AA^{+}) = A\frac{dA^{+}}{dx} + \frac{dA}{dx}A^{+}, \tag{7}$$

where A^{+} denotes the adjoint of A. Using (3) in (7) and averaging gives

$$\frac{d}{dx} < AA^{+} >= iK < AA^{+} > -i < AA^{+} > K^{+} + iM < h(x)AA^{+} > -i < h(x)AA^{+} > M^{+}. \tag{8}$$

The most straightforward approach in using the development in [3] to obtain the second moment transport equation consists of viewing $< AA^{+} >$ as an N^2-component vector, and the equation analogous to (5) involves an $N^2 \times N^2$ matrix. For our applications, N will be taken as 200, and the size of an $N^2 \times N^2$ matrix is prohibitive. Morozov and Colosi [4] have shown that the second moment transport equation can be rearranged and expressed in terms of only $N \times N$ matrices. After some effort the following result can be obtained:

$$\begin{aligned}\frac{d}{dx} < AA^{+} > &= iK < AA^{+} > -i < AA^{+} > K^{+} - MN < AA^{+} > - < AA^{+} > N^{+}M^{+} \\ &\quad + N < AA^{+} > M^{+} + M < AA^{+} > N^{+},\end{aligned} \tag{9}$$

where $< AA^{+} >$ is viewed as an $N \times N$ matrix.

DOZIER-TAPPERT APPROXIMATION

Even with the matrix size reduction enabled by the form given by (9), a straightforward numerical evolution of the second moment would take too much computation time with our parameters to qualify as an efficient solution method. For the example to be discussed here, the frequency is 3 kHz, the waveguide has a depth of 50 m, and the water sound speed is approximately 1500 m/s near the bottom. A sand bottom is modeled as an absorbing fluid (0.5 dB/λ) with a sound speed of 1600 m/s, giving a critical angle of about 20°. For these parameters there are 74 trapped modes. Using (9) and including just 74 modes would require advancing $74 \times 74 = 5476$ first-order coupled equations. And to show the effects of scattering at the surface into modes above the critical angle that are not trapped, it is useful to use more than 74 modes; in fact 200 modes have in used in the example that follows. The continuum modes are represented as closely spaced discrete modes by extending the computational domain 50 m into the sand bottom to a total depth of 100 m where a Dirichlet boundary condition is imposed. The mode complex eigenvalues and eigenfunctions are found with a LAPACK routine (SUBROUTINE ZGEEV).

Fortunately, an approximation, referred to here as the "Dozier-Tappert approximation," can be made that vastly reduces the computation burden with essentially no loss in accuracy for the surface scattering cases being considered. The off-diagonal elements in the matrix $< AA^{+} >$ represent cross-mode coherences. Dozier and Tappert [5],[6] in developing transport theory for acoustic forward scattering from internal waves argued that the cross-mode coherences should to a good approximation average out, and they therefore set the off-diagonal second moment terms to zero. Essentially the same approximation is made here. However, because the coherent field is significant in our case (but was not for Dozier and Tappert), the cross-mode coherences cannot be completely neglected if the spatial structure of the coherent field is to be replicated. Therefore, we use $< AA^{+} > \approx < A >< A^{+} > +$ diagonal term, where the diagonal term is found in the Dozier-Tappert approximation. This approximation effectively reduces the non-zero terms of $< AA^{+} >$ to an N-component vector leading to very efficient solutions. Comparisons with rough surface PE simulations that follow show that indeed there is very little loss of accuracy in making the Dozier-

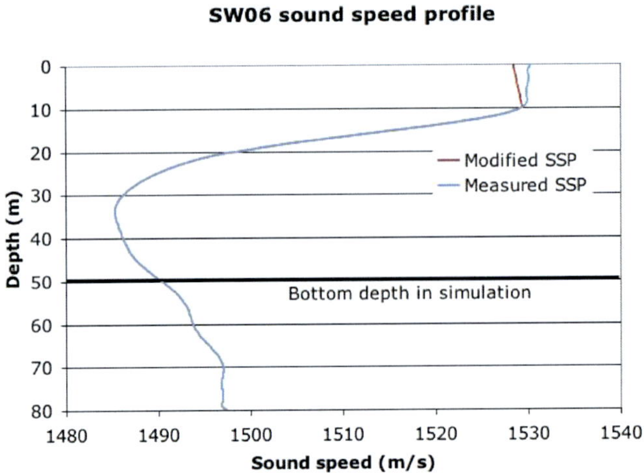

FIGURE 1. A sound speed profile (SSP) measured during the Shallow Water Experiment in 2006 (blue) and the modified profile (red from 0 to 10 m depth, blue otherwise) used in the simulation to give a more pronounced surface duct.

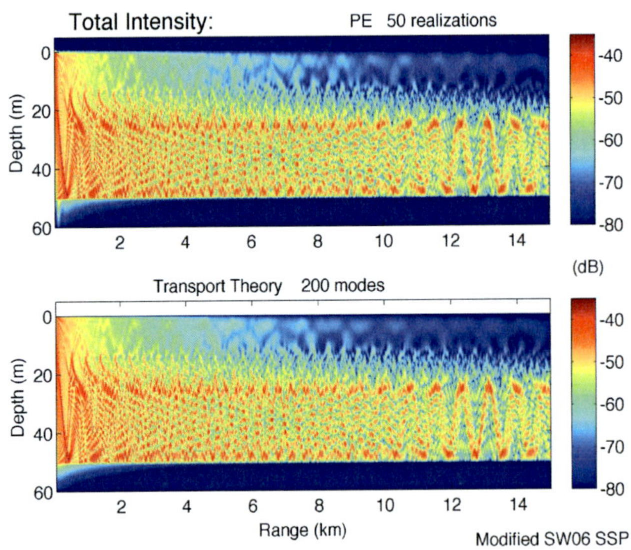

FIGURE 2. A comparison of the total intensity with the modified SW06 profile for PE (top) and transport

Tappert approximation. After making this approximation, the evolution equation given by (9) becomes identical to the alternate form obtained by working directly from Ref. [3], viewing $< \boldsymbol{AA}^{+} >$ as an N^2-component vector which leads to an $N^2 \times N^2$ matrix in the evolution equation, and then reducing down this form by making the Dozier-Tappert approximation.

Transport theory solutions have been obtained for both the first and second moments using an ODE solver for coupled first-order systems by Shampine and Gordon [7]. In [2] transport theory solutions for the scintillation index, a fourth moment, were also described under the assumption that the incoherent complex pressure obeys Gaussian statistics. In this special case the scintillation index can be represented in terms of first and second moments, and comparisons with rough surface PE simulations showed good agreement.

RESULTS

The accuracy of the transport theory method for shallow water propagation with surface forward scattering included will be shown for an example with a mild surface duct. The one-dimensional (1-D) roughness spectrum used in transport theory computations, and for generating realizations for PE simulations, is of a 1-D Pierson-Moskowitz form as given in [8]. For the example shown the surface waves are produced by a wind speed of 7.7 m/s (15 knots). A CW point source is located at the mid depth of 25 m, and a vertical (amplitude) beam pattern has been applied with a Gaussian shaded full width of 20° (to the 1/e points) and with the beam center aimed up at a 10° grazing angle. The effect of the beam pattern is to limit the vertical angular range of the starting field to a region well within the accuracy region of the rough surface PE simulation method.

Figure 1 shows a sound speed profile (blue) measured during the Shallow Water Experiment in 2006 (SW06), and a profile that was modified in the top 10 m (red) to give a more pronounced surface duct. The sound speed profile is cut off at the bottom depth of 50 m used in this example. With a source at a depth of 25 m, no energy would become trapped in the surface duct with a flat sea surface, but scattering from surface roughness can lead to energy being trapped. Both transport theory and rough surface PE will be used to model the energy trapped in the duct by surface scattering.

The total intensity (coherent plus incoherent) is compared in Figure 2 with the PE result at the top and the transport theory result at the bottom. In each case the PE results are obtained by averaging over the results for 50 surface realizations. As would be expected with the down-refracting sound speed profile of Figure 1, most of the propagated energy resides in the lower portion of the water column, though some does interact with the surface. Note the energy penetrating into the bottom in both the PE and transport theory plots at short range. Except for the first peak at a range of about 130 m, this energy has been scattered at the surface from below to above the critical angle and then readily penetrates into the bottom. This component is made visible with transport theory by using 200 modes in order to include the scattered energy at these higher angles. The lack of the peak at about 130 m in the penetrating component with transport theory is one consequence of using the Dozier-Tappert approximation.

Figure 3 gives the corresponding comparison for the coherent intensity. The coherent component that interacts with the surface can be viewed as simply reflecting from the surface (with some loss). Therefore, none of this component is trapped in the surface duct. Note the absence of energy penetrating into the bottom. With the choice made for the starting field, very little coherent energy can reach the bottom at angles above the critical angle. This supports the idea that the energy penetrating the bottom in Figure 2 had to first forward scatter from the surface.

Finally, Figure 4 shows the comparison for the incoherent energy. Energy trapping in the surface duct is clearly evident, showing the importance of trapping by surface scattering in this example. Since both rough surface PE and transport theory can account for any number of surface scatterings during propagation in the waveguide, depletion of energy from the duct by surface scattering will also be included in these results.

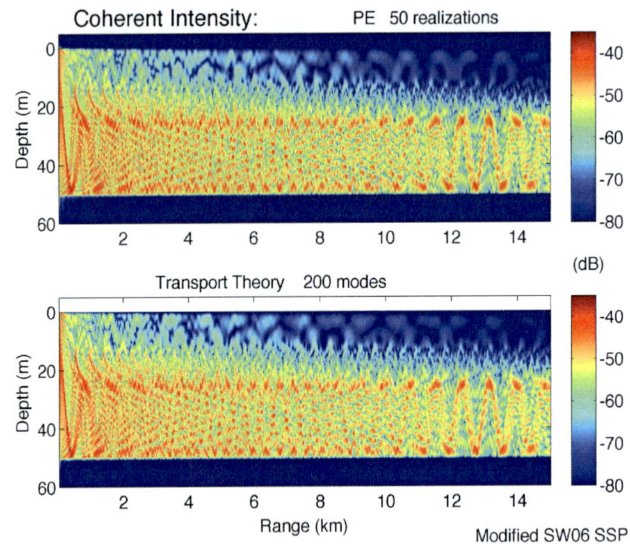

FIGURE 3. A comparison of the coherent intensity with the modified SW06 profile for PE (top) and transport theory (bottom).

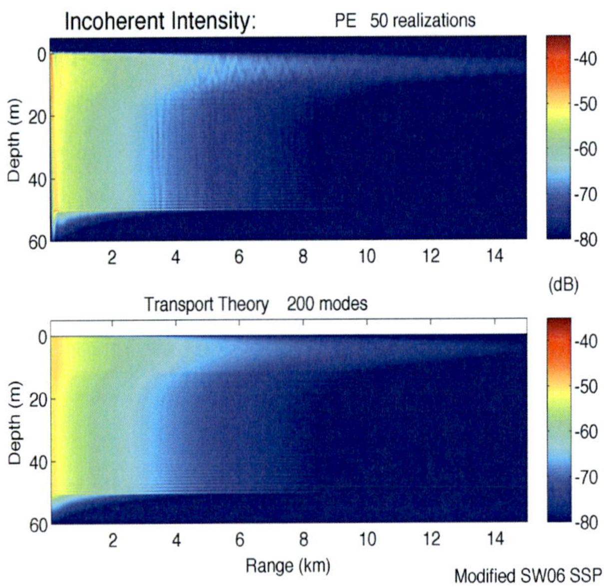

FIGURE 4. A comparison of the corresponding incoherent intensity for PE (top) and transport theory (bottom).

DISCUSSION

The comparisons shown illustrate the capability and accuracy of transport theory in treating effects of forward scattering in shallow water propagation. The computational efficiency of transport theory is also of special note. Typical computation times for modeling propagation to a range of 30 km is about 1 minute using typical office computer equipment, while the corresponding PE simulations with 50 surfaces can take about an hour. While the unperturbed propagation problem (that is, without surface roughness) is range independent in the example shown, the transport theory evolution equations have been advanced with an ODE solver, which can be expected to have application to range dependent problems as well.

ACKNOWLEDGMENTS

This work was supported by the U.S. Office of Naval Research.

REFERENCES

1. A. P. Rosenberg, *J. Acoust. Soc. Am.* **105**, 144-153 (1999).
2. E. I. Thorsos, F. S. Henyey, W. T. Elam, S. A. Reynolds, and K. L. Williams, "Modeling shallow water propagation with scattering from rough boundaries," Proceedings of High-Frequency Ocean Acoustics Conference, La Jolla, California, March 1-5, 2004, AIP Conference Proceedings 728, pp. 132-140.
3. N. G. Van Kampen, *Stochastic Processes in Physics and Chemistry,* Amsterdam: North-Holland, 1997, chapter XVI.
4. A. K. Morozov and J. A. Colosi, *Acoustical Physics* **53**, 335-347 (2007).
5. L. B. Dozier and F. D. Tappert, *J. Acoust. Soc. Am.* **63**, 353-365 (1978).
6. L. B. Dozier and F. D. Tappert, *J. Acoust. Soc. Am.* **64**, 533-547 (1978).
7. L. F. Shampine and M. K. Gordon, *Computer Solution of Ordinary Differential Equations: The Initial Value Problem,* San Francisco: W. H. Freeman and Co., 1975.
8. K. L. Williams, E. I. Thorsos, and W. T. Elam, *J. Acoust. Soc. Am.* **116**, 1975-1984 (2004).

Extension of Coupled-Modes Method to Waveguides with Elastic Bottom

Ning Wang, Hao Zhong Wang and Da Zhi Gao

Department of Physics
Ocean University of China, 238 Songling Rd. Qingdao, P.R.China

Abstract: In this paper, the coupled-modes method of acoustic propagation is extended to waveguides with smoothly varying fluid-elastic interfaces. To formulate the problem without recourse to the SH wave component from the ground up, we employ the $\{u_z,u_V\}$-formulation developed in seismology. The result is remarkably different from that for fluid-fluid interfaces; the boundary conditions imposed at fluid-elastic interfaces give rise to terms including unknown boundary values so that the differential system obtained is not closed. To make the system closed, additional compatibility and constraining conditions have to be imposed.

Keywords: coupled-modes, $\{u_z,u_V\}$-formulation , fluid-elastic interfaces .
PACS: 43.20

I. INTRODUCTION

Normal mode theory is an exact wave theoretical approach to the acoustic propagation problem in a horizontally layered waveguide. Pierce [1] extended the normal mode theory to handle gradual range-dependent environments, and derived a continuous system of differential equations for coupled modes. An alternate method often referred to as stepwise coupled modes [2] was proposed to calculate the range-dependent propagation problem by discretizing a range-varying waveguide into a number of range-independent sections and continue solutions of different sections. The standard continuous coupled-modes method is not complete for waveguides with density discontinuity boundary condition at the bottom interface [3]. Boundary condition correction was given by A. C. Fawcett [4] to improve the continuous coupled-modes method.

Modeling elasticity in the ocean bottom has gained importance in recent years because of renewed emphasis on shallow water acoustic propagations. Elasticity can be important, in particular, at low frequencies and for sources and receivers close to the ocean bottom; in these cases interface modes of propagation may dominate. There were a number of studies on the normal mode method with elastic bottoms, e.g., [5]. However, most studies on the coupled-modes method have been confined to liquid bottoms. The coupled-modes method may be adequate for considering the sound field in a water layer but not for the interface wave. The coupled-modes method may be adequate for considering the sound field in a water layer but not for the interface wave. A formulation for generic fluid-elastic environments is very complicated. In this paper, we report some of our recent results in this area .

CP1272, *Shallow-Water Acoustics, Second International Shallow-Water Acoustics Conference*
edited by J. Simmen, E. S. Livingston, J. X. Zhou, and F. H. Li

We sketch the geometry of the problem in Fig.1. The model that we are considering is a fluid-elastic waveguide in which the interfaces are range-dependent. For simplicity, sound speed and density profiles in the upper fluid layer are assumed constant and the underlying elastic layer is assumed a homogeneous isotropic half infinite. A source in the water column generates wave propagation. For a flat interface, the problem is a well-known classical problem, and a generalized Layleigh (P-SV) wave can be generated. It is noted that an elastic Love (SH) wave does nothing in the flat interface, because a source in the water makes no contribution to tangent displacement and stress components relevant to an SH wave. In the present problem, the Love modes make no contributions to the problem. SH wave/Love modes may be generated when P-SV propagates along an irregular interface, but it is second-order and we do not consider it here.

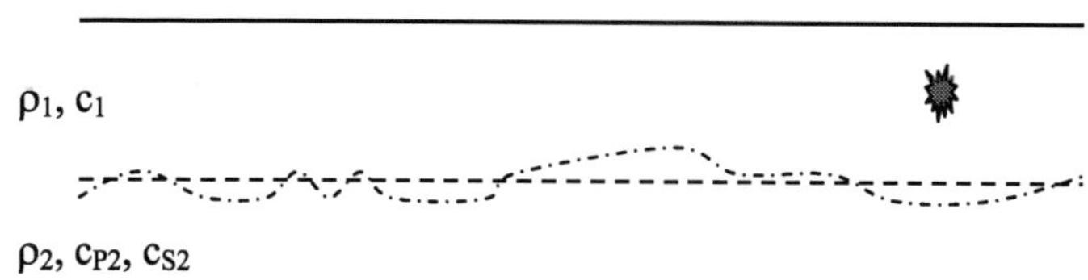

FIGURE 1. Geometry

Considering a three-dimensional problem with varying interface, the first difficulty is how to formulate the problem without recourse to the SH wave component from the ground up. This is a nontrivial problem. In the usual vector-potential or displacement vector approaches, a good gauge-fixed condition (being a good displacement vector-decomposition) can be choice, which helps us decompose and decouple the P-SV and SH components. However, for the present problem there is no known a priori geometric symmetry helpful for a gauge-fixed condition, and if we insist on using a vector-potential approach for example, the gauge-fixed condition has to be taken point by point. In other words, we could not determine a global decomposition which has a unified gauge-fixed condition. On physical ground, P-SV and SH components may couple at an interface, and the decomposition of a displacement vector into its P-SV and SH components is dependent on the geometry of both the incidence plane and the interface. For a spatially varying interface, it is impossible to have a global gauge-fixed condition. Thus, we have two choices; one begins with a fixed gauge condition and the other without recourse to the SH wave component from the ground up. In the first approach we have no known Love modes, but must keep three displacement components from the beginning, which is extremely difficult. Fortunately, a $\{u_z,u_V\}$-formulation developed in seismology can help us to resolve this problem. We shall introduce the $\{u_z,u_V\}$-formulation and reformulate the corresponding range-independent problem in Section 2.

In the $\{u_z,u_V\}$-formulation, the displacement vector field is replaced by a set of new field components, then governing equations and boundary conditions are transformed into new ones. There arises another difficulty; most well-developed tools for elastic wave problems, e.g., the representation theorem (Betti formula) and vectorial spherical function decomposition and explicit orthognality/completeness relation for normal modes were developed for usual vector-potential or displacement vector approaches, but we have little for the $\{u_z,u_V\}$-formulation. It is useful to have explicit expressions for orthognality/ completeness relation in the formulation, but as it is not possible to cover these in this paper, we prefer to bypass the problem to formulate our problem by analogy with the treatment for a fluid-fluid bottom waveguide given by A. C. Fawcett [4]. This analogy is discussed in Section 3, where a continuous system of partial differential equations of coupled modes is derived. It is remarkably different from the pure fluid-fluid bottom case (for simplicity, we call this case an acoustic case); the boundary corrections include unknown fields so that the differential system obtained is not closed. To make the system closed, additional compatibility and constraining conditions have to be imposed. All results obtained here are carefully verified to coincide exactly with the acoustical case in [4] when we take an elastic bottom to a liquid limit. This paper is a pure theoretical formulation; numerical code and its applications will be considered elsewhere.

II. $\{U_Z,U_V\}$-FORMULATION FOR NORMAL MODE THEORY

In this section, in the framework of $\{u_z,u_V\}$-formulation, we reformulate the normal mode theory for the range-independent fluid-elastic waveguide. The upper water column is taken to have constant depth H with sound speed c_1 and density ρ_1. There is the underlying homogeneous isotropic elastic half-infinite bottom with Lame constants, λ, μ and density ρ_2. We always assume the radiation boundary condition satisfied at infinite depth. It is easy to extend the treatment here to allow for depth-varying parameters and also attenuation in the elastic layer.

The problem we are considering is to solve the following boundary-valued problem:

$$\nabla^2\Theta+\frac{\omega^2}{c_1^2}\Theta(x,y,z)=0\ , \qquad 0\le z\le H \tag{1}$$

$$\rho_1\omega^2\Theta=p,\quad u_w=\nabla\Theta\quad , \tag{2}$$

where p denotes the sound pressure, Θ the displacement potential and satisfies the boundary conditions

$$\begin{aligned}&z=0,\quad \Theta=0;\\ &z=H,\quad \partial_z\Theta\big|_{z=H_-}=u_z\big|_{z=H_+}\quad -\rho_1\omega^2\,\Theta\big|_{z=H_-}=\sigma_{zz}\big|_{z=H_+},\\ &\qquad\qquad \sigma_{xz}\big|_{z=H}=\sigma_{yz}\big|_{z=H}=0\\ &z\to\infty,\quad \textit{radiating}-\textit{condition}\end{aligned} \tag{3}$$

For the elastic bottom, it is useful to introduce new elements

$$u_V\equiv\partial_x u_x+\partial_y u_y\,, \tag{4}$$

the displacement divergence in the $\{x,y\}$-plane and

$$\sigma_{Vz}\equiv\partial_x\sigma_{zx}+\partial_y\sigma_{zy}\,. \tag{5}$$

It is noted that elements introduced in eqs.(4) and (5) have different dimension with the usual displacement and stress components. In terms of these elements, the equations of motion describing P-SV waves in an isotropic horizontally stratified medium are decoupled from that of the SH wave, and can be rewritten compactly in the following form for $z\ge H$,

$$\partial_z u_z = (\lambda+2\mu)^{-1}\sigma_{zz} - (\lambda+2\mu)^{-1}\lambda u_V, \tag{6}$$

$$\partial_z u_V = \mu^{-1}\sigma_{Vz} - \nabla_{//}{}^2 u_z, \tag{7}$$

$$\partial_z \sigma_{zz} = -\omega^2 \rho_2 u_z - \sigma_{Vz}, \tag{8}$$

$$\partial_z \sigma_{Vz} = -(\lambda+2\mu)^{-1}\lambda\nabla_{//}{}^2(\sigma_{zz}) - \lambda_V \nabla_{//}{}^2 u_V - \omega^2\rho_2 u_V, \tag{9}$$

where $\nabla_{//}{}^2 \equiv \partial^2/\partial x^2 + \partial^2/\partial y^2$ denotes the {x,y}-Laplace operator and

$$\lambda_V = (\lambda+2\mu) - (\lambda+2\mu)^{-1}\lambda^2. \tag{10}$$

It is useful for later purposes to rewrite eqs.(6)-(9) to

$$\partial_z[\rho\alpha^2\partial_z u_z + \rho(\alpha^2-2\beta^2)u_V] + \rho\beta^2\partial_z u_V + \rho[\beta^2\nabla_{//}{}^2 + \omega^2]u_z = 0, \tag{11}$$

$$\partial_z[\rho\beta^2\partial_z u_V + \rho\beta^2\nabla_{//}{}^2 u_z] + \rho(\alpha^2-2\beta^2)\nabla_{//}{}^2\partial_z u_z + \rho(\omega^2 + \alpha^2\nabla_{//}{}^2)u_V = 0. \tag{12}$$

In the above equations, α and β define the P and SV wave speeds, respectively. These equations also satisfy for more general $\lambda(z)$, $\mu(z)$ and density $\rho_2(z)$. It is interesting to see that (x,y)-dependence of field components is separated completely from that on the z-coordinate. This allows us to make the standard variable separation. The equation governing depth functions can be obtained by the replacement of $\nabla_{//}{}^2 \to k^2$, k being the horizontal wave number. For $0 \le z \le H$, the eigenfunction for the n-th normal mode with eigen-number k_n satisfies

$$\frac{d^2}{dz^2}\hat{\Theta}_n + (\frac{\omega^2}{c_1^2} - k_n{}^2)\hat{\Theta}_n(z,k) = 0, \tag{13}$$

and for $z \ge H$, depth functions for eqs.(11)-(12) are determined by the following reduced second-order ordinary differential equations,

$$\partial_z[\rho_2\alpha^2\partial_z\hat{u}_z^n + \rho_2(\alpha^2-2\beta^2)\hat{u}_V^n] + \rho_2\beta^2\partial_z u_V^n + \rho_2[-\beta^2 k_n{}^2 + \omega^2]\hat{u}_z^n = 0, \tag{14}$$

$$\partial_z[\rho_2\beta^2\partial_z\hat{u}_V^n - \rho_2\beta^2 k_n{}^2\hat{u}_z^n] - \rho_2(\alpha^2-2\beta^2)k_n{}^2\partial_z\hat{u}_z^n - \rho_2(-\omega^2 + \alpha^2 k_n{}^2)\hat{u}_V^n = 0. \tag{15}$$

The boundary conditions in eq.(3) are converted to

$$\begin{aligned} & z=0, \quad \hat{\Theta}_n = 0; \\ & z=H, \quad \partial_z\hat{\Theta}_n\Big|_{z=H_-} = \hat{u}_z^n\Big|_{z=H_+} \qquad -\rho_1\omega^2\hat{\Theta}_n\Big|_{z=H_-} = \hat{\sigma}_{zz}^n\Big|_{z=H_+}, \\ & \qquad \hat{\sigma}_{Vz}^n\Big|_{z=H} = 0 \\ & z\to\infty, \quad radiating-condition \end{aligned} \tag{16}$$

The elastic wave propagation in a horizontally layered fluid-elastic environment using the above system has been widely studied cf. e.g. [6]. For more details on wave fields for a general layered fluid-elastic medium, we refer interested readers to [6] and [7]. Here we only need to keep in mind that there exists a set of finite or infinite discrete eigenvalues k_n, corresponding modes decay exponentially in the elastic bottom. We refer to these modes as normal modes.

III. DIFFERENTIAL EQUATIONS FOR COUPLED MODES

In this section, we extend the normal mode theory for fluid-elastic layered media to range-dependent environments, where the water column is allowed to vary *only* with depth *H(x)*. For simplicity, we consider only the interface being the x-coordinate dependent.

As argued by Fawcett, a correct treatment based on series approximation of modal summation has to be formulated in a function space sense. The basic idea is the following: first we have to distinguish two classes of field quantities; the first class is allowed to be directly approximated by the local modal summation and the second class with explicit z-derivation term is not allowed the similar approximation. The second step is to remove the z-derivation by applying partial integration to transform the second class to the first class. And last, apply the following local modal summation approximation

$$\Theta(r_{//},z) = \sum_n A_n(r_{//})\Theta_n(z,x) \quad , \tag{17}$$

$$u_z(r_{//},z) = \sum_n A_n(r_{//})u_z^n(z,x), \quad u_V(r_{//},z) = \sum_n A_n(r_{//})u_V^n(z,x) \tag{18}$$

Due to the range-dependent depth, local eigenvalue and normal mode functions are all functions of the x-variable. For simplicity, we will omit the hat symbol for depth functions in the following:

The boundary condition at the interface is now

$$\begin{gathered} z = H(x), \quad \partial_n \Theta\big|_{z=H^-} = u_n\big|_{z=H^+}, \quad -\rho_1\omega^2\,\Theta\big|_{z=H^-} = \sigma_{nn}\big|_{z=H^+}, \\ \sigma_{\mathrm{nt}}\big|_{z=H^+} = 0, \quad \sigma_{ny} = 0 \end{gathered} \tag{19}$$

where index n and t denote normal and tangent components with respect to the interface, respectively. In terms of {x,y,z}-coordinate system, (19) equals to

$$\begin{aligned} &[\sigma_{xx}n_x n_x + 2\sigma_{xz}n_x n_z + \sigma_{zz}n_z n_z] = \sigma_{nn}(H^+) = -\rho_1\omega^2\Theta(H^-), \\ &[-\sigma_{xx}n_x n_z + \sigma_{zx}(n_x n_x - n_z n_z) + \sigma_{zz}n_x n_z] = \sigma_{nt} = 0, \\ &\sigma_{zy}n_z + \sigma_{xy}n_x = \sigma_{ny} = 0, \quad z = H(x), \\ &u_n = u_z n_z + u_x n_x = \partial_n\Theta = \partial_z\Theta n_z + \partial_x\Theta n \end{aligned} \tag{20}$$

Quantities with explicit z-derivation terms, e.g., $\partial_z u_{z,V}$, etc, are not allowed to be approximated directly, but σ_{xy}, u_z and u_V, etc. can be approximated. To proceed to the second step, we first rewrite (11)-(12) to

$$\begin{aligned} &\partial_z[\rho_2\alpha^2\partial_z u_z + \rho_2(\alpha^2 - 2\beta^2)u_V] + \rho_2\beta^2\partial_z u_V + \rho_2[-\beta^2 k_m{}^2 + \omega^2]u_z \\ &+ \rho_2\beta^2[\nabla_{//}{}^2 + k_m{}^2]u_z = 0 \end{aligned} \quad , \tag{21}$$

$$\begin{aligned} &\rho_2(\alpha^2 - \beta^2)(\nabla_{//}{}^2 + k_m{}^2)\partial_z u_z + \rho_2\alpha^2(\nabla_{//}{}^2 + k_m{}^2)u_V \\ &+ \partial_z[\rho_2\beta^2\partial_z u_V - \rho_2\beta^2 k_m{}^2 u_z] - \rho_2(\alpha^2 - 2\beta^2)k_m{}^2\partial_z u_z + \rho_2(\omega^2 - \alpha^2 k_m{}^2)u_V = 0 \end{aligned} \quad , \tag{22}$$

calculating the following integrals,

$$\int_H^\infty dz u_z^m \times (21) + \int_H^\infty dz \, u_V^m / k_m^2 \times (22)\text{'}$$

where $1/k_m^2$ is inserted to make the two terms the same dimension. By partial integrations, transforms quantities of the second class into the first class, and after a straightforward but complicated calculation yield

$$\begin{aligned} &\sum_n \langle\langle w_m, w_n \rangle\rangle_1 (\nabla_{//}^2 + k_m^2)A_n + \langle\langle w_m, \nabla_{//}^2 w_n \rangle\rangle_1 A_n + \langle\langle w_m, \nabla w_n \rangle\rangle_1 \bullet \nabla_{//} A_n \\ &+ [u_z^m \sigma_{zz} - \sigma_{zz}{}^m u_z]\Big|_H^\infty + [\frac{u_V^m}{k_m^2}\sigma_{VZ} - \frac{\sigma_{VZ}{}^m}{k_m^2}u_V]\Big|_H^\infty - \sum_n \int_{-\infty}^{+\infty} dk_y [\frac{u_V^m}{k_m^2}\rho_2(2\beta^2 - \alpha^2)u_z^n(\nabla_{//}^2 + k_m^2)A_n]\Big|_H^\infty = 0 \end{aligned} \tag{23}$$

where

$$<< w_1, w_2 >>_1 \overset{def}{=} \int_H^\infty dz\{u_z^1 \rho_2 u_z^2 \beta^2 + \frac{u_V^1}{k_1^2}\rho_2 \alpha^2 u_V^2 - \frac{\partial_z u_V^1}{k_1^{\ 2}} u_z^2 \rho_2 (\alpha^2 - \beta^2)\}\cdot$$

Finally, calculating $\omega^2 \int_0^H dz \Theta_m \times (1) + (23)$ yields

$$\begin{aligned}
&\sum_n ((w_m, w_n))_1 (\nabla_{//}^{\ 2} + k_m^{\ 2}) A_n + ((w_m, \nabla_{//}^{\ 2} w_n))_1 A_n + ((w_m, \nabla w_n))_1 \bullet \nabla_{//} A_n \\
&+ [u_z^m \sigma_{zz} - \sigma_{zz}^{\ m} u_z]\Big|_H^\infty + [\frac{u_V^m}{k_m^{\ 2}} \sigma_{VZ} - \frac{\sigma_{VZ}^{\ \ m}}{k_m^{\ 2}} u_V]\Big|_H^\infty + \rho_1 \omega^2 \{\Theta_m \frac{\partial \Theta}{\partial z} - \frac{\partial}{\partial z}\Theta_m(z)\Theta\}\Big|_0^{H-}, \qquad (24) \\
&- \sum_n \int_{-\infty}^{+\infty} dk_y [\frac{u_V^m}{k_m^{\ 2}} \rho_2 (2\beta^2 - \alpha^2)(\nabla_{//}^{\ 2} + k_m^{\ 2}) u_z^n A_n]\Big|_H^\infty = 0
\end{aligned}$$

with

$$((w_1, w_2))_1 \overset{def}{=} \omega^2 \int_0^H dz \Theta_1 \Theta_z dz + \int_H^\infty dz\{u_z^1 \rho_2 u_z^2 \beta^2 + \frac{u_V^1}{k_1^2}\rho_2 \alpha^2 u_V^2 - \frac{\partial_z u_V^1}{k_1^{\ 2}} u_z^2 \rho_2 (\alpha^2 - \beta^2)\}\cdot$$

We have the following relation

$$<< w_m, w_n >>_1 = ((w_m, w_n)) - << w_m, w_n >>_2 + [\frac{\rho_2 (2\beta^2 - \alpha^2)}{k_m^{\ 2}} u_V^m u_z^m]\Big|_H^\infty, \qquad (25)$$

where

$$\begin{aligned}
- << w_m, w_n >>_2 &\overset{def}{=} \int_H^\infty dz\{\rho_2 \alpha^2 \frac{u_V^m}{k_m} u_V^n [\frac{1}{k_m} - \frac{1}{k_n}] + \rho_2 (2\beta^2 - \alpha^2)[\frac{1}{(k_n + k_m)}(\frac{u_V^m}{k_m}\partial_z u_z^n + \frac{u_V^n}{k_n}\partial_z u_z^m)\cdot \\
&- \frac{2u_V^m}{(k_m + k_m)k_m}\partial_z u_z^n] + \rho_2 \beta^2 [\frac{1}{(k_n + k_m)}[\frac{u_z^m \partial_z u_V^n}{k_n} + \frac{u_z^n \partial_z u_V^m}{k_m}] - \frac{2\partial_z u_V^m}{(k_m + k_m)k_m} u_z^n]\}
\end{aligned}$$

On substituting (25) into (24), the boundary term in (25) just cancels the last boundary term in (24), and we have

$$\begin{aligned}
&\sum_n [((w_m, w_n)) - << w_m, w_n >>_2](\nabla_{//}^{\ 2} + k_m^{\ 2}) A_n \\
&+ [((w_m, \nabla_{//}^{\ 2} w_n)) - << w_m, \nabla_{//}^{\ 2} w_n >>_2] A_n + [((w_m, \nabla_{//} w_n)) - << w_m, \nabla_{//} w_n >>_2] \bullet \nabla_{//} A_n \qquad (26) \\
&+ [u_z^m \sigma_{zz} - \sigma_{zz}^{\ m} u_z]\Big|_H^\infty + [\frac{u_V^m}{k_m^{\ 2}} \sigma_{VZ} - \frac{\sigma_{VZ}^{\ \ m}}{k_m^{\ 2}} u_V]\Big|_H^\infty + \rho_w \omega^2 \{\Theta_m \frac{\partial \Theta}{\partial z} - \frac{\partial}{\partial z}\Theta_m(z)\Theta\}\Big|_0^{H-} = 0
\end{aligned}$$

Some remarks are now in order:

- The matrix defined by elements $<<w_m,w_n>>_2$ is off-diagonal and vanishes in the fluid limit. The second boundary term in eq.(26) also vanished in the fluid limit, so eq.(26) for coupled modes coincides exactly with that in [4] in this limit.
- The matrix before the $(\nabla_{//}^{\ 2} + k_m^{\ 2}) A_n$, in contrast to the acoustic case, in general, is not the identity matrix.
- The derivation above is straightforward but not the most compact.

To obtain the final form, we need to transform all the boundary values in (26) to the first class quantities. By using the boundary conditions in eq.(20), the boundary terms (BTs) are transformed to

$$BTs = \sigma_{zz}^{m}[\partial_x \Theta(H^-) - u_x(H^+)]H'(x) \qquad (27)$$
$$-2u_z^m \sigma_{xz}(H^+)[n_x n_z + n_x^2 - n_z^2] - \frac{u_V^m}{k_m^2}\sigma_{Vz}(H^+)$$

The first term has the same form as that of the acoustic case, but the second term is a term due to the elastic bottom. The term vanishes in the exact horizontally stratified waveguide, but can not be naively neglected which has the same order as the first term with respect to *H'(x)*. It is however noted that there exist unknown boundary values $u_x(H^+)$, $\sigma_{xz}(H^+)$ and $\sigma_{Vz}(H^+)$, and they are all *H'(x)* order. The differential equation system for coupled modes becomes unclosed. From their defining relations, to make the system closed, we need u_x and u_x should these be the same? to determine these boundary values. Recall that in the exact range independent case, $\{u_x,u_y\}(x,y,z)$ can be recovered from u_V by the following relations[1]

$$u_x = \sum \frac{\partial_x A_n(x,y;k_n)}{-k_n^2} u_V^n(z), u_y = \sum \frac{\partial_y A_n(x,y;k_n)}{-k_n^2} u_V^n(z),$$
$$u_V = \sum_n A_n(x,y;k_n) u_V^n(z). \qquad (28)$$

For the range-dependent case, we show in Appendix-A that the following ansatz gives the first order approximation in *H'(x)*,

$$u_x = \sum_n \frac{\partial [A_n(x,y)]}{[-k_n(x)^2)]\partial x} u_V^n(z,x), u_y = \sum_n \frac{\partial [A_n(x,y)]}{[-k_n(x)^2)]\partial y} u_V^n(z,x) \cdot \qquad (29)$$

Substituting (29) into (27) by combination with (26), we have a closed differential system about unknown fields A_n.

We have thus formulated the coupled-modes problem for a fluid-elastic waveguide. No such formulation exists in the current literature to our knowledge. We believe that the coupled-modes system presented in this paper is useful to both numerical and theoretical studies.

ACKNOWLEDGMENTS

We would like to think Prof. S.E. Yang for leading us to consider this problem. Thanks also to Prof. S. C. Piao for his interestin this problem.

APPENDIX-A

In this appendix we show the ansatz (29). First, assume the following generic ansatz,

$$u_x = \sum_n \frac{\partial [B_n(x,y)]}{[-k_n(x)^2)]\partial x} u_V^n(z,x), u_y = \sum_n \frac{\partial [C_n(x,y)]}{[-k_n(x)^2)]\partial y} u_V^n(z,x), \qquad \text{(A-1)}$$

where $B_n(x,y;k)$, $C_n(x,y;k)$, n=1,2,... are two unknown functions to be determined. As eq.(A-1) must be compatible with eq.(4), we require the following compatibility condition

[1] It is worthwhile to note that in **sect.2.1.2 [6]** a similar formula is given for the case of a cylindrical coordinate system. Eq. (28) can be verified directly by using the defining relation for u_V.

$$\partial_x u_x + \partial_y u_y$$
$$= \sum_n \frac{\partial}{\partial x}\left[\frac{\partial_x [B_n(x,y)]}{\partial x}\frac{u_V^n(z,x)}{[-k_n(x)^2)]}\right] + \frac{\partial^2 [C_n(x,y)]}{\partial y^2}\frac{u_V^n(z,x)}{[-k_n(x)^2)]} \cdot \qquad \text{(A-2)}$$
$$= u_V = \sum_n A_n(x,y) u_V^n(z,x)$$

This condition is not enough to determine both $B_n(x,y)$ and $C_n(x,y)$! To determine the unknown boundary fields, we further require the following constraining condition (the third boundary condition of eq.(20)) satisfying

$$\sigma_{ny} = \sigma_{zy} n_z + \sigma_{xy} n_x$$
$$= \mu\{\sum_n H'[\frac{1}{(-k_n(x)^2)}\frac{\partial^2 B_n(x,y)}{\partial x \partial y} u_V^n(z,x) + \frac{\partial}{\partial x}(\frac{\partial C_n(x,y)}{\partial y}\frac{u_V^n(z,x)}{(-k_n(x)^2)})] + \cdot \qquad \text{(A-3)}$$
$$[\frac{\partial C_n(x,y)}{[-k_n(x)^2)]\partial y}\frac{\partial}{\partial z} u_V^n(z,x) + \frac{\partial A_n(x,y)}{\partial y} u_z^n(z,x)\}_{z=H(x)} = 0$$

Emphasize that (A-3) is a constraint condition. To determine completely C_n in terms of A_n and B_n, we require eq.(A-3) satisfying term-by-term

$$H'[\frac{1}{(-k_n(x)^2)}\frac{\partial^2 B_n(x,y)}{\partial x \partial y} u_V^n(z,x) + \frac{\partial}{\partial x}(\frac{\partial C_n(x,y)}{\partial y}\frac{u_V^n(z,x)}{(-k_n(x)^2)})] +$$
$$[\frac{\partial C_n(x,y)}{[-k_n(x)^2)]\partial y}\frac{\partial}{\partial z} u_V^n(z,x) + \frac{\partial A_n(x,y)}{\partial y} u_z^n(z,x)\}_{z=H(x)} = 0 \qquad \text{(A-4)}$$

This is a partially differential equation of C_n for given A_n and B_n. To the first order in *H'*, we solve (A-4) as follows

$$C_n(x,k_y)$$
$$\cong A_n(x,k_y) + H'\{\frac{k_n(x)^2}{\partial_z u_V^n}[\frac{1}{(-k_n(x)^2)}\frac{\partial B_n(x,k_y)}{\partial x} u_V^n(z,x) + \frac{\partial}{\partial x}(A_n(x,k_y)\frac{u_V^n(z,x)}{(-k_n(x)^2)})]\}_{z=H(x)} \qquad \text{(A-5)}$$

Assuming the above forms, the unknown coefficients C_n can be expressed by A_n and B_n, and boundary values $u_x(H^+)$, $u_x(H^+)$, $\sigma_{xz}(H^+)$ and $\sigma_{Vz}(H^+)$ in eq.(26) now can be determined by their defining relations. Similarly, by the boundary condition $\sigma_{nt} = 0$ we can show

$$B_n(x,y) = A_n(x,y) + o(H').$$

REFERENCES

[1] A. D. Pierce, "Extension of the method of normal modes to sound propagation in an almost stratified medium," J. Acoust. Soc. Am. **37**, 19-27 (1965).

[2] R. B . Evans, "A coupled modes solution for acoustic propagation in a waveguide with stepwise depth variations of a penetrable bottom," J. Acoust. Soc. Am. **74**, 188-195 (1983).

[3] S. R.. Rutherford and K . E. Hawker," Consistent coupled mode theory of sound propagation for a class of non-separable problems," J. Acoust. Soc. Am. **70**, 554-564 (1981).

[4] A. C. Fawcett, " A derivation of the differential equations of coupled-mode propagation," J . Acoust.Soc. Am. **92,** 290-295 (1992).

[5] E. K. Westwood, T. C. Tindle and N. R. Chapman, "A normal mode model for fluid-elastic ocean environments," J. Acoust. Soc. Am. **100**, 3631-3645 (1996).

[6] B. L. N. Kennett, *Seismic Wave Propagation in Stratified Media*, Cambridge U.P., Cambridge, UK, (1983).

[7] W. M. Ewing, W. S. Jardetsky, & F. Press, *Elastic Waves in Layered Media*, McGraw-Hill, New York. (1957).

Temporal Coherence of Normal Modes in an Ocean Waveguide

T.C. Yang
Naval Research Laboratory
Washington, DC 20375

Abstract. The coherence time of acoustic signals is derived using coupled mode equation to the first order of sound speed perturbation squared. It is found that mode coupling induced by internal waves in a waveguide changes the temporal coherence of the acoustic field. For a typical shallow water environment, numerical results show that the coherence time of acoustic field decreases as -3/2 power of frequency and -1/2 power of range at high frequencies in accord with experimental data. Ignoring the mode coupling, the coherence time of acoustic field decreases as -1 power of frequency and -1/2 power of range, same as that predicted for acoustic rays.

Keywords: Temporal coherence, coupled normal modes, linear internal waves, shallow water
PACS: 43.30.Re

I. INTRODUCTION

Acoustic signals propagating through a dynamic ocean fluctuate with time. Temporal fluctuations can be characterized by the temporal coherence of the signal, defined as the auto-correlation of the signal normalized by its power. The time it takes for the signal to become uncorrelated is referred to as the signal coherence time. The signal coherence time provides a time scale of the (de-)correlation of the signal and is a function of range and signal frequency. In an underwater acoustic waveguide, signal coherence loss is predominantly caused by signal scattering from the ocean inhomogeneities and surface waves. Generally speaking, as range increases, more scattering occurs and hence more coherence loss. As frequency increases, the range over wavelength ratios increases and hence more coherence loss. As a consequence, the signal coherence time will decrease with range and signal frequency. The exact relationship is a subject of great interest as it impacts signal processing in many practical applications.

In deep water, signal coherence loss is attributed predominantly to linear internal waves. Munk and Zachariasen [1] calculated the signal Coherence time for refractive rays (in deep ocean) using the Rytov approximation. Dashen, Flatté, and Reynolds [2] and Esswein and Flatté [3] derived similar results using the path integral approach. The results suggest that the signal coherence time of acoustic rays should decrease as -1 power of frequency and -1/2 power of range. In shallow water, the signal propagation is more appropriately described by normal modes as shown in Fig. 1a, where mode dispersion is significant [4]. However, there are no calculations for the signal coherence time using normal modes. Signal coherence time has been measured in various shallow water previously [5,6,7]. (These measurements were collected from source and receivers below the thermocline so that the effect of the rough surfaces on the temporal coherence is minimal.) The results suggest that the signal coherence time decreases as -3/2 power of frequency and -1/2 power of range (Fig. 1b). It was previously shown that mode

CP1272, *Shallow-Water Acoustics, Second International Shallow-Water Acoustics Conference*
edited by J. Simmen, E. S. Livingston, J. X. Zhou, and F. H. Li
2010 American Institute of Physics 978-0-7354-0822-7/10/$30.00

coupling is significant when internal waves are present [4,8]. Could the -3/2 frequency dependence be attributed to mode coupling?

This paper presents the coupled mode calculation of temporal coherence in an ocean waveguide with linear internal waves. The effect of internal waves is calculated to the first order of sound speed perturbation squared. Similar results were also derived using path integral [9]. It is shown that mode coupling has a significant effect on the signal coherence time. Numerical results for a typical shallow water environment show that the coherence time of the acoustic field for a fixed source to an array of receivers (otherwise known as the matched field coherence [5,7]) decreases as -3/2 power of frequency and -1/2 power of range with (full) mode coupling. One also finds that if one ignores the mode coupling, the coherence time of the acoustic field would decrease as -1 power of frequency and -1/2 power of range, same as that predicted for acoustic rays. In the real world, mode coupling can be induced by internal waves, turbulence and/or surface waves. The fast decrease of the signal coherence time with frequency, as observed in shallow water over a wide frequency range (Fig. 1b), could plausibly be explained by mode coupling.

This paper is organized as follows. In Sec. II, the coupled mode formulation of the temporal coherence is derived following Creamer [10]. Numerical calculation of the signal coherence time in a typical shallow water environment is given in Sec. III. A short summary is given in Sec. IV.

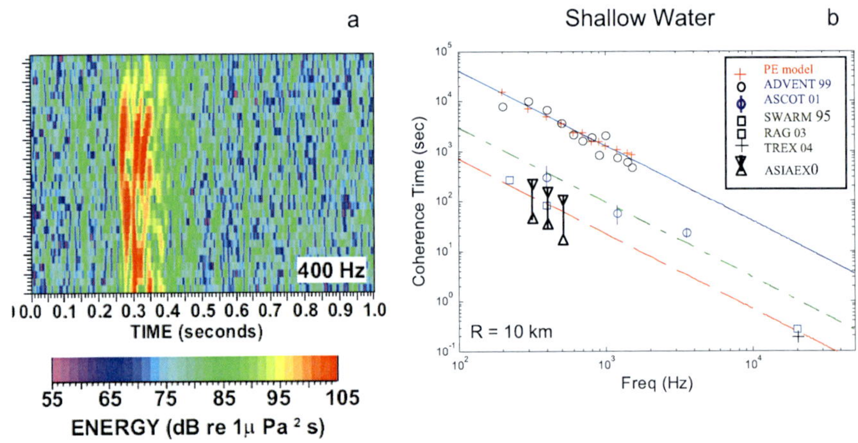

FIGURE. 1 (a) Signal intensity versus receiver depth and delay time in a typical shallow water exhibit mode arrivals (see Fig. 32 of Ref. 4). (b) Shallow water signal coherence time versus frequency. Measurements were obtained from various experiments listed in the legend. See ref. 4,5, and 6. The lines display the –3/2 power frequency dependence for different internal wave strengths.

II. THE COUPLED MODE CALCULATION

The coupled mode equation is well known in the literature [see, for example, 10,11]. Consider propagation in the (r, z) plane. Following Creamer [10], we express the acoustic field as a summation of normal modes,

$$P(r,z,t)=\sum_{n=1}^{M}A_n(r,t)\phi_n(z) \tag{1}$$

where $\phi_n(z)$ is the mode depth function, and express the mode amplitude, $A_n(r)$, in terms of

$$A_n(r,t)=\frac{\psi_n(r,t)e^{il_n r}}{\sqrt{k_n r}} \tag{2}$$

with $l_n = k_n + i\alpha_n$, where k_n the mode wavenumber and α_n is the mode attenuation coefficient. The mode amplitudes satisfy the following coupled mode equation,

$$\frac{\partial\psi_n(r,t)}{\partial r}=-i\sum_{m=1}^{M}\overline{v}_{nm}(r,t)\psi_m(r,t) \tag{3}$$

where

$$\overline{v}_{nm}=v_{nm}e^{-il_{nm}r},\quad l_{nm}=l_n-l_m=-l_{mn}, \tag{4}$$

and

$$v_{nm}=\frac{k_0^2}{\sqrt{k_n k_m}}\int dz\mu(r,z;t)\phi_n(z)\phi_m(z). \tag{5}$$

where $k_0=2\pi f/C_0$ and μ denotes the fractional change of the sound speed, $\mu(r,y,z,t)=\frac{\delta C(r,y,z,t)}{\overline{C}}$. The temporal correlation of the acoustic field is given by

$$<P^*(r,z,t+\tau)P(r,z,t)>=\sum_{n=1}^{M}\sum_{m=1}^{M}\phi_n(z)\phi_m(z)<\psi_n^*(r,t+\tau)\psi_m(r,t)>\frac{e^{-i(l_n^*-l_m)r}}{\sqrt{k_n k_m}r} \tag{6}$$

The matched field correlation is given by integrating the correlation over a large aperture vertical array. One finds

$$\begin{aligned}\int dz<P^*(R,z,t+\tau)P(R,z,t)>&=\sum_{n=1}^{M}\sum_{m=1}^{M}[\int dz\,\phi_n(z)\phi_m(z)]<\psi_n^*(R,t+\tau)\psi_m(R,t)>\frac{e^{-i(l_n^*-l_m)R}}{\sqrt{k_n k_m}R}\\&=\sum_{m=1}^{M}<\psi_m^*(R,t+\tau)\psi_m(R,t)>\frac{e^{-2\alpha_m R}}{k_m R}\end{aligned} \tag{7}$$

For small τ, the mode coherence time can be defined as

$$\begin{aligned}&<\psi_m^*(R,t+\tau)\psi_m(R,t)>-<\psi_m^*(R,t)\psi_m(R,t)>\\&\simeq-<\psi_m^*(R,t)\psi_m(R,t)>\tau^2/\tau_m^2=-\phi_m^2(z_s)\tau^2/\tau_m^2,\end{aligned} \tag{8}$$

where for the unperturbed field, the mode amplitude is given by that at the origin, $\psi_m(R,t)=\psi_m(0,t)=\phi_m(z_s)$. The matched field coherence time is defined as

$$\begin{aligned}&\int dz < P^*(R,z,t+\tau)P(R,z,t) > -\int dz < P^*(R,z,t)P(R,z,t) > \\ &\simeq -\frac{\tau^2}{\tau_{1/e}^2}\int dz < P^*(R,z,t)P(R,z,t) >\end{aligned} \tag{9}$$

Using Eq. (8), one finds

$$\tau_{1/e}^{-2} = \frac{\sum_{m=1}^{M}\tau_m^{-2}\phi_m^2(z_s)\dfrac{e^{-2\alpha_m R}}{k_m R}}{\sum_{m=1}^{M}\phi_m^2(z_s)\dfrac{e^{-2\alpha_m R}}{k_m R}} \approx \frac{\sum_{m=1}^{M}\tau_m^{-2}\phi_m^2(z_s)}{\sum_{m=1}^{M}\phi_m^2(z_s)} = \sum_{m=1}^{M}\tau_m^{-2}\phi_m^2(z_s) \tag{10}$$

where we made the approximation that mode wavenumbers l_n are equal, and used the mode closure condition, $\sum_{m=1}^{M}\phi_m^2(z_s)=1$. One finds that *the inverse square of the matched field coherence time is a weighted sum of the inverse square of the mode coherence times.*

To calculate the mode coherence time, we use the coupled mode equation (3) . Let $\gamma_{nm}(r,t',t)=\psi_n^*(r,t')\psi_m(r,t)$, one finds using Eq. (3) that

$$\frac{\partial \gamma_{nm}(r,t',t)}{\partial r} = \frac{\partial \psi_n^*(r,t')}{\partial r}\psi_m(r,t) + \psi_n^*(r,t')\frac{\partial \psi_m(r,t)}{\partial r} = -i\sum_{j,k=1}^{M}\rho_{nm,ij}(r,t',t)\gamma_{ij}(r,t',t), \tag{11}$$

where $\rho_{nm,ij}(r,t',t)=\delta_{ni}\overline{v}_{mj}(r,t)-\delta_{mj}\overline{v}_{ni}^*(r,t')$. Expressing $\boldsymbol{\gamma}$ as a vector of length M^2, whose element is given by γ_{nm}, *n,m* =1,…M, and $\boldsymbol{\rho}$ as a M^2 x M^2 matrix, whose (*nm, jk*)–th element is given by $\rho_{nm,jk}$, Eq. (11) can be expressed as [10]

$$\frac{\partial \boldsymbol{\gamma}(r,t',t)}{\partial r} = -i\boldsymbol{\rho}(t',t)\boldsymbol{\gamma}(r,t',t). \tag{12}$$

The solution to Eq. (12) is given by

$$\begin{aligned}\boldsymbol{\gamma}(R,t',t) &= \hat{R}\left[\sum_{m=1}^{M}\exp(-i\int_0^R dr\boldsymbol{\rho}(r,t',t))\right]\boldsymbol{\gamma}(0,t',t) \\ &= \boldsymbol{\gamma}(0,t',t) - i\hat{R}\left[\int_0^R dr\boldsymbol{\rho}(r,t',t)\right]\boldsymbol{\gamma}(0,t',t) - \frac{1}{2}\hat{R}\left[\int_0^R dr\boldsymbol{\rho}(r,t',t)\right]^2\boldsymbol{\gamma}(0,t',t) \\ &= \boldsymbol{\gamma}(0,t',t) - i\hat{R}\left[\int_0^R dr\boldsymbol{\rho}(r,t',t)\right]\boldsymbol{\gamma}(0,t',t) - \int_0^R dr\int_0^r dr'\boldsymbol{\rho}(r,t',t)\boldsymbol{\rho}(r',t',t)\boldsymbol{\gamma}(0,t',t)\end{aligned} \tag{13}$$

where $\hat{R}$ is a range ordering operator on the content within the [...]. Recall $\psi_m(0,t) = \phi_m(z_s)$ for a point source, hence $\gamma_{nm}(0,t',t) = \phi_n(z_s)\phi_m(z_s)$. We then find

$$
\begin{aligned}
<\gamma_{nm}(R,t',t)> &= \phi_n(z_s)\phi_m(z_s) - \sum_{p=1}^{M}\sum_{k=1}^{M}\int_0^R dr \int_0^r dr' <\bar{v}_{mp}(r,t)\bar{v}_{pk}(r',t)> \phi_k(z_s)\phi_n(z_s) \\
&+ \sum_{j=1}^{M}\sum_{k=1}^{M}\int_0^R dr \int_0^r dr' <\bar{v}^*_{nj}(r,t')\bar{v}_{mk}(r',t)> \phi_j(z_s)\phi_k(z_s) + \sum_{j=1}^{M}\sum_{p=1}^{M}\int_0^R dr \int_0^r dr' <\bar{v}^*_{nj}(r',t')\bar{v}_{mp}(r,t)> \phi_j(z_s)\phi_p(z_s) \\
&- \sum_{l=1}^{M}\sum_{j=1}^{M}\int_0^R dr \int_0^r dr' <\bar{v}^*_{nl}(r,t')\bar{v}^*_{lj}(r',t')> \phi_j(z_s)\phi_m(z_s)
\end{aligned}
\tag{14}
$$

where $<>$ denotes ensemble average, and we have used $<\bar{v}_{ij}(r,t)> = 0$, for $i, j = 1,...M$. Assuming that the temporal correlation is only a function of the time difference $\tau = t' - t$, $<\gamma_{nm}(r,t',t)> = <\gamma_{nm}(r,\tau)>$, and noting that the first, and last term in Eq. (14) are independent of τ, one finds

$$
\begin{aligned}
&<\gamma_{nm}(R,\tau)> - <\gamma_{nm}(R,0)> \\
&= \sum_{j=1}^{M}\sum_{k=1}^{M}\int_0^R dr \int_0^R dr' [<\bar{v}^*_{nj}(r,t+\tau)\bar{v}_{mk}(r',t)> - <\bar{v}^*_{nj}(r,t)\bar{v}_{mk}(r',t)>] \phi_j(z_s)\phi_k(z_s)
\end{aligned}
\tag{15}
$$

Using Eq. (5), one finds

$$
<v^*_{nj}(r,t+\tau)v_{mk}(r',t)> = \frac{k_0^4}{\sqrt{k_n k_j k_m k_k}} \iint dz dz' <\mu^*(r,z;t+\tau)\mu(r',z';t)> \phi_n(z)\phi_j(z)\phi_m(z')\phi_k(z')
\tag{16}
$$

For linear internal waves, the sound speed perturbation is related to the internal wave displacement by

$$
\mu(r,y,z,t) = \frac{\delta C}{\bar{C}} = GN^2(z)\eta(r,y,z,t) = H(z)\eta(r,y,z,t) \tag{17}
$$

where $N(z)$ is the Brunt-Väisälä buoyancy frequency, η is the internal wave displacement, G is a smooth function of depth, depending on the derivatives of the salinity and temperature profile and $H(z) \equiv GN(z)$. The particle displacement caused by internal waves can be represented as superposition of internal wave (IW) modes. In order to carry out the calculation analytically, we shall assume a two dimensional IW model [11,12] as follows

$$
\eta(r,z;t) = \sum_{l=1} \int dk_r b_l(k_r) W_l(z) e^{-ik_r r + i\omega t} \tag{18}
$$

where $W_l(z)$ is the IW mode depth function (of the l-th mode), k_r is the IW wave number, and ω is the IW mode frequency which is related to the mode wave number by the mode dispersion relation. Since the spectral amplitudes $b_l(k_r)$ are complex Gaussian random variables with zero mean and independent of each other,

$$<b_j^*(k_r)b_l(k_{r'})>=\delta_{jl}\delta(k_r-k_{r'})<|b_j(k_r)|^2>, \tag{19}$$

one finds

$$<\eta(r',z';t)\eta(r,z;t+\tau)>=\sum_l\int dk_l<|b_l(k_l)|^2>e^{-ik_l(r-r')}e^{+i\omega\tau}W_l(z)W_l(z'), \tag{20}$$

where k_l denotes the horizontal wavenumber k_r of the l-th IW mode. Changing the integration variable from wavenumber to frequency using the IW mode dispersion relation for each IW mode, $k_l=k_r(\omega,l)$, and using Eq. (17), one finds,

$$<\mu^*(r',z,t)\mu(r,z',t+\tau)>=\int d\omega\sum_l|\overline{a}_l(\omega)|^2H(z)W_l(z)H(z')W_l(z')e^{-ik_l(r-r')+i\omega\tau}, \tag{21}$$

where $|a_l(\omega)|^2\equiv\frac{dk_l}{d\omega}<|b_l(k_l)|^2>$ is the IW frequency spectral density for the l-th mode and. Substituting Eqs. (21) and (16) in Eq. (15), and setting n = m, one obtains

$$\begin{aligned}<\gamma_{mm}(R,\tau)>-<\gamma_{mm}(R,0)>=k_0^2\sum_{j=1}^{M}\sum_{i=1}^{M}\phi_j(z_s)\phi_i(z_s)\int_0^R dr\int_0^R dr'\int d\omega\\ e^{i[l_{nj}r-l_{mi}r'-k_l(r-r')]}[\cos(\omega\tau)-1]\sum_l|a_l(\omega)|^2F_{mj,l}F_{mi,l}\end{aligned}. \tag{22}$$

where

$$F_{mj,l}\equiv\frac{k_0}{\sqrt{k_mk_j}}\int dzH(z)W_l(z)\phi_m(z)\phi_j(z). \tag{23}$$

Let $\overline{r}=(r+r')/2$ and $\Delta r=(r-r')/2$, and assume a range-independent internal-wave spectrum, one finds

$$\begin{aligned}<\gamma_{mm}(R,\tau)>-<\gamma_{mm}(R,0)>=k_0^2\sum_{j=1}^{M}\sum_{i=1}^{M}\phi_j(z_s)\phi_i(z_s)G_{ji}\int d\omega\\ [\cos(\omega\tau)-1]\sum_l|a_l(\omega)|^2F_{mj,l}F_{mi,l}L_{mji}(\omega)\end{aligned}, \tag{24}$$

where, $L_{mji}(\omega)$ is given by

$$\begin{aligned}L_{mji}(\omega)&=\int_{-R/2}^{R/2}d(\Delta r)e^{i\Delta r[l_{nj}+l_{mi}-2k_l]}=R\text{sinc}[R(k_l-l_m+\frac{l_j+l_i}{2})]\\ &=2\pi\delta(k_l-k_m+\frac{k_j+k_i}{2}),\qquad \alpha=0,\quad R\to\infty\end{aligned}$$

and

$$G_{ji} \equiv \int_0^R d\bar{r} e^{l_{ji}\bar{r}} = \int_0^R e^{i(k_j - k_i)r} e^{-(\alpha_j - \alpha_i)r} dr$$

Let $\Delta k = k_j - k_i$ and $\Delta\alpha = \alpha_j - \alpha_i$, one finds, for $i \neq j$,

$$|G_{ji}|^2 = \frac{e^{2\Delta\alpha R} + 1 - 2e^{\Delta\alpha R}\cos(\Delta k R)}{(\Delta\alpha)^2 + (\Delta k)^2} = O(\Delta k^{-2}), \quad \text{for } R < (\alpha_i + \alpha_j)^{-1}. \text{ Thus}$$

$$G_{ji} = \int_0^R e^{-i(k_j - k_i)r} e^{(\alpha_j - \alpha_i)r} dr = \left[\begin{array}{ll} R & \text{for } i = j \\ O(\Delta k^{-1}) & \text{for } i \neq j, R < (\alpha_j + \alpha_i)^{-1} \end{array} \right. . \tag{25}$$

One finds that the condition $R < (\alpha_j + \alpha_i)^{-1}$ is satisfied in all practical conditions. For shallow water, α is typically of order of 10^{-5} m^{-1} or less. Realistic propagation range in shallow water is < 100 km, so the above condition is also satisfied. For propagation ranges much longer than the mode cycle distance, one finds $R >> \Delta k^{-1}$. In other words, G_{ji} behaves like a delta function, i.e., $G_{ji} \simeq R\delta_{ji}$. One then finds,

$$< \gamma_{mm}(R,\tau) > - < \gamma_{mm}(R,0) > \simeq k_0^2 \tau^2 R \sum_l \int d\omega |a_l(\omega)|^2 \omega^2 \sum_{i=1}^{M} |\phi_i(z_s) F_{mi,l}|^2 L_{mi}(\omega) , \tag{26}$$

where

$$L_{mi}(\omega) \equiv L_{mii}(\omega) \simeq 2\pi\delta(k_l - k_{mi}), \text{ for } \alpha = 0, \text{ and } R \to \infty. \tag{27}$$

where $k_{mi} \equiv k_m - k_i$. Using Eq. (26), the mode coherence time as defined by Eq. (8) is derived as

$$\phi_m^2(z_s)\tau_m^{-2} = \frac{k_0^2 R}{2} \int d\omega \sum_l |a_l(\omega)|^2 \omega^2 \sum_{i=1}^{M} |F_{mi,l}|^2 L_{mi}(\omega)\, \phi_i^2(z_s). \tag{28}$$

Using Eq. (10), the matched field coherence time is then

$$\tau_{1/e}^{-2} = \frac{k_0^2 R}{2} \int d\omega \sum_l |a_l(\omega)|^2 \omega^2 \sum_{m=1}^{M} \sum_{i=1}^{M} |\phi_i(z_s) F_{mi,l}|^2 L_{mi}(\omega). \tag{29}$$

Substituting Eq.(27) into Eq. (29) (for $\alpha \approx 0$, and large R), one finds

$$\tau_{1/e}^{-2} = \frac{k_0^2 R}{2} \sum_l \sum_{m=1}^{M} \sum_{i=1}^{M} |\phi_i(z_s) F_{mi,l}|^2 \omega_{mi}^2 |a_l(\omega_{mi})|^2. \tag{30}$$

where

$$\omega_{mi} = \omega(k_{mi}), \qquad \text{for } k_l = k_{mi}. \tag{31}$$

Equation (30) suggests that (1) mode coupling between the acoustic modes occurs when the IW wavenumber matches the difference of acoustic mode wavenumbers, and (2) the contribution to the temporal coherence (loss) comes from the high wavenumber spectrum as it is weighted by IW frequency squared. Note that the Garrett-Muck spectrum behaves like ω^{-2} for large ω, hence $\omega^2 |a_l(\omega)|^2$ can be approximated by a constant, which will be denoted by A_l. One then finds

$$\tau_{1/e}^{-2} = \frac{k_0^2 R}{2} \sum_l A_l \Gamma_l(z_s). \tag{32}$$

where

$$\Gamma_l(z_s) \equiv \sum_{m=1}^{M} \sum_{i=1}^{M} |\phi_i(z_s) F_{mi,l}|^2. \tag{33}$$

One notes $k_0^2 \sim f^2$ where f is the acoustic frequency. Numerical calculations in the next section indicates that $\Gamma_l(z_s) \sim f$ for high frequencies. In that case, the matched field coherence time $\tau_{1/e}$ will decrease as -3/2 power of frequency and -1/2 power of range.

III. NUMERICAL EXAMPLES

Numerical modeling is done using the environmental data from the SWARM95 experiment [4] which took place on the mid-Atlantic bight, on the New Jersey Continental Shelf, where internal waves have been extensively studied. Figure 3a shows a typical summer sound speed at that site. The spectrum of the diffused internal waves has been analyzed by Yang and Yoo based on SWARM95 data [13]. Figure 2b shows the buoyancy frequency measured during the SWARM95 experiment [13]. Based on that, one can calculate the IW mode depth function $W_l(z)$, as shown in [13]. Figure 2c plots the function, $F_l(z) = H^2(z) W_l^2(z)$, for four IW modes, l = 1, 2, 3 and 4, where $H(z)$ is determined primarily by the buoyancy frequency squared (Eq. 17). One notes that $F_l(z)$ peaks at depth around 10 m, where the buoyancy frequency is largest. The IW mode depth functions are calculated for wavenumber k = 0.0122 m^{-1}, corresponding to frequency of 122 cycles/day for the first mode. One finds that the mode depth functions vary slowly for internal wave frequency between 70 and 200 cycles/day where the most of the internal wave energy lies. Figure 3 shows the variation of the acoustic mode depth functions with respect to the acoustic frequency for 3 acoustic modes. Note that the acoustic mode depth functions are confined to the lower water column. Figure 4a plots $\Gamma_1(z_s)$ of Eq. (33) as a function of frequency, for three source depths: z_s = 20, 50 and 65 m, for the first IW mode, assuming a source-receiver range of 10 km as used in Fig. 1b. We note that the dependence on the source depth is minimal. Also plotted in Figs. 4a is $\overline{\Gamma}_1(z_S)$ for the same three source depths where mode coupling ($m \neq i$ in Eq. 33) is ignored. One finds that $\overline{\Gamma}_1(z_s)$ is practically independent of frequency. The same frequency dependence is also found for higher IW modes.

Figure 4a shows that the mode coupling term $\Gamma_1(z_s)$ increases approximately linearly with frequency (f> 1 kHz) for the SWARM95 environment. As a result, one expects $\tau_{1/e}$ to decrease like $f^{-1.5}$ as frequency increases, which was suggested by data (Fig. 1b). Figure 4b shows parabolic equation (PE) calculations of the signal coherence time, $\tau_{0.8}$ (the delay time when the temporal coherence drops to 0.8) for the SWARM95 environment [12]. Numerical simulation of temporal coherence is sensitive to the maximum wavenumber used in the IW spectrum. Figure 4b is obtained at a range of 20 km using $k_{max} = 1$ m^{-1}. (Ideally, $k_{max} \sim k_{signal}$.) One observes that the PE result is consistent numerically with a $f^{-1.5}$ frequency dependence as expected by theory.

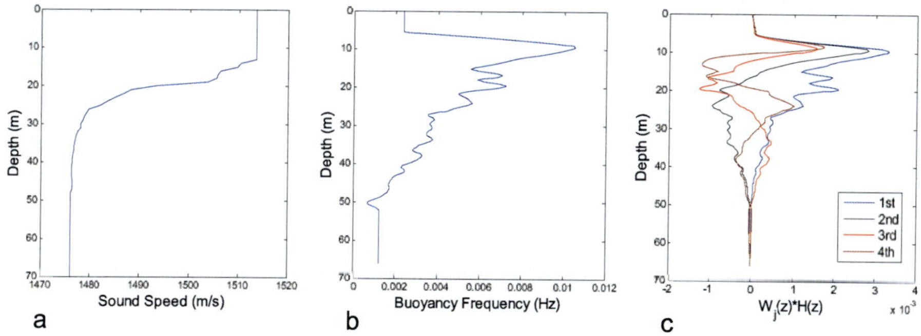

FIGURE. 2 (a) A typical sound speed profile measured during the SWARM95 experiment. (b) Buoyancy frequency versus depth measured during the SWARM 95 experiment. (c) F_l versus depth for l = 1, 2, 3, and 4.

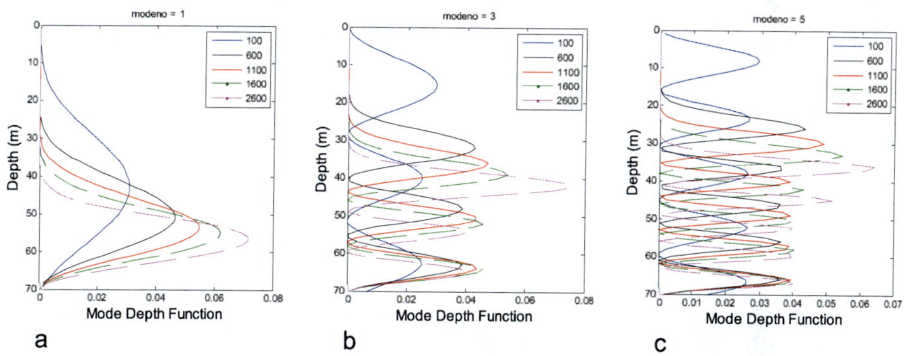

FIGURE. 3 Acoustic mode depth function in shallow water at various frequencies shown in legend in Hz: (a) first mode, (b) third mode and (c) fifth mode.

IV. SUMMARY

This paper provides a theoretical explanation for the -3/2 power frequency dependence of the signal coherence time as observed in several shallow experiments (at high frequencies). The increased temporal coherence loss is attributed to mode coupling.

ACKNOWLEDGMENT

This work is supported by the US Office of naval Research.

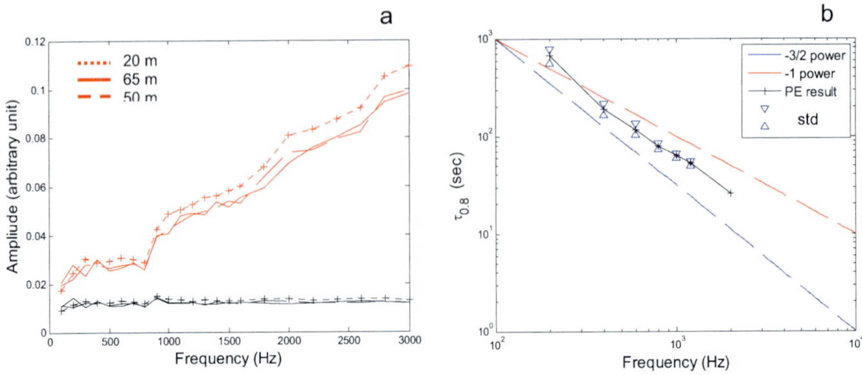

FIGURE. 4 $\Gamma_1(z_S)$ and $\overline{\Gamma}_1(z_S)$ versus frequency for three different source depths; the straight line is a linear fit as described in text (a). $\tau_{0.8}$ as a function of frequency, simulated with a PE code (b).

REFERENCES

1. W. H. Munk and F. Zachariasen, "Sound propagation through a fluctuating stratified ocean: Theory and observation," J. Acoust. Soc. Am. **59**, 818-838 (1976).
2. R. Dashen, S. Flatté, and S. Reynolds, "Path-integral treatment of acoustic mutual coherence functions for rays in a sound channel," J. Acoust. Soc. Am. 77, 1716–1722 (1985).
3. R. Esswein and S. M. Flatte, "Calculations of the phase-structure function from oceanic internal waves," J. Acoust. Soc. Am. **70**, 1387–1396 (1981).
4. J. Apel, M. Badiey, C. S. Chiu, S. Finette, R. Headrick, J. Kemp, J. Lynch, A. Newhall, M. Orr, B. Pasewark, D. Teilbuerger, A. Turgut, K. von der Heydt, and S. Wolf, "An overview of the 1995 SWARM shallow water internal wave acoustic scattering experiment," IEEE J. Ocean. Eng. **22**, 465–500 (1997).
5. T.C. Yang, "Measurements of temporal coherence of sound transmissions through shallow water," J. Acoust. Soc. Am, **120**, 2595-2614 (2006).
6. T.C. Yang, "Toward continuous underwater acoustic communications," Proc. of OCEANS 2008, Quebec, Canada (2008).
7. P. C. Mignerey and M. H. Orr, "Observations of matched field autocorrelation time in the South China Sea," IEEE J. Oceanic Eng. 29, 1280-1291(2004).
8. D. Tielbuerger, S. Finette, and S. Wolf, "Acoustic propagation through an internal wave field bounded by a shallow water waveguide," J. Acoust. Soc. Amer., 101, 789–808 (1996).
9. T.C. Yang, "Temporal coherence of normal modes," Pacific Rim Underwater Acoustic Conference, Xi'an, China (Sept. 2009).
10. D. B. Creamer, "Scintillating shallow water waveguides," J. Acoust. Soc. Am. **99**, 2825–2838 (1996). See Eq. (10a).
11. John A. Colosi, "Acoustic mode coupling induced by shallow water nonlinear internal waves: Sensitivity to environmental conditions and space-time scales of internal waves," J. Acoust. Soc. Am. **124**, 1452–1464 (2008).
12. K. B. Yoo, " Shallow water temporal coherence of broadband acoustic signal in the presence of internal waves," *IEEE J. Ocean. Eng.*, **30**, 391-405 (2005).
13. T. C. Yang and K. Yoo, "Internal waves spectrum in shallow water: Measurement and comparison with the Garrett-Munk model," IEEE J Oceanic Eng. **24**, 333-345 (1999).

Wave Propagation in Water-saturated Sand and Grain Contact Physics

Nicholas P. Chotiros and Marcia J. Isakson

Applied Research Laboratories, The University of Texas at Austin, Texas 78713-8029

Abstract. Measurements in sandy ocean sediments over a broad range of frequencies show that the sound speed dispersion is significantly greater than that predicted by the Biot-Stoll model with constant coefficients, and the observed sound attenuation does not seem to follow a consistent power law. The observations may be explained in terms of the Biot-Stoll model with frequency-dependent complex frame bulk and shear moduli that are governed by the grain-grain contact physics, and random motion at the grain level. In the case of water-saturated sands, the contact stiffness is dominated by squirt flow and viscous drag of a thin fluid film that permeates the contact area. Using this approach, the observed sound and shear wave speeds and attenuations may be modeled over a broad band of frequencies.

Keywords: underwater, sediment, poroelastic, dispersion.
PACS: 43.30.Ma, 43.30.Pc, 43.30.Vh

INTRODUCTION

The ocean sediment is often approximated as a lossy fluid or a viscoelastic solid. These approximations are relatively simple to model with the tools that are currently available. Measurements have shown that there are significant discrepancies between these approximations and real ocean sediments. This study is particularly concerned with sandy ocean sediments. Physically, it is a granular material that is saturated with sea water. It is a two-phase medium in which the sand grains form a loose skeletal frame permeated by sea water. The logical starting point would be a poroelastic model, such as the Biot model [1], which was reformulated by Stoll and applied to sandy sediments [2]. The theory accounts for the interaction between the solid and fluid phases. The sound wave propagation is governed by two coupled wave equations.

$$\mu\nabla^2 u + (H-\mu)\nabla(\nabla \bullet u) - C\nabla(\nabla \bullet w) = \rho\frac{\partial^2}{\partial t^2}u - \rho\frac{\partial^2}{\partial t^2}w \tag{1}$$

$$C\nabla(\nabla \bullet u) - M\nabla(\nabla \bullet w) = \rho_f\frac{\partial^2}{\partial t^2}u - \frac{c\rho_f}{\beta}\frac{\partial^2}{\partial t^2}w - \frac{F\eta}{\kappa}\frac{\partial}{\partial t}w \tag{2}$$

$$\rho = \rho_s(1-\beta) + \rho_f\beta \tag{3}$$

The terms u and w refer to solid and fluid particle displacements as illustrated in Fig. 1; ρ is the bulk density of medium, ρ_s and ρ_f the solid and fluid densities, β the porosity; H, C, M and μ are elastic moduli of the solid phase as defined in Ref. [2]; η,

CP1272, *Shallow-Water Acoustics, Second International Shallow-Water Acoustics Conference*
edited by J. Simmen, E. S. Livingston, J. X. Zhou, and F. H. Li

is the viscosity of the pore fluid; κ, F and c are the permeability, high-frequency correction function and tortuosity of the pore spaces.

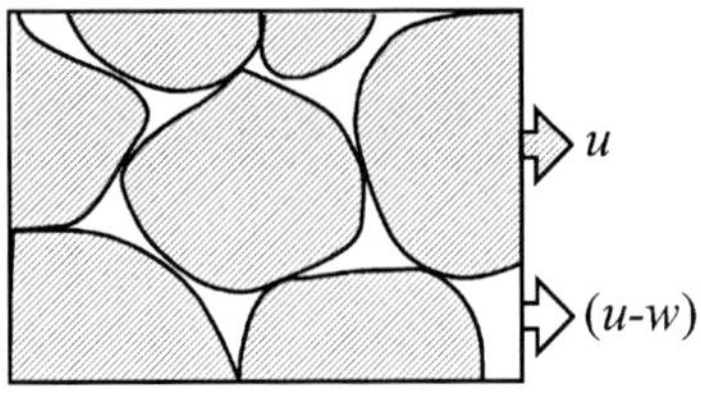

FIGURE 1. Illustration of particle displacements of the solid and fluid phases.

The Biot-Stoll model predicts two compressional waves, usually called the fast and slow waves. In ocean sediments, the fast wave is commonly identified with the sound wave. The slow wave is predicted to be highly attenuated, but its influence is indirectly evident in the reflection coefficient. The Biot-Stoll model predicts two frequency regimes, separated by the critical frequency [3], defined as,

$$f_c = \frac{\beta\eta}{\kappa\rho_f} \tag{4}$$

Below the critical frequency, the attenuation of both the sound and shear waves increase with the second power of frequency. Above this frequency, the attenuations continue to increase at a lower rate proportional to the half-power of frequency. The wave speeds show the greatest dispersion in the vicinity of the critical frequency, as illustrated in Fig. 2.

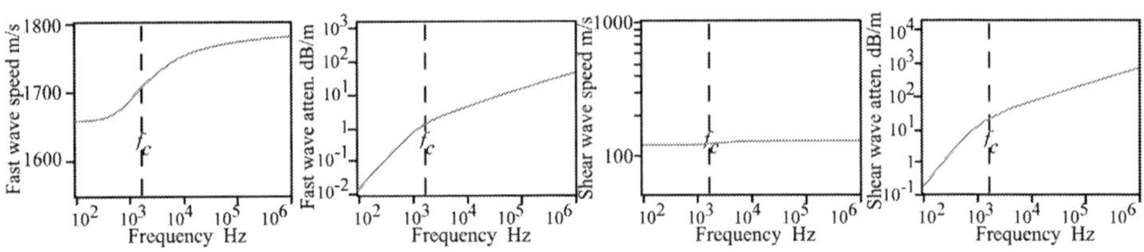

FIGURE 2. Example of sound and shear wave speeds and attenuations predicted by the Biot-Stoll model.

THEORETICAL MODELS

The frequency dependencies illustrated in Fig. 2 are not in total agreement with observations. The low-frequency dependence of attenuation on the second power of frequency is observed but not the high-frequency half-power dependence. The observed sound speed dispersion is also somewhat greater than the model prediction. To address these issues, an extension of the Biot-Stoll model, which includes the effects of contact squirt flow and viscous shear drag at the grain-grain contact (BICSQS), was developed [4]. The surface of the sand grain is rough to some degree, and mechanical contact is made at just a few asperities, as illustrated in the second panel in Fig. 3. These contact points are surrounded by a thin fluid film which is quite large in comparison with the solid contact area. The geometry is idealized in the third panel as a thin fluid film of uniform thickness surrounding a small central contact point. The hypothesis is that the mechanical properties of the contact are governed by both the elasticity of solid contact and the hydrodynamics of the fluid film.

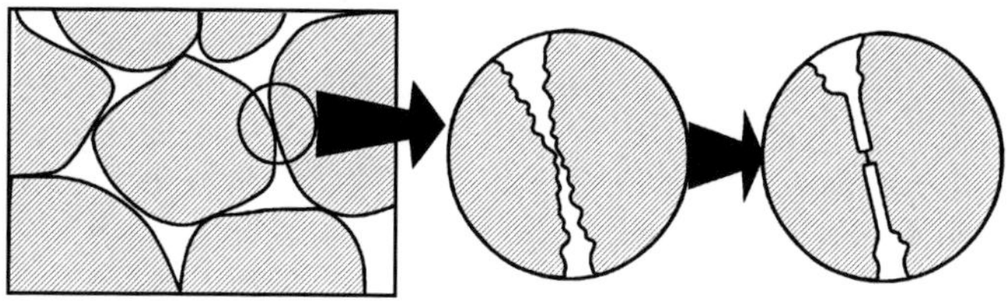

FIGURE 3. Illustration of the grain-grain contact model in the BICSQS model.

The normal S_n and shear S_t contact stiffnesses contain both a solid elastic component and a viscous fluid component.

$$S_n = k_c + \frac{c_y}{1 + i\dfrac{f_c}{f}} \; ; \; S_t = g_c + i g_c \frac{f}{f_\mu} \tag{5}$$

where k_c and g_c are the low frequency stiffnesses due to the solid connection; c_y is the high-frequency increment due to the fluid film; f_c and f_μ are the relaxation frequencies determined by the average dimensions of the fluid film. Since the elastic properties of the skeletal frame are dominated by the contact stiffnesses, its frame bulk K_b and shear μ moduli are predicted to have the same form,

$$K_b = K_c + \frac{C_y}{1 + i\dfrac{f_k}{f}} \; ; \; \mu = G_c\left(1 - i\frac{f}{f_\mu}\right) \tag{6}$$

The terms K_c and G_c are the low-frequency asymptotic values due to the solid part of the contacts only; C_y is the asymptotic increment at high frequencies. The frame shear modulus μ appears directly in Eqs. (1) and (2), and the frame bulk modulus K_b determines the values of the other elastic coefficients C, H and M. A comparison of the BICSQS model predictions to those of the Biot-Stoll model are shown in Fig. 4. The parameter values are from Table IV under the WJTTS column in Ref. [4].

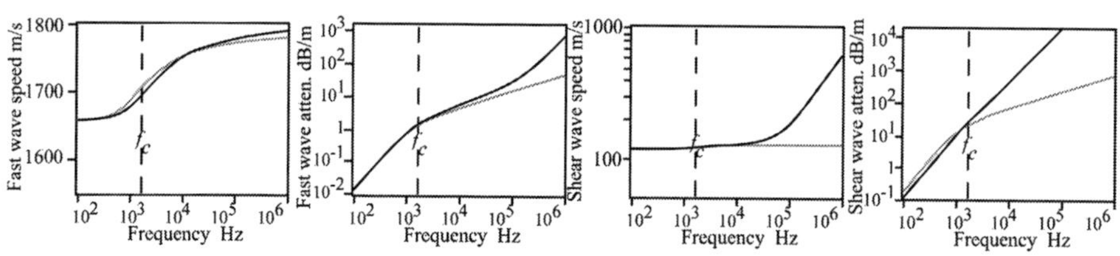

FIGURE 4. Example of wave speeds and attenuations predicted by the BICSQS model compared to the corresponding Biot-Stoll model.

EXPERMIMENTAL DATA

The models are compared with the measured sound speed and attenuation from the Sediment Acoustics Experiments of 1999 and 2004 (SAX99 and SAX04), and the Shallow Water Experiment of 2006 (SW06), as shown in Fig. 5. The BICSQS and Biot-Stoll models predict an attenuation that increases as the second power of frequency which is in agreement with the sediment attenuation inverted from propagation measurements in SW06 [5]. The BICSQS model is also consistent with direct higher frequency attenuation measurements from the SAX99 experiment in Ref. [6]. With respect to sound speed, both models show some degree of dispersion and are consistent with the measurements above about 10 kHz. However, at low frequencies, none of the models are able to give sound speed values that are low enough to match the measured values. The measured sound speeds shown are from a number of investigators at SAX99 [6] and from Hines, Osler, Scrutton and Lyons at SAX04 [7]. The Wood's Equation sound speed, which is the sound speed in a suspension of the same porosity, is the lower bound for all effective medium models. There are data points below the Wood's Equation lower bound from both SAX99 and SAX04, as shown in Fig. 5, that need to be addressed.

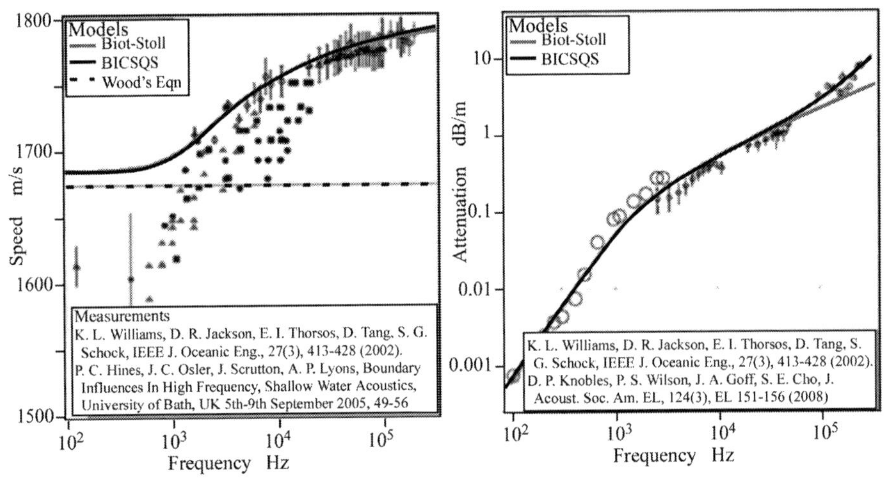

FIGURE 5. Comparison of measured sediment sound speeds from Refs. [6] and [7] and measured attenuations from [5] and [6] to the model predictions.

SOUND SPEED REDUCTION IN GRANULAR MEDIA

At this point it is instructive to review the modeling of sound speed in granular media. Given a plane acoustic wave of the form,

$$u = A_u \cos\left(\frac{\omega x}{c_p} - \omega t\right) \tag{7}$$

the potential and kinetic energies in the wave are given by,

$$E_{potential} = -\left(K + \frac{4}{3}G\right)\frac{u}{2}\frac{d^2u}{dx^2} = \left(K + \frac{4}{3}G\right)\frac{\omega^2}{2c_p^{\ 2}}A_u^2\cos^2\left(\frac{\omega x}{c_p} - \omega t\right) \quad (8)$$

$$E_{kinetic} = \frac{\rho}{2}\left(\frac{du}{dt}\right)^2 = \frac{\rho}{2}\omega^2 A_u^2 \sin^2\left(\frac{\omega x}{c_p} - \omega t\right) \quad (9)$$

In the lossless case, energy must be conserved. Therefore, given the Pythagorean relationship between the squares of the sine and cosine terms, the following solution is obtained for the sound speed,

$$c_p^2 = \frac{K + \frac{4}{3}G}{\rho} \quad (10)$$

In the case of a granular medium, due to random variations in the grain-grain contact stiffness, there will be additional random motions at the grain level, that will partition the kinetic energy. This is illustrated with the aid of Fig. 6. Panel (a) shows 5 spherical grains in their initial positions. Consider the grains within the pressure peak in a plane acoustic wave traveling from left to right. In certain effective medium models, such as the Biot-Stoll, uniform deformation is assumed, which implies that spherical grains become oval and the porosity remains constant as illustrated in (b). It is obvious that uniform deformation is a poor approximation for a granular medium because most of the strain takes place at the grain-grain contact, in accordance with the Hertz-Mindlin theory. The contacts are more or less uniformly distributed over the surface of each sphere due to geometrical constraints. If all the contacts have equal stiffness, then the contact forces are symmetric about a vertical line through the center of each sphere, as illustrated in (c). However, since the packing is random, it should follow that the contact stiffnesses must also be variable and there will an asymmetry of contact forces. This will cause random rotations and translations in directions that are orthogonal to the wave vector, as illustrated in (d). Then random movements will partition the kinetic energy of the acoustic wave. Finally, some of the weaker contacts may slip and slide which tends to enhance the random motions, as illustrated in (e).

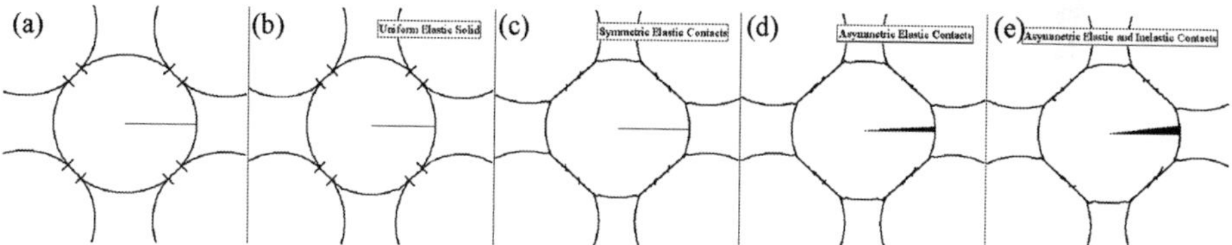

FIGURE 6. Illustration of different models of deformation in a granular medium (a) grains in their initial positions (b) uniform deformation, (c) symmetric elastic contacts, (d) asymmetric elastic contacts, and (e) asymmetric contacts with slip, in response to compression by a sound wave in the horizontal direction.

Consequently, the expression for sound speed will have an additional inertial term to account for the grain level energy due to random motion. This term will be referred to as the frame virtual mass, m_p.

$$c_p^2 = \frac{K + \frac{4}{3}G}{\rho(1 + m_p)} \tag{11}$$

An analysis of the micromechanics [8] shows that m_p for a dry granular medium in vacuo is given by the following expression.

$$m_p \approx \frac{\sigma^2\left(6N\left[R_{nt} - 1\right]^2 + 50N_s R_{nt}^2\right)}{5N^2(2R_{nt} + 1)^2} \tag{12}$$

where σ^2 is the scintillation index of the grain-grain contact stiffness, N the coordination number, N_s is the average number of contacts that do not slip, and R_{nt} is the ratio of the tangential to normal contact stiffness. For quartz grains, R_{nt} is 0.96, or almost equal to 1. In a random close-packing of spheres of equal size, N is approximately 9 [8]. It is postulated that the value of N_s is about half of N [9]. Finally, σ^2 is estimated to be 1 since contact stiffness is estimated to follow an exponential distribution [10].

In a water-saturated granular medium, the grain-grain contact stiffness has two components, solid and fluid, as quantified in Eq. (5). The solid component is as described above, and is expected to have a scintillation index of 1. The fluid component, which is due to the fluid film surrounding the solid contact, is expected to have a negligible scintillation index since fluid films have a tendency to redistribute and smooth out stress concentrations. Thus, the frame virtual mass of water-saturated sand m_{psat} is frequency dependent, as illustrated in Fig. 7, and it modifies the density of the solid frame [11].

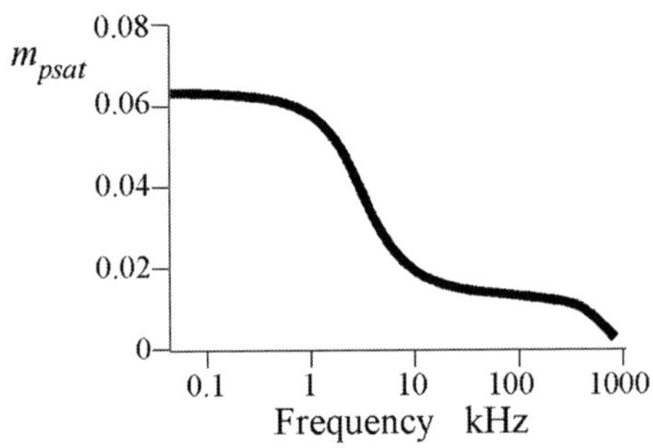

FIGURE 7. Illustration of the frame virtual mass of water-saturated sand.

At high frequencies, where the fluid component dominates, the effective value of m_{psat} is negligible. At low frequencies, where the solid component is dominant, the effective value of m_{psat} will be as given above for m_p, causing the effective inertia of the medium to be larger than the physical density, and allowing the sound speed to be lower than the Wood's Equation lower bound. Replacing ρ_s with $\rho_s(1+m_{psat})$ in Eq. (3), the resulting sound speed and attenuation curves from the modified BICSQS model are as shown in Fig. 8.

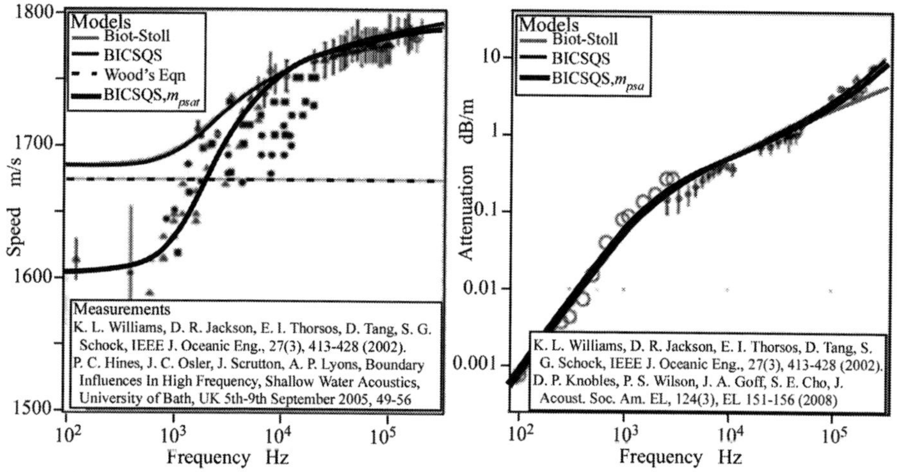

FIGURE 8. Comparison of measured sediment sound speeds from Refs. [6] and [7] and measured attenuations from [5] and [6] to the model predictions.

CONCLUSIONS

The observed low frequency attenuation in the sediment follows the second power of frequency consistent with the Biot-Stoll poroelastic model, but the observed sound speed dispersion requires a more complicated approach. The sound speed dispersion is due to three mechanisms: a relaxation process inherent in the Biot-Stoll model related to the pore fluid flow, a relaxation process due to the squirt flow of the fluid film at the grain-grain contacts, and the frequency dependence of the frame virtual mass. In conclusion, a poroelastic model that has been extended to include grain-grain contact physics, specifically squirt flow and viscous shear drag, and the micromechanical inertial effects due to randomness in the contact stiffness and the corresponding grain movement, fits well with the published experimental measurements.

ACKNOWLEDGMENTS

This work was sponsored by the United States Office of Naval Research, Ocean Acoustics Program.

REFERENCES

1. M. A. Biot, "Theory of Propagation of Elastic Waves in a Fluid-Saturated Porous Solid -- I. Low Frequency Range," *J. Acoust. Soc. Am.*, 28(2), 168-178 (1956).
2. R. D. Stoll, *Sediment acoustics*, New York, Springer-Verlag, 1989.
3. T. Yamamoto, "Acoustic propagation in the ocean with a poro-elastic bottom," *J. Acoust. Soc. Am.*, 73(5), 1587-1596 (1983).
4. N. P. Chotiros, M. J. Isakson, "A broadband model of sandy ocean sediments: Biot-Stoll with contact squirt flow and shear drag," J. Acoust. Soc. Am., 116(4), 2011-2022 (2004).

5. D. P. Knobles, P. S. Wilson, J. A. Goff, S. E. Cho, "Seabed acoustics of a sand ridge on the New Jersey continental shelf," J. Acoust. Soc. Am. EL, 124(3), EL 151-156 (2008).
6. K. L. Williams, D. R. Jackson, E. I. Thorsos, D. Tang, S. G. Schock, "Comparison of sound speed and attenuation measured in a sandy sediment to predictions based on the Biot Theory of porous media," IEEE J. Oceanic Eng., 27(3), 413-428 (2002).
7. P. C. Hines, J. C. Osler, J. Scrutton, A. P. Lyons, "Time-of-flight measurements of acoustic wave speed in sandy sediments from 0.6 – 20 kHz," Boundary Influences In High Frequency, Shallow Water Acoustics, N.G.Pace and P Blondel, (Eds), University of Bath, UK 5th-9th September 2005, 49-56 (2005).
8. G. Mavko, T. Mukerji, J. Dvorkin, The rock physics handbook, Cambridge, United Kingdom, Cambridge University Press, 1998.
9. R. Bachrach, J. Dvorkin, A. Nur, "Seismic velocities and Poisson's ratio of shallow unconsolidated sands," Geophysics, 65(2), 559-564 (2000)
10. C. H. Liu, S. R. Nagel, D. A. Schecter, S. N. Coppersmith, S. Majumdar, O. Narayan, T. A. Witten, "Force fluctuations in bead packs," Science, 269( ), 513-515 (1995).
11. N. P. Chotiros, "Wave speed dispersion in Fluid-Saturated Granular Media Due to Grain Contact Physics," Poromechanics IV: Proceedings of the Fourth Biot Conference on Poromechanics, H. I. Ling, A. Smyth, R. Betti Eds., 8-10 June 2009, Columbia University, New York, 455-460 (2009)

Observations of 0th Order Head Waves in the Yellow Sea

Peter H. Dahl[a] and Jee Woong Choi[b]

[a]*University of Washington, Applied Physics Laboratory and Dept. of Mechanical Engineering*
Seattle, Washington
[b]*Hanyang University, Dept. of Environmental Marine Science*
Ansan, Korea

Abstract. Measurements made as part of the 1996 Yellow Sea experiment, undertaken by China and the U.S. are revisited with focus on precursor arrivals. The distinction between a first-order head wave (zero gradient) and a zeroth-order interference head wave (non-zero, positive gradient) is highlighted, with the latter being much more energetic and the first- and zeroth-order terminology relating to their ray series classification. A parameter, ζ, named the Chekin parameter, is described, and an analysis of the potential range of ζ prior to carrying out geoacoustic inversion of data is suggested.

Keywords: head waves, precursor arrivals, sound speed gradient
PACS: 43.30.Cq, 43.30.Ma, 43.30.Pc

INTRODUCTION

In Choi and Dahl [1], the influence of a sound speed gradient in surficial sediments on the set of precursor arrivals commonly called head waves is analyzed. The distinction between a first-order head wave (zero gradient) and a zeroth-order interference head wave (non-zero, positive gradient) is highlighted, with the latter being much more energetic and the first- and zeroth-order terminology relating to their ray series classification. Figure 1 shows the two arrival types, with a single ray path identified for the first-order case, and multiple ray paths (refracted waves) Cn-waves (n=0,1,2,...) identified for the zeroth-order case. For the zeroth-order case, the amplitude spectrum mimics to a degree the source spectrum, $S(f)$, in contrast to the first-order case for which the spectrum goes as $S(f)/f$.

This paper summarizes and advances the discussion started in [1], continued in Dahl and Choi [2] for which measurements of zeroth-order head waves made in the Yellow Sea are presented and modeled the parabolic wave equation (PE) approximation, and in Choi and Dahl [3], for which a parameter first introduced by Chekin [4] is used to elucidate important properties of precursor arrivals.

CP1272, *Shallow-Water Acoustics, Second International Shallow-Water Acoustics Conference*
edited by J. Simmen, E. S. Livingston, J. X. Zhou, and F. H. Li

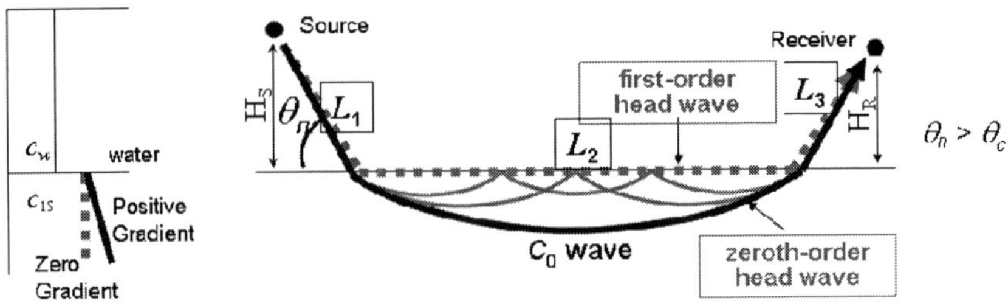

first-order head wave travel time: $$T = \frac{r}{c_{1S}} + \frac{(H_S + H_R)c_W}{\sqrt{1 - \left(\frac{c_W}{c_{1S}}\right)^2}} \left(\frac{1}{c_W^2} - \frac{1}{c_{1S}^2} \right)$$

zeroth-order (refracted) head wave (C_0 wave) travel time:

$$T = \frac{1}{a} \left| \ln \left(\frac{1 - \sin\theta_t}{1 + \sin\theta_t} \right) \right| + \frac{H_S + H_R}{c_W \sin\theta_0}$$

a : linear sound speed gradient coefficient
θ_t : transmitted grazing angle

FIGURE 1. Upper left: sound speed profiles showing positive gradient and zero gradient within the sediment, where c_{1s} is the sediment sound speed at the water-sediment interface. Upper Right: Corresponding first-order head wave (zero gradient) and zeroth order head wave (positive gradient). Travel times for these two arrivals are given based on sound speeds and length scales as shown in upper right figure.

THE YELLOW SEA EXPERIMENT

The measurements were made as part of the 1996 Yellow Sea experiment (YS96) conducted by China and the U.S. Several works published in the proceedings of the 1st International Conference on Shallow Water Acoustics in Beijing [5] provide experimental overview and quick-look results from YS96. The experimental site (37°N, 124°E) is located in the central Yellow Sea, roughly centered between the tip of the China's Shandong Peninsula and the landmass of Korea, in waters 75 m in depth (Fig 2). The YS96 head wave data set originates from the broad band propagation measurements made at ranges less than 1 km, on 22 August 1996. Explosive sources (38-g) set from the Chinese R/V *Shi Yan 2* were deployed to detonate at 7 and 50 m as the *Shi Yan 2* opened in range from the Chinese R/V *Shi Yan 3*, while the latter held station at the experimental site. These data are measured from a 16-element vertical line array (VLA), element spacing 4 m, deployed from the R/V *Shi Yan 3* (Fig. 3)

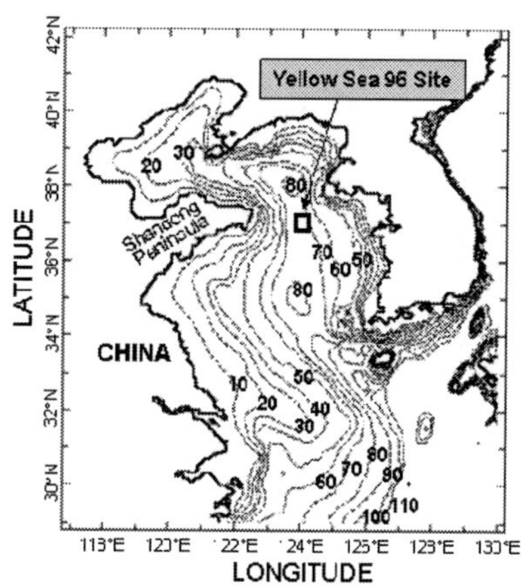

FIGURE 2. Location of the 1996 Yellow Sea experiment conducted by China and the U.S. including bathymetry contours in meters.

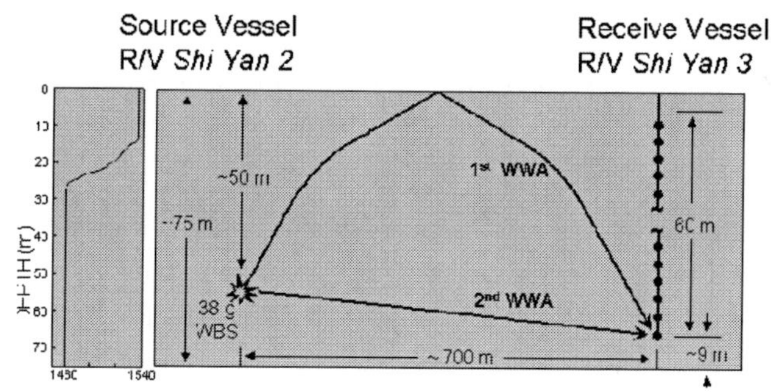

FIGURE 3. (Left plot) Sound speed versus depth as determined by the Conductivity- Temperature-Depth (CTD) casts from the R/V Shi Yan 3. (Right plot) Ray diagram for the two water-borne arrivals; first arrival interacts with the sea surface.

Precursor Arrivals and Travel Time Inversion

Precursor arrivals (PA) that arrive prior to any water borne arrival (Fig. 4) represent the key observation of interest. Figure 5 is a similar display showing all four data sets taken at ranges between 700 and 870 m. A travel time inversion is done using the equation for the C_0-wave travel displayed with Fig. 1, with result being c_{1s} = 1573 m/s and gradient a = 1.1 s^{-1} (see [2] for details and error bounds). A further confirmation of this inversion is shown in Fig. 6 which compares a ray-based estimate of vertical arrival angle versus range with estimates based on beam forming. For reference, the significantly lower, and range independent, vertical arrival angle for a 1st order head wave is shown by the horizontal line.

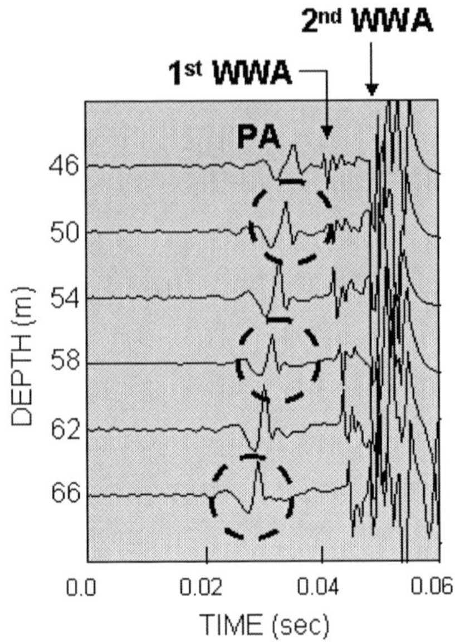

FIGURE 4. Precursor arrivals (three of which are circled) shown arriving prior to water-wave arrivals (see Fig. 3). The time axis is reduced time with 0.0 representing approximately 460 ms after source detonation.

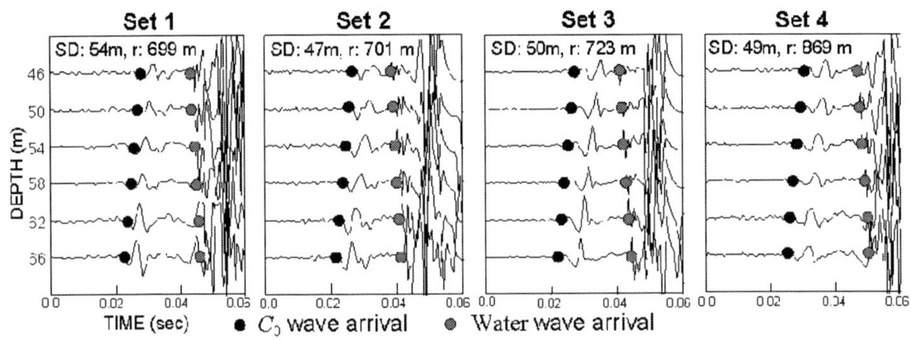

FIGURE 5. (Left plot, set 1) The arrival structure versus time and depth for the bottom six hydrophones for set 1. The closed (Co wave arrival) and open (water wave arrival) symbols reflect the final inversion result with similar display for sets 2-4.

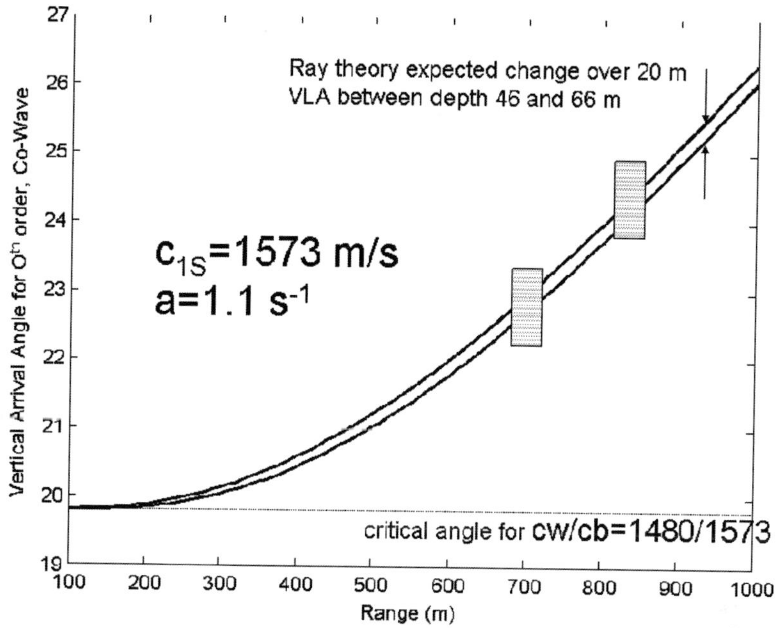

FIGURE 6. Ray-based estimates of vertical arrival angle (solid lines) versus range compared with estimates based on beam forming (boxes) at two ranges (size of boxes indicates uncertainty in estimates). The ray estimates are the sound speed and gradient indicated; for reference the range dependent vertical arrival angle for case zero gradient is shown by horizontal line.

THE CHEKIN PARAMETER

The Chekin parameter is discussed in [3] (including this proposed nomenclature) and is defined as follows:

$$\zeta = \frac{L_2(a/c_{1s})^{2/3}}{\lambda_1^{2/3}} \tag{1}$$

where L_2 and c_{1s} are defined in Fig. 1, λ_1 is wavelength in the sediment corresponding to c_{1s} and a is the sound speed gradient. For $\zeta << 1$ the precursor arrival assumes the properties of a 1st order (classical) head wave, while for $\zeta > 1$ the precursor arrival assumes the properties of a 0th order head wave, or C_0 wave [3]. The presence of gradient results in a significant change in the amplitude of the precursor wave and its spectral properties [3]. The amplitude effect is illustrated in Fig. 7 which is based on PE simulation [1]. The time series set given on the left is from an iso-speed seabed or zero gradient, and the set on the right is based on a gradient of 1 s^{-1}. The two sediment sound speeds (c_{1s}) differ slightly in order to match the vertical arrival angle (26.8°). The gradient case is described by $\zeta \approx 1.5$, and the ratio of 0th order to 1st order wave amplitude goes as $\pi\zeta^3$.

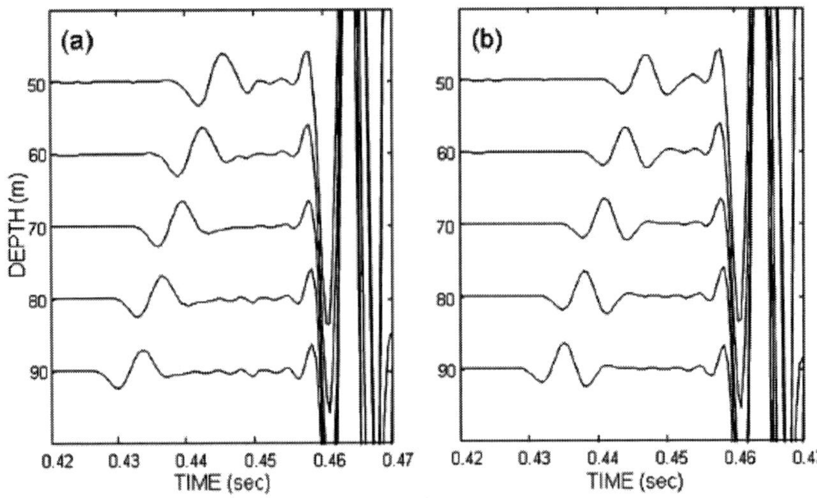

FIGURE 7. (a) PE simulated time series showing 1st order head wave arrivals as viewed from a vertical line array (VLA) for a Pekeris waveguide, and (b) 0th order head wave for equivalent gradient waveguide (equivalence defined as matching the vertical arrival angle). The water depth is 100 m, source depth is 50 m, and VLA range is 700 m. The 1st order head waves are amplified by a factor of 7 in the region to the left of 0.45 s; the amplitude of the water wave arrivals in each case are essentially equal and out of scale for this plot.

SUMMARY

The properties of the precursor (head wave like) arrival associated with a linear and positive sound speed gradient are governed by the Chekin parameter ζ; for $\zeta \ll 1$ the precursor arrival assumes the properties of a 1st order (classical) head wave, while for $\zeta > 1$ the precursor arrival assumes the properties of a 0th order head wave, or C_0 wave. [3]. The 1st and 0th order arrivals can display similar time arrival structure, however the amplitude of a 1st order head wave is significantly smaller. Furthermore, although in many cases it is possible to describe a geoacoustic model to predict the arrival times of 1st order head waves, that model will not correctly describe simultaneously observed travel time and arrival angle, and also observed amplitude relative to other arrivals if a gradient exists. An analysis of the potential range of Chekin parameter prior to carrying out a geoacoustic inversion of data is suggested.

ACKNOWLEDGMENTS

This work was funded by the U.S. Office of Naval Research, Code 321 Ocean Acoustics.

REFERENCES

1. J.W. Choi and P. H. Dahl, *J. Acoust. Soc. Am.* **119**, 3660-3668 (2006).

2. P. H. Dahl and J. W. Choi, *J. Acoust. Soc. Am.* **120**, 3525-3533 (2006).
3. J.W. Choi and P. H. Dahl, *J. Acoust. Soc. Am.* **122**, 146-150 (2007).
4. B. S. Chekin, *Izv., Acad. Sci., USSR, Earth Phys*. Ser. **3**,. . **122**, 143-147 (1965).
5. See several works in: "*Shallow-water Acoustics*" edited by R. Zhang and J. Zhou, China Ocean Press, Beijing, 1997.

Seabed attenuation in water saturated sands on New Jersey continental shelf in 50-3000 Hz band

D. P. Knobles[a]

[a]*Applied Research Laboratories, The University of Texas at Austin*
P.O. Box 8029
Austin, Texas 78713

Abstract. The Shallow Water 2006 ocean acoustics experiment on the New Jersey continental shelf was designed in part to determine the frequency dependence of sound speed and attenuation for a sandy sediment in the 50-3000 Hz band. Two acoustic arrays in L-geometries were positioned about 25 km apart on a ridge of coarse sand. Segments of narrowband and impulsive acoustic data were analyzed for the information they contain on the seabed geoacoustic structure and the frequency dependence of the top sediment layer attenuation. The results clearly demonstrate a non-linear frequency dependence of the absorption.

Keywords: Shallow water acoustics, seabed geoacoustics, attenuation
PACS: 43.30.Ma, 43.30.Pc, 43.20Jr, and 43. 30Ky.

INTRODUCTION

In 1980 Hamilton noted that the absorption loss associated with water-saturated sands below 1 kHz was an issue that needed to be addressed with carefully controlled experiments [1]. His remark was part of a discussion on the frequency dependence of attenuation in marine sediments, i.e., the value of the exponent n for $A(\mathrm{f}) = \alpha\,(\mathrm{f}/1000)^n$ where A, α, and f have units of dB/m, dB/m @ 1 kHz, and Hz, respectively. For clays and muddy sediments Hamilton noted that experimental evidence indicated n =1 below 1 kHz. For sandy sediments that are commonly found in shallow littoral ocean environments, Hamilton noted that some researchers suggested n =2, which would be consistent with a Biot-Stoll theory of low-frequency (< 1 kHz) acoustics for water-saturated sandy sediments [2-7]. Both prior to and after the paper by Hamilton, reports in the literature suggested the frequency dependence of the absorption in sandy sediments to be described with $n \cong 1.8$ for f < 1 kHz [see for example 8-11]. There is not a consensus in the scientific community on the attenuation in sandy sediments, due in large measure to the scarcity of measurements. Some researchers argue that the frequency dependence of the attenuation for water-saturated sands is linear at all frequencies, and that the observed frequency dependence of the loss is caused by other mechanisms [12-14]. In all discussions of this debate, it is of critical importance to keep in mind not only the sediment type but also the frequency band. For example, in the Biot-Stoll theory for water-saturated sand, the exponent changes smoothly from n =2 at the low frequencies to n = 0.5 at the higher frequencies. The frequency where

CP1272, *Shallow-Water Acoustics, Second International Shallow-Water Acoustics Conference*
edited by J. Simmen, E. S. Livingston, J. X. Zhou, and F. H. Li

the transition occurs depends on a number of parameters such as porosity, grain size, and permeability.

This paper reports on an experiment designed in part to infer the seabed attenuation over a large frequency band (50-3000 Hz) for a unique shallow water area where the surface seabed is known to be a coarse water-saturated sand with an average grain size of about ϕ=1.0. Continuous wave (CW) sources, that produced tonals in the 50-3000 Hz band, were towed over a 20 km track. From these data the frequency dependence of the attenuation was deduced. Further, a tropical storm passed through the experimental area. The acoustic measurements and sea surface measurements during the storm provide sufficient information to test the validity of the frequency dependence of the attenuation obtained from the long-range transmission loss measurements.

DESCRIPTION OF EXPERIMENT

Figure 1 shows the geographical location of part of an acoustic experiment (known as Shallow Water 2006 or SW06) performed in August-September 2006 on the New Jersey continental shelf. The line connecting the positions of two acoustic arrays lies approximately along the 70 m depth bathymetry contour in the middle of a *sand ridge* encompassing an area about 180 km^2 in which the first 2-5 m of sediment is a uniform coarse sand.

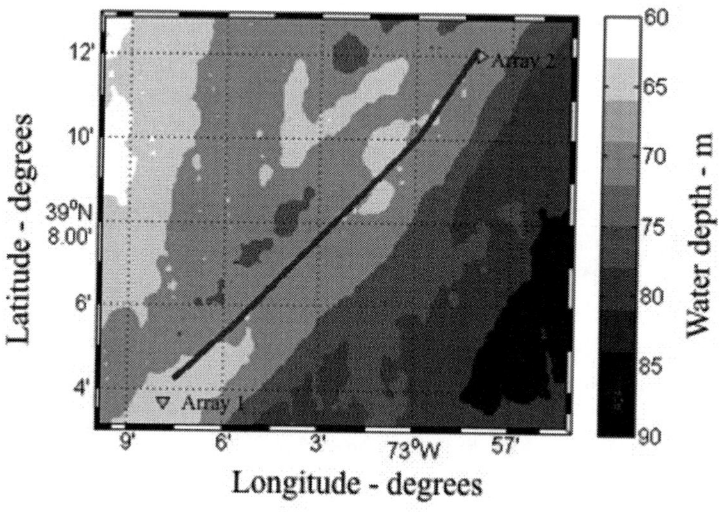

FIGURE 1. Locations of two acoustic arrays and a propagation track in the sand ridge portion of the SW06 experiment.

The unique properties of the sand ridge provide an ideal experimental setting in which to make a large bandwidth (50-3000 Hz) study of the attenuation in water-saturated sand. The sediments to the southeast at the 80 m depth contour are less ideal for such a study because of their reportedly large clay content [15]. Figure 1 also shows a track where the RV KNORR towed a J-15-1 (50-900 Hz band) source and an ITC 250 (1-4 kHz band) source that produced narrowband CW lines. The track over which these sources were towed starts near Array 2 in the NW direction and passes within about 100 m of Array 1. Further, along this track broadband light bulb implosions and combustive sound source (CSS) were deployed.

The seabed layering structure of the region was mapped by CHIRP bottom reflection measurements from two surveys prior to the acoustic experiment. Figure 2 shows the seabed layering structure between Array 1 and Array 2. For most of the track the surface sediment is a sand layer with an average thickness of about 3 m. Inversion analyses suggest that the sound speed in this layer is about 1650 m/s [16]. In the proximity of Array 1 the sediment beneath the sand layer is known as the outer shelf wedge and is composed of thin layers of clay in addition to sand. This causes the average sound speed to about 60 m/s less than the speed in the sand layer. Beneath the sand layer starting about 5 km from Array 1, is a layer comprised of consolidated sand with a higher sound speed (~ 1720 m/s).

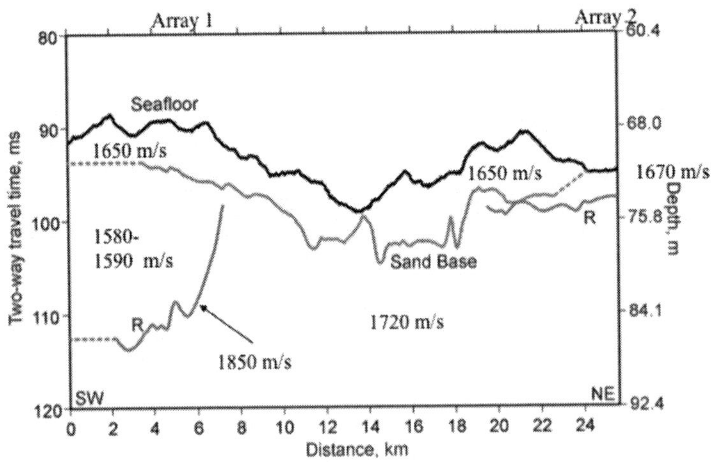

FIGURE 2. Seabed layering structure between Array 1 and Array 2.

ANALYSIS OF ACOUSTICAL MEASUREMENTS

Short-ranged CW tow data collected on Array 2 in the 50-250 Hz band and the received time series in the 50-325 Hz band from a CSS impulsive source collected at Array 1 were used to infer average parameter values for parameters such as the sound speed in the top sediment layer, the layer thickness, the layer density, the sound speed in the second sediment layer, etc. [16]. For frequencies above 50 Hz, the geoacoustic structure (not including the attenuation) of the seabed on the sand ridge is reasonably well described by a half-space with a sound speed and density of 1650 m/s and 1.85 g/cc, respectively. For a bottom water sound speed of 1495 m/s, the sound speed ratio of the surface sediment is about 1.104, a value that is consistent with a water-saturated sand classified as coarse with a mean value for the grain size ϕ=1.

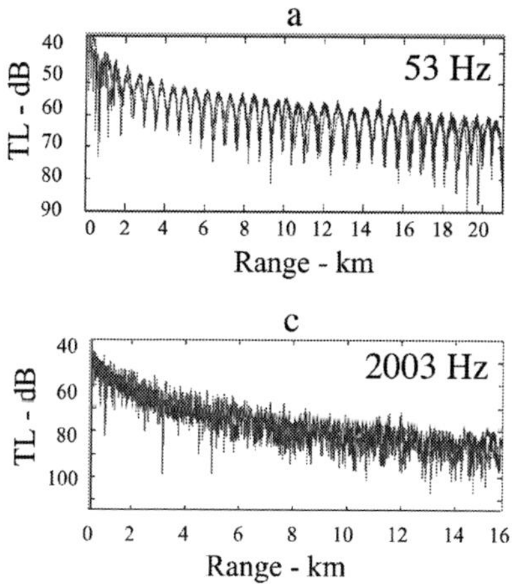

FIGURE 3. TL model (red)-data (blue) comparisons at 53 and 2003 Hz after attenuation optimization.

The results of the geoacoustic inversions performed at Array 1 and Array 2, the measured bathymetry and sub-bottom layering along the propagation track shown in Fig. 2, and the measured SSPs near the two arrays during the time of the multi-frequency source tows give an approximate representation of the range-dependent waveguide. For the 1-3 kHz and the 50-900 Hz data, the range over which there was adequate SNR was about 15 and 20 km, respectively. Absorption in the water column is included with a standard Thorp attenuation relationship. A parabolic equation model [17] then found the value for the sediment attenuation in the first sediment layer that gave the optimal fit to the measured TL versus range for each frequency in the 50-3000 Hz band. Figure 3 shows comparisons of modeled and measured TL at 53 and 2003 Hz. Matching both the measured TL levels and overall range modal interference structure were important considerations in estimating the attenuation.

Figure 4 shows the inferred attenuation values for the first sediment layer as a function of frequency. For purposes of comparison Fig. 4 includes several predictions of the Biot theory [2-3]. The specific Biot calculations are based on the formulation of Stoll and Kan [7]. The solid black line is the prediction using the Biot parameters listed in Table I of Ref. 18, which were derived from measurements made during the Sediment Acoustics Experiment in 1999 (SAX99). The solid blue line is derived from the SAX99 parameters but with a lower bound estimate (β=0.363) for the porosity at the New Jersey shelf location [19]. The two dotted lines are derived from the SAX99 parameters but with upper and lower bound estimates for the permeability, κ=1.0x10-10 m^2 and κ=1.0x10-11 m^2, respectively. Permeability was not measured at the New Jersey shelf site, but these estimates were derived from sediment physical models that relate permeability to porosity and grain size [20]. Core samples for the New Jersey site yielded a mean grain size ϕ=1. For frequencies f below about 2 kHz the dependence of the inferred attenuation is approximately proportional to $f^{1.85}$. These results are consistent with other studies that have inferred a non-linear frequency dependence of the attenuation from analyses of forward propagation data [8-11, 21]. While the data are not inconsistent with the Biot predictions above 2 kHz, one cannot draw any definitive conclusions concerning the frequency exponent because of the sparseness of data there the observed dependence is about $f^{1.1}$.

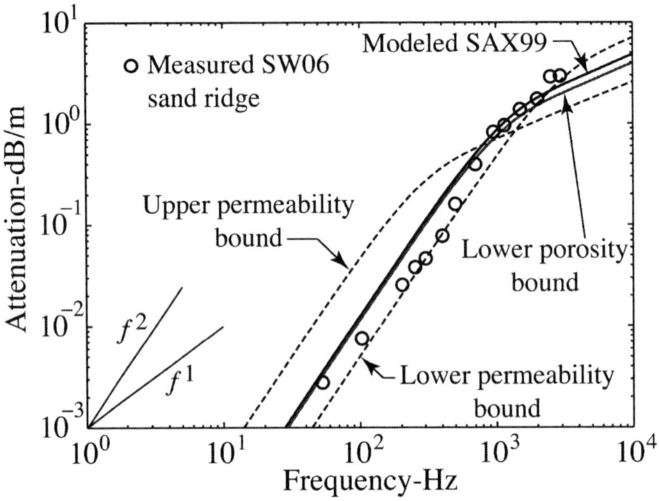

FIGURE 4. Inferred attenuation in top sediment layer on sand ridge

Figure 5 shows an omni-directional spectrogram from a hydrophone on the horizontal segment of Array 1 during a time period when tropical storm Ernesto passed through the experimental area [22]. The wind speed, measured on an Air Sea Interaction Spar (ASIS) buoy positioned about 150 m from Array 1, is superimposed on the spectrogram. The ambient noise levels are clearly well correlated with the wind speed. The wind-generated noise extends above 400 Hz for wind speeds above 3 knots. Below 400 Hz it is difficult to separate the distant shipping and wind components of ambient noise in shallow water. It was observed (not shown here) that the averaged ambient noise levels in the 200-1000 Hz band at the Array 2 location were within about 1 dB of the ambient noise levels measured at the Array 1 site.

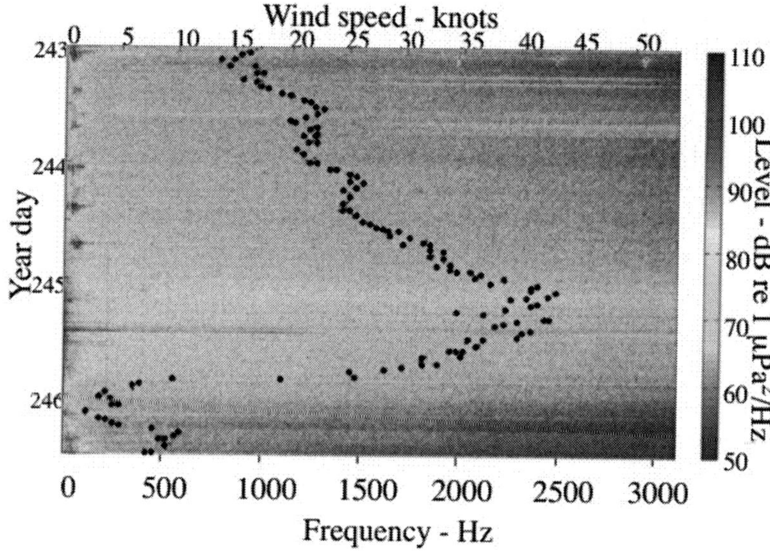

FIGURE 5. Spectrogram measured on Array 1 for tropical storm Ernesto

In shallow water, the wind-generated noise spectrum should be sensitive to the frequency dependence of the seabed attenuation. This hypothesis is tested by examining modeled and measured difference spectra [22]. A difference spectrum is the difference between the ambient noise levels for a site of interest, for example the New Jersey shelf location, and the ambient noise level for a reference deep-water location. In this manner the shape of the difference spectrum does not depend on wind source level. However, evaluating the difference spectrum at high wind speeds results in greater accuracy due to the asymptotic behavior of noise levels with increasing wind speed [22]. The difference in ambient noise at two locations for a specific frequency, f, and wind speed, V, is

$$\Delta AN_{1,2}(f,V) = AN_1(f,V) - AN_2(f,V) \quad . \tag{1}$$

Assuming that the effective source levels for a specific wind speed are independent of site, $\Delta AN_{1,2}(f,V)$ is approximately independent of V. Further, define for site j

$$L_j(z_0,z_j,f) = 10\,Log_{10}(2\pi\int_0^\infty dr\, r 10^{-TL_j(r,z_0,z_j,f)/10}) \quad . \tag{2}$$

L_j represents the ambient noise level at site j for frequency f at depth z_j with 0 dB source strength per unit area uniformly distributed in the x-y plane at depth z_0. TL_j represents the modeled transmission loss as a function of range from a reference point below the sea surface to the receiver. It is assumed that the waveguide possesses azimuthal symmetry. Finally, define

$$\Delta L_{1,2}(f) = L_1(z_0,z_1,f) - L_2(z_0,z_2,f) \quad . \tag{3}$$

The premise of in this study is that

$$\Delta AN_{1,2}(f) = \Delta L_{1,2}(f) \quad . \tag{4}$$

The location of the *Church Opal* (CO) experiment [23] will be used as the deep-water reference site. As discussed in Ref. 23, the environment is well understood, and the transmission loss in Eq. 2 can thus be accurately modeled. Figure 6 compares the measured $\Delta AN_{SW06,CO}$ and modeled $\Delta L_{SW06,CO}$ values. The integrals in Eq. 2 were evaluated numerically. The $\Delta AN_{SW06,CO}$ values lie approximately between 6 and 8 dB in the 400-3200 Hz band, and the $\Delta L_{SW06,CO}$ values also lie between about 6 and 8 dB. In this sense the modeled results are in good agreement with the measured values. Figure 6 includes the result of making the frequency dependence of the attenuation for the SW06 seabed linear with α=0.55 dB/m at 1 kHz. An attenuation that depends linearly with frequency causes the modeled $\Delta L_{SW06,CO}$ to have a spread of about 6 dB over the 500-3000 Hz band, whereas the spread of measured $\Delta AN_{SW06,CO}$ values is only about 2 dB. Assuming that the sediment sound speed values deduced from the previous analysis [16] are approximately correct, a nonlinear frequency dependence of the sediment attenuation appears to be the reason that the $\Delta AN_{SW06,CO}$ values have such a weak frequency dependence in the 500-3000 Hz band.

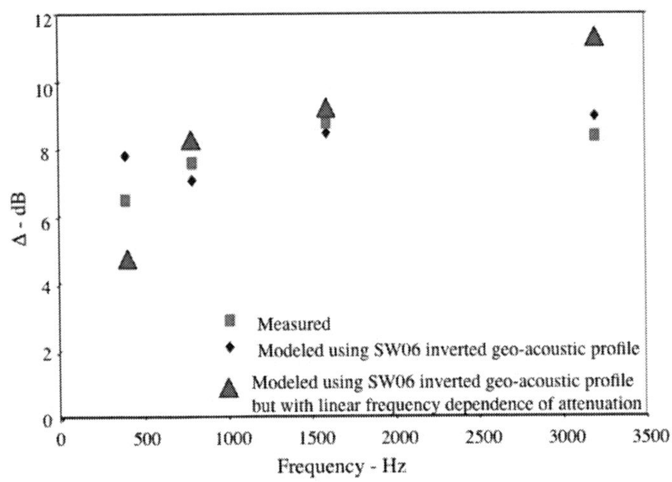

FIGURE 6. Mode-data comparison of frequency dependence of wind generated noise difference spectra

SUMMARY

Both broadband and narrowband acoustic measurements in the 50-3000 Hz band made on the New Jersey continental shelf during the Shallow Water 2006 (SW06) Experiment were analyzed for the information they contain on the frequency dependence of the attenuation in the seabed. The array positions and propagation track were on a sand ridge. Geoacoustic inversions produced values for the sound speeds of the surface sand layer that were between 1650 and 1710 m/s, and that were consistent with the existing geophysical data. For frequencies below about 2 kHz the attenuation values varied approximately as $f^{1.8}$. The results of an analysis of wind-generated noise due to a tropical storm at the same location where the transmission loss measurements were made are in agreement with the inferred value for the frequency exponent below 2 kHz.

REFERENCES

1. E. L. Hamilton, "Geoacoustic modeling of the seafloor," *J. Acoust. Soc. Am.* **68**, 1313-1340 (1980).
2. M. A. Biot, "Theory of propagation of elastic waves in a fluid-saturated porous solid. I. Low-frequency range," *J. Acoust. Soc. Am.* **26**, 168-178 (1956).
3. M. A. Biot, "Theory of propagation of elastic waves in a fluid-saturated porous solid. II. Higher frequency range," *J. Acoust. Soc. Am.* **28**, 179-191 (1956).
4. M. A. Biot, "Generalized theory of acoustic propagation in porous dissipative media," *J. Acoust. Soc. Am.* **34**, 1254-1264 (1962).
5. R. D. Stoll, "Acoustic waves in ocean sediments," *Geophysics* **42**, 715-725 (1977).
6. R. D. Stoll, "Theoretical aspects of sound transmission in sediments," *J. Acoust. Soc. Am.* **68**, 1341-1350 (1980).
7. R.D. Stoll and T.K. Kan, "Reflection of acoustic waves at a water-sediment interface," *J. Acoust. Soc. Am.* **70**, 149-156 (1981).
8. L. Bjorno, "Finite-amplitude wave propagation through water-saturated marine sediments," Acustica **38**, 195-200 (1977).
9. J. M. Hovem and G.D. Ingram, "Viscous Attenuation of sound in saturated sand," *J. Acoust. Soc. Am.* **66**, 1807-1812 (1979).
10. A.D. Pierce and W.M. Carey, "Low-frequency attenuation of acoustic waves in sandy/silty marine sediments," *J. Acoust. Soc. Am.* **124**, EL308-EL312 (2008).
11. J. X. Zhou, X. Z. Zhang, and D. P. Knobles, "Effective geo-acoustic model for sandy seabottoms in shallow water," *J. Acoust. Soc. Am.* **125,** 2647-2666 (2009).
12. M. J. Buckingham, "Wave propagation, stress relaxation, and grain-to-grain shearing in saturated, unconsolidated marine sediments," *J. Acoust. Soc. Am.* **108**, 2796-2815 (2000).
13. M. J. Buckingham, "Compressional and shear wave properties of marine sediments: comparisons between theory and data," *J. Acoust. Soc. Am.* **117**, 137-152 (2005).
14. M. J. Buckingham, "On pore-fluid viscosity and the wave properties of saturated granular materials including marine sediments," *J. Acoust. Soc. Am.* **122**, 1486-1501 (2007).
15. Y. M. Jiang and N. R. Chapman, "Measurement of Low-Frequency Sound Attenuation in Marine Sediment," *IEEE J. Ocean. Eng.* **35**, 70-78 (2010).
16. D. P. Knobles, P. S. Wilson, J. A. Goff, and S. E. Cho, "A seabed acoustics experiment on a sand ridge on the New Jersey continental shelf," *J. Acoust. Soc. Am.* **124**, EL151-EL156 (2008).
17. M. D. Collins, "A split-step Pade solution for the parabolic equation method," *J. Acoust. Soc. Am.* **93**, 1736-1742 (1993).
18. K. L. Williams, D. R. Jackson, E. I. Thorsos, D. J. Tang, and S. G. Schock, "Comparison of sound speed and attenuation measured in a sandy sediment to predictions based on the Biot theory of porous media," *IEEE J. Ocean. Eng.* **27**, 413-428 (2002).

19. J. A. Goff, B. J. Kraft, L. A. Mayer, S. G. Schock, C. K. Sommerfield, H. C. Olson, S. P. S. Gulick, and S. Nordfjord, "Seabed characterization on the New Jersey middle and outer shelf: Correlatability and spatial variability of seafloor sediment properties," *Mar. Geol.* **209**, 147-172 (2004).
20. S. G. Schock, "A method for estimating the physical and acoustic properties of the sea bed using chirp sonar data," *IEEE J. Ocean. Eng.* **29**, 1200-2017 (2004).
21. S. M. Dediu, W. L. Siegmann, and W. M. Carey, "Statistical analysis of sound transmission results obtained on the New Jersey continental shelf," *J. Acoust. Soc. Am.* **122**, EL23-EL28 (2007).
22. D. P. Knobles, S. M. Joshi, and R. D. Gaul, H. C. Graber, and N. J. Williams, "On the nature of wind-driven ambient noise in a shallow water environment with a sandy seabed," *J. Acoust. Soc. Am.* **124**, EL157-EL162 (2008).
23. R. D. Gaul, D. P. Knobles, J. Shooter, and A. F. Wittenborn, "Ambient noise analysis of deep ocean measurements in the Northwest Pacific," *IEEE J. Ocean. Eng.* **32**, 497-512 (2007).

ACKNOWLEDGMENTS

The author wishes to acknowledge the ARL:UT engineering team and the crew of the RV KNORR for their assistance in obtaining the acoustic measurements.

Data–Model Comparisons for Sea Surface Waves from the ASIAEX East China Sea Experiment

Jie Yang [a)], Ji-Xun Zhou [b)], and Peter H. Rogers [b)]

[a)] *Applied Physics Laboratory, University of Washington, Seattle, WA 98105, USA.*
[b)] *School of Mechanical Engineering, Georgia Institute of Technology, Atlanta, GA 30332, USA.*

Abstract. During ASIAEX, surface directional wave spectra were recorded for nine consecutive days. Data are analyzed in this paper to show the characteristics of the surface wave field in the East China Sea. Three surface spectrum models, Pierson–Moskowitz, JONSWAP, and Toba, are evaluated and compared with the ASIAEX RMS surface wave height data. The data–model comparisons show that the JONSWAP–Mitsuyasu model is the best fit to the experimental results. Model performance can be improved by a combined model, JONSWAP–Mitsuyasu for U / c_p <0.8 and Toba-Donelan for U / c_p >0.8. The preliminary results suggest that locally measured wave age can be used to find the most appropriate surface model for acoustic prediction purposes.

Keywords: Wind-generated sea surface waves; air-sea interactions; sea surface spectrum.
PACS: 43.30.Re; 92.60.Cc; 92.60.Gn; 92.10.Hm.

INTRODUCTION

The scattering of sound from wind-generated surface waves has been studied extensively since the early 1950s [1]. Sound propagation attenuation has been reported due to the interaction of sound and the rough surface [2], [3]. In shallow water, this attenuation may depend on other properties of the waveguide, such as the seabed. A quantitative description of the sea surface is necessary to study its acoustical effect on shallow water sound propagation.

In coastal areas, the development of a surface wave field is complicated, depending on the wind and its variation, duration and fetch, as well as on the coastal geometry [4]. As a random process, the correlation and its Fourier transform pair, the surface wave spectrum, have been studied widely [5]–[8].

Past studies focused on spectral models of surface gravity waves, especially the low-frequency domain. There are two main types of surface spectra: they differ mainly in the slope of their rear face. The first category includes the Pierson–Moskowitz [5] and JONSWAP [6] models with an f^{-5} proportionality for frequency beyond the spectral peak. The other form, proposed by Toba [7], has an f^{-4} frequency dependence. The PM model has a simple form but is only valid for a fully developed sea, which requires steady winds of long fetch and duration. Both the JONSWAP and Toba spectra made modifications to the PM form based on the high-frequency formulation of Phillips [8].

Since 1990 efforts have been made to build a surface spectrum that is valid over all wavenumbers with application to the area of electromagnetic scattering [9]–[11]. The

CP1272, *Shallow-Water Acoustics, Second International Shallow-Water Acoustics Conference*
edited by J. Simmen, E. S. Livingston, J. X. Zhou, and F. H. Li

new form has combined low- and high-wavenumber regimes corresponding to surface gravity and capillary waves. The low-wavenumber regime takes a form similar to the existing JONSWAP or Toba–Donelan models [12]. The spectral forms at high wavenumbers are based on the work of Phillips, Kitaigorodskii and Plant [13]–[15].

During the Asian Seas International Acoustics EXperiment (ASIAEX) in the East China Sea (ECS), nine days of consecutive wind velocity and directional wave spectrum data were recorded from 29 May to 6 June 2001 [16]. These data are analyzed here to show the characteristics of surface waves in shallow water and used to compare different surface spectrum models with application to acoustic prediction.

This paper is organized as follows. Sections II and III describe the experiment and data characteristics, respectively. Section IV presents a comparison of the ASIAEX data with different surface spectrum models. A summary is given in Section V.

ASIAEX ECS EXPERIMENTAL OBSERVATIONS

ASIAEX was a multinational scientific project conducted in both the East China Sea (ECS) and South China Sea (SCS). Here we focus only on the ECS experiment, conducted from 27 May to 9 June 2001 in the boxed area defined by 28° to 30° N and 126° 30' to 128° E (Fig. 1). The experimental site has a nominal water depth of 104 m with a variation of less than 10 m over a range of 30 km [16].

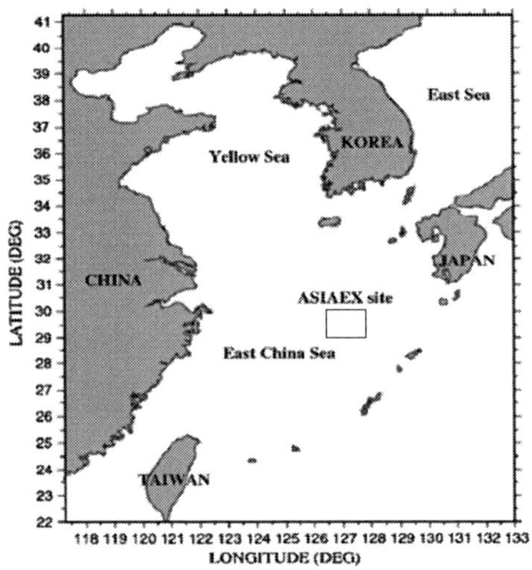

FIGURE 1. Experimental site of the ASIAEX in the East China Sea.

Wind speed and direction were recorded every half hour from the R/V *Melville* IMET station, about 14 m above the sea surface. Figure 2 shows the nine-day record of wind speed and direction. One interesting feature to note is the sharp decrease in wind speed at the beginning of day 2 and day 9, which may be due to a 180° wind direction change. The ASIAEX data provide valuable information on surface wave field response to changing wind fields.

Many researchers ([17]–[20]) use "wave age" to quantify the development of surface wave fields. Using the deep water dispersion relation $\omega^2 = gk$, wave age is defined as U_r/c or U_r/c_p with U_r the wind velocity at a reference height, c the surface wave phase velocity, and c_p the phase velocity at the peak frequency. The magnitude of wave age is a measure of wind input. For a young developing sea, its wave age is more than unity. Over time the young sea will become fully developed with U/c_p in the range 0.8–1.0. Energy input from the wind ceases as the wave speed exceeds the wind speed. Wave age decreases as the fully developed sea gradually relaxes to old waves. Figure 3 shows that for the conditions during ASIAEX, wave age was 0.1–1.6. Over the 9-day observation period, a fully developed sea was rarely observed and only for very short periods, although the wind speed changed significantly from 1 to 11 m/s. Heitsenrether and Badiey considered this to be characteristic of surface wave fields in a coastal region which depend on fetch and swell [22].

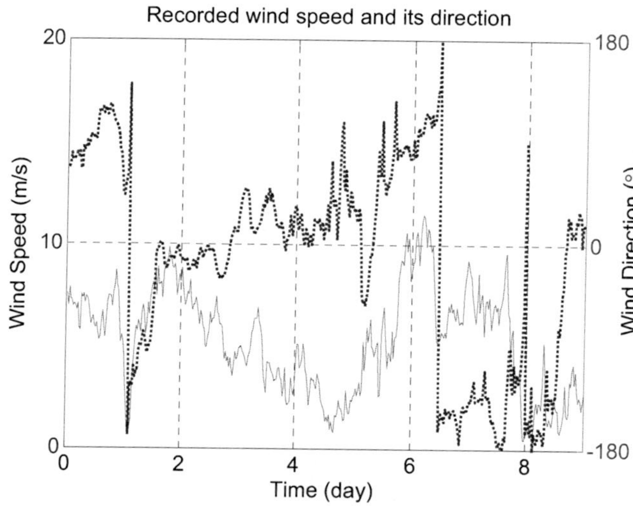

FIGURE 2. Recorded wind speed and direction for nine consecutive days at the ASIAEX site starting 00:00 May 29, 2001. Solid line: wind speed; dotted: wind direction.

The variability of the recorded spectrum complicates direct determination of the spectral peak, so we used a technique introduced by Young [21] to calculate f_p:

$$f_p = \frac{\int_0^\infty f F^4(f)\, df}{\int_0^\infty F^4(f)\, df}, \tag{1}$$

where $F(f)$ is the surface frequency spectrum.

Directional surface wave spectra were recorded by Dahl [22] using a 0.9-m diameter TRIAXYS directional wave buoy 500 m from the *Melville*. It measured wave height variance spectra in 0.005-Hz bins from 0.03 Hz to 0.64 Hz, and in 3-degree directional bins. Spectra were computed every 0.5 h based on a 20-min averaging time.

The non-directional spectrum shown in Fig. 4 is the integration of the directional wave spectrum over all angles. It is apparent that surface wave energy is highly concentrated in the frequency range 0.1–0.4 Hz and spectral level is in direct response to the magnitude of the wind speed.

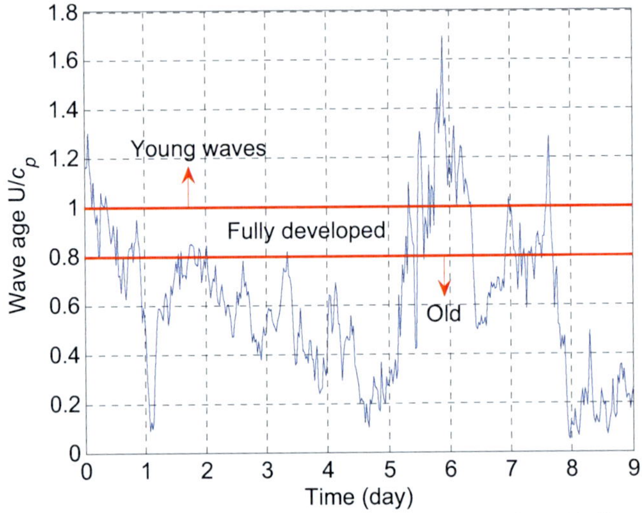

FIGURE 3. Time history of wave age during ASIAEX. Wave age above 1 indicates young waves.

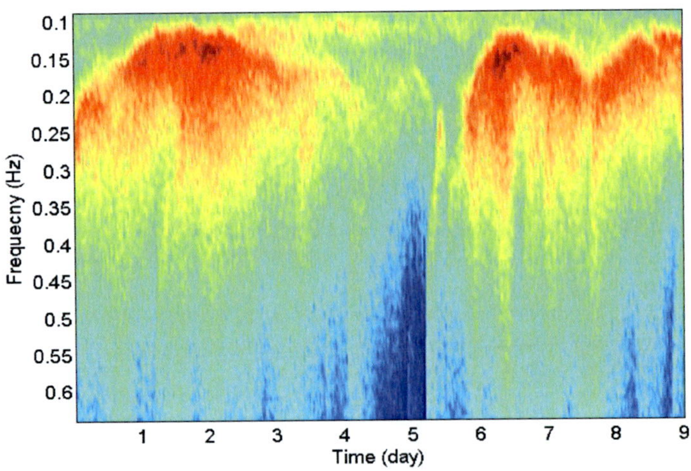

FIGURE 4. Color plot of 9-day non-directional spectrum level starting May 29, 2001(from [16]).

CHARACTERISTICS OF WIND-GENERATED WAVE FIELDS DURING ASIAEX

The ASIAEX data were analyzed to show the characteristics of the wind-driven sea surface wave field in the ECS according to their wave age. Over the 9-day record young or mature sea were observed 26% of the time. The coexistence of wind driven sea and swell was observed about 27% of the time—from 12:00 June 1 to 20:00 June 3, which results in a double-peaked spectrum.

Non-directional Surface Spectra

The surface frequency spectrum usually takes a unimodal form as in Fig. 5(a), left. The well-defined peak falls in the range of 0.1–0.3 Hz and f_p decreases with increasing wind speed. High–frequency spectra respond to wind forcing more rapidly and hence have a flatter rear face. For a steady or gently changing wind field, the spectral peak usually aligns well with the wind direction [Fig. 5(a), right].

Figures 5(a), (b), and (c) correspond to young waves, a fully developed sea, and old waves with wave ages U/c_p= 1.1, 0.92, and 0.41, respectively. The three cases show decreasing wind speed and hence with decreasing intensity level as shown in the figure.

A bimodal structure of the frequency spectrum was observed from 12:00 June 1 to 20:00 June 3 [Fig. 5(c)]. The first peak persisted throughout the period at the same position at 0.11 Hz, which corresponds to a swell component. The color plot of a twin peak case indicates this as well [Fig. 5(c), right]. Bimodal structure has been observed in coastal regions by other researchers ([18]–[20]).

Response of Directional Wave Spectra to Veering Wind Fields

Gradual Change in Wind Direction

The ASIAEX data contain a 10-h period of gradually changing wind direction on day one (05:00–15:00 29 May). The wind speed was 6–8 m/s and wind direction changed 50° in about 10 h. Figure 6 (1)–(4) shows the evolution of the directional spectra due to the mild direction change; the spectral peaks align with the wind direction most of the time. Our observation agrees with Young [23] that the whole spectrum rotates with the wind if the wind direction change is less than 60°. Different frequencies were also studied (results not shown here) and it was found that their spectral peaks follow the gradual wind direction change quite well for frequencies higher than 0.25 Hz.

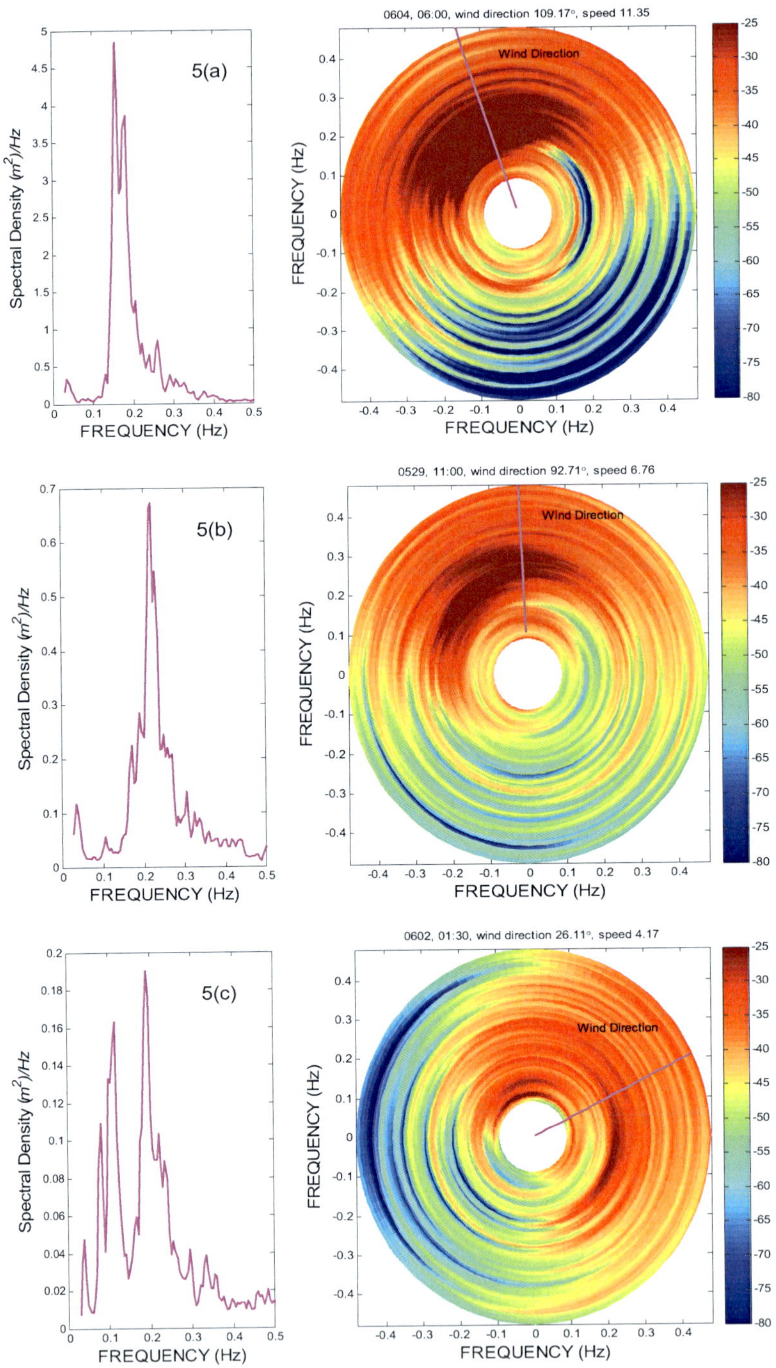

FIGURE 5. (a) young waves, (b) a fully developed sea, and (c) old waves.

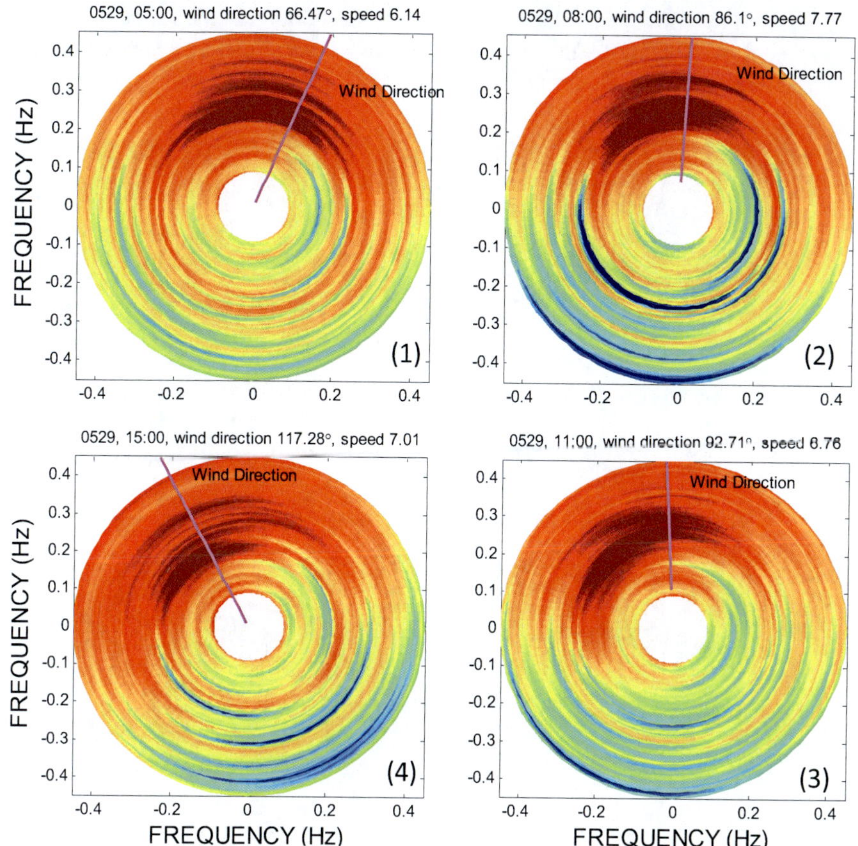

FIGURE 6. Response of surface waves to a 50° wind direction change from 05:00–15:00 May 29, 2001. Clockwise from the upper left (1)–(4) show the alignment between directional spectral and wind direction over this period. Same color bar as Fig. 5.

Sudden Change in Wind Direction

Two periods of sudden wind direction change were also recorded: 01:00–03:00 May 30 and 08:30–12:00 June 4 (days two and seven). The first period had a 150° direction change in 2 h and the second had a 100° variation in 3.5 h. The latter is chosen as an example.

Figure 7 (1) and (2) shows the directional spectrum at 08:30 and 12:00 with a 100° wind direction change. The quick response in the high-frequency range is anticipated while the rather rapid shift in the low-frequency range is surprising. In about 3.5 h both the high- and low-frequency components show an obvious movement toward the wind direction. Young [23] reports that a second peak will form in the new wind direction if the sudden change is greater than 90°. This was observed in the ASIAEX data [Fig. 7(3) and (4)]. It took about 5 h for a second peak to form in the new wind direction. The old peak, which remains in Fig. 5(4), will die out gradually. Figure 7(5) demonstrates how low and high frequencies respond to the sudden change. Frequencies above 0.2 Hz exhibit a clear response to the direction change while the lowest frequency (0.15 Hz) shows no response.

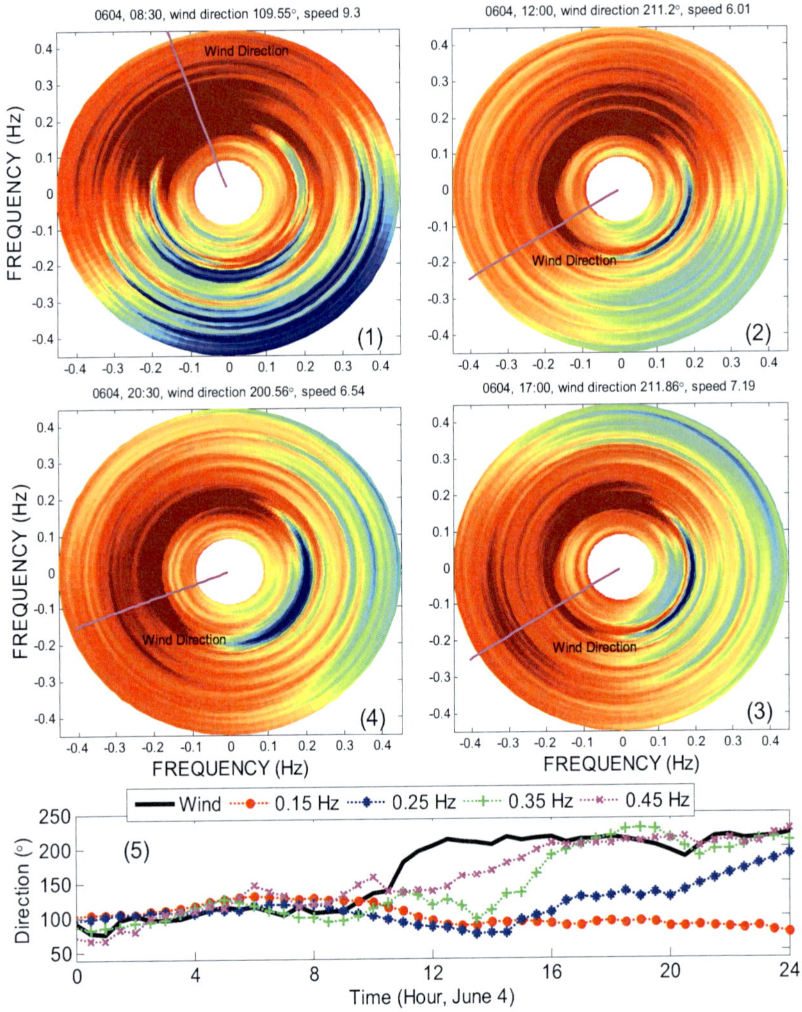

FIGURE 7. (1)–(2): response of surface waves to 100° wind direction change in 3.5 h 08:30–12:00 June 4, 2001. (3)–(4): new spectral peak formed in the new wind direction 5 h after the sudden change. (5) thick line: wind direction; dots, stars, pluses and crosses represent directions of the spectral peaks at 0.15, 0.25, 0.35, and 0.45 Hz, respectively. Same color bar as Fig. 5.

ASIAEX DATA COMPARED WITH MODELS

Surface Spectrum Models

For surface gravity waves, the two types of surface spectrum models are based on the work from Pierson & Moskowitz [5] and Phillips [8].

The original Pierson–Moskowitz surface spectrum model (PM) [5] has been used most widely due to its mathematical simplicity:

$$S(f)=\frac{\alpha g^2}{(2\pi)^4 f^5}\exp\left(-\frac{\beta g^4}{(2\pi)^4 f^4 U^4}\right)=\frac{\alpha g^2}{(2\pi)^4 f^5}\exp\left(-\frac{5}{4}\left(\frac{f_p}{f}\right)^4\right). \tag{2}$$

Here, $\alpha = 0.0081$, $\beta = 0.74$, U is the wind speed at 19.5 m, g is the gravitational acceleration, and f_p is the spectral peak frequency defined as $f_p = 0.14g/U$. The PM model is only valid for fully developed seas, i.e., wave age $U/c_p 0.8–1$.

In coastal areas, fully developed sea states are rarely reached due to limited fetch and swell [24] (also see Fig. 3 of this paper). Based on Phillips' formulation in high-frequency range [13], Hasselmann et al. [6] propose the JONSWAP (Joint North Sea Wave Project) spectrum for fetch-limited conditions:

$$S(f)=\frac{\alpha g^2}{(2\pi)^4 f^5}\exp\left(-\frac{5}{4}\left(\frac{f_p}{f}\right)^4\right)\gamma^{\exp\left(-\frac{(f-f_p)^2}{2\sigma^2 f_p^2}\right)}. \tag{3}$$

In (2), α, γ, and σ are power-law functions of the non-dimensional parameter υ with $\upsilon = f_p \times U/g$. υ and the wave age U/c_p only differ by a factor of 2π. f_p is now fetch-dependent: $f_p = 3.57g/(U\xi^{0.33})$, in which ξ is non-dimensional fetch defined as $\xi = gx/U^2$, where x is the actual fetch. Using the same formula, Mitsuyasu et al. [25] report a different set of relations between α, γ, σ and υ with less scatter. All of these results are summarized in Table 1 [4].

Toba [7] proposes the f^{-4} form:

$$S(f)=\frac{\alpha g^2}{(2\pi)^4 f_p f^4}\exp\left(-\frac{5}{4}\left(\frac{f_p}{f}\right)^4\right)\gamma^{\exp\left(-\frac{(f-f_p)^2}{2\sigma^2 f_p^2}\right)}. \tag{4}$$

The slope of the spectrum changes for $f_p^{-1} f^4$ instead of f^5 as in the PM and JONSWAP models. This formulation has been supported in the literature both experimentally and theoretically ([12], [26]). Donelan's results are based on a data set with wave age in the range 0.83–4.6 (Table I).

Recently, unified surface spectrum models containing both gravity and capillary waves have been developed ([9]–[11]). Elfouhaily et al. [9] and Plant [11] utilize Donelan's results in the low-frequency regime while Lemaire [10] takes the JONSWAP form. They have distinct forms at high wavenumbers. Because the

ASIAEX data only contain frequencies up to 0.64 Hz, the inclusion of [9]–[11] is merely for the purpose of completeness. Performance comparison between these models can be found in [11].

TABLE 1. Surface model parameters. (All parameters in the form of y = amp.* υ^{power}.)

Model		PM	JONSWAP–Hasselmann	JONSWAP–Mitsuyasu	TOBA–Donelan	ASIAEX
U/c_p range		$0.8 \sim 1$	$0.88 \sim 5$	< 0.82	$0.83 \sim 5$	0.1~1.6
α	Amp.	Const. = 0.0081	0.036	0.033	0.0165	0.011
	Power		0.87	0.86	0.55	0.465
γ	Amp.	-	4.12	4.42	1.7, if $\upsilon < 0.159$;	4.560
	Power		0.32	0.43	6.489+6log υ, o.w.	0.257
σ_a	Amp.	-	0.54	Const. = 0.07		0.270
	Power		-0.32		$\sigma = 0.08+1.29*10^{-3}\upsilon^{-3}$	0.578
σ_b	Amp.	-	0.078	Const. = 0.09		0.075
	Power		-0.16			-0.061

Model Comparisons Using ASIAEX Data

The surface spectrum models studied here have distinct wave age ranges based on the data pools with which each model was built. To compare the ASIAEX data with the existing models, data are first divided into three categories according to wave age: $U/c_p < 0.8$, $0.8 < U/c_p < 1$, and $U/c_p > 1$. The ASIAEX data contain 432 spectra and 318 of them have $U/c_p < 0.8$. As the occurrence of a fully developed sea was brief (see Fig. 3), we focus on young waves with $U/c_p \geq 1$. Among 50 young wave spectra, eight of them included a swell component and hence were excluded.

Surface wave height is an important parameter for acoustic predictions. It is thus important to know which model gives the best wave height predictions. The RMS wave height was calculated using the four models and compared with the ASIAEX data. The JONSWAP–Mitsuyasu model exhibits the best fit (Fig. 8), due to the fact that over 70% of the ASIAEX data have low wind input (wave age < 0.8), which coincides with Mitsuyasu's presumed wave ages. Examining Figures 3 and 8 shows that the Mitsuyasu model deviates from the experimental curve under three conditions: (1) $U/c_p<0.6$; (2) $U/c_p>1$; (3) existence of swell.

Each surface model has its own wave age range, which can be regarded as each model's "validity region." In that case, RMS wave height would be better predicted using the JONSWAP–Hasselmann or TOBA–Donelan to model ASIAEX data for wave age $U/c_p>1$. One remaining question here is how to include swell in the surface model because none of the surface models accounts for a swell component.

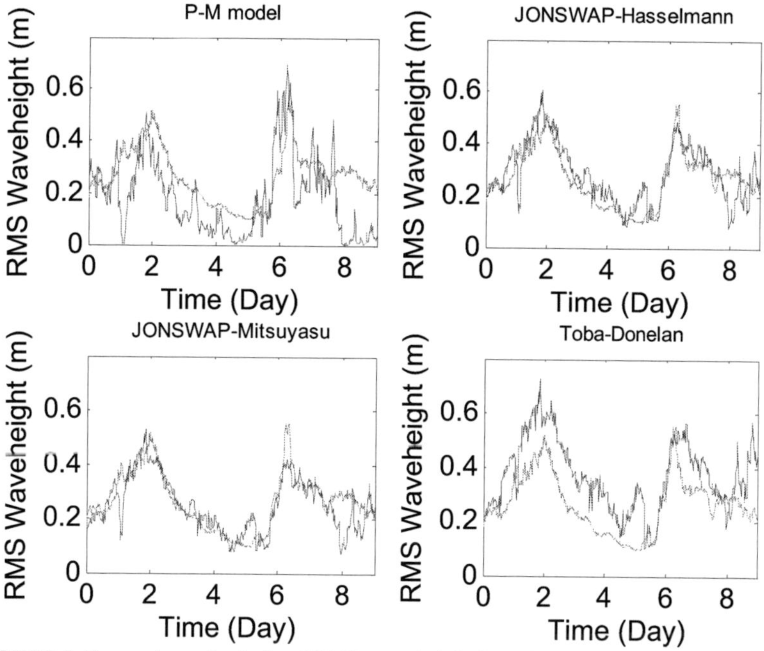

FIGURE 8. Comparison of calculated RMS wave height from four models with the ASIAEX data. Red: experimental RMS; blue: model RMS.

The ASIAEX experimental site was 300 km from the coast in all directions so a fetch-limited condition may not apply. The two spectrum functions (3) and (4) contain five parameters α, γ, σ_a, σ_b and rear face slope n. Curve fitting techniques were applied to find the slope n of the rear face. Curve fitting starts from f_p to the highest frequency of 0.64 Hz (Fig. 9). The average slope n is –4.08 based on 42 spectra from young waves. Data seem to favor the TOBA–Donelan form with f^{-4} frequency dependence.

The prediction of the RMS wave height is separated into two regions: $U/c_p<0.8$ and $U/c_p>0.8$, with the Mitsuyasu model used for the first region and the Donelan model for the second. The RMS comparison between experimental data and the combined model shows that the combined Mitsuyasu–Donelan model is in better agreement with the experimental curve when wave age > 0.8 around day 6 and 7 than the Mitsuyasu model alone.

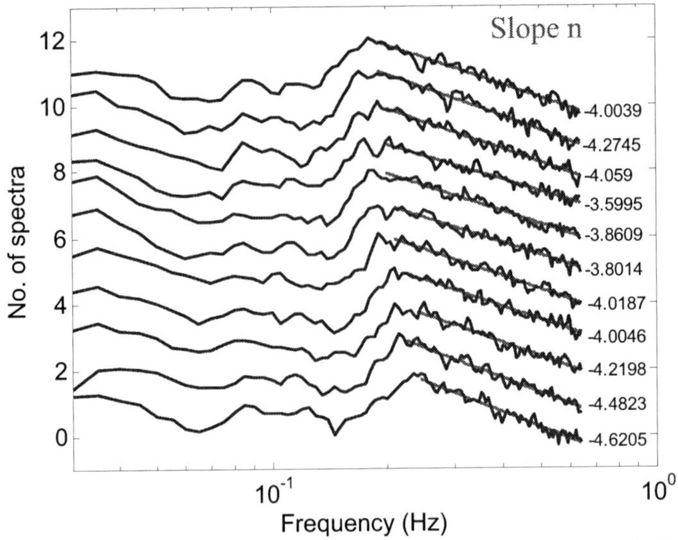

FIGURE 9. Rear face slope fitting using12 spectra recorded consecutively from 20:00 June 3 to 01:30 June 4. Fitting starts from f_p to 0.64 Hz.

Now the prediction of the RMS wave height is separated into two regions: U/c_p<0.8 and U/c_p>0.8 with the Mitsuyasu model used for the first region and the Donelan model for the second. Figure 10 shows the RMS comparison between experimental data and the combined model. The blue curve which corresponds to the combined Mitsuyasu-Donelan model is seen to be in better agreement with the experimental curve when wave age > 0.8 around day 6 and 7 than the Mitsuyasu model alone.

To summarize, the best RMS wave height prediction comes from a combined model in the frequency range $f \le 3f_p$:

$$S(f)=\begin{cases} \dfrac{\alpha g^2}{(2\pi)^4 f^5}\exp\left(-\dfrac{5}{4}\left(\dfrac{f_p}{f}\right)^4\right)\gamma^{\exp\left(-\frac{(f-f_p)^2}{2\sigma^2 f_p^2}\right)} & for\ U/c_p<0.8 \\ \dfrac{\alpha' g^2}{(2\pi)^4 f^4 f_p}\exp\left(-\left(\dfrac{f_p}{f}\right)^4\right)\gamma'^{\exp\left(-\frac{(f-f_p)^2}{2\sigma^2 f_p^2}\right)} & for\ U/c_p\ge 0.8 \end{cases} \quad (5)$$

where α, γ, and α', γ', correspond to Mitsuyasu and Donelan models, respectively.

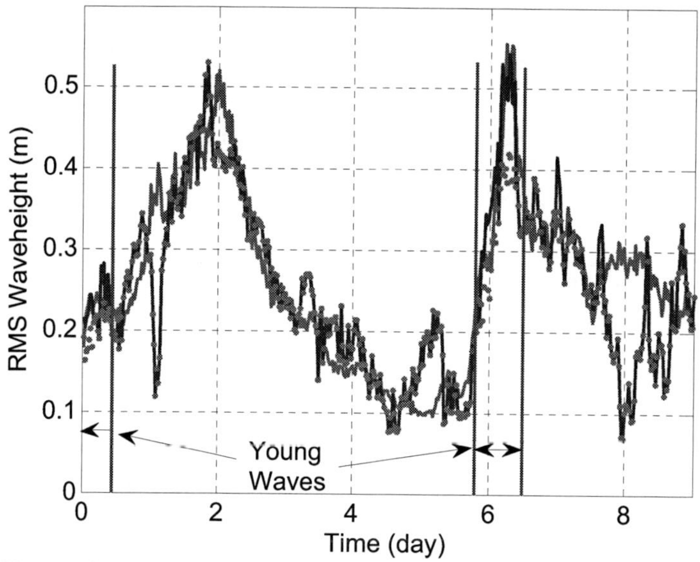

Figure 10. A combined Mitsuyasu–Donelan model. Red: experimental RMS; blue: combined model RMS; magenta dots: Mitsuyasu model.

SUMMARY AND DISCUSSIONS

The ASIAEX surface directional wave spectra have been analyzed. The non-directional surface wave spectra usually display a unimodal shape. Bimodal structure is also observed with the lower frequency peak corresponding to a swell component, usually from a distant wind field, and the higher one to wind-generated waves. The spectral peak frequency shows a tendency to decrease with increasing wind speeds.

The directional spectra demonstrate apparent response to wind direction change. For a gradual change, the whole spectrum rotates with the wind field. A second peak is generated in the new wind direction if the sudden wind direction change exceeds 90°. High frequencies respond more rapidly to wind variations; in one case approximately 5 h were needed to catch up with the sudden direction change.

Existing surface spectrum models were compared with the ASIAEX data to determine which model gives the best prediction of the RMS surface wave height in the coastal region. The JONSWAP–Mitsuyasu model has a better fit to the ASIAEX RMS wave height data because both the model and the data have similar wave age ranges. All models deviate from the experimental data when the wind input measure, U/c_p, is small or when a swell component is present. Models with a higher wave age range (Donelan) can be used to treat corresponding parts of the data. A combined Mitsuyasu–Donelan model can make a better RMS surface wave height prediction than any of the models alone.

The results of this work may be beneficial for acoustic propagation and scattering modeling; wave age can be used to find the most appropriate surface model to use for acoustic predictions. The conclusions, however, are based on a limited 9-day directional surface wave spectrum data set and need to be confirmed by more observations. The inclusion of a swell component in the model remains to be determined.

ACKNOWLEDGMENT

This research was supported by the Office of Naval Research. The authors thank P. Dahl at the Applied Physics Laboratory, University of Washington for providing the surface wave data from ASIAEX ECS.

REFERENCES

1. C. Eckart, *J. Acoust. Soc. Am.*, **25**, 566-570 (1953).
2. W. A. Kuperman and F. Ingenito, *J. Acoust. Soc. Am.*, **113**, 1178-1187 (1977).
3. D. E. Weston and P. A. Ching, *J. Acoust. Soc. Am.*, **86**, 1530-1545 (1989).
4. I. R. Young, *Wind Generated Ocean Waves*, Elsevier Ocean Engineering Book Series, Vol. 2, 1999.
5. W. J. Pierson and L. Moskowitz, *J. Geophys. Res.*, **69**, 5181-5190 (1964).
6. K. Hasselmann, D. B. Ross, P. Muller, and W. Sell, *J. Phys. Oceanogr.*, **6**, 200-228 (1976).
7. Y. Toba, *J. Oceanogr. Soc. Japan*, **29**, 209-220 (1973).
8. O. M. Phillips, *J. Fluid Mech.*, **4**, 426-434 (1958).
9. T. Elfouhaily, B. Chapron, K. Katsaros, and D. Vandemark, *J. Geophys. Res.*, **102**, 781-796 (1997).
10. D. Lemaire, P. Sobieski, and A. Guissard, *IEEE Trans. Geosci. Remote Sens.*, **37**, 1038-1051, (1999).
11. W. J. Plant, *J. Geophys. Res.*, **107**, 3-1~3-21 (2002).
12. M. A. Donelan, J. Hamilton, and W. H. Hui, *Philos. Trans. R. Soc. Lond. A*, **315**, 509-562 (1985).
13. O. M. Phillips, *J. Fluid Mech.*, **156**, 505-531 (1985).
14. S. A. Kitaigorodskii, *The Physics of Air-Sea Interacti*ons, Keter Press, Jerusalem, 1973.
15. W. J. Plant, *J. Geophys. Res.*, **91**, 735-749 (1986).
16. P. Dahl, R. Zhang, J. H. Miller, L. R. Bartek, Z. Peng, S. R. Ramp, J.X. Zhou, C.-S. Chiu, J. F. Lynch, J. A. Simmen, and R. C. Spindel, *IEEE J. Ocean. Eng.*, **29**, 920-928 (2004).
17. D. E. Hasselmann, M. Dunckel, and J. A. Ewing, *J. Phys. Oceanogr.*, **10**, 1264-1280 (1980).
18. J. H. Allender, J. Albrecht, and G. Hamilton, *J. Phys. Oceanogr.*, **13**, 1519-1525 (1983).
19. D. Masson, *J. Phys. Oceanogr.*, **20**, 1876-1885 (1990).
20. S. D. Smith, R. J. Anderson, W. A. Oost, C. Kraan, N. Maat, J. DeCosmo, K. B. Katsaros, K. L. Davidson, K. Bumke, L. Hasse, and H. M. Chadwick, *Boundary-Layer Meteorol.*, **60**, 109-142 (1992).
21. I. R. Young, *Ocean Eng.*, **22**, 669-686 (1995).
22. P. H. Dahl, *J. Acoust. Soc. Am.*, **115**, 589-599 (2004).
23. I. R. Young, S. Hasselmann, and K. Hasselmann, *J. Phys. Oceanogr.*, **17**, 1317-1338 (1987).
24. R. M. Heitsenrether and M. Badiey, *AIP Conf. Proc.*, **728**, 214-221 (2004).
25. H. Mitsuyasu, F. Tasai, T. Suhara, S. Mizuno, M. Onkusu, T. Honda, and K. Rukiski, *J. Phys. Oceanogr.*, **10**, 286, 296 (1980).
26. S. A. Kitaigorodskii, *J. Phys. Oceanogr.*, **13**, 816-827 (1983).

Physical Parameters for Four Seabed Geoacoustic Models from Low-Frequency Measurements

Ji-Xun Zhou and Xue-Zhen Zhang

School of Mechanical Engineering, Georgia Institute of Technology, Atlanta, GA 30332-0405, USA.
State Key Laboratory of Acoustics, Institute of Acoustics, Chinese Academy of Sciences, Beijing, China
E-mail: jixun.zhou@me.gatech.edu

Abstract. Low-frequency (LF) sound speed and attenuation in sandy/silty seabottoms have recently been analyzed and summarized from long-range field measurements in shallow water [Zhou, Zhang and Knobles, *J. Acoust. Soc. Am.*, **125**, 2847-2866, 2009]. Field measurements conducted at 20 locations in different coastal zones around the world were analyzed. The seabed attenuations, inverted from different acoustic field characteristics, exhibit similar magnitude and nonlinear frequency dependence below 2000 Hz. The present paper shows that both LF field-derived sound speed and attenuation in the bottom can be equally well described by the Biot-Stoll model, the Buckingham VGS model, the Chotiros-Isakson BICSQS model, and the Pierce-Carey model. Possible ranges of the physical parameters for each of these four seabed acoustic models are estimated based on the LF field-inverted sound speed and attenuation. These four models with different physical mechanisms in consideration may result in almost identical non-linear frequency dependence of sound attenuation in sandy/silty seabottoms.

Keywords: Acoustics of sediments, seabed acoustics models, seabed geo-acoustic inversion.
PACS: 43.30.Ma, 43.30.Pc, 43.20Jr, and 43. 30Ky.

INTRODUCTION

An accurate solution to the wave equation in shallow-water waveguides requires accurate seabed acoustic parameters to define a bottom boundary condition. Sound propagation losses in shallow water (SW) are generally dominated by bottom reflection and scattering losses. Thus, the seabed has been the *king* of the SW acoustics problems. Much progress has been made in our understanding of seabed acoustics through significant efforts on seabed acoustic modeling and geoacoustic inversion. However, there remain important unanswered questions and a scarcity of high-quality basic research data sets. For example, the debate over the sound speed dispersion and the frequency dependence of sound attenuation in sandy bottoms has persisted for decades. There is a lack of high-quality wideband data of sediment sound speed and attenuation that covers the low- to high-frequency transition band (from low hundred of Hz to several kHz). Many geoacoustic inversion papers have been published with different algorithms. However, these inversion methods have yet to produce a fully populated wideband data set of bottom sound speed and attenuation for practical sonar applications and sediment acoustic modeling.

CP1272, *Shallow-Water Acoustics, Second International Shallow-Water Acoustics Conference*
edited by J. Simmen, E. S. Livingston, J. X. Zhou, and F. H. Li

LF sound speed and attenuation in sand and silt seabottoms have recently been analyzed and summarized by Zhou et al. from long-range field measurements in shallow water [1]. Field measurements conducted at 20 different locations were analyzed. The present paper shows that the LF field-derived bottom sound speed and attenuation can be equally well described by the Biot-Stoll model [2-8], the Pierce-Carey model [9], the Buckingham VGS model [12-14], and the Chotiros-Isakson BICSQS model [15]. The LF field-derived data in [1] are used to estimate possible ranges of physical parameters in these four physics-based seabed geoacoustic models for those LF experimental sites.

SOUND SPEED AND ATTENUATION IN SANDY/SILTY SEABED FROM LF FIELD MEASUREMENTS IN 50-2000 HZ

The sound speed and attenuation in sand and silt seabottoms, inverted from the LF field measurements with different methods in a frequency range of 50-2000 Hz, have been analyzed and summarized in [1]. The measurements were conducted at 20 locations in different coastal zones around the world. The LF field-inverted sound speed and attenuation combined with the SAX99/04 and other high-frequency measurements are shown in figures 1 and 2. The LF sound speed ratios at the water-sediment interface are in the 1.04-1.08 range. The sound attenuations in sandy/silty bottoms exhibit similar magnitudes and a similar nonlinear frequency dependence. The average bottom attenuation can be expressed by $\alpha = (0.37 \pm 0.01) f^{(1.80\pm0.02)}\ dB/m$ in 50-1000 Hz range, where the frequency f is in *kHz*. The LF field-inverted data appear to smoothly dovetail into the higher frequency data from the SAX 99/04 experiments and other mid- and high-frequency measurements. They also match Hamilton's prediction on sound attenuation very well around 1 kHz. These data will be used in the present paper to estimate the physical parameters for physics-based geoacoustic models for sandy/silty seabottoms.

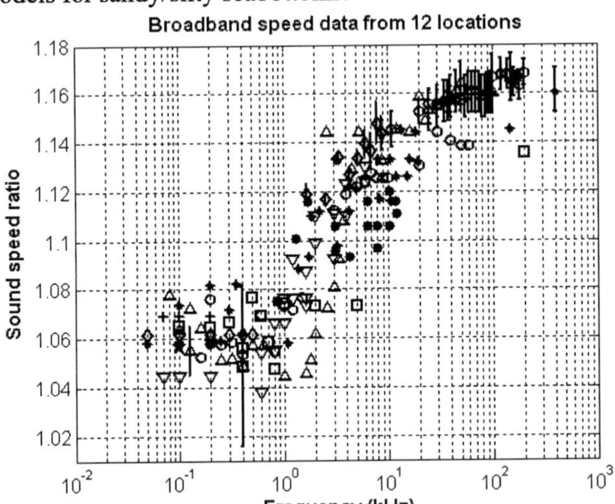

FIGURE 1. Sound speed ratio at the bottom-water interface. [1]

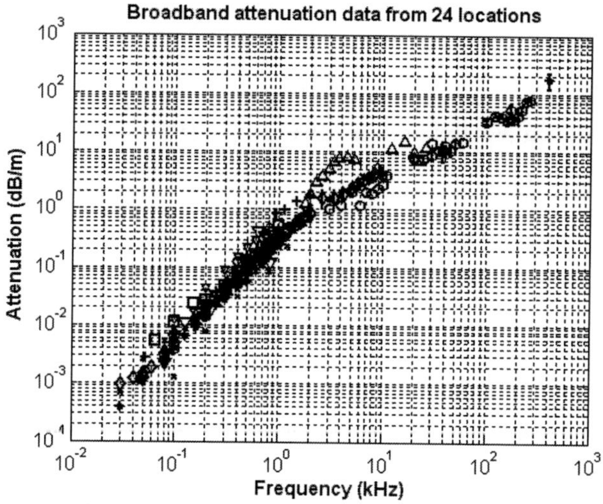

FIGURE 2. Sound attenuation in sandy/silty seabottoms [1].

EFFECTIVE GEOPHYSICAL PARAMETERS FOR FOUR SEABED ACOUSTIC MODELS

1. Biot-Stoll model

In the Biot-Stoll theory, acoustic wave propagation in an isotropic saturated porous medium (such as in water-saturated sediments) is described by a pair of coupled differential equations. The solutions for the equations are obtained in terms of the potentials ϕ_s, ϕ_f, ψ_s and ψ_f [7]. The displacement vector of the skeletal frame **u** and the displacement vector of the fluid relative to solid **W** are defined as **u** $=\nabla\phi_s+\nabla\times\psi_s$; **W**$=\nabla\phi_f+\nabla\times\psi_f$. In this way, Biot's equations become:

$$\nabla^2(H\phi_s - C\phi_f) = \frac{\partial^2}{\partial t^2}(\rho\phi_s - \rho_f\phi_f) \tag{1}$$

$$\nabla^2(C\phi_s - M\phi_f) = \frac{\partial^2}{\partial t^2}(\rho_f\phi_s - m\phi_f) - \frac{F\eta}{\kappa}\frac{\partial\phi_f}{\partial t} \tag{2}$$

Where ρ_f is the pore fluid density; η is the pore fluid viscosity; κ is the hydraulic coefficient of permeability; ρ is the total mass density, $\rho = \beta\rho_f + (1-\beta)\rho_s$, where β is the porosity, ρ_s is the density of grain. The parameter $m = c\rho_f/\beta$ accounts for the inertia of fluid flow with respect to the macroscopic pressure gradient, where c is tortuosity. The factor F is a viscosity correction to account for the frequency dependence of viscous losses in the oscillating flow within the sediment pore, and is given by

$$F(\varepsilon)=\frac{1}{4}\frac{\varepsilon T(\varepsilon)}{[1-(2T(\varepsilon)/i\varepsilon)]}, \quad T(\varepsilon)=\frac{ber'(\varepsilon)+bei'(\varepsilon)}{ber(\varepsilon)+bei(\varepsilon)} \tag{3}$$

where ber and bei are the real and imaginary parts of the Kelvin function of the first kind of order zero, ber' and bei' are their derivatives. The argument $\varepsilon = a\sqrt{\omega\rho_f/\eta}$, in which ω is the angular frequency. The parameter a is the "pore size parameter". It is related to the permeability and porosity by the Kozeny-Carman equation, often expressed by $a=\sqrt{8c\kappa/\beta}$ in which the constant c depends on the shape and size of the pore spaces.

The expressions for the Biot moduli in (1-2) can be expressed in terms of the measurable sediment properties as follows:

$$H=\frac{(K_r-K_b)^2}{D-K_b}+K_b+\frac{4\mu}{3} \tag{4}$$

$$C=\frac{K_r(K_r-K_b)}{D-K_b} \tag{5}$$

$$M=\frac{K_r^2}{D-K_b} \tag{6}$$

$$D=K_r\left[1+\beta\left(\frac{K_r}{K_f}-1\right)\right] \tag{7}$$

where K_r is the bulk modulus of an individual sediment grain and K_f is the bulk modulus of the pore fluid. The sediment frame bulk and shear moduli K_b and μ are complex to account for dissipation losses at grain contacts. For a harmonic plane wave of $\exp(i\omega t-ikx)$ traveling through a sediment medium, Biot's equations for the scalar potentials reduce to

$$-Hk^2\phi_s+Ck^2\phi_f=-\rho\omega^2\phi_s+\rho_f\omega^2\phi_f \tag{8}$$

$$-Ck^2\phi_s+Mk^2\phi_f=-\rho_f\omega^2\phi_s+m'\omega^2\phi_f \tag{9}$$

where the virtual mass density m' is given by $m'=m-(iF\eta/\kappa\omega)$. In (8) and (9), k is the wave number in the sediment. The determinant of the coefficients in (8) and (9) must be zero.The resulting equation for the admissible solutions for ϕ_s and ϕ_f is

$$\begin{vmatrix} Hk^2-\rho\omega^2 & \rho_f\omega^2-Ck^2 \\ Ck^2-\rho_f\omega^2 & m'\omega^2-Mk^2 \end{vmatrix}=0 \tag{10}$$

The wave numbers of two compressional waves can be found by solving (10). The solutions for complex velocities of the fast (slow) compressional waves *V1 (V2)* can be expressed as:

$$\begin{pmatrix} V_1 \\ V_2 \end{pmatrix}=\left[\frac{2(HM-C^2)}{(Hm'+\rho M-2C\rho_f)\mp[(m'H-\rho M)^2+4\rho_f H(\rho_f M-m'C)+4\rho C(m'C-\rho_f M)]^{1/2}}\right]^{1/2} \tag{11}$$

The attenuation in sediment is $\alpha = 8.686\,\mathrm{Im}(k)\ dB/m$. In the basic Biot-Stoll theory, it is largely three parameters that control the fluid motion that results in the nonlinear dispersion that is important in the coarser granular sediments: the permeability, the porosity, and the tortuosity which defines the necessary amount of "added mass". Our data/model comparisons show that the LF sound velocity ratio is most sensitive to the porosity of the sediment, and is less sensitive to the permeability and tortuosity; the LF sound attenuation is most sensitive to the permeability of sediment, and is less sensitive to the porosity and tortuosity. The dispersion is sensitive to the tortuosity. Thus, these three physical parameters can be easily obtained by finding values that provide the best match between the Biot-Stoll model and the sound speed and attenuation data.

In this paper, a model that provides the best match to the upper boundary of the LF sound speed and attenuation data in Figs. 1&2 will be referred to as Model A; Model B provides the best match to the lower boundary of the data. On the basis of a simple visual fit to the LF field-inverted sound speed and attenuation data, we obtained physical parameters for the Biot-Stoll models (model A and model B) that are given in Table I [1]. Two numerical curves from these two models are plotted in Fig. 3 and Fig. 4 by the black solid and dashed lines, respectively. That is, for those LF experimental sites with sandy/silty seabottoms, the sediment porosity is in a range of $0.395-0.450$, the permeability is in a range of 0.5×10^{-11} - 2.5×10^{-11}, the tortuosity is in 1.25-1.5 range. The other 10 Biot parameters are assumed the same as in Williams [16] from SAX99.

2. Pierce-Carey model

By examining locally averaged stresses and taking into account "the bowing of the fluid velocity profile in the interstices when the overall local region is being accelerated" and the counteraction of this bowing via viscosity, Pierce and Carey derived a low-frequency sound attenuation α_{lf} in sandy/silty marine sediments (Nepers per unit length):

$$\alpha_{lf} = \tau\omega^2/2c_0 = \beta_{PC} d^2 \beta(1-\beta)^2(\rho_s - \rho_f)^2 \omega^2 /(2\eta\rho c_0) \qquad (12)$$

where d is the mean grain diameter, ρ is the mass density, c_0 is given by a Wood's equation:

$$c_0 = 1\Big/\sqrt{[\beta/K_f + (1-\beta)/K_s][\beta\rho_f + (1-\beta)\rho_s]} \qquad (13)$$

From the Effective Density Fluid Model (EDFM), in which the frame moduli are set to zero[10], Williams identified that the non-dimensional parameter in the Pierce-Carey model, β_{PC}, can be expressed in terms of physical parameters : $\beta_{PC} = \kappa/(\beta d^2)$, and the time constant by several other parameters [11]:

$$\tau = \kappa(1-\beta)^2(\rho_s - \rho_f)^2/(\rho\eta) \qquad (14)$$

The high frequency limit for (12) and (14) is determined by Biot's characteristic frequency $f_c = \beta^2\eta/(2\pi\kappa\rho_f)$ or Williams' f_h [11].

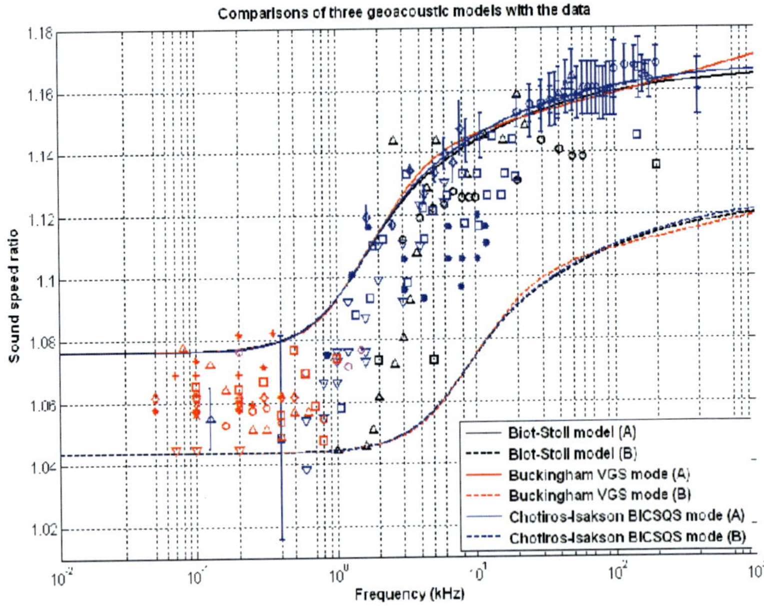

FIGURE 3. Data-model comparisons for sound speed in sandy/silty sea bottoms.

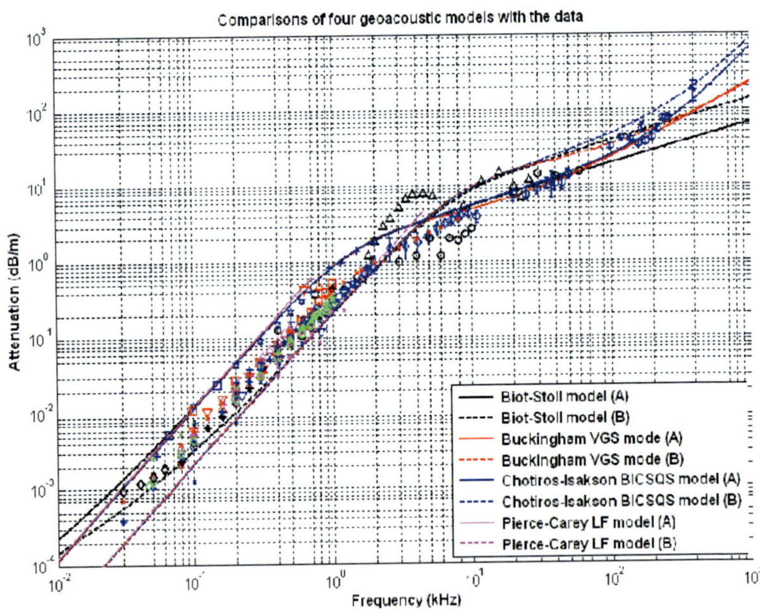

FIGURE 4. Data-model comparisons for sound attenuation in sandy/silty sea bottoms

TABLE 1. Parameters of the Biot-Stoll model [1]

Symbols (units)	Historical	SAX99	Mode A	Model B
Bulk properties				
#1. Porosity β	0.36-0.47	0.385 (0.415)	**0.39**	**0.45**
#2. Grain density $\rho_s\ (kg/m^3)$	2650	2690	2690	2690
#3. Fluid density $\rho_f\ (kg/m^3)$	1000	1023	1023	1023
#4. Grain bulk modulus $K_s\ (Pa)$	(3.6-4.0)E+10	3.2E+10	3.2E+10	3.2E+10
#5. Fluid bulk modulus $K_f\ (Pa)$	(2.0-2.3)E+09	2.395E+09	2.395E+09	2.395E+09
Fluid motion				
#6. Viscosity $\eta\ (kg/m-s)$	0.001	0.00105	0.00105	0.00105
#7. Permeability $\kappa(m^2)$	(10.0-0.65)E-11	2.50 E-11	**2.5E-11**	**0.5E-11**
#8. Pore size $a\ (m)$	(1.00-5.00)E-05			
#9. Tortuosity of sediment α_0	1.00-1.25	1.35 (1.12)	**1.25**	**1.50**
Frame response				
#10. Frame shear modulus $\mu_0\ (Pa)$	(2.61-11.9)E+07	2.92E+07	2.92E+07	2.92E+07
#11. Shear log decrement δ_s	0.00-0.15	0.0616	0.0616	0.0616
#12. Frame bulk modulus $K_0\ (Pa)$	(4.36-43.6)E+07	4.36E+07	4.36E+07	4.36E+07
#13. Bulk log decrement δ_b	0.00-0.15	0.0477	0.0477	0.0477

From the LF-field sound attenuation shown in Fig.2, we have estimated a range of the time constant τ in the Pierce-Carey model: $\tau_A = 11.92\times10^{-6}s$, $\tau_B = 1.06\times10^{-6}s$. These two time constants can also be calculated from (14) using the relevant 7 Biot parameters for Model A and Model B listed in Table I. Two bottom attenuation curves from (12) for model A and Model B are plotted in Fig. 4 by magenta solid and dashed lines, respectively.

3. Buckingham VGS model

Considering grain-to-grain contacts in marine sediments, Buckingham [12-14] developed a grain-shearing (GS) model to correlate the extensive HF data sets, published by Hamilton and others, and the physical properties of the sediments. The GS theory has recently been extended to a new version (designated as the VGS theory) to include the effects of the viscosity of the molecularly thin layer of pore fluid separating contiguous grains.[14] The viscosity mechanism (lubrication) in the molecularly thin layer constitutes a different physical phenomenon from the motion of the pore fluid relative to the "frame" in the Biot theory. The VGS theory yields the following expressions for sound speed and attenuation in the form of simple algebraic equations:

$$c_p = \frac{c_0}{\mathrm{Re}[1+\chi(j\omega T)^n g(\omega)]^{-1/2}}, \qquad 0 < n < 1 \tag{15}$$

$$\alpha_p = -\frac{\omega}{C_0}\mathrm{Im}[1+\chi(j\omega T)^n g(\omega)]^{-1/2}, \quad 0 < n < 1 \tag{16}$$

$$\chi = \frac{\gamma_p + (4/3)\gamma_s}{\rho_0 c_0^2}, \tag{17}$$

$$g(\omega) = \left(1 + \frac{1}{j\omega\tau_0}\right)^{-1+n} \tag{18}$$

where the attenuation α_p is in Np/m, γ_p and γ_s the compressional and shear rigidity coefficients, respectively, n the material exponent (strain-hardening index), χ the grain shearing coefficient, and τ_0 a viscoelastic time constant. The best match between the LF field-derived sound speed and attenuation and theoretical predictions by (15-16) results in the following ranges for the physical parameters in the VGS model: $n = 0.06$; $\gamma_{pA} = 3.7\times10^8$, $\gamma_{pB} = 2.8\times10^8$; $\gamma_{sA} = 2.0\times10^7$, $\gamma_{sB} = 1.0\times10^7$ (these 3 parameters-n, γ_p, γ_s are close to Buckingham's [14]); $\tau_{0A} = 9.0\times10^{-5}\,s$, $\tau_{0B} = 1.7\times10^{-5}\,s$. Two curves for the VGS models (A&B) are plotted in Fig. 3 and Fig. 4 by red solid and dashed lines, respectively. For matching the Biot-Stoll model and the LF field-inverted sound speed data, a speed that is little higher than c_0 is needed in (15-17); the speed is about 0.6%-0.7% higher than c_0 from the Wood's equation of (13).

4. Chotiros and Isakson BICSQS model

The extant published experimental data in water-saturated sands show that the sound speed dispersion is often greater than that predicted by the Biot-Stoll model with constant coefficients. In order to explain this, Chotiros and Isakson [15] have developed the BICSQS model that is an extension of the Biot-Stoll model. It uses a physical model of the grain to grain contact, which includes squirt flow and shear drag to calculate the frame moduli. The effective frame bulk and shear moduli, K_b and μ, are expressed as a function of frequency by

$$K_b = K_c + K_y/[1 + i(\omega_k/\omega)] \tag{19}$$

$$\mu = G_c[1 - i(\omega/\omega_\mu)] \tag{20}$$

$$K_c = [2\sigma/(1-2\sigma) + 2/3]G_c \tag{21}$$

Where K_c and G_c are the asymptotic frame bulk and shear moduli at the LF limit, K_y is the difference between the asymptotic HF and LF values of the bulk modulus. $\omega_k (= 2\pi f_k)$ and $\omega_\mu (= 2\pi f_\mu)$ are the bulk and shear relaxation angular frequencies.

The complex frame moduli in the Biot-Stoll model are computed using (19-20), i.e., the real and imaginary parts of the frame bulk and shear moduli in the Biot-Stoll model are replaced by computed values using four independent parameters of f_k, f_μ, G_c and K_y. K_c is related to G_c by LF frame Poisson's ratio, σ.

The LF field-derived sound speed and attenuation require the following physical parameters for the BICSQS model: $f_k = 4800\ Hz, f_\mu = 100\,kHz$; $G_c = 3.2\times10^7$; $K_{yA} = 5.0\times10^7$, $K_{yB} = 1.0\times10^7$. (Other parameters for model A and model B are same as those in Table I). Two curves for the BICSQS models (A&B) are plotted in Fig. 3 and Fig. 4 by blue solid and dashed lines, respectively.

SUMMARY

Four seabed geo-acoustic models are briefly introduced in this paper. They are: the Biot-Stoll model [2-8], the Pierce-Carey model [9], the Buckingham VGS model [12-14], and the Chotiros-Isakson BICSQS model [15]. The LF field-inverted data of sound speed and attenuation in sandy/silty sea bottoms, obtained from different sites with different field characteristics [1], are used to invert for the physical parameters of these four geoacoustic models for the LF experimental sites. Figure 3 and figure 4 show that these geoacoustic models, which represent different physical mechanisms, may predict almost identical sound speeds and attenuations at low- to mid-frequencies using a set of the physical parameters in the relevant models. That is, the LF sound speed and attenuation in sandy/silty seabottoms, obtained from the LF field measurements, may be equally well matched by these four seabed geoacoustic models when the input physical parameters of these seabed acoustic models are appropriately inverted.

ACKNOWLEDGMENTS

This work was supported by ONR OA322 and CAS. The authors thank Dr. Nicholas Chotiros, Dr. Allan Pierce and Dr. Michael Buckingham for reading the draft manuscript and helpful discussions on seabed geo-acoustic models.

REFERENCES

1. J. X. Zhou, X.Z. Zhang and D.P. Knobles, "Low-frequency geoacoustic model for the effective properties of sandy seabottoms," *J. Acoust. Soc. Am.* **125,** 2847-2866 (2009).
2. M. A. Biot, "Theory of propagation of elastic waves in a fluid-saturated porous solid. I. Low-frequency range," *J. Acoust. Soc. Am.* **26**, 168-178 (1956).
3. M. A. Biot, "Theory of propagation of elastic waves in a fluid-saturated porous solid II. Higher frequency range," *J. Acoust. Soc. Am.* **28**, 179-191 (1956).
4. M. A. Biot, "Generalized theory of Acoustic propagation in porous dissipative media," *J. Acoust. Soc. Am.* **34**, 1254-1264 (1962).
5. R. D. Stoll, "Acoustic waves in ocean sediments," *Geophysics* **42,** 715-725 (1977).

6. R. D. Stoll, "Theoretical aspects of sound transmission in sediments," *J. Acoust. Soc. Am.* **68**, 1341-1350 (1980).
7. R.D. Stoll and T.K. Kan, "Reflection of acoustic waves at a water-sediment interface," *J. Acoust. Soc. Am.* **70**, 149-156 (1981).
8. D.R. Jackson and M.D. Richardson, *High-Frequency Seafloor Acoustics*, Springer, 2007, pp.285-307.
9. A.D. Pierce and W.M. Carey, "Low-frequency attenuation of acoustic waves in sandy/silty marine sediments," *J. Acoust. Soc. Am.* **124**, EL308-EL312 (2008).
10. K.L. Williams, "An effective density fluid model for acoustic propagation in sediments derived from Biot theory,' *J. Acoust. Soc. Am.* **110**, 2276-2281 (2001).
11. K.L. Williams, "Sand acoustics: The effective density fluid model, Pierce/Carey expressions, and inferences for porous media modeling," *J. Acoust. Soc. Am.* **125**, EL164-EL170 (2009).
12. M. J. Buckingham, "Wave propagation, stress relaxation, and grain-to-grain shearing in saturated, unconsolidated marine sediments," *J. Acoust. Soc. Am.* **108**, 2796-2815 (2000).
13. M. J. Buckingham, "Compressional and shear wave properties of marine sediments: comparisons between theory and data," *J. Acoust. Soc. Am.* **117**, 137-152 (2005).
14. M. J. Buckingham, "On pore-fluid viscosity and the wave properties of saturated granular materials including marine sediments," *J. Acoust. Soc. Am.* **122**, 1486-1501 (2007).
15. N. P. Chotiros and M. J. Isakson, "A broadband model of sandy ocean sediments: Biot-Stoll with contact squirt flow and shear drag," *J. Acoust. Soc. Am.* **116**, 2011-2022 (2004).
16. K. L. Williams, D. R. Jackson, E. I. Thorsos, D. J. Tang, and S. G. Schock, "Comparison of sound speed and attenuation measured in a sandy sediment to predictions based on the Biot theory of porous media," *IEEE J. Oceanic Eng.* **27** (3), 413-428 (2002).

Experimental observations of internal wave influence on acoustic propagation in shallow water

Mohsen Badiey

Ocean Acoustics Laboratory
University of Delaware, Newark, DE 19716

Abstract. Over the last three decades several observational and theoretical studies have established the measureable effects of anisotropy in shallow water environment and its effects on the acoustic wave propagation. In this paper, the influence of shallow water internal waves on propagation of acoustic waves is revisited and verified by new experimental data. Results using a towed source show feasibility for remote mapping of the internal waves by underwater acoustics.

Keywords: Shallow water acoustics, internal waves, intensity fluctuations, anisotropy.
PACS: 43.30.Re,43.30.Es,43.30.Dr,43.30.Bp

INTRODUCTION

The first comprehensive study of shallow water acoustics dates back to 1948 where Ewing, Worzel, and Pekeris conducted a detailed study of explosive sounds in the ocean [1]. Since that time, several important investigations over six decades have brought the state of our understanding to where it is in the present day. Shallow water acoustics for many years was studied predominantly for the effect of the sea bottom on the propagation and scattering of sound. However, in the last twenty years, the effects due to the water column have become of interest. We have begun to understand the complex relationship between the frequency dependent acoustic wave propagation and scattering due to the combined effects of the water column and the seafloor. The most interesting relationships summarize the intensity and temporal fluctuations of sound by the anisotropic variability that can be related to the sound field index of refraction in the ocean both in the water column and the sea floor. To have a better understanding of the issues in the shallow water regions continued studies are necessary in order to utilize the important shallow water regions.

A summary of theoretical studies in the past few years is given in an accompanying paper in this proceeding [2]. Here we provide a view on the recent experiment observations in recent years. Table 1 summarizes reported observations from 1991 to 2009. Different parameters of the waveguide including the angle between the acoustic propagation track and the internal wave front α, as well as the acoustic wave

CP1272, *Shallow-Water Acoustics, Second International Shallow-Water Acoustics Conference*
edited by J. Simmen, E. S. Livingston, J. X. Zhou, and F. H. Li

frequency, and the resulting intensity fluctuations due to the interaction with the internal waves are listed. This table shows the relationship between the angle and the intensity fluctuation plus the mechanism of propagation for each case.

TABLE 1. Recent experimental reports on the influence of shallow water internal waves on acoustic propagation *.

Reported by	**Year**	**Waveguide**			**IW Parameter**			**Acoustic Signal**	**Fluctuations**	
		Location	**L (km)**	**D (m)**	**N (cph)**	**A (m)**	α	**Frequency (Hz)**	**dB**	**Mechanism**
Zhou et al.[3]	1991	Yellow Sea	28	40	Hypothesized		>45°	100-1000	> 20	MC
Rubenstein et al.[4]	1991	Wash. Coast	18.5	150	10	10	10-15°	400	3	AD
Rubenstein[5]	1999	Gulf Mexico	30	185	15-20	10	30°	240	2	MC
Badiey et al.[6]	2002	NJ Shelf	15	70	10-15	12	5°	30-160 & LFM 50-250	6-7	HR
Badiey et al.[6]	2002	NJ Shelf	19	70-100	10-15	12	35-40°	30-160 & LFM 50-250	2-3	MC
Frank et al.[7]	2004	NJ Shelf	15	70	10-15	12	35-40°	30-160 & LFM 50-250	2-3	MC
Badiey et al.[8]	2005	NJ Shelf	15	70	10-15	12	5°	30-160	6-7	HR
Badiey et al.[9]	2007	NJ Shelf	15	70	10-15	12	5°	30-160	6-7	HR
Luo et al.[11]	2008	NJ Shelf	20	70	10-15	12	3-5°	270-330	7	HR
Katsnelson et al.[12]	2009	NJ Shelf	20	70	10-15	12	35-40°	30-160	2-3	MC

*L=source-receiver range, D=water depth, N=Brunt-Visala frequency, α=angle between IW front and acoustic track, A=IW amplitude, MC=Mode coupling, AD=Adiabatic, HR=Horizontal refraction.

Based on recent field observations of the sound intensity fluctuations a theory was developed to summarize different regimes of acoustic propagation in the waveguides [9]. This theory that goes back to the work of Burridge et. al. on horizontal rays and vertical modes [10], utilizes the variation of the angle between the propagation track and the front of internal waves (α) as a determining factor for refraction effects, mode coupling, or adiabatic regimes.

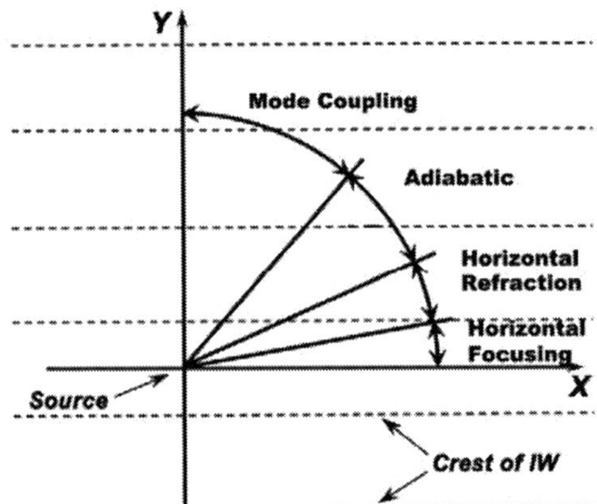

FIGURE 1. Schematic diagram summarizing the propagation mechanisms with respect to the angle (α) between internal wave front (dashed lines) and acoustic track between source and receiver [9].

To examine the validity of this theory, experiments were carried out with fixed and towed sound sources and a vertical-horizontal array (VHLA) during the Shallow Water experiment in 2006 (SW06) [11,13]. The following sections provide a summary of two specific geophysical times in which simultaneous environmental and acoustic measurement were made.

ENVIRONMENTAL AND ACOUSTIC MEASUREMENT

The SW06 experiment was conducted on the New Jersey continental shelf during summer 2006 with participation of several institutions [13]. During the experiment fixed sound sources transmitted acoustic signals on designated time schedule while thermistor chains at different locations were measuring the water column temperature profile continuously. Two fixed acoustic source-receiver tracks included along-shelf and cross-shelf configurations. Also a towed source (J15) was deployed from R/V Sharp in various positions relative to the receiving VHLA in order to measure the angular dependence of sound field with respect to the propagation direction of internal waves. Here we discuss fixed source data for the along-shelf configuration and towed source data for different angles with the internal wave front. Figure 2 shows the along-shelf fixed source-receiver position (dashed line). Along this track, water column temperature was measured at four positions (at the source, at the receiver, and at two mid points, SW20 and SW32 (see Fig. 2)). The colored sectors designated by M1 to M5 in this figure represent the location of R/V Sharp equipped with surface radar imaging at different positions. In these sectors the towed source was transmitting chirp signals in frequency band of 50-450 Hz for 15 minutes. Each of the colored segments also shows a time window that the acoustic track and the internal wave front had a fixed angle. Similarly, F1 through F7 sectors represent the location of the surface ship with respect to the transmission times of the fixed source (NRL300). Using the surface radar images from two ships (R/V Sharp and R/V Oceanous), the surface expressions of IW were recorded. During F1 to F7 time windows when the NRL300 was transmitting the towed source was silent.

Temperature profiles at four points along the acoustic track for an IW that originated and propagated towards the coast during August 17, and 18, 2006 is shown in Fig. 3. Accurate measurement of the position and speed of the internal waves is possible via surface ships as well as the thermistor chains and are used to generate the 3D sound speed profile in the water column [11].

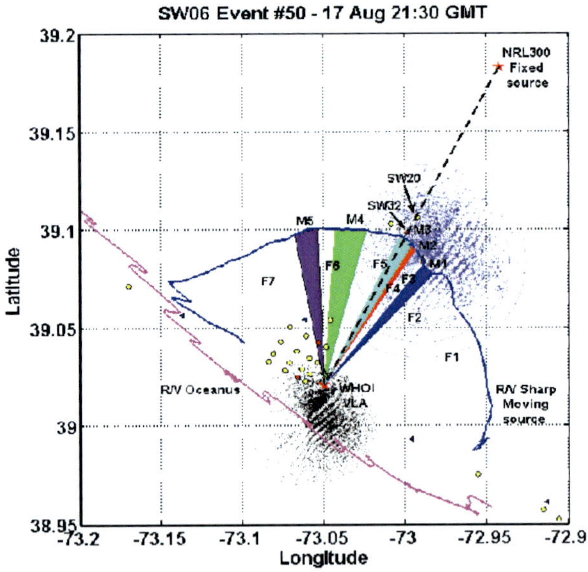

FIGURE 2. Experimental configuration for fixed and towed source-receiver positions. The blue and purple line tracks show the ship tracks for R/V Sharp and Oceanus during their coverage of the internal wave while propagating from south vest to the northeast direction [15].

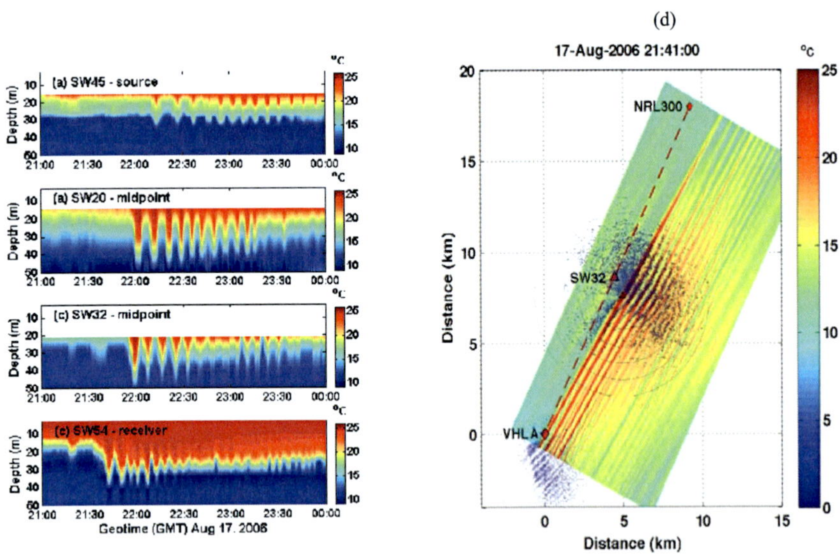

FIGURE 3. (a) - (c) Measured water column temperature at four points along the track. (b) Reconstructed 3D temperature field from thermistors and the surface radar images on R/V Sharp and R/V Oceanous [15].

Radar images have been correlated with the subsurface internal waves in several studies for example [14]. Combined with the thermistor chain data, they can aid with the detailed 3D modeling of the environment. Images shown in Fig. 4 are from both research vessels at 21:36 GMT on Aug. 17, 2006 showing the surface impressions and the curvature of the internal waves during its propagation. Internal wave distribution in horizontal plane can be assessed by carefully examining radar images combined with the depth distribution of the temperature between source and receiver (in this case from points SW45, SW20, SW32, and SW54 shown in Fig. 2). An environmental model for this distribution has been developed starting with the assumption of horizontal IW front and is being advanced to the curved fronts [15].

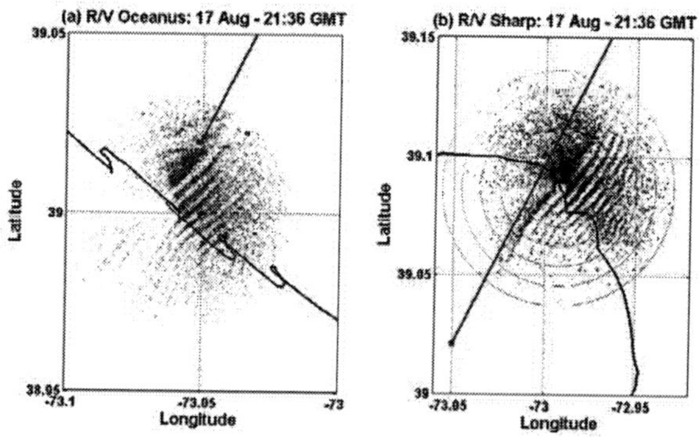

FIGURE 4. Radar images on (a) R/V Oceanous, and (b) R/V Sharp during August 17, 2006 at 21:36 GMT showing the front of the internal wave with respect to the fixed source-receiver track. See Fig. 3(d) for relative position of these two radar images with respect to the approaching IW front to the acoustic track.

Fixed source and fixed receiver configuration

The fixed source was moored about 10.5 m above the sea floor at 39° 10.957' N, 72° 56.575' W, and transmitted 2.0489 second Linear Frequency Modulating (LFM) signals from 270 to 330 Hz. Transmissions continued for 7.5 minutes and then repeated every half hour. A vertical and horizontal receiver array (Shark VHLA) on mooring SW54 was located at 39° 01.252' N, 73° 02.983' W, about 20.2 km south of the NRL300 source. The vertical part of the receiver array consisted of 16 hydrophones with 3.5 m spacing and was positioned in the water column from 13.5 to 77.75 m below the surface. The horizontal part of the array consisted of 32 hydrophones on the seafloor with spacing of 15 m, providing 478 m of horizontal aperture. The sampling rate of the array was 9.766 kHz. The acoustic track and the locations of the source and receiver are shown in Fig. 2. The water depth along the acoustic track was about 80 m [11,13].

Figure 3 shows the received acoustic signals during geotimes F1 through F4 (also shown as sectors with the same letters in Fig. 2). We calculate the total intensity integrated over the depth

$$I(T) = \int_0^H I(z,T)dz \tag{1}$$

where $I(z,T) = (\frac{1}{\rho c})\int_{\tau}^{\tau+\Delta\tau} p^2(z,T,t)dt$ is the intensity of the signal arrivals integrated over a pulse length $\Delta\tau$ at a given depth z, where p is acoustic pressure, ρ is water density, and c is sound speed. During F1, the IW has not reached the acoustic track and therefore the integrated acoustic intensity received on the vertical and horizontal arrays remain relatively constant (see middle panel in Fig. 5). Minimal modal fluctuations are observed in F1 sector. As the IW continues to approach the acoustics track (i.e. during F2 sector), the intensity of the signal increases, but still no major interferences occurs among different modes. This trend continues for another 30 minutes and for the first few minutes in F3 sector where an interesting interference pattern starts to occur. Depth dependent intensity continues to increase to the end of F3. This is the geotime window (or sector) that the IW is approaching the acoustic track, but has not yet crossed it. When the IW reaches the acoustic track, the angle between the IW front and the acoustic track is about 5°. This provides a condition for horizontal refraction. The value of *I (T)* shows dramatic increase after a couple minutes into the F3 geotime. The peak to peak goes up to about 5 dB. Hence a clear transition is observed between the adiabatic and horizontal focusing. Between F3 and F4 there is a gap because the acoustic source was not transmitting signals. The conditions are ripe for a high intensity peak at this geotime that is related to convergences of the horizontal rays and the corresponding vertical modes. When the IW front enters the acoustic track and the overlap occurs, (i.e. during the first few minutes of the F4 sector), the intensity starts to decrease first leading to a focusing event a few minutes later the F4 geotime (sector) region. The depth- averaged intensity (1) shows dramatic increase of more than 10 dB (see Fig. 5 (d)). Dynamic behavior of the waveguide is clearly shown by observing the acoustic intensity on the VLA and the HLA arrays. The data can be modeled in detail because of combined radar and thermistor data in range.

Bottom panels in Fig. 5 show the acoustic intensity for F1 through F4 registered on the horizontal array. The horizontal lines in the HLA plots for F1 and F2 correspond to the modes of the waveguide observed on the HLA during the quiescent time (i. e. no internal wave activity). During the sector F3, at about 21:33 GMT, these lines start to become sloped. This is the geotime when the first effect of the refraction is observed. However, the IW front is only approaching the acoustic track and has not yet crossed it. An interesting interference pattern on the VLA plot in F3 indicates multiple refraction in the horizontal plane and modal interference in vertical plane [16]. The depth-integrated intensity during this period continues to increase. In F4 sector, the internal wave has just passed the acoustic track. (No transmission occurred for 30 minutes between F3 and F4 due to the pre set transmission schedule). The intensity fluctuation patterns in the beginning of the F4 sector lead to the strong focusing event at about 22:05 GMT [16]. This transition is also observed in the HLA portion of the data shown in the bottom panels of F3 and F4 sectors. The sloped intensity modulations correspond to the motion of an interference pattern registered on the

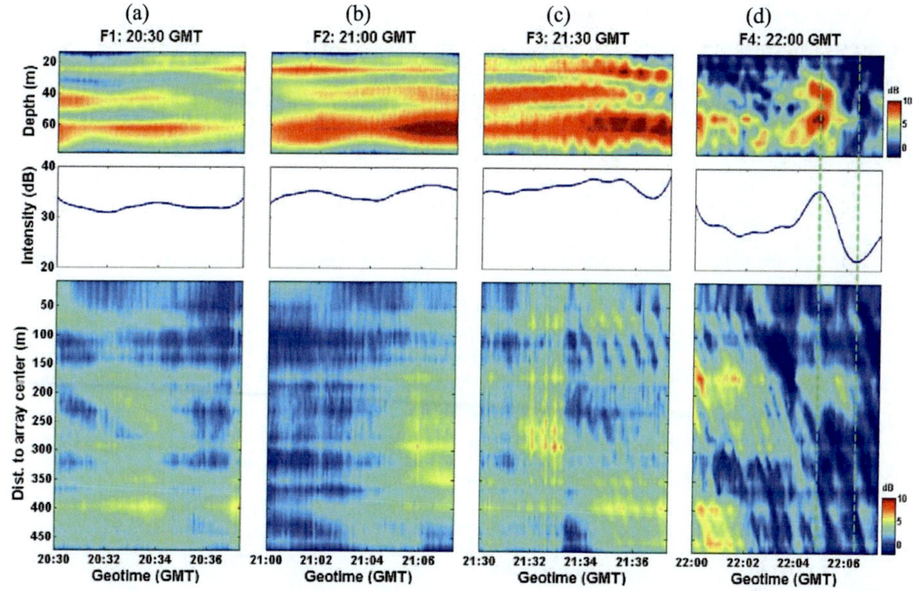

FIGURE 5. Measured acoustic intensity on the VLA and HLA respectively during two different transmission times on August 17, 2006, (a and c) from 20:30:00 to 20:37:20 GMT and (d and f) from 22:00:00 to 22:07:20 GMT. Plots shown in (b) and (d) depict the integrated intensity over the VLA. Right hand panels show evidence of strong intensity variation leading to a focusing event shown by dashed vertical lines [16].

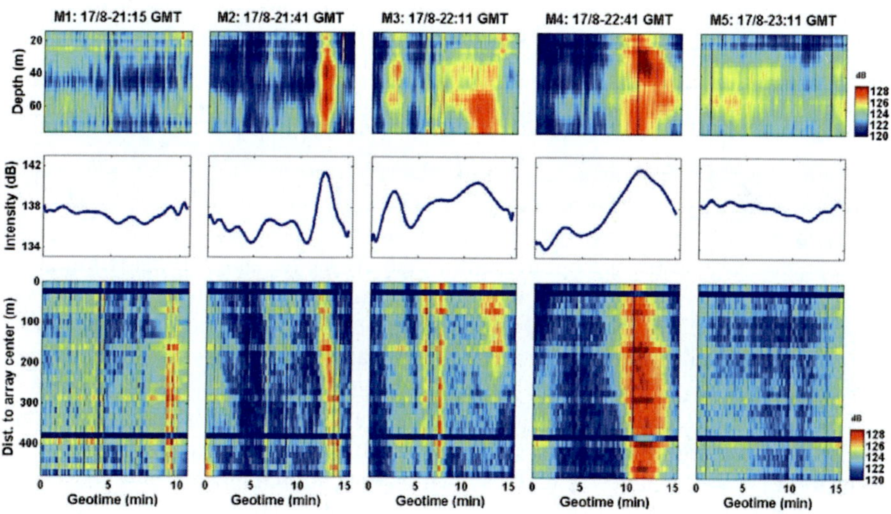

FIGURE 6. Measured acoustic intensity on the VLA and HLA respectively during five different transmission times on August 17 2006 from 21:15 to 23:26 GMT. Each column designated from (a) through (e) shows the geotime window (or position) of the source when the J15 transmitted acoustic signals.

horizontal array, is the result of motion of the IW. The velocity of this interference pattern is estimated as 1 to 1.5 m/s from the acoustic data. This is close to the velocity of the internal waves (measured from shipboard radar to be about 1 m/s). Further detailed modeling is needed to fully describe the behavior of the waveguide during this transition.

Moving source and fixed receiver configuration

For the moving source scenario, R/V Sharp was positioned on the leading edge of the IW and moved with the speed of its propagation (about 1.0 m/s). During the times when the fixed sound source (NRL300) was not transmitting, the J15, which was placed about 15 m below R/V sharp, transmitted 50 to 450 Hz chirp signals for up 15 minute in duration. This transmission schedule created five data sets during geotime zones labeled M1 to M5 in Fig. 2. Towed source transmissions during the SW06 experiment has provided a unique and novel technique to study the propagation mechanisms as acoustic signals traveled through an internal wave train for different angles of α shown in Fig. 1.

In Figure 6, the top panels from left to right show the received acoustic intensity on VLA in transmission geotime zones M1 to M5. Middle panels show the depth-integrated intensity and the bottom panels show the intensity registered on the horizontal array, similar to the case for the fix source position shown earlier. At about 21:15 GMT (sector M1), a portion of an IW front reached the receiver while the major part was still outside the acoustic track between the moving source R/V Sharp and the VLA. From 21:41 GMT, the IW intersected the acoustic track. Intensity fluctuations of about 7~dB were observed on the VLA in geotime sectors M2, M3, and M4. The acoustic wave propagation mechanism shows horizontal focusing (Fig. 6), which agreed with our proposed theory. When the source (R/V Sharp) traveled into the large angle region (sector M5), only weak (~2 dB) fluctuations were recorded on the receiver, confirming that horizontal refraction was no longer the dominant mechanism.

While M1 and M5 show a very small intensity variation in geotime, M2, M3, and M4 show strong peaks. The middle panels from (a) to (e) show several peaks in the averaged intensity over depth. In (b) and (c) columns, these regions correspond to the small angle while in (d) in contradiction to the theory the intensity variation is large while the angle is also large. To show this point, we plot these intensities on a circular pattern shown in Fig. 7. The large intensity around 25° does not correspond to the small angle prescribed by the theory. This unusual feature shown in the acoustic data is not clearly resolved at this time and it may be attributed to other factors beyond the effects of the internal waves such as bathymetric effects and needs additional research to explain. Acoustic intensity received on the HLA is also shown here. The modal behavior of the channel is clearly shown in the bottom panels of Fig. 6 and is corresponding to the focusing events shown in panels (b) to (d) on the top panels for the VLA.

SUMMARY AND CONCLUSIONS

Oceanography of shallow water regions can be highly inhomogeneous and anisotropic due to the presence of nonlinear internal waves and other features such as temperature fronts and eddies. Understanding the shallow water acoustic propagation in the presence of complex oceanography requires detailed field observation combined with theoretical research. The key point in connecting the theory and modeling to the field observations is accurate measurement and reconstruction of the environmental parameters for input to the acoustic propagation models. In an effort to study the three-dimensional acoustic wave propagation in shallow water regions, a combined field observation and theoretical advancement [2] have been presented. The SW06 experiment has provided high quality acoustic and environmental data to investigate the acoustic wave interaction with internal waves. Preliminary analysis of the acoustic data and observations of radar images and temperature records shows that during the passage of an IW event, acoustic wave propagation in the horizontal plane results in significant acoustic intensity variation. This observation confirmed our previously reported results [6] and agrees with the recently reported theory of sound propagation through IW [7]. Future work includes detailed modeling of the environmental conditions and considerations of the effects due to wave curvature and additional features of the internal waves on acoustic mode and frequency content effects [17,18].

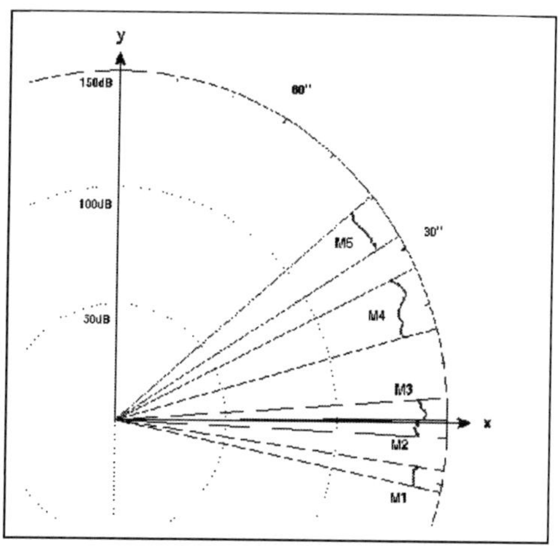

FIGURE 7. Integrated acoustic intensity on the vertical array during M1 to M5 segments of geotime (or ship positions) showing the variations of the intensity from the mean value (~ 130 dB). The fluctuation of the intensity for small angles is due to the refraction and agrees well with the theory. Variation of intensity shown in M4 (~25 °) is not clearly resolved in this case.

ACKNOWLEDGMENTS

The author is grateful to many colleagues and graduate students who participated in various aspects of this work. Special thanks are due to Drs. Ellen Livingstone and Bob Headrick for their support and encouragement during this study. Many thanks also are due to Drs. Jeff Simmen and Jixun Zhou for organizing the scientific workshop in Shanghai. Continued leadership and support of the Office of Naval Research (ONR321OA) during the years in developing this scientific field is most appreciated.

REFERENCES

1. J. L. Worzel, M. Ewing, and C. L. Pekeris, "Propagation of sound in the ocean," Geological Society of America, Memoir 27, (1948).
2. B. G. Katsnelson, and M. Badiey,"Theory and modeling the horizontal refraction in shallow water," Paper in this proceeding, (2010).
3. J. Zhou, X. Zhang, and P. H. Rogers, *J. Acoust. Soc. Am.* 90, 2042-2054 (1991).
4. D. Rubenstein and M. N. Brill, in Ocean Variability and Acoustic Propagation, Ed. by J. Potter and A. Warn-Varnas (Kluwer Academic, Dordrecht, 215–228, (1991).
5. D. Rubenstein, IEEE J. Ocean. Eng. 24, 346, (1999).
6. M. Badiey, Y. Mu, J. F. Lynch, J. R. Apel, and S. N. Wolf, *IEEE J. Ocean. Eng.***27**,117-129 (2002).
7. S. D. Frank, M. Badiey, J. Lynch,W. L. Siegmann, *J. Acoust. Soc. Am.* **118**(2), 723-734 (2005).
8. M. Badiey, B. Katsnelson, J. Lynch, S. Pereselkov, W. Siegmann, *J. Acoust. Soc. Am.* **117**(2), 613-625, (2005).
9. M. Badiey, B. Katsnelson, J. Lynch, S. Pereselkov, *J. Acoust. Soc. Am.* **122**(2), 747-760, (2007).
10. R. Burridge, G. Weinberg, "Horizontal rays and vertical modes" in *Wave propagation and underwater acoustics*, edited: J.B. Keller and J.S. Papadakis, Springer-Verlag, 1977, pp. 86–150.
11. J. Luo, M. Badiey, E. Karjadi, B. Katsnelson, A. Tskhoidze, J. Lynch, J. Moum, *J. Acoust. Soc. Am.* **124**(3), 66, (2008).
12. B. G. Katsnelson, V. Grigorev, M. Badiey, & J. F. Lynch, *J. Acoust. Soc. Am.* **126**(1), 41, (2009).
13. A. E. Newhall, T. F. Duda, J. D. I. K. von der Heydt, J. N. Kemp, S. A. Lerner, S. P. Liberatore, Y.-T. Lin, J. F. Lynch, A. R. Maffei, A. K. Morozov, A. Shmelev, C. J. Seller, and W. E. Witzell, WHOI technical report No. WHOI-2007-04, (2007).
14. R. A. Kropfli, L. A. Ostravski, T. P. Stanton, et al., *J. Geophys. Res.* **104**(C2), 3133-3148, (1999).
15. J. Luo, M. Badiey, Y.-T. Lin, *J. Acoust. Soc. Am.*, **127**(3), Pt. 2, p.1786, (2010).
16. M. Badiey, J. Lynch, Y.-T. Lin, B. katsnelson, *J. Acoust. Soc. Am.*, **127**(3), Pt. 2, p.1974, (2010).
17. J. Lynch, Y.-T. Lin, T. Duda, and A. Newhall, IEEE, J. Oceananic Engr., **35**(1), (2010).
18. Y.-T. Lin, T. F. Duda, and J. F. Lynch, *J. Acoust. Soc. Am.* 126(4), 1752-1765, (2009).

Nonlinear Internal Wave Behavior near the South of Hai-Nan Island and Simulating the Effect on Acoustic Propagation

Tao Hu and Li Ma

Key Laboratory of Underwater Acoustic Environment, Chinese Academy of sciences, Beijing, 100190, China

Institute of Acoustics, Chinese Academy of sciences, Beijing, 100190, China

Abstract. An internal wave observation experiment was performed near the south of Hai-Nan Island in the South China Sea in July 2004. Three vertical thermistor arrays were moored to estimate internal wave propagation direction and velocity. A nonlinear internal wave packet was observed in this experiment. It appeared at flood tide time of wee hours. Computation indicated that the nonlinear internal wave packet's velocity was 0.54m/s and its propagation direction was northwest. From its propagation direction, we estimated that the nonlinear internal wave packet was generated near Xi-Sha Islands. The dnoidal model of KdV(Korteweg-deVries) equation was used to simulate the waveform of thid nonlinear internal wave. Measured data shows the crest interval of nonlinear internal waves was shorter when they propagated. In the last section of this paper we simulate a nonlinear internal wave packet's effect on sound propagation and analyzed mode coupling led by the nonlinear internal wave packet.

Keywords: Nonlinear internal wave, Propagation direction, Velocity, Mode couple
PACS: 43.30.Re

INTRODUCTION

An ocean current flowing over a topographic feature such as a sill or continental shelf in a stratified flow can produce nonlinear internal waves of tidal frequency and has been studied by many researchers [1]. The nonlinear internal wave was observed in the northern South China Sea, near the northeast of Hai-Han Island, during Asian International Acoustics Experiment (ASIAEX)[2,3]. Data obtained during ASIAEX shows that nonlinear internal waves in the northern South China Sea had a 12 hour tide period, and were generated near Batan Island in the Luzon Strait. In July 2004, an internal wave observation experiment (IWOE 2004) was performed near the south of Hai-Han Island in the South China Sea, and a nonlinear internal wave packet was observed during the experiment.

CP1272, *Shallow-Water Acoustics, Second International Shallow-Water Acoustics Conference*
edited by J. Simmen, E. S. Livingston, J. X. Zhou, and F. H. Li

This paper has three parts. The first part shows the internal waves' behavior during IWOE 2004. The second part gives the simulation results for acoustic propagation through the nonlinear internal wave packet. Last is the conclusion of paper.

I. INTERNAL WAVE OBSERVATION DURING IWOE 2004

Figure 1 shows the position of this experiment. Three vertical thermistor arrays were used to measure internal waves. Figure 2 shows the moored site of three thermistor vertical arrays, named A, B and C. The three arrays built up a triangle, and distance of each side was about 1000meters. Sampling frequency of the thermistors was 30 seconds. Using those thermistor arrays, the internal wave's speed and propagating direction can be estimated.

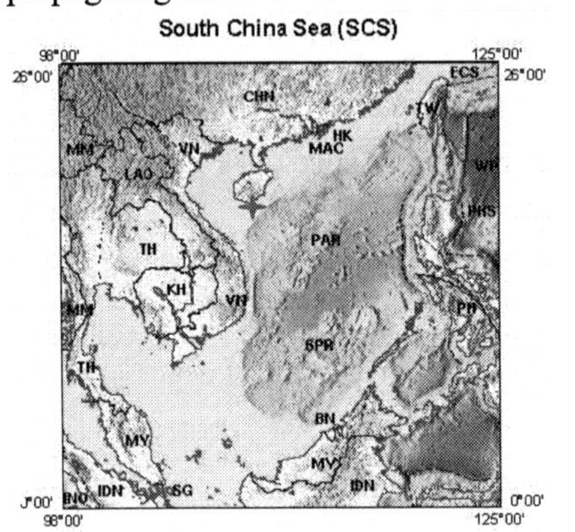

FIGURE 1. Experiment site is shown by "✦"

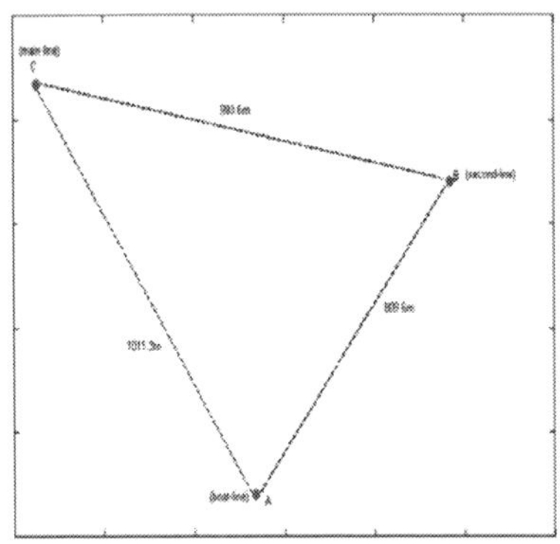

FIGURE 2. Site of three thermistor arrays

Figure 3 shows internal wave measurements obtained by array A. Though observation of internal waves lasted only 30 hours, this time covered a diurnal tide period. From figure 3, a very strong internal wave was observed. Figure 4 gives the expansion of the internal wave in figure 3. From figure 4, the internal wave was composed of some nonlinear wave packets. The first several wave packets had regular shapes, but the following wave packets' shapes were aberrant and very different from the first several wave packets. Figure 5 shows the nonlinear internal wave packet observed by array B. Figure 6 shows the nonlinear internal wave packet obtained by array C. The nonlinear internal wave packet of array B was earliest to observe and the nonlinear internal wave packet of array C was 30 minutes later than that of array B. The nonlinear internal wave packet of array A was five minutes later than that of array B.

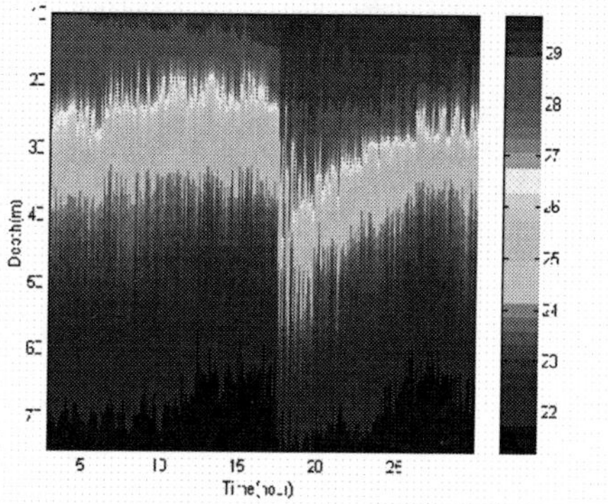

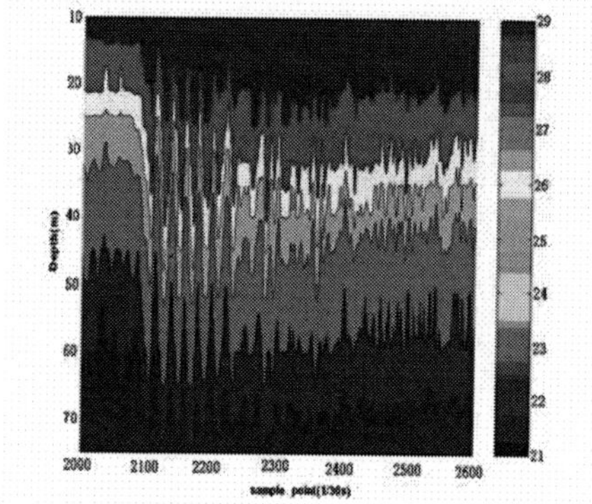

FIGURE 3. Temperature profile data measured by array A

FIGURE 4. Nonlinear internal wave packet measure by array A

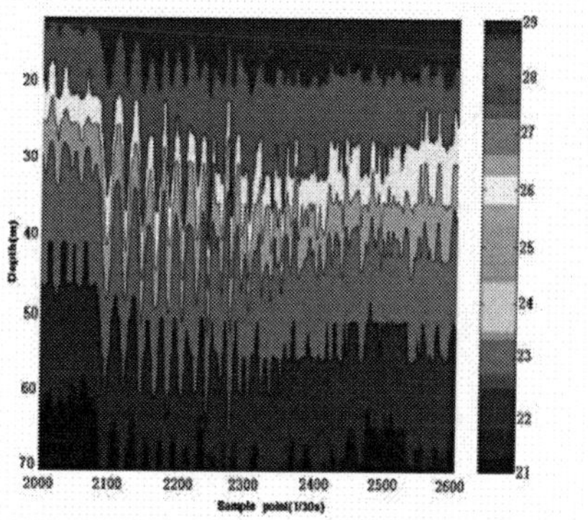

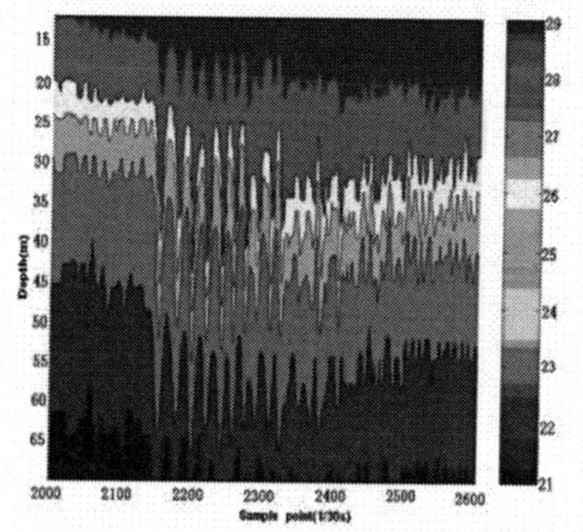

FIGURE 5. Nonlinear internal wave packet measured by array B

FIGURE 6. Nonlinear internal wave packet measured by array C

Using the time of the nonlinear internal wave's arrival at three measurement arrays, and knowledge of geometry, the nonlinear internal wave's speed was estimated. Calculation showed that the nonlinear internal wave's speed was 0.54m/s and that it propagated to the northwest. We estimated that the nonlinear internal wave packet was generated near Xi-Sha Islands.

Figure 7 shows the depth time series measured during experiment by the pressure sensor in the moored thermistor array. From the depth curves, we knew the tide information of the experiment site. The nonlinear internal wave packet appeared at the time marked by the red line. This time was low tide.

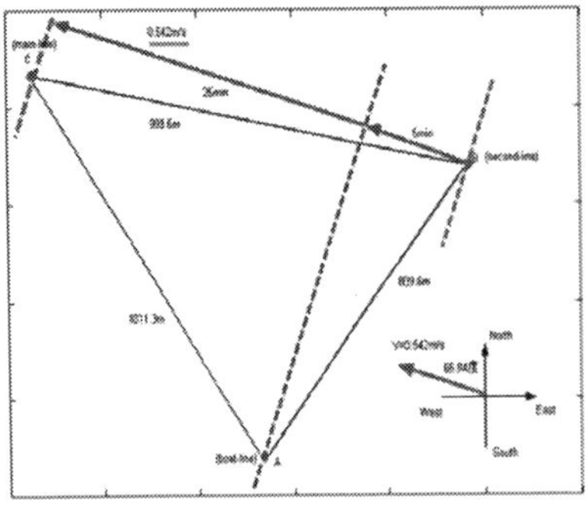

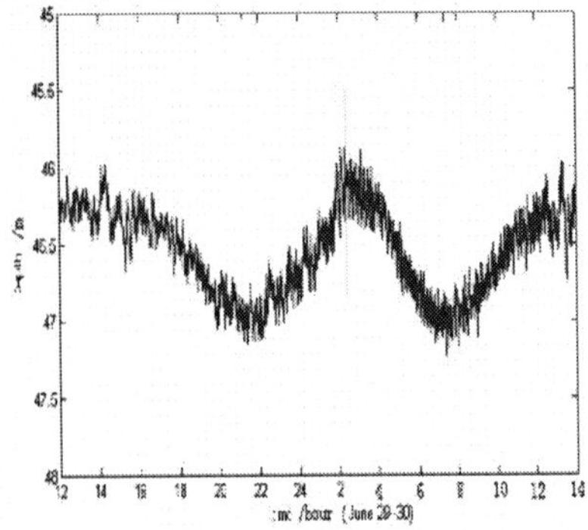

FIGURE 7. Estimating internal wave's speed using time of arrival at different observation positions

FIGURE 8. Depth data of pressure sensor in vertical thermistor array

Figure 9 shows measured temperature fluctuation of array A at 30.7m during the experiment. At the time the nonlinear internal wave packet appeared, the temperature changed strongly, and appeared periodically to fluctuate.

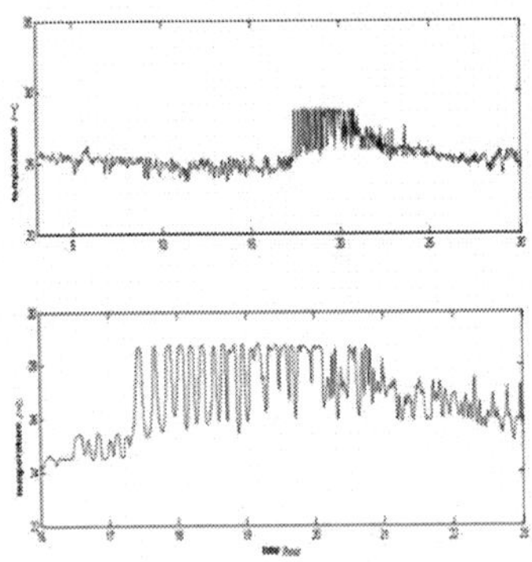

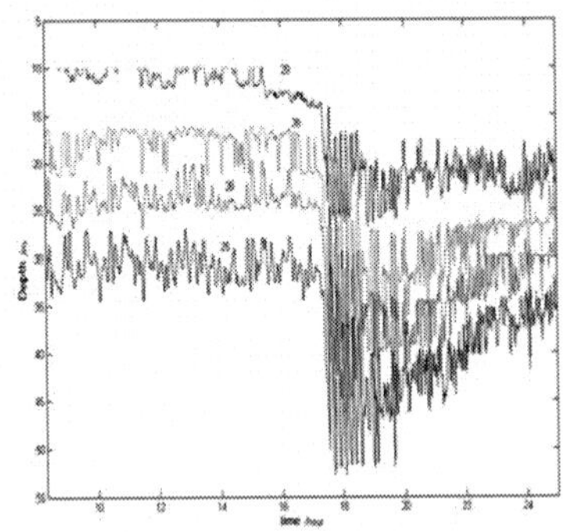

FIGURE 9. The temperature time series at 30.7m depth

FIGURE 10. Isothermal line time series measured by array A

Figure 10 gives the isothermal line of 29, 28, 26, and 25°C fluctuation. From the isothermal line, the nonlinear internal wave packet's shape was approximated. Figure11 compares the shapes of two nonlinear internal wave packet measured by array B and by array C. The nonlinear internal wave packet of array C propagated a longer distance than that of array B. The red solid curve was the internal wave of array C and the blue dashed curve was the internal wave of array B. Observing the first several wave crests, we found that the nonlinear internal wave crest interval of array C was shorter than that obtained at array B. This meant that the nonlinear internal wave crest interval would be shorter during the internal wave's propagation. Figure 12 shows that using the dnoidal model of the KdV equation simulated the isothermal line of 25 °C. The fit of the simulated curve and the real measured curve was very coherent.

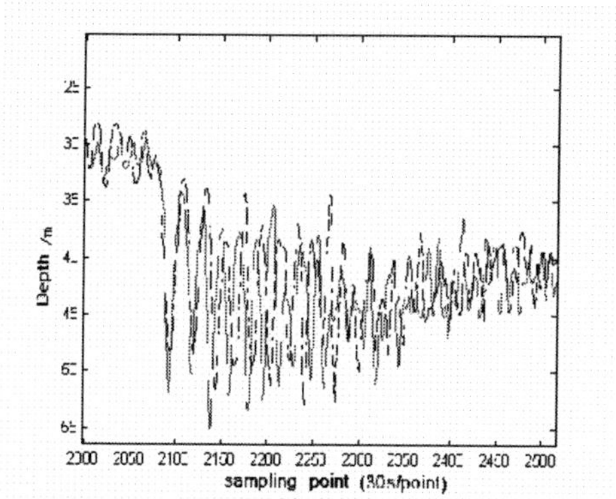

FIGURE 11. 25 °c Isotherms of array C and array B (dash line is array B, solid line is array C)

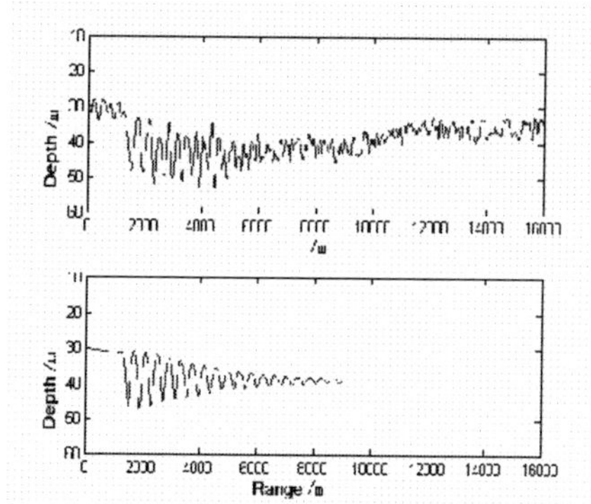

FIGURE 12. Comparison between $dn^2(x,t)$ solution and Isotherms at 25 °c of array B

II. SIMULATING NONLINEAR INTERNAL WAVE'S EFFECT ON ACOUSTIC PROPAGATION

Propagation characterized by such an exchange between modes is generally called coupled mode propagation [4]. Mode coupling by moving wave packets will probably always induce greater temporal acoustic field variability. The acoustic field variability effect of packet coupling depends strongly on modal composition, distance between packet and receiver, acoustic waveguide character, etc [4].

In this part, we simulated the nonlinear internal wave's effect on acoustic propagation. To simplify the calculation, only 4 wave crests were used to simulate the nonlinear internal wave packet, and we assumed that the waveform of the nonlinear internal wave was steady when it moved. Figure 13 shows the simulated nonlinear internal wave. Figure 14 shows the parameters of the environment; the sound source depth was 5m and the receiver depth was 80m. The water depth was 82m.

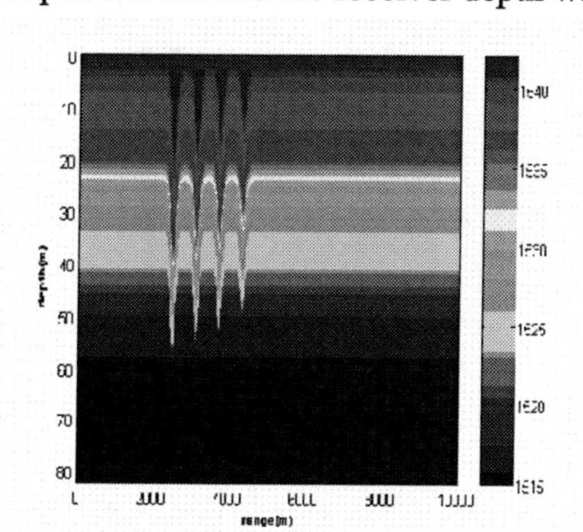

FIGURE 13. Sound speed profile vs. range with nonlinear internal wave

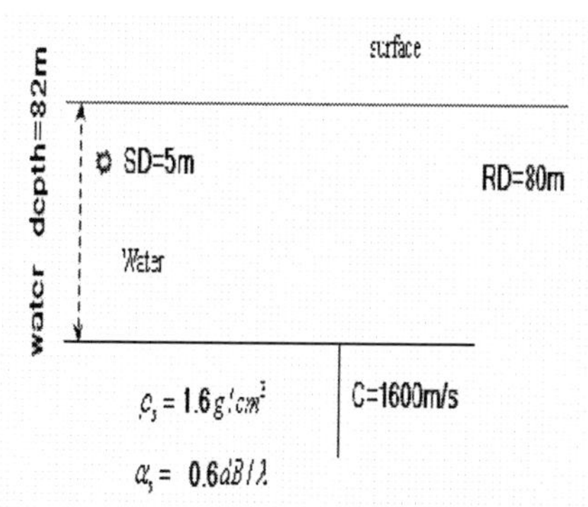

FIGURE 14. Parameters of environment

The problem was addressed by performing monochromatic 400Hz propagation simulations. A version of the COUPLE97 code was used, utilizng methods published by R.B.Evans. We calculated five cases. The first case assumed an environment of no nonlinear internal wave. The remaining cases assumed environments in which nonlinear internal waves existed and the distance between the nonlinear internal wave and the sound source were respectively 200 meters, 2100 meters, 5100 meters, and 7100 meters. Figure 15 shows the incoherent acoustic transmission loss vs. range. In figure 15, the lowest solid line was the case wherein the nonlinear internal wave didn't exist, and the upper four dashed lines were cases in which nonlinear internal waves existed. From figure 15, two points were known. First, the transmission loss with nonlinear internal waves was less than the transmission loss without nonlinear internal waves. Second, the shorter the distance between internal wave and sound source, the smaller the transmission loss was.

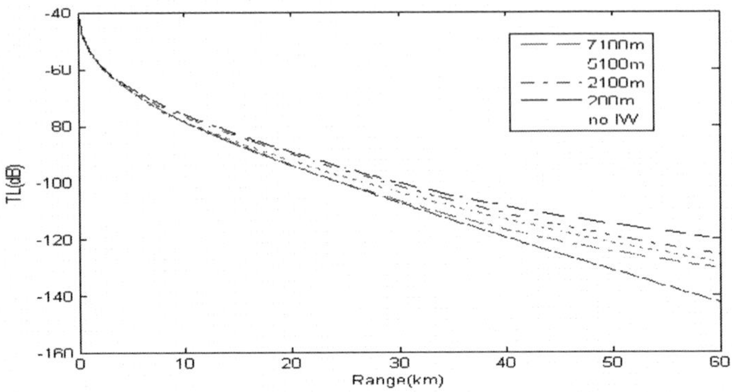

FIGURE 15. Incoherent transmission loss curve vs. range with and without nonlinear internal wave

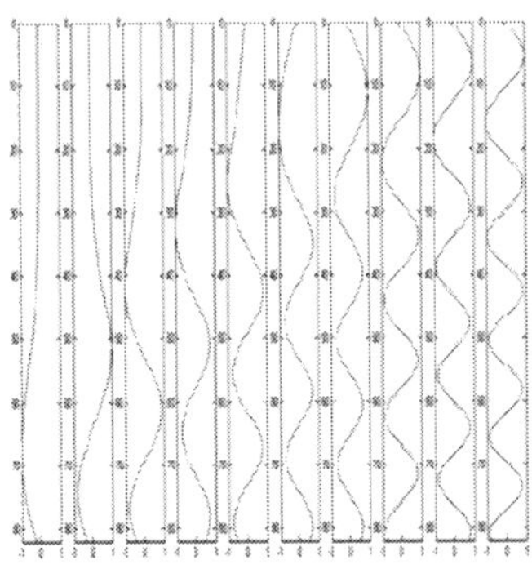

FIGURE 16. First ten mode shapes

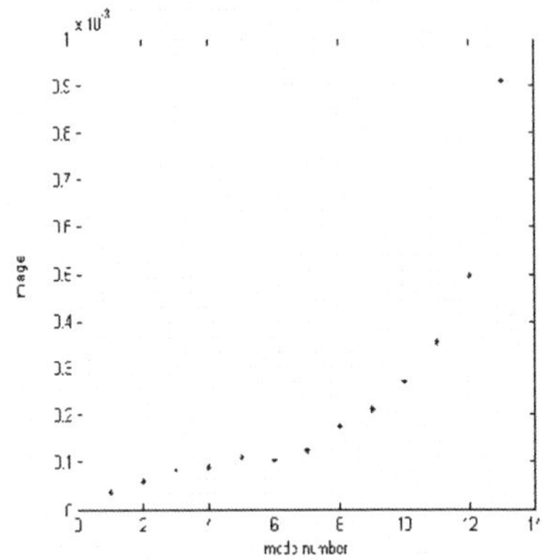

FIGURE 17. The image part of mode eigenvalues

Why did this phenomenon appear? Figure 16 shows the first ten mode shapes and figure 17 shows the image parts of mode eigenvalues. In our calculation, the sound source depth was 5meters, which was near the zero point of lower-order mode shapes and the maximum point of higher-order mode shapes, so the source excited strong higher-order modes and weak lower-order modes.

The red curve in figure 18 shows amplitude transmission loss vs. range for modes 1,

2, 3, 8, 9, and 10 in a case with a nonlinear internal wave and a distance of 200 meters the nonlinear internal wave and the sound source. The blue point curve gives the case of no nonlinear internal wave. From figure 18, the first three mode amplitudes were weak in the first 200m distance because source was shallow. When modes propagated into areas where a nonlinear internal wave existed, the first three mode amplitudes increased and the mode 8, 9 and 10 amplitudes decreased. This meant that the energy of the higher-order modes coupled into the lower-order modes. From figure 17, the higher-order modes had more transmission loss than the lower-order modes when they propagated. Nonlinear internal waves made the energy of higher-order modes coupled to lower-order modes, so in received depth of a low-order mode shape's maximum, the sound field's energy was enhanced.

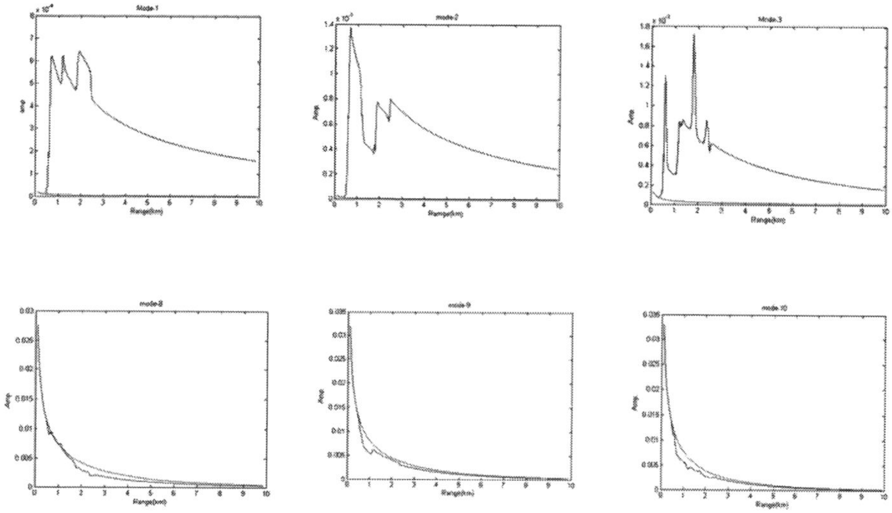

FIGURE 18. Mode amplitude vs. range with and without nonlinear internal wave

III. CONCLUSION

This Internal wave observation experiment in 2004 shows that a nonlinear internal wave packet was found, and it appeared just one time during a 30 hour experiment. Computation indicated that the nonlinear internal wave packet's velocity was 0.54m/s, and from its propagation direction we estimated that it was generated near Xi-Sha Islands. The dnoidal model of KdV(Korteweg-deVries) equation was used to simulate the waveform of the nonlinear internal wave packet. Measured data shows that the crest interval of the nonlinear internal wave packet would be shorter when it propagated. We simulated the nonlinear internal wave packet's effect on sound propagation and mode coupling of sound propagation. We found that the nonlinear wave packet made the energy of higher-order modes couple to lower-order modes, so in received depth of low-order mode shape's maximum, the sound field's energy was enhanced.

ACKNOWLEDGMENTS

This work was supported by National Natural Science Fund (No.10874199) and Innovative Fund of Chinese Academe of Science (No.CXJJ-09-S11).

REFERENCES

1. J.R.Apel, J.R.Holbrook, A.k.Liu, and J.Tsai, "The Sulu Sea internal soliton experiment," J.phys. Oceanogr., Vol.15, 1985, pp.1625-1651.
2. Timothy, Dude E, James E ,Lynch, et al. "Internal tide and nonlinear internal wave behavior at the continental slop in the northern south china sea". IEEE J. Oceanic Eng. Vol.29, NO.4, 2004, pp.1105-1129.
3. Antony K.Liu, Stevrn R. Ramp, et al. "A case study of internal solitary wave propagation during ASIAEX 2001". IEEE J. Oceanic Eng, Vol.29, No.4, 2004, pp.1144-1156.
4. Timothy F.Duda, "Acoustic mode coupling by nonlinear internal wave packets in a shelf break form area," IEEE J. Oceanic Eng, Vol.29, No.1. 2004, pp.118-215.

Effect of tidal internal wave fields on shallow water acoustic propagation

Ju Lin, Huan Wang, and Junping Sun

College of Information Science and Technology, Ocean University of China, Qingdao 266100, P. R. China

Abstract. Internal waves are one of the most pronounced oceanic phenomenons to the oceanographer. During past decades much effort has been made to investigate the effect of internal waves on shallow water acoustic propagation. Even though many field observations, such as SWARM '95, have provided fruitful information about the relation between internal waves and acoustic propagation, it is necessary to conduct more numerical simulations due to their extensive feasibility. In this study, the shallow water internal wave environment is constructed by using a non-hydrostatic ocean model, the open boundary forcing is set by considering single or several internal wave modes at the M2 tidal frequency. In order to show the mode coupling caused by the internal wave field more clearly, the acoustic starting field with different single normal modes is adopted. The acoustic simulation can be used to check whether a specific combination of internal wave modes is related to the mode coupling, and which mode pair will be affected. The combination of internal wave modes can be separated into several groups. Even though the internal wave fields are different among every case in each group, the acoustic field structure and the mode coupling are similar. Different acoustic normal mode coupling occurs due to the different combinations of internal wave mode forcing. When the parameters of internal wave mode are modified gently, the acoustic mode coupling becomes quite different. It is interesting and important to investigate the sensitivity of acoustic fields to the variability of the internal mode combination.

Keywords: Internal wave, acoustic propagation, non-hydrostatic ocean model.
PACS: 92.10.hj, 43.30.Bp, 92.05.Bc.

INTRODUCTION

Internal waves (IWs) often occur at interfaces between water layers of different density created by tidal currents, surface waves, currents over abrupt topography, or wind blowing, etc[1]. The IWs play an important role in water mixing and sediment transport. Internal solitary waves (ISW), a kind of nonlinear phenomenon, have been observed in the relatively shallow seas such as the Yellow Sea, the Luzon Strait at the boundary of the Pacific and the South China Sea.

Zhou et al.[2] reported the anomalous, frequency-dependent transmission loss of sound due to strong internal solitons. Many ocean acoustic transmission experiments have shown that travelling sound can be greatly damped by shallow water internal waves. The SWARM '95 experiment that was conducted off the New Jersey coast in 1995[3] showed an intensity fluctuation of about 7 dB. Rubenstein[4] showed relatively small sound variability from experimental results obtained in the Gulf of Mexico, but that variability still had a strong correlation with internal solitons.

CP1272, *Shallow-Water Acoustics, Second International Shallow-Water Acoustics Conference*
edited by J. Simmen, E. S. Livingston, J. X. Zhou, and F. H. Li

In addition to fruitful field experiments, many researchers have focused on the simulation study of acoustic scattering by internal waves. By considering a simple soliton model, Zhou et al[2] concluded that the anomalous transmission loss was due to strong acoustic mode coupling caused by an interaction between acoustic waves and internal waves (Resonant interaction). Lamb[5] and Finette et al.[6] constructed the sound speed profile by applying the non-hydrostatic and hydrostatic ocean models, respectively. The tidal forcing is set up along the open boundary and the temperature oscillation is simulated by considering the near-realistic topography. But the internal wave can be decomposed into many modes, and the first mode is normally dominant in the ocean. The sound speed field is perturbed quite differently by different internal wave modes. It is important and interesting to study the effect of the internal wave mode on the fluctuation of sound transmission. And furthermore it is useful to understand the acoustic normal mode coupling or sound intensity variability caused by IWs.

The purpose of this paper is to reconstruct the sound speed structure by using a two-dimensional non-hydrostatic ocean model, in which the different internal wave modes are combined as open boundary forcing, and to investigate the corresponding variability of single acoustic normal mode propagation. The arrangement of this paper is as follows: We describe a non-hydrodynamic ocean model and the setup of an open boundary condition. The simulation results are presented to show the effect of different internal wave modes on sound transmission. We conclude with a summary and discussion.

OCEAN MODEL

During the past several decades many numerical models have been developed to simulate the ocean dynamistic phenomenon, such as POM[7], ROMS[8], etc. These are hydrodynamic models, which often traditionally make the hydrostatic approximation and neglect non-hydrostatic pressure and subsequently, vertical momentum. Hydrostatic aproximations can not be made at open boundaries[9] and steep slopes[10], where strong vertical velocities occur. Unlike linear waves, the evolution of nonlinear internal waves is inherently related to non-hydrodynamic pressure. In order to model the development and propagation of internal waves well, non-hydrostatic pressure must be considered in a numerical model.

Marshall et al.[11,12] developed a non-hydrostatic model which is a z-coordinate finite-volume model used for studying small- to basin-scale processes in the ocean and atmosphere. Legg[13] applied this model to investigate the internal wave reflection from concave and convex slopes, and Legg[14] examined the generation of internal waves and dissipation by tidal flow over different topographies. In this paper, such a non-hydrostatic model is adopted to simulate the propagation and evolution of internal waves in the coastal ocean, and the results are used to construct the sound speed field. The simple description of the incompressible Boussinesq equations written in the z-coordinate mode of the model is given as follows:

The momentum equation with Boussinesq assumption is

$$\frac{D\vec{V}_h}{Dt} + f\vec{k} \times \vec{V}_h + \frac{1}{\rho}\nabla_z p = \vec{F} \tag{1}$$

$$\varepsilon_{nh}\frac{Dw}{Dt} + g + \frac{1}{\rho}\frac{\partial p}{\partial z} = \varepsilon_{nh}F_w \tag{2}$$

The continuity equation for an incompressible fluid is

$$\frac{1}{\rho}\frac{D\rho}{Dt} + \nabla_z \cdot \vec{V}_h + \frac{\partial w}{\partial z} = 0 \tag{3}$$

where the three-dimensional velocity vector is $\vec{V} = (\vec{V}_h, w) = (u, v, w)$, $\vec{V}_h = (u, v)$ are the two-dimensional horizontal velocity components, w is the vertical velocity component. ε_{nh} is the non-hydrostatic parameter. f is the Coriolis parameter, p is pressure, ρ is in-situ density, g is the constant gravitational acceleration, and $\vec{F}, F_w$ are corresponding forcing fields.

In the model the spatial discretization is carried out using the finite volume method, which treats the irregular topography by using orthogonal curvilinear grids and shaved cells. The 2-D elliptic equation is solved and the hydrostatic pressure is obtained through integration of the hydrostatic relation vertically. The non-hydrostatic pressure is calculated by introducing its perturbation to the 3-D elliptic equation. The model treats vertical diffusion and viscosity implicitly in time using a backward method.

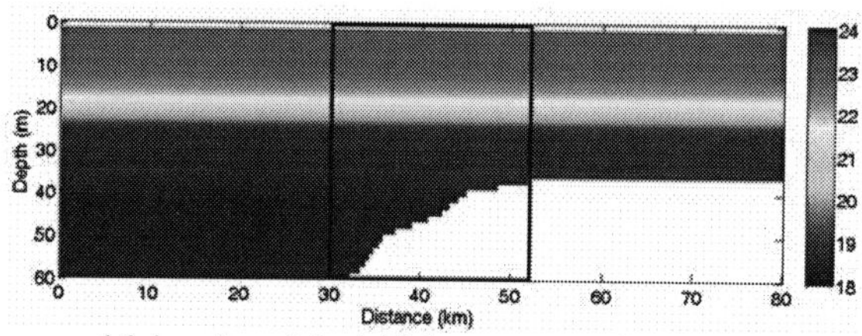

FIGURE 1. The ocean model domain and the initial temperature distribution. The rectangular area is the acoustic computation domain.

In this preliminary study the 2-D non-hydrostatic simulations are operated over a coastal sea situation with a range of 80 km. The topography, initial temperature and salinity field in the simulation come from the experiment of AEYSFI 2005 (Fig. 1). In the domain there is a flat bottom 60 m deep between 0-32 km, a sloping region exists between 32-51 km, and a plateau extends to the right at a depth of 35 m. The initial salinity field is set to 35 psu homogeneously over the whole horizontal and vertical field. The stratification is dependent on the temperature with the same profile in the horizontal direction, and only modified by the depth variation. A thin thermocline appears between 15-20 m water depth. The horizontal grid resolution is set non-uniformly between 25-200 m according to the variable topography, and higher resolution is adopted along the sloping region. In the vertical direction the domain is divided into 40 equally spaced levels with 1.5m thickness. At the bottom boundary, the no-slip condition is applied for the velocity field, i.e. the velocity parallel to the topography is set to zero. The no-stress condition is applied at the surface, and the free surface condition is chosen.

In order to effectively investigate the relation between IWs and sound propagation, only an internal wave forcing is set at the offshore boundary. The velocities and buoyancy anomalies are calculated according to the internal wave equations. The various combinations of modes 1-4 are considered in the forcing form (Table 1). The forcing frequency is set to the M2 tidal frequency. The Orlanski radiation condition is applied at the open boundary to avoid the internal wave being re-reflected back into the domain.

Table 1. Configuration of internal wave forcing at the offshore open boundary

Amplitude of velocity (m/s)		Mode 1	Mode 2	Mode 3	Mode 4
Group 1	Case 1	0.08	0.08	0.0	0.0
	Case 2	0.08	0.05	0.03	0.02
	Case 3	0.08	0.05	0.03	0.0
	Case 4	0.08	0.08	0.08	0.0
	Case 5	0.08	0.0	0.08	0.08
Group 2	Case 6	0.08	0.08	0.0	0.08
	Case 7	0.08	0.08	0.08	0.08
Group 3	Case 8	0.08	0.0	0.0	0.0

In the whole ocean model domain, the sloping subset region with a range of 22 km is chosen to be the acoustic computation domain. The evolution and the transition of internal waves is very clear in that region, therefore the acoustic normal mode coupling is expected to be stronger. The sound speed field is calculated by using the simulation results of the ocean model.

SIMULATION RESULTS

Evolution of Internal Wave Fields

In order to simulate the progress of internal waves effectively, the mean value of temperature is removed from the vertical profile. The time step is set to be 5 s to guarantee stability. To simulate the evolution of IWs, eight numerical experiment cases are conducted by applying different forcing. Each case is run for a spin-up of three days, and then the steady simulation results of the fourth day are chosen as input data for acoustic computation.

Figure 2 shows four temperature field snapshots of internal wave propagation over the slope region for Case 8, where only mode 1 is considered as open boundary forcing. As the internal wave propagates shoreward, the wave form is changed and wave properties are transformed. During the propagation, an internal solitary wave (ISW) is generated. The shallower the water is, the stronger the ISW is. The waveform of the ISW is distorted by the shoaling effect, which is caused by the varying topography. The amplitude of the ISW becomes high and the wavelength becomes short. As the energy existing in the leading wave is dissipated and the trailing oscillations are generated, the number of ISWs in the wave packet is increased. Another phenomenon is that the ISW polarity is reversed near the range of 43 km, where the upper layer depth and lower layer depth are almost equal; The ISW is reversed from a depressed type to an elevation type.

When the other internal wave modes are included at the open boundary, the simulated temperature structure is similar, but the temperature variability becomes strong, the number of ISWs in the wave chain is increased, and the point at which the ISW polarity is reversed is shifted seaward.

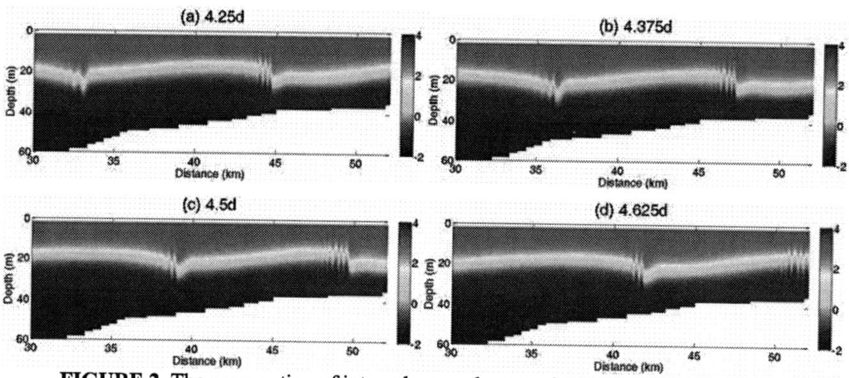

FIGURE 2. The propagation of internal wave shoreward with waveform evolution at time (a) 4.25d, (b) 4.375d, (c) 4.5d and (d) 4.625d (Case 8).

Variability of Acoustic Field

The sound speed field is constructed by deriving the simulation result of a non-hydrostatic ocean model, and a parabolic equation (PE) acoustic transmission model For3D is adopted to simulate the acoustic field under the aforementioned ocean environment, after assuming the temperature field is 'frozen' during sound transmission. The source frequency is set between 300-900 Hz, and for simplicity the source is located at depth 30 m, where it is under the thermocline. A single-mode starting field is specified in the PE model. Figure 3 shows the spatial-temporal variability of transmission loss for Case 8 (the 300 Hz mode 2 starting field), and the time dependence of acoustic propagation is clearly seen. The structures of the acoustic field are related to the source frequency, source depth and internal wave structure, especially the position of the ISW group where the acoustic normal mode coupling occurs strongly. In order to check the acoustic field variability more clearly, mode filtering is performed for the whole computation domain at a given time.

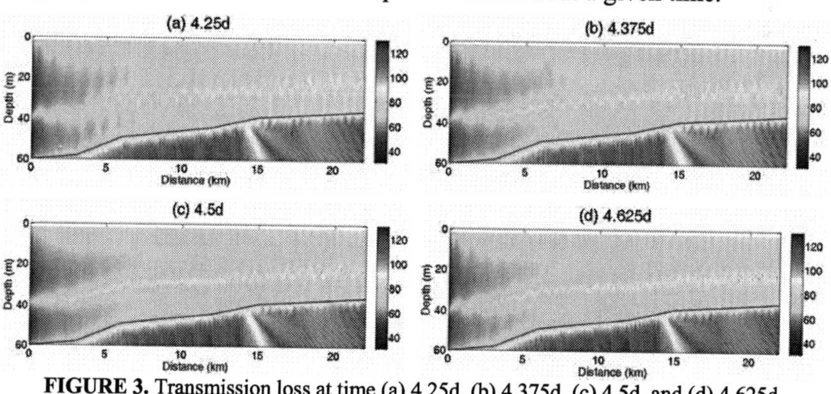

FIGURE 3. Transmission loss at time (a) 4.25d, (b) 4.375d, (c) 4.5d, and (d) 4.625d by specifying a 300Hz mode 2 starting field (Case 8).

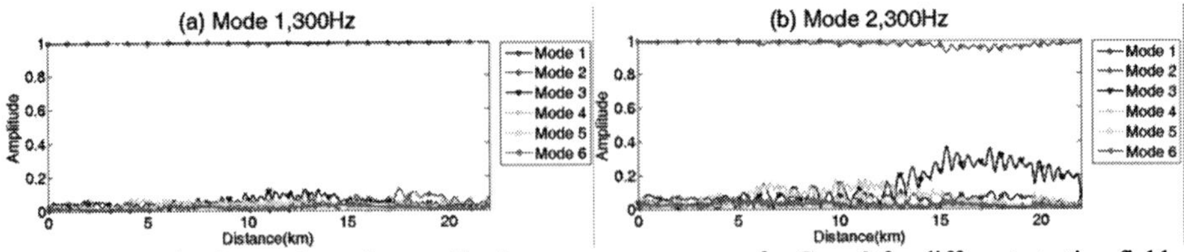

FIGURE 4. The former 6 modes amplitude versus range curves for Case 0 for different starting field, (a) mode 1, (b) mode 2. Source frequency is 300Hz.

The initial temperature and salinity field is also applied to simulate the acoustic transmission (Case 0), which can be considered as the original field. The normalized mode amplitude curves are shown in Fig. 4. When the mode 1 starting field is specified, the mode coupling is very weak and the amplitude of mode 1 remains fairly constant. The weak coupling occurs for mode pairs 1/3 at a range of 10 km and for mode pairs 1/2 at a range of 18 km. When the starting field is composed of mode 2, the coupling becomes a little stronger for mode pairs 2/3 at the 12 km range.

For Case 0, the normal mode coupling occurs only due to the stratification caused by the varying topography. If a different internal mode combination is set as external forcing, the couplings should become quite different. The corresponding simulations are shown in Fig. 5 and 6.

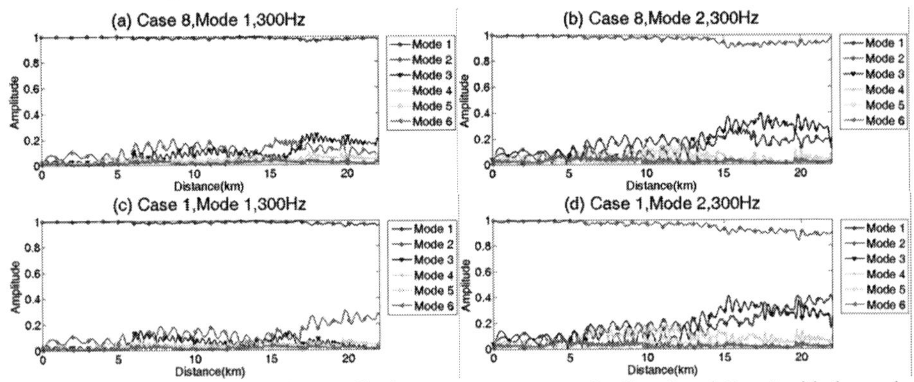

FIGURE 5. The former six modes amplitude versus range curves for Case 8 and Case 1 with the mode 1 (left panels) and the mode 2 (right panels) starting field at time 4.375d. Source frequency is 300Hz.

After comparing the simulation for different situations, it is found that the temperature structures among some cases are similar qualitatively, even though the combination of IW mode is quite different, therefore all eight cases can be separated into three groups (see Table 1 for details). Only one representative case is chosen from each group for comparison. Figure 5 shows the distribution of the mode amplitudes. For both Group 1 and 3, the amplitudes of original mode 1 or mode 2 remain almost constant, and are decreased a little as the depth becomes less. When acoustic signals reach the first ISW packet at a range of about 6 km, the different mode coupling situation occurs. While it is the mode 1 starting field, the mode coupling occurs for mode pairs 1/2 for both groups, but mode 3 is also stimulated for Group 3. When acoustic signals pass through the second ISW packet, after mode coupling the amplitude of mode 2 is increased gradually for Group 3. For the mode 2 starting field, the coupling structure is similar for both groups. The energy is transferred from mode 2 to mode 1 at the first ISW trains area, and mode 3 is stimulated at the second ISW packet. The amplitude of modes 1-3 are always antiphasic.

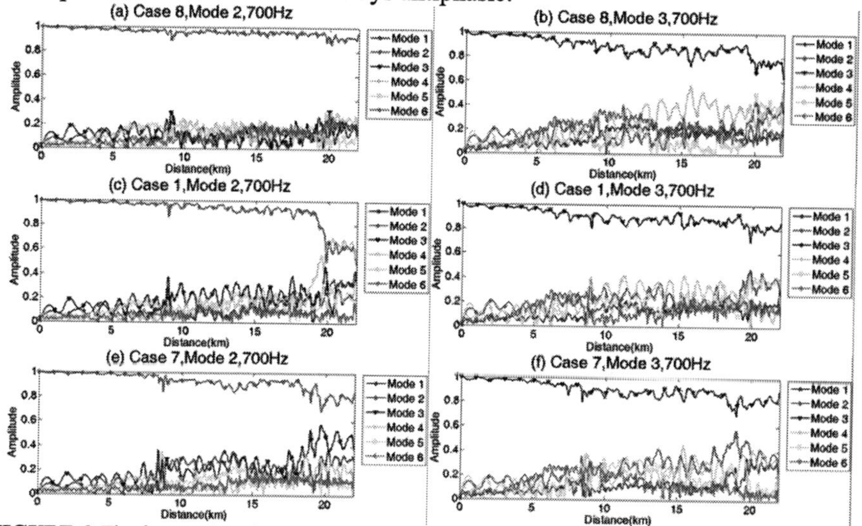

FIGURE 6. The former 6 modes amplitude versus range curves for three cases with the mode 2 (left panels) and the mode 3 (right panels) starting field at time 4.0d. Source frequency is 700Hz.

When an internal wave is propagated shoreward continually, its waveform is changed along with the corresponding water depth; of course, the mode coupling is also altered (Figure 6). Whether the starting field is mode 2 or mode 3, the original mode amplitude is decreased with distance from the source, but is still dominant. In Fig. 6, the most dramatic phenomenon is the transparent resonance[15] and the mode coupling is only limited among the first ISW packet area. In the first half of the coupling area, the energy is transferred from mode 2 (all three groups) to mode 3 (Group 2 and Group 3) or mode 3 and mode 4 (Group 1); in the second half of the area, the direction of energy transfer is reversed. When acoustic signals just transmit outside the ISW group area, the corresponding mode energy is recovered. While among the second ISW packet area, the transparent resonance is not obvious, but the strong multi-mode coupling occurs. As for Group 3 and the mode 3 starting field, the acoustic energies are exchanged among modes 2-5.

From the above simulation results, in general it can be seen that strong mode coupling takes place in the second ISW packet region. When compared with those in the first ISW group area, the amplitude of the second ISW packet is large, the wavelength is short, and the horizontal gradient of sound speed profile becomes large. The higher the source frequency is, the more complicated mode coupling is, due to the fact that more normal modes are generated for high frequency signals.

SUMMARY AND DISCUSSION

When different mode combinations of internal wave forcing are set at open boundary (as listed in Table 1), two ISW packets exist among the sloping area. The first one is the depressive type, the second one is the elevation type. The number of ISWs is increased in the second group, i.e. the energy existing in the leading wave is dissipated and the trailing oscillations are generated.

Under such environments, a single-mode starting field is applied in the PE model, and the acoustic propagation results can be separated into three groups. Within the same group, even though the combinations of internal wave modes are quite different, the structures of the acoustic field are almost the same qualitatively. Among different groups, there is not a large discrepancy in the sound speed field, but the acoustic normal mode coupling is quite different. When the combination of internal wave modes is the same, but the amplitude of each mode velocity changes gently, the acoustic field variability is changed. The coupling intensity and mode pair are dependent on the internal wave group and acoustic source frequency, and of course also dependent on the source depth (this issue is not discussed in the present study). The 'transparent resonance' phenomenon exists, especially for higher source frequencies and higher single mode starting fields.

The relation between the internal wave mode and acoustic normal mode coupling is as yet inadequately understood, and more theory and numerical simulations are needed.

ACKNOWLEDGMENTS

This work is supported by the National High Technology Research and Development Program of China (Grant No. 2006AA09Z114), the Ministry of Science and Technology of China (National Key Program for Developing Basic Science Grant No. 2007CB411803 and No. 2005CB422307) and The Scientific Research Foundation for the Returned Overseas Chinese Scholars, State Education Ministry of China (Acoustic Monitoring Technology of Shallow Water Dynamic Environment)

REFERENCES

1. J. R Apel, L. A. Ostrovsky, Y. A Stepanyants, J. F Lynch, *J. Acoust. Soc. Am* 121(2), 695-722 (2007).
2. J. Zhou, X. Zhang, and P. H. Rogers, *J. Acoust. Soc. Am.* **90**, 2042–2054 (1991).
3. M. Badiey, Y. Mu, J. F. Lynch, J. R. Apel, and S. N. Wolf, *IEEE J. Ocean. Eng.* **27**, 117–129 (2002).
4. D. Rubenstein, *IEEE J. Ocean. Eng.* **24**, 346–357 (1999).
5. K. Lamb, *J. Geophys. Res.(Ocean)* **99**, 843–864 (1994).
6. Finette S, Oba R, Shen C, and Evans T, *J. Acoust. Soc. Am.* **121(5)**, 2575-2590 (2007).
7. A.F., Blumberg and G.L. Mellor, "A description of a three-dimensional coastal ocean circulation model," in Three-Dimensional Coastal Ocean Models, edited by N. S. Heaps, American Geophysical Union, Washington D.C. 1987, pp. 1-16.
8. D.B. Haidvogel, H.G. Arango, K. Hedstrom, A. Beckmann, P. Malanotte-Rizzoli and A.F. Shchepetkin, *Dynamics of Atmospheres and Oceans* **32**, 239–281 (2000).
9. A. Mahadevan, J. Oliger and R. L. Street, *J. Phys. Oceanogr.* **26**, 1868-1880 (1996).
10. D. A. Horn, J. Imberger, G. N. Ivey, and L. G. Redekopp, *J. Fluid Mech.* **467**, 269-287 (2002).
11. J. Marshall, C. Hill, L. Perelman, A. Adcroft, *J. Geophys. Res.(Ocean)* **102**, 5733–5752 (1997).
12. J. Marshall, A. Adcroft, C. Hill, L. Perelmam, C. Heisey , *J. Geophys. Res.(Ocean)* **102**, 5753–5766 (1997).
13. S. Legg and A. Adcroft, *J. Phys. Oceanogr.* **33**, 2224-2246 (2003).
14. S. Legg, A. Adcroft and KMH. Huijts, *Deep Sea Research, Part II-Topical Studies in Oceanography* **53**, 140-156 (2006).
15. J. Preiseg and T. F. Duda, *IEEE J. Ocean. Eng.* **22**, 256–269 (1997).

Sound Velocity Fluctuation due to Linear Internal Waves in North Yellow Sea

Kun Liu and Shihong Zhou

Hangzhou Applied Acoustic Research Institute, Hangzhou, China

Abstract. The data measured by a thermal chain array covering depths of 4 to 46.6 meters in the North of the Yellow Sea in September 2008 is analyzed in this paper. Results show that the water particle vertical displacement of 15 meters occurs due to internal waves. The vertical displacement power spectrum of isotherm curves is conducted and the spectrum attenuation coefficients are between -1.93 and -2.19. The mean spectrum attenuation coefficient is -2.05, which is consistent with the G-M internal wave spectrum with power law of -2. With these, the sound velocity fluctuation $\delta c / c$ caused by internal waves is predicted at different ranges, depths and times respectively. The results show that mean $\delta c / c$ is about $10^{-4}$and varies strongly within the thermocline layer at the depth of 20-35 meters. It will be useful for deterministic modeling and stochastic modeling in a random sea.

Keywords: Internal wave. Sound velocity fluctuation.

THEORY OF INTERNAL WAVE

An internal wave is a gravity wave in the sea; its amplitude in deep water is higher than that of surface. It has a great influence on sound transmission and is also a random process of temporal and spatial variability that is further complicated in shallow water.

The movement of internal waves is conducted by inertial frequency ω_i and Brunt-Väisälä frequency $N(z)$. It can be proved that the internal wave frequency ω must be between ω_i and $N(z)$. In the sea, the internal wave which can be propagated satisfies: $\omega_i < \omega < N(z)$.

The vertical velocity of an internal wave can be expressed as a linear superposition of the modes:

$$w_{jk}(x, y, z, t) = W_{jk}(z) \exp[i(kr - \omega_{jk} t)] \tag{1}$$

Where k is the horizontal wave number, j is the mode number. $W(z)$ is the amplitude of the internal wave at a depth of z meter and satisfies [1] [2]:

$$\frac{\partial^2 W_{jk}}{\partial z^2} + \frac{N^2(z) - \omega_{jk}^{\ 2}}{\omega_{jk}^{\ 2} - \omega_i^2} k^2 W_{jk} = 0 \tag{2}$$

Boundary conditions: $W(z) = 0$ at surface and bottom.

Where $\omega_i = (2cycles/day)\sin(latitude)$, the latitude of experiment is 38°31.4124'N, so ω_i=0.0519cph, Brunt-Väisälä frequency is:

CP1272, *Shallow-Water Acoustics, Second International Shallow-Water Acoustics Conference*
edited by J. Simmen, E. S. Livingston, J. X. Zhou, and F. H. Li

$$N^2(z) = -\frac{g}{\rho(z)}\frac{\partial \rho(z)}{\partial z} \tag{3}$$

and expression (3) satisfies: $\int_0^H \left[N^2(z)-\omega_i^2\right]W^2(z)dz = 1$.

The density gradient $\partial\rho(z)/\partial z$ is controlled by temperature gradient $\partial T(z)/\partial z$ and salt gradient $\partial S(z)/\partial z$:

$$\frac{1}{\rho(z)}\frac{\partial \rho(z)}{\partial z} = -a\frac{\partial T(z)}{\partial z} + b\frac{\partial S(z)}{\partial z} \tag{4}$$

where a=0.13 × 10^{-3}(1/℃) and b=0.80 × 10^{-3}(1/‰) are temperature expansive coefficient and salt compressive coefficient respectively. In shallow water, temperature gradient has much more contribution than salt gradient. Therefore equation (4) can be written as: $N^2(z) = agdT(z)/dz$.

The relation between vertical displacement ξ and the vertical velocity of an internal wave is: $W = d\xi/dt$ .So the vertical displacement of an internal wave also can be written as [3]:

$$\xi(x,y,z,t) = \mathrm{Re}\{\sum_j \int dk_1 \int dk_2 A_j(k_1,k_2)W_{jk}(z)\exp[i(k_1x+k_2y-\omega_{jk}t)]\}$$
$$k^2 = k_1^2 + k_2^2 \tag{5}$$

Where $A_j(k)$ is an independent complex Gaussian random variable, satisfies:

$$\begin{gathered} \langle A_j(k_1,k_2)\rangle = 0 \\ \langle A_j(k_1,k_2)A_j^*(k_1',k_2')\rangle = P_j(k_1,k_2)\delta_{jj'}\delta(k_1-k_1')\delta(k_2-k_2') \end{gathered} \tag{6}$$

and where $P_j(k_1,k_2)$ is the internal wave spectrum.

CALCULATION OF SOUND FLUCTUATION IN SEA

From equation (5), at a certain horizontal position, the water particle vertical displacement due to an internal wave is:

$$\xi(z,t) = \frac{1}{2\pi}\int_{\omega_i}^{N_{max}} \sum_j A_j(\omega)W_j(\omega,z)\exp(-i\omega t)d\omega \tag{7}$$

In fact, the normal mode function of an internal wave varies slowly with frequency, and equation (7) can be simplified as:

$$\xi(z,t) \approx \sum_j \left[\int A_j(\omega)\exp(-i\omega t)d\omega\right]W_j(z) = \sum_j \alpha_j(t)W_j(z) \tag{8}$$

Or be rewritten as matrix form [4]:

$$\boldsymbol{\xi} = \mathbf{W}\,\mathbf{a} + \mathbf{n} \tag{9}$$

Where $\mathbf{W}_{ij} = W_j(z_i)$, $\boldsymbol{\xi}$ is the water particle vertical displacement matrix at different times calculated by isotherm at different depth, $\mathbf{n}$ is the measurement error. By taking the inverse of the matrix in equation (9), the normal mode coefficient $\alpha_j(t)$ of an internal wave at different times can be obtained, and the internal wave spectrum estimation $\left|A_j(\omega)\right|^2$ can also be analyzed and obtained.

The water particle vertical displacement $\xi(r,z,t)$ can be calculated from equation (5). According to a theory of W. H. Munk, the sound velocity fluctuation expression is about:

$$\delta c(r,z,t) = -\bar{c}(z)N^2(z)\frac{\eta}{g}\xi(r,z,t) \tag{10}$$

Where η and g are, respectively, 24.5 and gravity acceleration.

RESULT OF SEA EXPERIMENTAL DATA

An internal wave observation experiment was performed in the North of the Yellow Sea in September 2008. In the experiment a thermal chain array is used which covers a depth of 4-46.6 meters and the sample interval is 1 minute. The temperature variability versus depth and time is shown in FIGURE 1. The thermocline layer is at the depth of 20-30 meters. FIGURE 2 provides the Brunt-Väisälä frequency $N(z)$ variability versus depth and time, and FIGURE 3 shows the average variability of $N(z)$. The variability of isotherms of 14℃, 15℃, 16℃, 17℃, 18℃ and 19℃ versus time is provided in FIGURE 4, the result shows that the Maximum of vertical displacement of isotherm curves is about 15 meters due to internal waves.

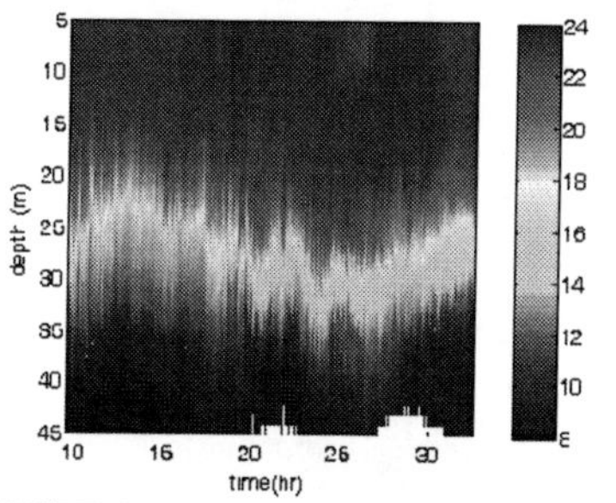

FIGURE 1. Temperature variability versus depth and time.

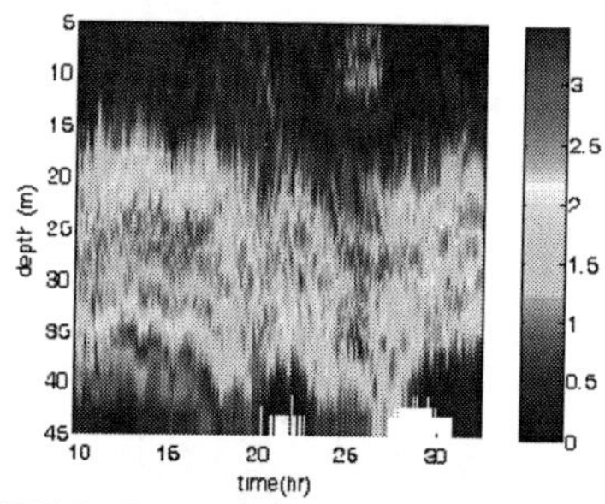

FIGURE 2. Brunt- Väisälä frequency variability versus depth and time.

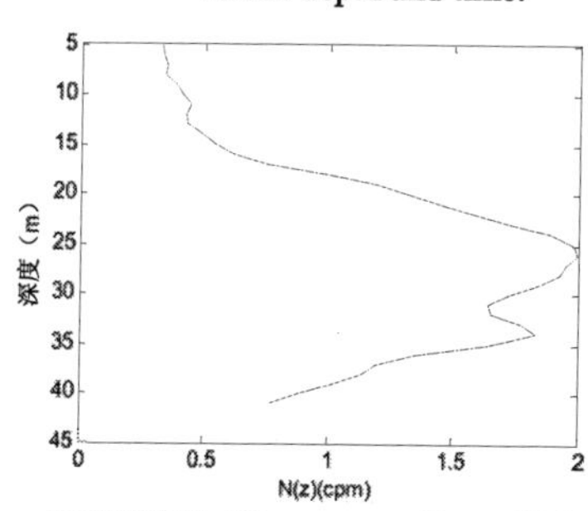

FIGURE 3. The average Brunt-Väisälä frequency variability versus depth

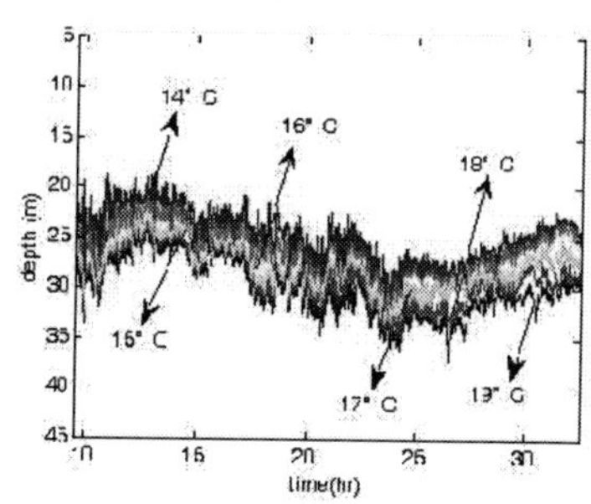

FIGURE 4. The variability of isotherms of 14℃, 15℃, 16℃, 17℃, 18℃ and 19℃ versus depth.

The water vertical displacement power spectrum of isotherm curves in FIGURE 4 is depicted in FIGURE 5. It shows the spectrum attenuation coefficients are between -1.94 and -2.19, and the mean spectrum attenuation coefficient is -2.05, which is consistent with the G-M internal wave spectrum with power law of -2. So the estimation expression of internal wave power spectrum can be written as:

$$\left|A_j(\omega)\right|^2 \approx 0.85\omega^{-2}(j^2+1)^{-2} \tag{11}$$

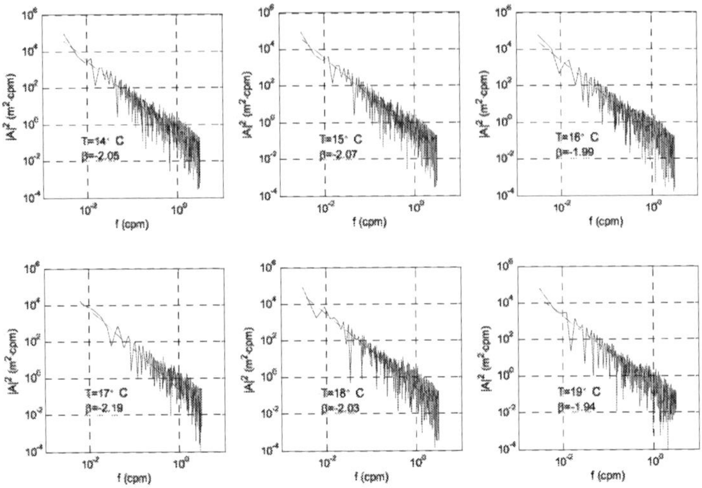

FIGURE 5. The vertical displacement power spectrum of isotherm curves.

According to equations (8), (10) and (11), the sound velocity fluctuation at different depths, ranges and times is predicted. FIGURE 6 shows variability of $\delta c/c$ versus range and depth at different times, the result shows that the mean $\delta c/c$ is about 10^{-4} and varies strongly at the depth of 20-35 meters thermocline layer.

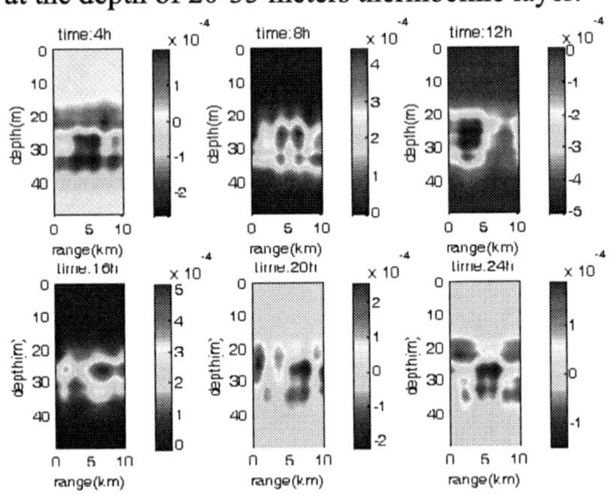

FIGURE 6. Variability of $\delta c/c$ versus range and depth at different times

With the result of $\delta c/c$ at different ranges, depths and times, the deterministic modeling and stochastic modeling in a random sea can be studied in future work.

CONCLUSION

Within this paper, the sound velocity fluctuation due to internal waves is studied with linear internal wave theory. The sound velocity fluctuation $\delta c / c$ at different ranges, depths and times are obtained from the data of an experiment in the North of the Yellow Sea in September 2008. It is vital to study deterministic modeling and stochastic modeling in a random sea in the future.

REFERENCES

1. S. Flatte, W. Munk et al., Sound transmission through a fluctuating ocean. Cambridge: Cambridge Univ. Press, 1979.
2. W. H. Munk and F. Zachariasen, Sound propagation through a fluctuating stratified ocean: Theory and observation, J. Acoust. Soc. Am, 59(4), 818-838, 1976.
3. Dozier L B, Tappert FD, Statistics of normal mode amplitudes in a random ocean: I theory, J. Acoust. Soc. Am, 63(2), 352-365, 1978.
4. Yan Jin, Zhang Renhe, Zhou Shihong et al, Characteristics of the internal waves and their effects on the sound transmission in the midst of the Yellow Sea, Chinese Journal of Acoustics, 18(1), 47-55, 1999.

The Effects of Internal Waves on the Signal Temporal Correlation Length in the South China Sea

Yun Ren, Lixin Wu, and Zhenglin Li

State Key Laboratory of Acoustics, Institute of Acoustics, Chinese Academy of Sciences, Beijing 100190

Abstract. The temporal correlation of sound transmission is simulated using a two-dimensional advective frozen-ocean model using the temperature data of a temperature sensor array deployed on the propagation path in the South China Sea (SCS) experiment 2009. The relationships of temporal correlation length, source-receiver range, and sound speed standard deviation (STD) caused by internal waves (including both linear internal waves and solitons) are numerically investigated. It is found that the temporal correlation length is −1.2 power dependent on source-receiver range and −0.9 power dependent on sound speed STD.

Keywords: internal waves, temporal correlation length, source-receiver range, sound speed STD
PACS: 43.30

INTRODUCTION

The ocean is a dynamic medium because of the existence of the tide, internal waves, and so on. As a result, acoustic signals propagating through the ocean fluctuate with time. The rate of signal fluctuation is often described by the signal temporal correlation length. The signal temporal correlation length is useful for many practical applications, such as detection of narrow band acoustic signals, matched field processing, and adaptive signal processing.

The practical need has prompted many experimental and theoretical studies of the temporal correlation of acoustic signals. P. C. Mignerey studied the influence of internal waves on the signal temporal correlation in an environment along a shelf break using the 2001 ASIAEX experimental data [1]. It is concluded that, nonlinear internal wave packets propagating in the acoustic path shorten the signal temporal correlation length. T. C. Yang analyzed acoustic data from three experiments and found the relationships of temporal correlation length, source-receiver range, and the sound speed STD [2]-[3]. And G. H. Ji numerically investigated the relationship based on oceanographic data from two shallow water experiments with the existence of linear internal waves [4].

In this paper, the relationships of temporal correlation length, source-receiver range, and the sound speed STD caused by both linear internal waves and solitons are numerically studied with the environmental data from the SCS experiment 2009.

CP1272, *Shallow-Water Acoustics, Second International Shallow-Water Acoustics Conference*
edited by J. Simmen, E. S. Livingston, J. X. Zhou, and F. H. Li

DEFINITIONS AND MEASUREMENT TECHNIQUES

The signal temporal correlation and the temporal correlation length are usually used to describe the rate of the signal fluctuation with time. If a well-defined signal is radiated repeatedly by an underwater source and received by a hydrophone at times t and t+△T, and the two received signals are denoted as p1(t) and p2(t+△T), the temporal correlation is given by

$$\rho(t,\tau) = \max_{\tau} \frac{\int_{-\infty}^{+\infty} p_1(t')p_2(t'-\tau)dt'}{\sqrt{\int_{-\infty}^{+\infty} p_1^{\ 2}(t')dt' \int_{-\infty}^{+\infty} p_2^{\ 2}(t')dt'}}, \tag{1}$$

where τ is the delay time.

The signal temporal correlation length τ0.707 is commonly defined as the delay time at which the temporal correlation drops to 0.707,

$$\rho(t,\tau_{0.707}) = 0.707. \tag{2}$$

For deep water, earlier theoretical studies have suggested that Doppler spread ν should increase linearly with the amplitude of the internal waves γ, the acoustic frequency f, and the square-root of the source-receiver range R [5],

$$\nu = \gamma f \sqrt{R}. \tag{3}$$

And the signal temporal correlation length is related to the Doppler spread by [2]

$$\nu = \sqrt{\frac{-\ln(0.707)}{2\pi^2}} \tau_{0.707}^{\ -1} = 0.13\tau_{0.707}. \tag{4}$$

So the signal temporal correlation length decreases as the inverse linear power of the amplitude of the internal waves and the inverse quadratic power of the source-receiver range.

In this paper, the sound speed STD δc represents the amplitude of the internal waves. And it is supposed that signal temporal correlation length τ0.707, sound speed STD δc, and source-receiver range R have a relationship in shallow water that is similar to that in the deep sea. The relationship is

$$\tau_{0.707} = lR^m(\delta c)^n. \tag{5}$$

The values of l, m, and n can be derived from the simulation data.

THE SCS EXPERIMENT 2009

The experiment was conducted near the SCS shelf break during June 2009. FIGURE 1 shows the site of the experiment. There is a temperature sensor array at point T in the left figure (point R4 in the right figure). The right figure zooms in the area in the red frame of the left figure. The water depth along the track from O to R5 varies from 1320 m to 620 m (as shown in FIG. 2). FIGURE 4 depicts the one-day temperature measurements from June 24 15:40 to June 25 16:40. Both the large amplitude solitons and linear internal waves can be observed from FIG. 4. During the temperature measurements, a CTD cast was done at point R4, and the measured salinity is shown in FIG. 3.

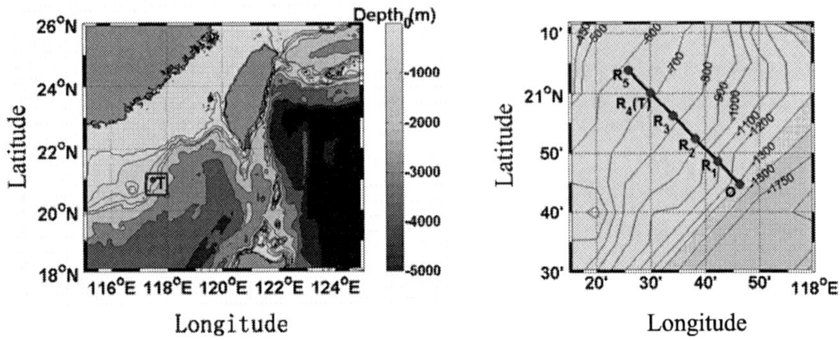

FIGURE 1. The site of the SCS experiment 2009.

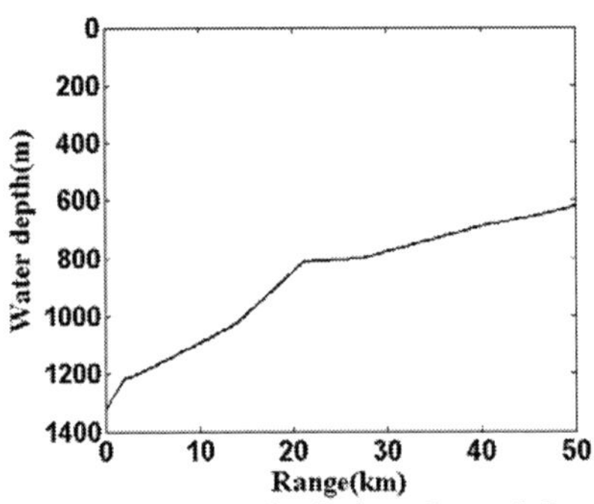

FIGURE 2. Water depth along the track from O to R5.

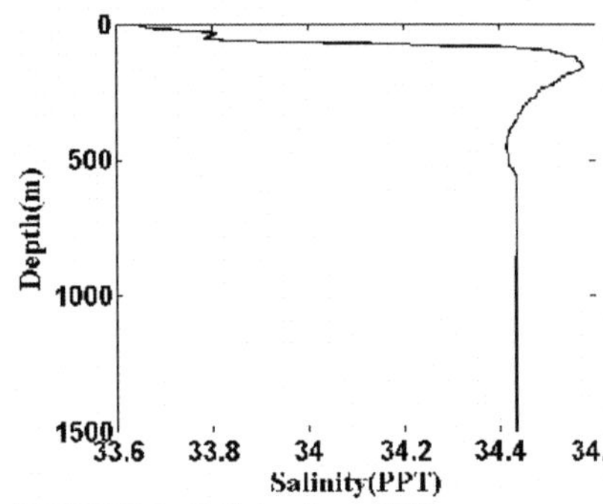

FIGURE 3. Salinity from CTD measurement at point R4.

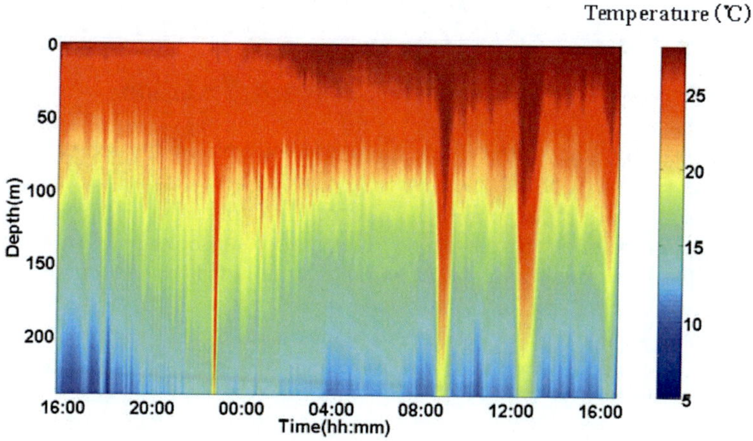

FIGURE 4. One-day temperature measurement at point R4.

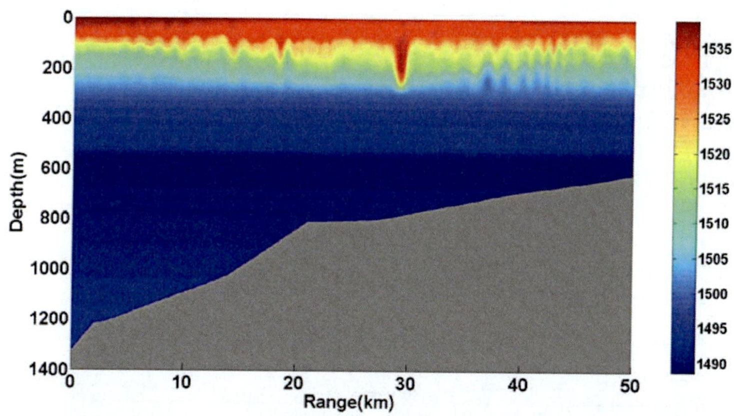

FIGURE 5. Converted sound speed profile from the temperature data.

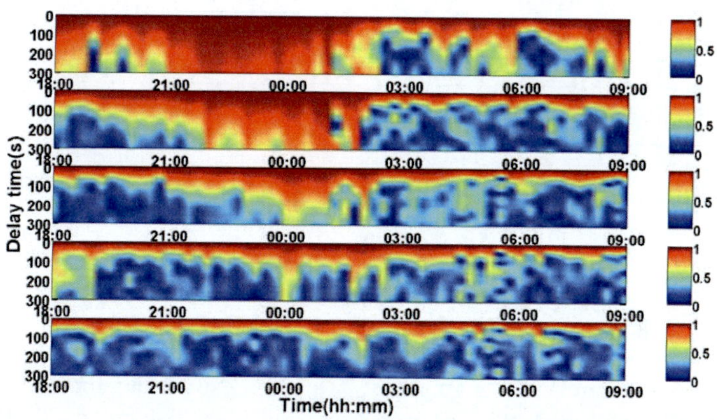

FIGURE 7. Temporal correlations as a function of time and delay time at 400Hz.

DATA ANALYSIS

A. Sound Speed Standard Deviation

To numerically study the effects of internal waves on the temporal correlation length, it is necessary to estimate the sound speed as a function of range from the temperature measurements of FIG. 4. Salinity has less effect on sound speed than temperature. Therefore, it is reasonable to let the profile shown in FIG. 3 represent the salinity of the experiment area. Assuming temporal variety in temperature is the result of internal waves passing the temperature sensor array at range r0 with a constant speed, sound speed profiles at time t and range r can be converted from the temperature profile measured at time t0 by using the linear transformation

$$r = r_0 + a(t - t_0), \tag{6}$$

where a is the advection speed.

As an example, FIG. 5 shows the converted sound speed profile varying with depth and range from the temperature data of FIG. 4 with the time period from 18:00 on June 24 to 9:00 on June 25, where a =1.5 m/s. A large amplitude soliton can be seen at the range of 30 km. The sound speed STD at five points R1, R2, R3, R4, and R5 in FIG. 1 were obtained for different times, and are shown in FIG. 6, where the range of each point from O is 10, 20, 30, 40, and 50 km, respectively.

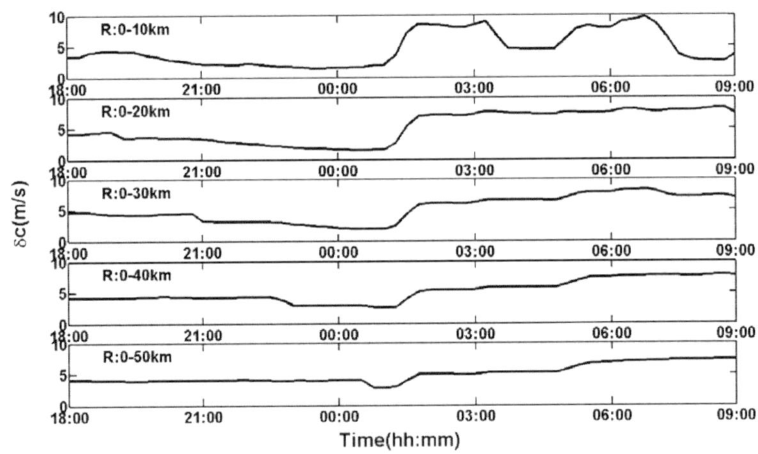

FIGURE 6. Sound speed STD of five ranges.

B. Analysis of Temporal Correlation Length

In the temporal correlation length analysis, a 400 Hz source is supposed to be put at point O and at 329.5m in depth. The propagating signals at five points: R1, R2, R3, R4, and R5 at different times can be calculated by the RAM PE [6] model with the converted sound speed profiles. In the acoustic field calculation, the bottom parameters are obtained from the Hamilton equations [7]-[8]. The relationship of the sediment sound speed versus sediment depth is

$$V(z) = V_0 + 1.304z - 0.741z^2 + 0.257z^3, \tag{7}$$

where V is the sound speed (km/s), V_0 is the value on the water–bottom interface (km/s), and z is the depth from that interface(km). The attenuation coefficient of sediment k_p (dB/kHz/m) versus the depth z (m) is

$$\begin{aligned} k_p(z) &= k_{p0} + k_{p0} \cdot z^{-\frac{1}{6}}, && z \le 200\,\text{m}; \\ k_p(z) &= k_p(200) - 1.4\times10^{-4} z + 0.03, && z > 200\,\text{m}. \end{aligned} \tag{8}$$

And the sediment density ρ (g/cm^3) can be calculated from the sediment sound speed V (km/s),

$$\begin{aligned} \rho &= 1.135V - 0.19, && z \le 500\,\text{m}; \\ \rho &= 0.917 + 0.744V - 0.08V^2, && z > 500\,\text{m}. \end{aligned} \tag{9}$$

In this paper, we let V0=1.646 km/s and kp0=0.34 dB/kHz/m. From Eq. (7), (8), and (9), the properties of the sediments versus sediment depth can be obtained.

Then temporal correlation and temporal correlation length can be obtained from Eq. (1) and Eq. (2). FIGURE 7 and FIG. 8 show the temporal correlation and temporal correlation length at points R1, R2, R3, R4, and R5 (from above to bottom), respectively. Comparing FIG. 7 and FIG. 8 with FIG. 6, it can be seen that the times with long temporal correlation length match those with a small sound speed STD and short range.

To obtain the relationship of the temporal correlation length, source-receiver range, and sound speed STD quantitatively, the empirical relationship as shown in Eq. (5) is estimated by fitting the simulation data shown in FIG. 8. In the linear fitting, Eq. (5) can be converted to the form of

$$\lg \tau_{0.707} = \lg l + m \lg R + n \lg(\delta c), \tag{10}$$

where l, m, and n can be obtained with a linear fit. The linear fit can be performed with three steps. First, for each range R of those five different ranges, lg$l$+$m$lg$R$ is considered as a whole in the linear fit. Then five values of lg$l$+$m$lg$R$ and n can be obtained. Second, in order to omit effects of R, Eq. (10) is changed to the form of $\lg\tau_{0.707}-(\lg l+m\lg R)=n\lg(\delta c)$, and n can be obtained from the linear fit as shown in FIG. 9. Finally, by using the five values of lg$l$+$m$lg$R$ obtained in the first step, we can get l and m. FIGURE 10 shows the comparison of temporal correlation length for different ranges with the fitted results. In the end, we obtained l=3.8×10^6, m=−1.2, and n=−0.9.

That means the temporal correlation length is −0.9 power dependent on the sound speed STD and −1.2 power dependent on the source-receiver range.

$$\tau_{0.707} = 3.8 \times 10^{6} R^{-1.2} (\delta c)^{-0.9} \tag{11}$$

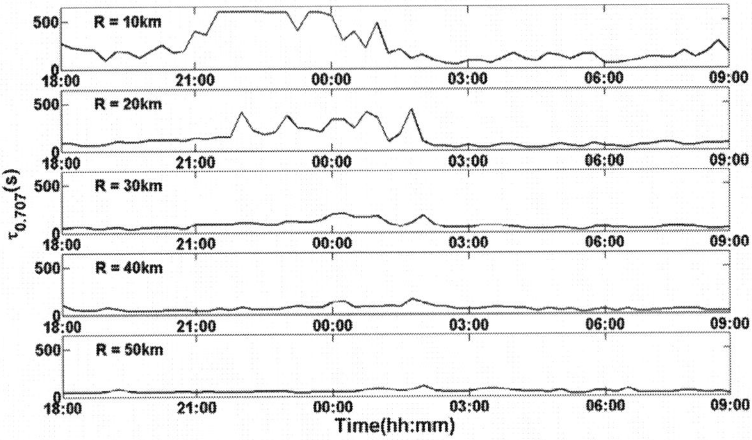

FIGURE 8. Temporal correlation lengths for different times corresponding to FIGURE 7.

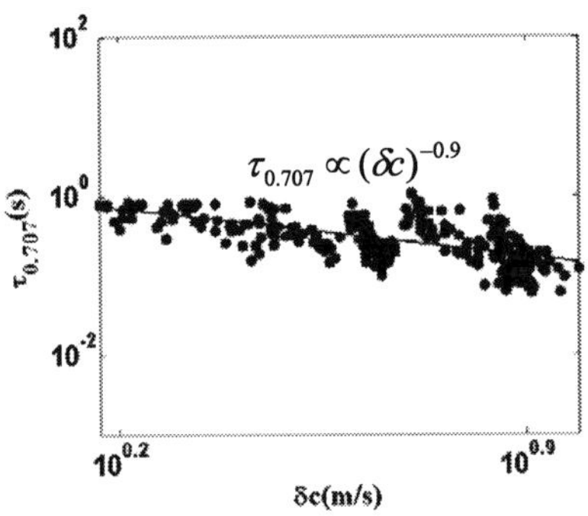

FIGURE 9. The relationship between temporal correlation length and sound speed STD.

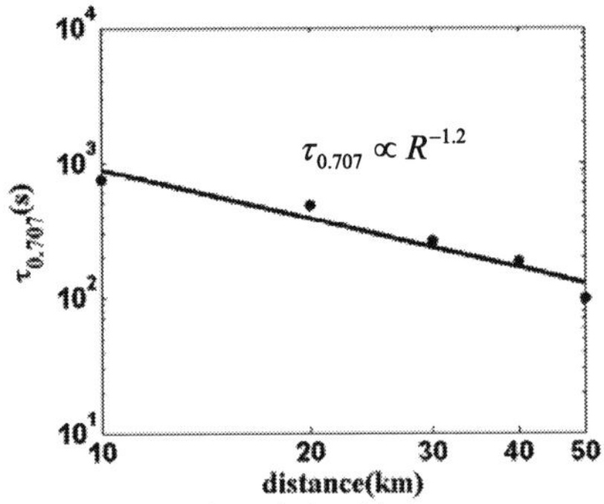

FIGURE 10. The relationship between temporal correlation length and source-receiver range.

As shown in Eq. (11), the temporal correlation length is −1.2 power dependent on R, which is different from the −0.5 power dependence on R obtained by T. C. Yang [2] and G. H. Ji [4]. It is mainly caused by two circumstances. The first is the existence of the large amplitude solitons in the experiment, while T. C. Yang and G. H. Ji only considered environments with linear internal waves. The second is the water depth variation along the propagation track. In order to validate the water depth variation, a situation with a flat bottom was analyzed and showed a −0.9 power dependence on R.

CONCLUSIONS

The SCS experiment 2009 was conducted near the shelf break, and both linear internal waves and strong solitons were observed in the experiment. The effects of source-receiver range and sound speed STD caused by internal waves on the signal temporal correlation length were numerically investigated by using environmental measurements of the experiment. The results suggest that the temporal correlation length is −0.9 power dependent on sound speed STD and −1.2 power dependent on source-receiver range.

REFERENCES

1. P. C. Mignerey and M. H. Orr, "Observations of matched-field autocorrelation time in the South China Sea," *IEEE Journal of Oceanic Engineering*, 29(4), 1280-1291 (2004).
2. T. C. Yang, "Measurement of temporal coherence of sound transmissions through shallow water," *J. Acoust. Soc. Am.*, 120(5), 2595-2614 (2006).
3. T. C. Yang, "Temporal coherence of sound transmissions in deep water revisited," *J. Acoust. Soc. Am.*, 124(1), 113-127 (2008).
4. G. H. Ji and Z. L. Li, "Coherence-time of matched-field processing in shallow water in the presence of linear internal waves," *Chin. Phys. Lett.*, 26(9), 0943011-0943014 (2008).
5. W. H. Munk and F. Zachariasen, "Sound propagation through a fluctuating stratified ocean: Theory and observation," *J. Acoust. Soc. Am.*, 59, 818–838 (1976).
6. M. D. Collins, "A split-step Pade solution for parabolic equation method," *J. Acoust. Soc. Am.*, 93(4), 1736-1742 (1993).
7. E. L. Hamilton, "Sound velocity-density relations in sea-floor sediments and rocks," *J. Acoust. Soc. Am.*, 63(2), 1366-1377 (1978).
8. E. L. Hamilton, "Geoacoustic modeling of the sea floor," *J. Acoust. Soc. Am.*, 68(5), 1313-1340 (1980).

Geo-Acoustic Doppler Spectroscopy: A Novel Acoustic Technique For Surveying The Seabed

Michael J. Buckingham

Marine Physical Laboratory, Scripps Institution of Oceanography
University of California, San Diego, La Jolla, CA, 92093-0238, USA

Abstract. An acoustic inversion technique, known as Geo-Acoustic Doppler Spectroscopy, has recently been developed for estimating the geo-acoustic parameters of the seabed in shallow water. The technique is unusual in that it utilizes a low-flying, propeller-driven light aircraft as an acoustic source. Both the engine and propeller produce sound and, since they are rotating sources, the acoustic signature of each takes the form of a sequence of narrow-band harmonics. Although the coupling of the harmonics across the air-sea interface is inefficient, due to the large impedance mismatch between air and water, sufficient energy penetrates the sea surface to provide a useable underwater signal at sensors either in the water column or buried in the sediment. The received signals, which are significantly Doppler shifted due to the motion of the aircraft, will have experienced a number of reflections from the seabed and thus they contain information about the sediment. A geo-acoustic inversion of the Doppler-shifted modes associated with each harmonic yields an estimate of the sound speed in the sediment; and, once the sound speed has been determined, the known correlations between it and the remaining geo-acoustic parameters allow all of the latter to be computed. This inversion technique has been applied to aircraft data collected in the shallow water north of Scripps pier, returning values of the sound speed, shear speed, porosity, density and grain size that are consistent with the known properties of the sandy sediment in the channel.

Keywords: geo-acoustic inversion, sediment, aircraft sound.
PACS: 43.20.Bi, 43.30,Bp,43.30.Es.

INTRODUCTION

It has long been known that the sound from large propeller-driven aircraft such as the Lockheed Orion P-3 and the de Havilland Twin Otter is detectable underwater [1-3]. In some of our early experiments, conducted off the coast of southern California, north of Scripps Pier, we demonstrated that the sound produced by a propeller-driven light aircraft with a single piston engine can also be detected beneath the sea surface, on a hydrophone either in the water column or buried in the seabed [4,5].

The sound of a light aircraft is created predominantly by the engine and the propeller, both of which are rotating sources, and hence each produces a sequence of narrow-band tones, or harmonics, at frequencies that are multiples of the fundamental. Due to the horizontal motion of the aircraft, the harmonics experience significant Doppler shifts in frequency, upwards on approach and downwards on departure, which gives rise to a fore-aft asymmetry in the field. On penetrating the sea surface, each harmonic excites a set of normal modes in the shallow-water waveguide. These modes are all Doppler frequency-shifted, the degree of shift decreasing with increasing mode

CP1272, *Shallow-Water Acoustics, Second International Shallow-Water Acoustics Conference*
edited by J. Simmen, E. S. Livingston, J. X. Zhou, and F. H. Li

number, a phenomenon that may be understood from consideration of Snell's law along with the fact that the grazing angle of a modal equivalent ray increases with increasing mode number.

The amplitudes and Doppler-shifted frequencies of the normal modes depend on several factors, including the nature of the seabed. Since the seabed influences the spectral properties of the modal arrivals at a hydrophone in the water column (or, for that matter, buried in the sediment), it follows that the modes provide the basis of an inversion technique for estimating the geo-acoustic parameters of the bottom. A brief description of the inversion technique is given below.

DOPPLER-SHIFTED MODES AND RAYS

The inversion for the sediment properties relies on a comparison between the spectrum of the underwater arrivals associated with a particular engine or propeller harmonic and a similar spectrum computed from an acoustic propagation model of the air-sea-sediment waveguide [6]. An example of such a computed spectrum is shown in Fig. 1, in this case for an airborne source with an unshifted frequency of 88 Hz, a horizontal, unaccelerated speed of 50 m/s and an altitude of 50 m above the sea surface. The receiver in this example was taken to be in the water column directly beneath the source track at a depth of 53 m; and the shallow-water channel itself was as specified by Zhang and Tindle [7], with a depth of 54 m, sound speeds of 1500 and 1600 m/s, respectively, in the water column and sediment, a sediment attenuation of 0.3125 dB/m/kHz, and a water-to-sediment density ratio of 0.8.

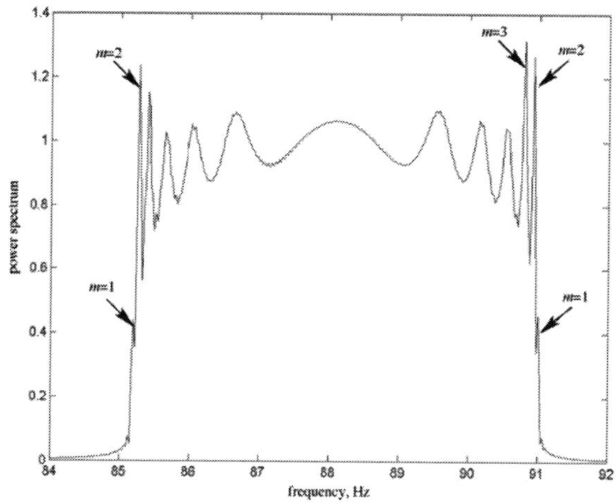

FIGURE 1. High-resolution computed power spectrum showing up- and down-shifted modal peaks, identified by the mode number, m. Reprinted from [6].

The high-resolution spectrum in Fig. 1 shows the frequency spread (approximately 85 to 91 Hz) in the arrival structure from the 88 Hz source tone. Three up-shifted (approach) modal peaks appear at frequencies of 91.007, 90.921 and 90.773 Hz and two down-shifted (departure) modal peaks are present 85.189 and 85.273 Hz. These are all true, propagating modes. Strictly, the third down-shifted mode is cut-off, although there is a sharp, mode-like peak at 85.419 Hz, which is actually a virtual mode [8,9], as are the remaining, broader peaks in Fig. 1.

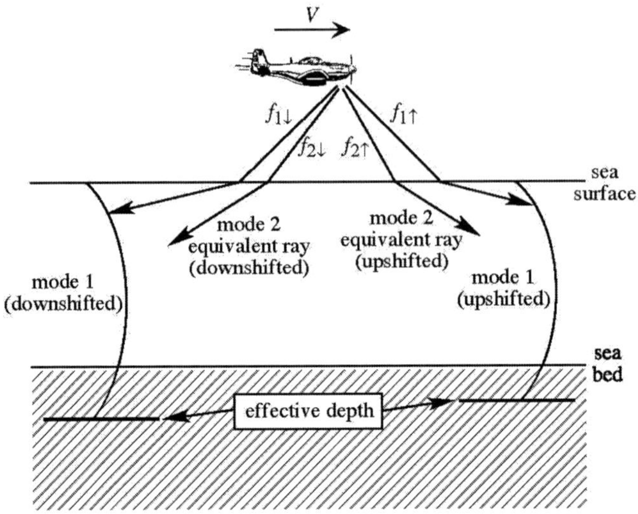

FIGURE 2. Ray schematic of sound from a monotonic moving airborne source coupling into normal modes in the channel. To avoid clutter, the second mode shape has been omitted. Reprinted from [6].

The Doppler-shifted modal structure in Fig. 1 may be interpreted in terms of equivalent modal rays, as illustrated in Fig. 2. If f_0 is the unshifted frequency of the source, then the frequency of an airborne ray is

$$f = \frac{f_0}{1-(V/c_1)\cos\alpha}\ , \tag{1}$$

where V is the speed of the source, c_1 is the speed of sound in air, and α is the declination of the ray below the horizontal. Such a ray will couple into the equivalent ray of a normal mode when the two rays satisfy Snell's law, as shown in Fig. 2. Clearly, the shallowest rays, fore and aft of the source, have the greatest Doppler shifts and couple into the lowest mode, with successively steeper rays exciting modes 2, 3 and so on.

The mode shapes are governed by the effective depth [10,11] of the channel, which varies with the shifted-frequencies of the individual modes: the effective depth is greatest for the lowest (m = 1) down-shifted mode and least for the lowest (m = 1) up-shifted mode. In general, the effective depth is given by the frequency-dependent expression

$$H = h + \frac{1 \mp \beta_2}{b_{23} k_2 \sin(\alpha_c)} \tag{2}$$

where h is the channel depth, β_2 is the Mach number of the source relative to the speed of sound in seawater, k_2 is the acoustic wavenumber in the water column, b_{23} is the water-column to sediment density ratio, and α_c is the critical grazing angle of the seawater-sediment interface. Where there is a choice of sign in Eq. (2), the minus (plus) sign corresponds to the approach towards (departure from) the sensor station. Clearly, the effective depth depends on frequency, through k_2, as well as the properties of the sediment, notably the sediment sound speed embedded in the critical grazing angle α_c.

Since spectra of the type shown in Fig. 1 depend on the effective depths associated with individual up- and down-shifted modes, such spectra may be inverted to return the sound speed in the sediment. The remaining geo-acoustic parameters may then be determined using the theoretical correlations developed in recent analyses of wave propagation in saturated, unconsolidated granular materials [12,13].

EXAMPLE OF AN INVERSION

Moving-source data have been collected in several locations, including the coastal region north of Scripps Pier, southern California, a shallow-water area off the western tip of Kauai, and the Gulf of Mexico near Fort Walton Beach, Florida. Various types of single-engine light aircraft have been used as acoustic sources, some with two- and others with three-blade propellers; and in all cases acoustic sensors were co-located (horizontally) in the air, the water column and the sediment. The aircraft flew race-track patterns at constant speed, passing directly over the sensor station once per circuit in straight and level flight. The position, altitude and speed of the aircraft were monitored using onboard GPS.

Fig. 3 shows the water-borne spectrum (dotted line) of the first engine harmonic (unshifted frequency ≈ 79.5 Hz) of a Diamond Star DA40 with a three-blade propeller. The spectrum was taken north of Scripps Pier with a hydrophone lying on the seabed at a depth of 14.4 m. The sediment at this location is known to be a fine sand whose geo-acoustic properties are well characterized as the result of an extensive survey of the area performed recently by Hogarth *at al.* [14].

A pair of modal peaks appears in the measured spectrum in Fig. 3, on either side of the unshifted frequency, close to 82 Hz (up-shifted) and 76 Hz (down-shifted), and the two members of the pair correspond to the first and second propagating normal modes. A computed spectrum (light solid line) based on a three-layer propagation model, atmosphere-seawater-sediment, each with isotropic acoustic properties, yields a spectrum showing just the first modal peaks on either side of the unshifted frequency, which match the corresponding peaks in the data nicely. However, the three-layer model fails to account for the presence of the second mode in the observations.

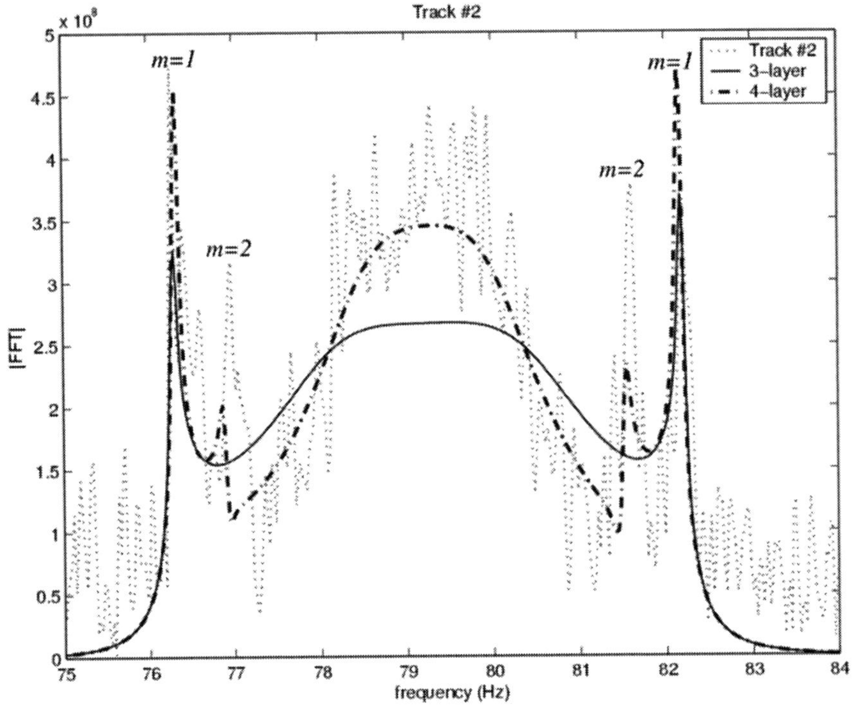

FIGURE 3. Measured spectrum from a hydrophone on the seabed and two theoretical spectra computed from a three-layer and a four-layer model of the air-sea-sediment waveguide.

According to the recent survey of the area [14], the fine sand layer is about 10 m thick, overlying indurated sand (bedrock) in which the sound speed is approximately 2,200 m/s. By including a fourth layer in the acoustic propagation model to represent the high-speed substrate, the heavy dash-dot line in Fig. 3 is obtained. Now modes 1 and 2 can be seen in the computed spectrum at positions on either side of the unshifted frequency that agree reasonably well with the observed modal peaks. The geo-acoustic parameter values inferred from the computed four-layer spectrum in Fig. 3 are as follows: compressional wave speed = 1634 m/s; compressional wave attenuation = 0.12 dB/m/kHz; shear wave speed = 94.2 m/s; shear wave attenuation = 38.8 dB/m/kHz; bulk density = 2019 kg/m^3; mean grain diameter = 83 micrometer; critical grazing angle = 21.8°; sound speed ratio = 0.12. All these values are consistent with the known properties of the sediment. For instance, Hogarth *et al.* [14] report a sediment sound speed of 1640 m/s, which is less than 0.5% higher than the value obtained from the inversion of the aircraft signature.

CONCLUDING REMARKS

The geo-acoustic inversion technique outlined above relies on a high-speed airborne source to produce a measurable separation of Doppler shifted modal peaks in the underwater spectral signatures of an individual transmitted tone. It is possible that vehicles other than aircraft, for instance, hovercraft, could act as sound sources with sufficient Doppler shift to allow a similar inversion technique to be employed. Indeed, given its proximity to the sea surface, a hovercraft may provide better acoustic coupling, air-to-water, than an aircraft, although the physical disturbance of the sea surface by a hovercraft could lead to complications in interpreting the modal spectrum, not to mention other complexities, including its acoustic signature.

Be that as it may, the use of a light aircraft as a sound source for underwater acoustic experiments in general and geo-acoustic inversions in particular seems to have some potential. From a budgetary point of view, it is interesting to note that a light aircraft costs about $200 per hour or less, compared with a typical oceanographic research ship, which has a price tag of several tens of thousands of dollars per day.

ACKNOWLEDGMENTS

The research reported here was supported by the Office of Naval Research under Grant No. N00014-04-1-0063, sponsors Drs. Ellen Livingston and Robert Headrick, Ocean Acoustics Code 321.

REFERENCES

1. R. J. Urick, "Noise signature of an aircraft in level flight over a hydrophone in the sea," J. Acoust. Soc. Am. **52**, 993-999 (1972).
2. H. Medwin, R. A. Helbig, and J. D. Hagy, Jr., "Spectral characteristics of sound transmission through the rough sea surface," J. Acoust. Soc. Am. **54**, 99-109 (1973).
3. W. J. Richardson, C. R. Greene, Jr., C. I. Malme and D. H. Thomson, *Marine Mammals and Noise*, Academic Press, New York, 1995.
4. M. J. Buckingham, "Acoustical remote sensing of the sea bed using propeller noise from a light aircraft," in *Sounds in the Sea: from Ocean Acoustics to Acoustical Oceanography*, 581-597, edited by H. Medwin, Cambridge University Press, Cambridge, 2005.
5. M. J. Buckingham, E. M. Giddens, and F. Simonet, "Inversion of the propeller harmonics from a light aircraft for the geoacoustic properties of marine sediments," in *Acoustic Sensing Techniques for the Shallow Water Environment: Inversion Methods and Experiments*, 257-263, edited by A. Caiti, N. R. Chapman, J-P. Hermand and S. M. Jesus, Springer, Dordrecht, 2006.
6. M. J. Buckingham and E. M. Giddens, "Theory of sound propagation from a moving source in a three-layer Pekeris waveguide," J. Acoust. Soc. Am. **120**, 1825-1841 (2006).

7. Z. Y. Zhang and C. T. Tindle, "Complex effective depth of the ocean bottom," J. Acoust. Soc. Am. **93**, 205-213 (1993).
8. C. T. Tindle, A. T. Stamp, and K. M. Guthrie, "Virtual modes and the surface boundary condition in underwater acoustics," J. Sound Vib. **49**, 231-240 (1976).
9. C. T. Tindle, A. T. Stamp, and K. M. Guthrie, "Virtual modes and mode amplitudes near cutoff," J. Acoust. Soc. Am. **65**, 1423-1428 (1979).
10. D. E. Weston, "A Moiré fringe analog of sound propagation in shallow water," J. Acoust. Soc. Am. **32** (6), 647-654 (1960).
11. M. J. Buckingham, "Array gain of a broadside vertical line array in shallow water," J. Acoust. Soc. Am. **65**, 148-161 (1979).
12. M. J. Buckingham, "Wave propagation, stress relaxation, and grain-to-grain shearing in saturated, unconsolidated marine sediments," J. Acoust. Soc. Am. **108**, 2796-2815 (2000).
13. M. J. Buckingham, "On pore-fluid viscosity and the wave properties of saturated granular materials including marine sediments," J. Acoust. Soc. Am. **122** (3), 1486-1501 (2007).
14. L. J. Hogarth, J. Babcock, N. W. Driscoll *et al.*, "Long-term tectonic control on Holocene shelf sedimentation offshore La Jolla, California," Geol. **35** (3), 275-278 (2007).

Geoacoustic Inversion in a Spatially and Temporally Variable Shallow Water Environment

N.R. Chapman and Yong-min Jiang

School of Earth and Ocean Sciences, University of Victoria, Victoria, BC Canada V8W3V6.

Abstract. This paper describes matched field inversion of data from the Shallow Water 06 (SW06) experiment that was carried out in a shallow water environment in which the water column sound speed profile was spatially and temporally variable. The profile was characterized by a variable thermocline between relatively stable upper and lower homogeneous layers. The inversion accounted for the variable water column sound speed profile by parameterizing the profile using empirical orthogonal function (EOFs), and including the EOFs as parameters in the inversion. Two approaches were used to obtain information about the sound speed variability. The first used only the variation in the thermocline, whereas the second used EOFs for the full water column. The inversions showed that consistent estimates of the geoacoustic model parameters could be obtained with either of the two approaches. (Work supported by the Ocean Acoustics Team at Office of Naval Research)

Keywords: Geoacoustic inversion, matched field processing, sound speed variability.
PACS: 43.30.Pc,43.60.Pt

INTRODUCTION

Matched field geoacoustic inversion (MFI) has been successfully applied to experimental data in various different range-independent and range-dependent ocean environments.[1] Most of the inversion studies to date were carried out by assuming that the water column sound speed profile (SSP) is both range independent and time invariant. However, spatial and temporal variations of the oceanic SSP can arise due to internal waves, eddies and tidal currents that generate local inhomogeneities in the water at different spatial and temporal scales. These uncertainties in the water sound speed along the propagation path cause acoustic field fluctuations that reduce the signal coherence at the receiver and degrade sonar performance in applications for underwater communication and source localization. Spatial and temporal variations of the oceanic SSP will also contaminate geoacoustic inversion results.[2-5]

In this paper we investigate the impact of temporal and spatial variations of the water column SSP on geoacoustic inversion. In some cases of relatively weak variation of the environmental parameters, it is possible to use a single sound speed profile to account for the temporally and spatially varying SSP along the propagation path. We investigate the performance of this approach of using a range-independent SSP in geoacoustic inversions of data from the Shallow Water 2006 (SW06) experiments.[6]

CP1272, *Shallow-Water Acoustics, Second International Shallow-Water Acoustics Conference*
edited by J. Simmen, E. S. Livingston, J. X. Zhou, and F. H. Li

We assume that the SSP is represented by an effective range-independent profile that is based on empirical orthogonal functions (EOFs) generated from information from different oceanographic stations in the vicinity. Two approaches for generating the EOFs are compared. The first approach used a limited set of SSPs that were measured from the source ship at the time of the signal transmissions. This approach includes only the relevant variation of the ocean SSP during the experiment. It also models only the variations in the thermocline, which was the portion of the SSP where the greatest variation occurred. Inversions based on this approach were compared with inversions for which the EOFs were constructed from a more extensive set of SSPs from other environmental moorings nearby to capture a wider degree of sound speed variation. The first approach requires fewer EOFs in generating an effective sound speed profile. This provides a distinct advantage of computational efficiency in MFI, since fewer model parameters are needed for calculating the replica acoustic fields.

THE SW06 EXPERIMENT

The New Jersey shelf-break location (Figure 1) was specifically chosen to provide a site where internal waves were known to propagate in the water column. Additional complicating effects due to the shelf break front, a salinity/density front and current that meanders near the shelf edge, and the frequent presence of warm-core eddies shed from the Gulf Stream were also present. Experiments were carried out in SW06 to study the interaction of sound with the ocean bottom over a broad frequency band from 50 Hz to 20 kHz[6]. In this paper, we focus on the low frequency (< 1000 Hz) experiments for geoacoustic inversion.

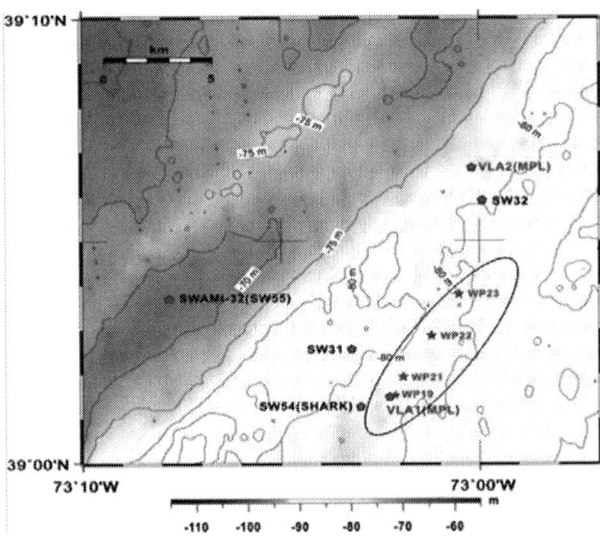

FIGURE 1. Bathymetry of the experimental site and the positions of the source, acoustic arrays, and environmental array moorings.

The data were recorded on a bottom-moored vertical line array (VLA) deployed near the shelf break from the research vessel R.V. KNORR in about 79 m water depth. The VLA consisted of 16 hydrophones equally spaced at 3.75 m, with the bottom-most sensor 8.2 m above the sea floor. The sampling rate at each sensor was 2000 samples/s, and the data were stored on a hard drive in an underwater pressure case at the base of the array. Depth of the array was monitored by a pressure sensor that was located ~0.5 m below the top-most hydrophone. Tilt and compass direction were also monitored at the same location on the array.

A J–15 sound projector was towed from the R.V. KNORR while transmitting low frequency CW (continuous wave) signals in one of two modes: either four tones from 53 to 253 Hz, or five tones from 303 to 953 Hz. This paper reports inversions of data obtained on a single radial track from the VLA out to ~8 km along one of the chirp survey lines. The source ship stopped at stations of 1, 3 and 5 km and transmitted the CW tone sequences for about five minutes at a source depth of 30 m while maintaining position at each site using a dynamic positioning system. The water depth was range independent out to about 3 km, and gradually increased to about 83 m over the remainder of the track.

Water Sound Speed Profile Parameterization

Sound speed profiles in the water were measured at each station using a CTD, and the variation in the profiles is shown in Fig. 2(a). Substantial SSP variations are evident at depths from 10 to 50 m over the 4 h time period and 5 km range of the experiment. The source depth of ~30 m was in the middle of the thermocline, and it is evident that the water column sound speed changed considerably at this depth over the observation time. Since the SSPs in the small sample set were almost identical from 0–10 m and from 50–79 m, only the middle parts (10–50 m) were used in the EOF analysis. This procedure added reasonable constraints for generating an effective ocean SSP.

Additional information about the SSPs was obtained from the Woods Hole Oceanographic Institution (WHOI) environmental array moorings deployed over the experimental region. The pentagon symbols in Fig. 1 show the environmental array moorings (SW31, SW32, and SW54) that were closest to the source track. SSPs recorded at those three locations on the day of the experiment are shown in Fig. 2(b1) to Fig. 2(b3). The SSPs were sampled in time at a rate of 30 s. In Fig. 2, the sample time was color coded to represent the SSPs measured at different times. Figure 2(b4) shows all of the SSPs collected at the three locations during the 4 h time period that the tones were transmitted. Notable differences in the thermocline are seen from the SPPs measured at SW54 and SW31/SW32. The differences can be characterized in terms of the sound speed gradient and the depth of the thermocline.

EOFs were generated from small and large SSP sample sets to compare the effect of the statistics of SSP observations on the uncertainties of the estimated parameter values in the inversions. The small SSP sample set contained only the SSPs collected from CTD measurements on the source ship at the experimental stations. The large SSP set contained the SSPs recorded on the SW31, SW32, and SW54 environmental array moorings within the entire 4 hour time period of the experiment, and all of the samples in the small SSP sample set. In total, there were 5 SSPs in the small sample set and 905 SSPs in the large sample set.

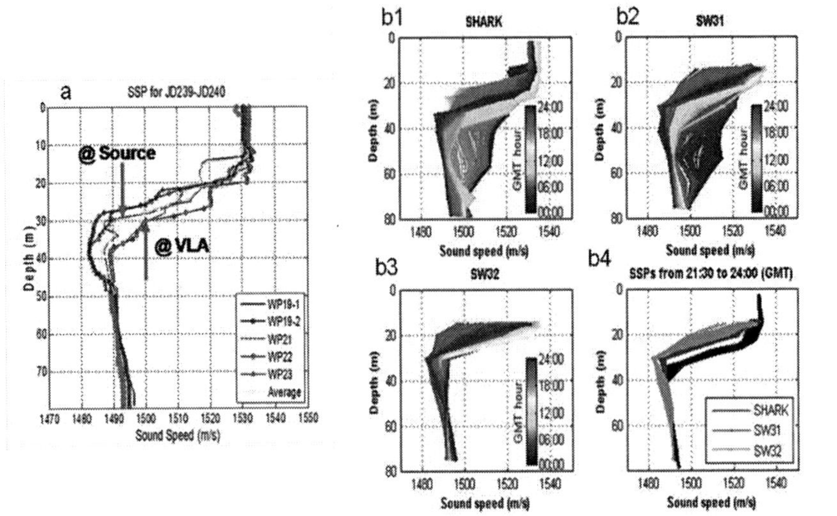

FIGURE 2. (a) CTD casts measured at the source stations during the experiment. (b) Oceanographic observations recorded on WHOI environmental arrays during the day of the experiment.

The method for deriving the EOF representation from the SSP data is described in Jiang and Chapman[7]. The number of EOFs required to reconstruct an individual SSP to a prescribed degree of misfit was determined for each of the two approaches. Figure 3 shows a comparison of the misfit ratio for the small and large SSP sample sets. For the large sample set, 8 EOFs were needed to reach over 90% of the misfit energy, whereas only 4 EOFs were required to reach over 95% of the misfit energy for the small sample set.

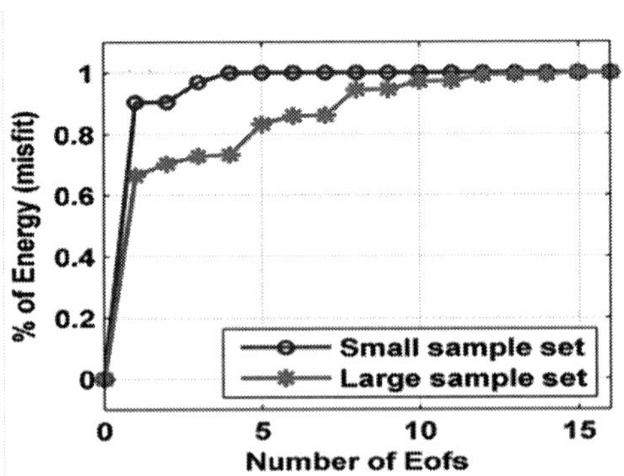

FIGURE 3. Comparison of the percentage of misfit energy over total misfit versus the number of EOFs used to recover an individual SSP for the small and large SSP sample sets.

GEOACOUSTIC INVERSION

The details of the Bayesian MFI method are described in Jiang and Chapman.[7] We focus attention here on the impact of correlated data errors in MFI, and the estimation of an appropriate data error covariance matrix. Data error information is one of the most important factors in Bayesian inversion since wrong assumptions about the data errors may bias the uncertainties. Data errors include measurement errors and theory errors. Theory errors arise from reasons such as inappropriate model parameterization and inaccurate acoustic theory. For example, in matched field inversions, theory errors on the adjacent hydrophones of the VLA could be correlated. Hence, it is not generally true that the data error covariance matrix has identical independent distribution especially for cases of low frequencies and relatively small hydrophone.

While the variance of the measurement errors can be reduced by averaging over multiple data samples, we cannot assume that the theory errors can be reduced by the same means. However, information about the theory errors may be extracted by changing the acoustic signal propagation conditions, such as by varying the source-receiver range[7], changing the source depth or receiver depth, or simply maintaining the source and receiver geometry in a highly variable ocean environment. The dominant parameters in the proposed geoacoustic model should produce the acoustic replica that fits the data with small uncertainties under different sound propagating conditions.

In this experiment, the differences between the measured and the modeled data from different sound transmission conditions in the rough sea state during the experiment was sufficient to provide theory error information. The source depth variations within the strong thermocline caused changes in the sound propagation conditions in the waveguide that resulted in variations of the acoustic field. The data error covariance matrix was constructed from multiple 2-s data samples that contained these variations.

Geoacoustic Model

A high resolution sub–bottom survey revealed well-resolved structure down to about 30 m below the sea floor, most prominently showing an interface known as the 'R' reflector at about 20 m. This interface, which is pervasive in the region, is overlain with alternating layers of sand and mud. Based on the geophysical surveys, a simple geoacoustic model of one layer over a half space was chosen to represent the sea bottom for all the three sites.

Three groups of parameters are inverted: (i) 4 geometric parameters including water depth, source depth, range, and array tilt; (ii) 9 geoacoustic parameters including sediment depth, sediment *p*-wave sound speed and density at top and bottom, the constant and the exponent of frequency dependence of the attenuation of the sediment, and *p*-wave sound speed and density of the lower half space; and (iii) 4/8 EOF coefficients to describe the water column SSP, where 4 and 8 are the number of EOF coefficients to be inverted for the small and large SSP sample sets, respectively. Due to the spatial variability of the geoacoustic parameters over the range and the sediment depth, it is expected that the parameter estimates from the inversion are effective ones that are averaged over range and depth.

Inversion Results

One dimensional (1D) marginal distributions of the estimated geoacoustic parameters are shown in Fig. 4 for inversions with the 1-km range data using the two different sets of SSP observations. The most sensitive parameters, including the sediment sound speed, density and depth and half space sound speed, are very well resolved and consistent between both approaches. Similar consistency between the two approaches was found for the geometrical parameters (range, water depth, source depth and array tilt).[7] The EOF coefficients were also sensitive parameters, and were well estimated in the inversions.[7] The close agreement in the model parameter estimates suggests that the restriction of using a limited SSP sample set and fitting only the variation in the thermocline did not degrade the inversion performance. This enabled a significant improvement in the efficiency of the inversion, since the computation time of the inversion was greatly reduced.

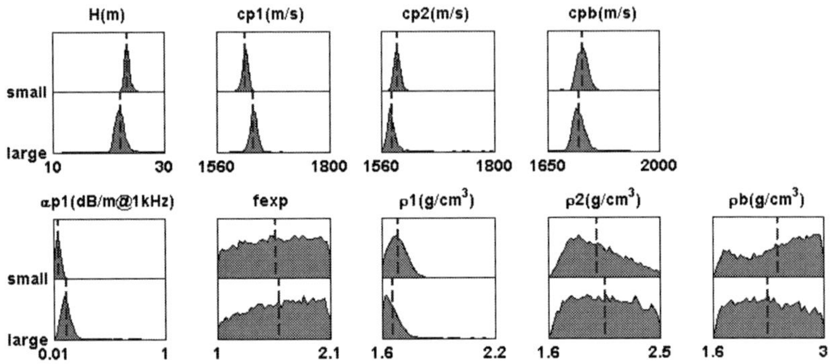

FIGURE 4. Comparison of the geoacoustic parameter estimates obtained by using small and large ocean SSP observation samples. The dashed lines indicate the mean values of the estimates..

The correlations between the EOF and the most sensitive geometric/geoacoustic parameters are displayed in Fig. 5 in terms of two dimensional marginal distributions for the small SSP sample set inversions. It is evident that the water column SSP has a strong effect on the geometric parameters, as expected, but there is little or no correlation with the sediment sound speed. This result provides further support for our assumption that a single range-independent SSP is adequate for geoacoustic MFI in this case of relatively weak variability in the SSP. However, this approach is limited if the environmental variation along the propagation path is more severe.[7]

CONCLUSIONS

The results presented here show that the impact of a dynamic ocean environment on can be mitigated by using a single range-independent SSP in matched field geoacoustic inversions. Furthermore, the SSP can be generated from a reduced sample set of SSP observations that characterize the variability at the time of the experiment. Some caution is needed in using this as a general approach, since the application is restricted to weak variability that does not cause significant mode coupling.

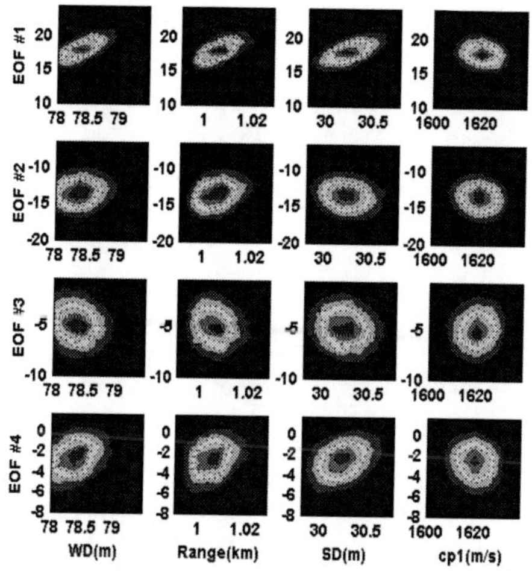

FIGURE 5. 2D marginal distributions of EOFs and water depth, range, source depth, and sediment sound speed for the inversion of 1-km range data with the small SSP sample set.

ACKNOWLEDGMENTS

This work is supported by the Office of Naval Research Ocean Acoustic Team under Grant No. N00014-05-1-0264. The authors thank William Hodgkiss, Peter Gerstoft, John Goff, David Knobles and Arthur Newhall for helpful assistance with various aspects of the SW06 data.

REFERENCES

1. N. R. Chapman, S. Chin-Bing, D. King, and R.B. Evans, "Benchmarking geoacoustic inversion methods for range-dependent waveguides," IEEE J. Ocean. Eng. **28**, 320–330 (2003).
2. M. Siderius, P.L. Nielsen, J. Sellschopp, M. Snellen, and D. Simons, "Experimental study of geo-acoustic inversion uncertainty due to ocean sound-speed fluctuations," J. Acoust. Soc. Am. **110**, 769–781 (2001).
3. M. Snellen, D.G. Simons, M. Siderius, J. Sellschopp, and P.L. Nielsen, "Evaluation of the accuracy of shallow water matched field inversion results," J. Acoust. Soc. Am. **109**, 514–527 (2001).
4. C.-F. Huang, P. Gerstoft, and W.S. Hodgkiss, "Effect of ocean sound speed uncertainty on matched-field geoacoustic inversion," J. Acoust. Soc. Am. **123**, EL162–EL168 (2008).
5. Y.-M. Jiang and N.R. Chapman, "Bayesian geoacoustic inversion in a dynamic shallow water environment," J. Acoust. Soc. Am. **123**, EL155–EL161 (2008).
6. D.J. Tang, J. Moum, J. Lynch, P. Abbot, R. Chapman, P. Dahl, T. Duda, G. Gawarkiewicz, S. Glenn, J. Goff, H. Graber, J. Kemp, A. Maffei, J. Nash, and A. Newhall, "Shallow Water '06—A joint acoustic propagation/nonlinear internal wave physics experiment," Oceanogr. **20**, 156–167 (2007).
7. Y.-M. Jiang and N.R. Chapman, "The impact of ocean sound speed variability on the uncertainty of geoacoustic parameter estimates", J. Acoust. Soc. Am., **125**, 2881-2895, (2009).

Estimating marine sediment properties in a temporally varying water column environment

Yong-Min Jiang and N. Ross Chapman

School of Earth and Ocean Sciences, University of Victoria, Victoria, BC V8W 3P6, Canada

Abstract. This paper presents geoacoustic inversion of broadband data collected at a site on the New Jersey continental shelf in the Shallow Water 2006 experiments. The experimental geometry of multiple source depths, vertical line array and short distance between the source and receiver was used to obtain a wide-angle coverage of the sea bottom reflection. The sound speed profile in the water was a typical coastal one that had a strong thermocline in the middle part of the water column. During the data collection, the variability of the water column sound speed was evident. To mitigate the effect of the variable water environment on the geoacoustic inversion, a multi-step inversion was carried out. The geometric parameters such as the water depth, source and receiver range were inverted first, then the water column sound speed profile was determined by inverting the coefficients of its empirical orthogonal function representations, and finally the sediment sound speed and layer thickness were inverted by using the parameters estimated in the first two steps. Furthermore, a method of estimating the sediment attenuation and its frequency dependence by using this short range experimental geometry is developed. The uncertainty of the attenuation estimate due to measurement is examined by utilizing the variances at different frequency bands.

Keywords: Travel time geoacoustic inversion, hybrid optimization, sediment sound attenuation, frequency dependence of attenuation, sound speed profile, empirical orthogonal function.
PACS: 43.30.Pc, 43.60.Pt

INTRODUCTION

Sound propagation in the shallow water environments is strongly influenced by the sea bottom, with which the sound has multiple interactions. Motivated by the demands of predicting sound field and acoustic energy distributions over long range in the littoral regions, geoacoustic inversion has been an active research topic.

As is often the case on the continental shelves, sound propagation is also highly affected by the variable ocean environments due to internal waves, fronts, eddies and diurnal tides etc. The impact of a variable ocean environment on sound propagation is greater at higher frequencies because the acoustic signal is more sensitive to the variations in the water column sound speed profile (SSP) [1]. The complicated temporally /spatially varying ocean environments clearly bring more challenges to geoacoustic inversion.

This paper describes a geoacoustic inversion study of high frequency, broadband signals in a time varying oceanic environment [2] [3](© [2010] IEEE). A multi-step inversion scheme was applied to invert the bottom parameters sequentially. First, effective water column sound speed profile and experimental geometry are estimated

CP1272, *Shallow-Water Acoustics, Second International Shallow-Water Acoustics Conference*
edited by J. Simmen, E. S. Livingston, J. X. Zhou, and F. H. Li

to mitigate their effects on sea bottom property estimates; second, the sea bottom sound speed and sediment layer thickness are inverted; third, the frequency dependence of the sediment p-wave attenuation is obtained by evaluating the ratio of the reflections from the sea bottom and a sediment layer, and then the attenuations in dB/m/kHz and in dB/λ are computed by using the estimates inverted in the previous steps.

THE EXPERIMENT AND DATA

The data analyzed here were collected on August 31, 2006 near the shelf break of New Jersey Continental Shelf during the Shallow Water Experiments 2006 (SW06) [4]. The acoustic signals were transmitted from the research vessel *Knorr* and recorded on a vertical line array (VLA) developed by Marine Physical Laboratory, Scripps Institution of Oceanography. *Knorr* was dynamically positioned at (39°1.520'N, 73°2.109'W) during the experiment and the VLA was anchored at (39°1.4771'N, 73°2.259'W). According to the surveys in the vicinity, the experimental site was on top of the low-speed clay outer shelf sediment wedge [5]. Underneath was a subsurface seismic reflector (the so called "R"-reflector), which is ubiquitous at SW06 site [5]-[7]. The depth of the "R"-reflector is about 20 m under the seafloor based on a sediment sound speed of 1600 m/s [8].

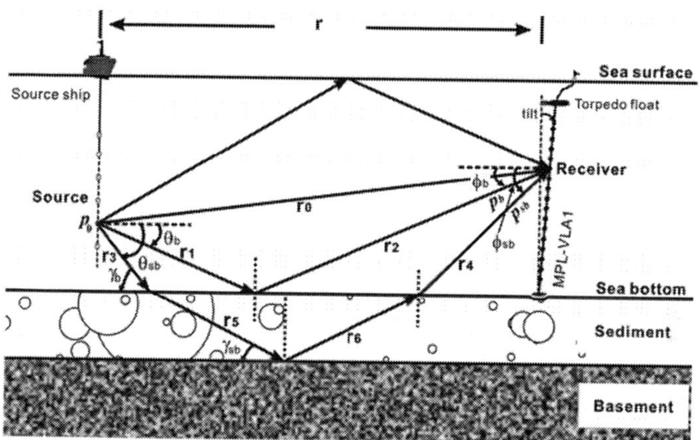

FIGURE 1. Experimental geometry and typical eigen rays [3].

Experimental Geometry

Figure 1 sketches the experimental setup. The distance between the source and the VLA was 230 m. The water depth was around 79 m. During the experiment, the source was lowered in the water at depths of 15 to 65 m in 10 m intervals. The source was held about 5 minutes at each depth. There were 16 hydrophones evenly spaced at 3.75 m on the VLA. According to the data acquired by the tilt/pressure/temperature (TPT) sensor attached to the VLA, the array leaned to one direction with a tilt of 2°±0.6°. The source lowering experiment lasted 45 minutes.

Oceanic Environmental Measurements

Two sets of oceanic environmental observations were obtained. One is conductivity -temperature-depths (CTD) measured at the source position (shown in Fig.2(a)), and the other one is the temperature measurement by the TPT on the VLA at a rate of one sample every 30 seconds, shown in Fig.2(b). The nominal depth of the TPT is 14.05 m.

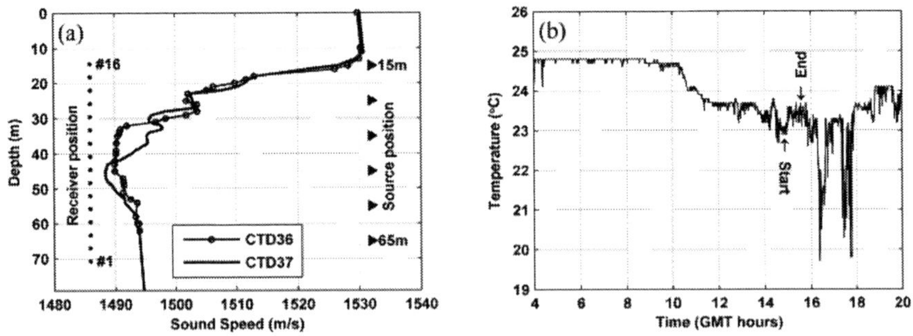

FIGURE 2. Oceanic environment measurements. (a) SSPs derived from CTD casts at source position [2], and (b) temperature measurements at receiver position [2].

There were two CTDs obtained for the source lowering experiment, one was taken before the 15 m source lowering (CTD36) and the other one was taken immediately after the acoustic data acquisition of the 65 m source (CTD37). Both SSPs derived from the CTD casts have different, well-resolved microstructures between the depths of 20 to 60 m. Notable differences in the thermocline structure suggest oceanic environmental variations in time. The 'Start' and 'End' arrows in Fig.2(b) mark the beginning and the ending times of the source lowering experiment. The temperature variations measured at the VLA position over this period of time support the hypothesis of temporally varying oceanic activities observed at the source position.

Acoustic Observations

Linear frequency modulation (LFM) pulses were transmitted from the source at a rate of one pulse per second. The LFM pulse duration was 1 s over a frequency band from 1.5 to 4.5 kHz. To assist in identifying different arrivals such as surface reflected (SR), bottom reflected (BR), surface–reflected–bottom-reflected (SRBR) and bottom–reflected–surface-reflected (BRSR), the received pulses received at the VLA were matched filtered and then aligned with respect to direct arrivals.

Severe fluctuation in the amplitude of direct arrival is observed in the signals at most of the hydrophones, especially for the ones whose depths are shallower than 55 m and for source depths are from 15 to 45 m. Accompanied by the large amplitude fluctuation is broadening of the signal duration for the source-receiver pairs within the thermocline. The main reason is that the sound travels within/through the thermocline region and is affected by the SSP microstructures. Consequently the direct arrival often splits into more than one path so that the arrival times at the same hydrophone are different. Large amplitude fluctuation in the SR arrival is also observed. If the pulses are aligned with respect to the direct arrival, large travel time fluctuation of the SR is evident. The fluctuation in the SR is mainly caused by the rough sea surface and the depth variation of the acoustic source. The amplitude fluctuation of the BR arrival is relatively small and so is the travel time fluctuation. A strong reflection from a sub-bottom layer is observed in the received signals from the 25 to 65 m sources. An example of the preprocessed signals for source depth at 45 m is shown in Fig. 3. The sub-bottom reflected (SUBR) signal, which is identified as reflections from the 'R'-reflector, is evident at hydrophones 8 to 14.

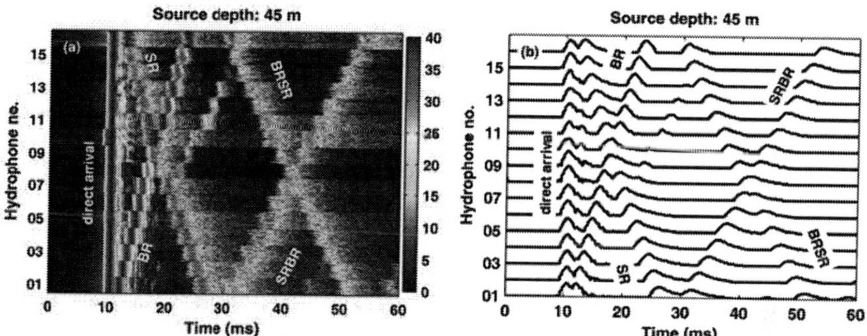

FIGURE 3. Matched filtered chirp pulses received by the hydrophones on the VLA for the source at 45 m. (a) One minutes signals stacked vertically (the unit of the colorbar is dB), and (b) The waveform display of the received signals averaged over one minute[3]

GEOACOUSTIC INVERSION APPROACH

Multi-Step Inversion Scheme

Due to the significant amplitude fluctuation in the direct arrivals, the approaches that involve amplitude information of direct arrival, such as making use of the bottom reflection coefficient versus grazing angle, are excluded. Only the relative travel time of the SR, BR and SUBR arrivals, and the amplitude of BR and SUBR arrivals are used in this inversion study.

The travel times of different arrivals are determined by the experimental geometry as well as the water column SSP. Since the simulations using different measured water column SSPs indicate that the impact of the oceanic SSP on the travel time is not negligible [2], two important facts need to be addressed in the inversion: a) the effects of the time varying oceanic SSP and the experimental geometry on the sub-bottom parameter estimates, and b) the correlation between the water column SSP and the experimental geometry. A multi-step inversion scheme (MSIS) was applied to account for the complications mentioned above:

1) Use the SSP derived from CTD37, which was measured immediately after the 65 m source transmitting, and the relative travel times of the SR and BR arrivals with respect to the direct arrival measured when source at 65 m, to

invert for the experimental geometry including water depth (WD), the range between the source and the VLA (RNG), source depth (SD) and the VLA tilt.

2) For the acoustic data for the source depths from 25 to 55 m: The water depth and the source-receiver range are fixed at the values that are inverted in the previous step, since it is reasonable to assume that WD and RNG will not change significantly within 45 minutes. Again, we use relative travel time of SR and BR with respect to the direct arrival to invert for oceanic SSP, SD and the VLA tilt.
3) The experimental geometry and the water column SSP are fixed at the values inverted in the previous steps. The relative travel times of the SUBR with respect to the BR arrival at multiple hydrophones are used to invert for the sub-bottom sound speed and the layer thickness.
4) For the acoustic data for source depths of 35 and 45 m. The relative amplitudes of the BR and SUBR arrivals at different sub-frequency bands are used to estimate the sediment sound attenuation.

Travel Time Inversion

The first three steps in MSIS are done by nonlinear travel time inversion through optimizations. In steps 1) and 2), the matched filtered pulses are aligned with respect to the direct arrival, and in step 3), the pulses are aligned with respect to the BR path. The pulses that increase the standard deviation (STD) of the travel time are excluded so that the STDs of the travel times are within the time resolution of the transmitted signal, which is 0.3ms.

The search algorithm is a hybrid optimization algorithm – adaptive simplex differential evolution (ASDE)[2]. ASDE embeds the local search algorithm downhill simplex into the global search algorithm differential evolution to retain the advantages of both algorithms and consequently improve the search efficiency.

Since the bathymetry between the source and the VLA is almost flat with only 0.6 m variation over 230 m of range, range independence is assumed in the analysis. A range-independent ray tracing model is used as acoustic forward model. The model searches for the ray parameters for the eigenrays connecting the source and receiver. The performance of the ray tracing performance in predicting travel times has been benchmark against a normal mode acoustic model.

The geoacoustic parameterization is based on the resolvable sub-bottom reflections in the received signals. It depends on the time resolution of the signal transmitted and the strength of the reflections from the sediment fine structure. Since a strong reflection from a sub-bottom layer is observed at multiple hydrophones on the VLA for the 25 to 65 m sources, a one sediment layer over a half-space is used as the geoacoustic model in the inversion.

To mitigate the impact of the oceanic SSP variation on the travel time geoacoustic inversion, a range-independent oceanic SSP is inverted before the inversion of the sediment properties at each source depth. The oceanic SSP is parameterized in terms of empirical orthogonal functions (EOFs). Since the SSPs measured either in the vicinity of the experimental site or near the time period the experiment being conducted will provide the most relevant information [1] [2], a total of 10 CTDs measured within 5 km from the VLA on August 31, 2006 was used in the EOF analysis.

The objective function for the travel time inversion is based on comparing the relative ray travel time differences between the measured and modeled ones. In step 1) and 2) of MSIS, the relative ray travel time is the travel time difference of SR and the direct arrivals, and the difference of BR and the direct arrivals. In step 3) of MSIS, the relative ray travel time is the travel time difference of the SUBR and BR paths. The objective function is written as:

$$E(\mathbf{m}) = \sum_{l=1}^{N_L} \left\{ \frac{1}{K_{HL}} \sum_{k=1}^{K_{HL}} \left(t_{l,k} - T_{l,k}(\mathbf{m}) \right)^2 \right\} \tag{1}$$

where:

$E(\mathbf{m})$: mismatch to be minimized by the optimization algorithm;

$\mathbf{m}$: M-element vector of model parameter to be estimated;

N_L : number of layers. N_L=2 in steps 1) and 2) of MSIS and N_L=1 in step 3);

K_{HL} : number of hydrophones' signals to be used in the inversion. K_{HL}=16 in steps 1) and 2). In step 3), it depends on the number of hydrophones that contain resolvable SUBR.

$t_{l,k}$: measured relative travel time at layer l and hydrophone k.

$T_{l,k}(\mathbf{m})$: modeled relative travel time at layer l and hydrophone k.

The mismatch takes any positive value, with 0 indicating a perfect match.

Sediment Sound Attenuation Estimation

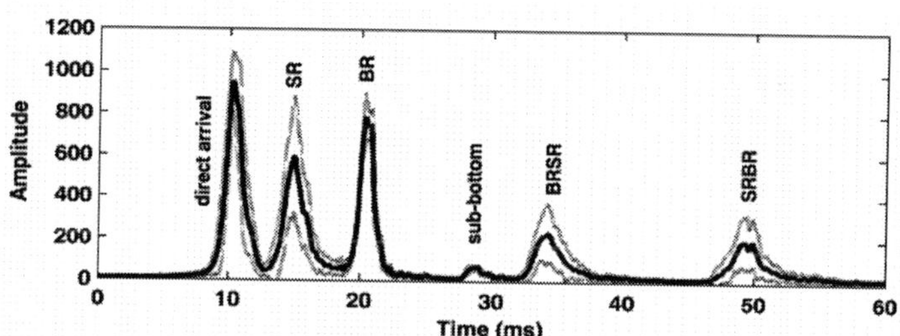

FIGURE 4. Averaged matched filtered chirps received by one the hydrophones on the VLA for the source at 45 m. The dark solid line in the middle is the mean amplitude and the two light dotted lines are mean +/- standard deviation [3]

Typical single bounce paths between the source and the receiver are shown in Fig. 1. The paths considered in this analysis are the bottom (i.e., r_1-r_2) and sub bottom (i.e. r_3-r_5-r_6-r_4) paths. The matched filtered signal averaged over one minute at one of the hydrophones on the VLA for the 45 m source is shown in Fig. 4. It is seen that the STDs of the BR and SUBR paths are much smaller than the other paths.

If we assume water sound attenuation is negligible over the short path lengths in the chirp frequency band, and the reflection and transmission coefficients of the sound at the interface of the sea bottom and sub-bottom for ray paths with grazing angles greater than the critical angle are frequency independent, then the amplitude ratio of BR to SUBR $\Delta P(f)$ can be written as [3]:

$$\Delta P(f) = (r_5 + r_6) \cdot \alpha_{sb}^{(f)} \cdot f + B \tag{2}$$

or

$$\Delta P(f) = (r_5 + r_6) \cdot \alpha_{sb}^{(f_0)} \cdot (f / f_0)^{\beta} + B \tag{3}$$

where f is frequency in kHz, α_{sb}^{f} is sediment sound attenuation in dB/m·kHz, $\alpha_{sb}^{f_0}$ is the sediment sound attenuation in dB/m at a reference frequency f_0, here 1 kHz, β is the exponent of the frequency dependence of the attenuation, and B is a positive constant. Eqs. (2) and (3) represent linear and nonlinear frequency dependence of the attenuation respectively.

The amplitudes of the sea bottom and sub bottom reflections at different frequencies were obtained from the compressed pulses filtered by a series of band pass filters (total eight sub-bands). The STD of $\Delta P(f)$ at each frequency band was obtained by averaging the signals over one minute. In order to determine the frequency dependence of the attenuation from the data, a Monte Carlo approach was carried out by fitting every possible combination of $\Delta P(f)$ according to the data fluctuation distributions over eight frequency bands. Gaussian distribution of $\Delta P(f)$ fluctuation at each frequency was assumed in the analysis. The consistency of the estimates and their 95% credibility interval from the data of different source - receiver combinations were used to determine the frequency dependence of the attenuation.

When the frequency dependence of the attenuation is determined, the actual values of α_{sb}^{f} and $\alpha_{sb}^{f_0}$ are computed by using $(r_5 + r_6)$ and sediment sound speed which were inverted in the previous steps in MSIS.

Inversion Results

The sediment sound speed and layer thickness estimates are shown in Table 1. Overall, the estimates for difference source depths are consistent. The combination of sediment sound speed and layer thickness matches the chirp seismic reflection data well in terms of two way travel time from the sea floor to the 'R'-reflector [5] [7].

TABLE 1. Summary of sediment sound speed and layer thickness from travel time inversion [2].

Source Depth (m)	Sound Speed (m/s)	Layer Thickness (m)
25	1609	21.5
35	1597	20.6
45	1600	22.3
55	1601	22.3
65	1584	20.4

Only four pairs of source – receiver data were chosen to estimate sound attenuation. One of the reasons [3] is because of the amplitude fluctuation at these hydrophones is smaller. The frequency dependence of the attenuation derived from the data is linear over the LFM frequency band. The attenuation estimates from linear fitting of the data from the four source-receiver pairs are self consistent, with values from 0.090 to 0.103 dB/m·kHz, or from 0.14 to 0.17 dB/λ.

SUMMARY

The experimental geometry using a vertical line array and a depth varying source at short range provides a simple and effective method for wide angular coverage of reflections from the ocean bottom, and also high quality single bounce reflection data from the sea floor and sub bottom interfaces. A multi step inversion scheme was applied to mitigate the effects of time varying ocean environment on the experimental geometrics and geoacoustic property estimates. A practical method of estimating the sediment sound speed and its frequency dependence by using this short range geometry is also developed.

ACKNOWLEDGMENTS

This work is supported by the Office of naval Research Ocean Acoustic Team. The authors would like to thank Dr. William S. Hodgkiss and Dr. Peter Gerstoft, for the acoustic data, Dr. David Knobles for the navigation data, and Dr. John Goff for the chirp seismic reflection data.

REFERENCES

1. Y.-M. Jiang and N. R. Chapman, "The impact of ocean sound speed variability on the uncertainty of geooacoustic parameter estimates ", *J. Acoust. Soc. Amer.*, **125**, pp2881-2895, 2009.
2. Y.-M. Jiang, N. R. Chapman and P. Gerstoft, "Estimation of geoacoustic properties of marine sediment using a hybrid differential evolution inversion method", *IEEE J. Ocean. Eng.* **35(1)**, pp59-69, 2010.
3. Y.-M. Jiang and N. R. Chapman, "Measurement of low-frequency sound attenuation in marine sediment", *IEEE J. Ocean. Eng.* **35(1)**, pp70-78, 2010.
4. D. J. Tang, J. Moum, J. Lynch, P. Abbot, R. Chapman, P. Dahl, T. Duda, G. Gawarkiewicz, S. Glenn, J. Goff, H. Graber, J. Kemp, A. Maffei, J. Nash, and A. Newhall, "Shallow Water '06—A joint acoustic propagation/nonlinear internal wave physics experiment", *Oceanography*, **20(4)**, pp. 156–167, 2007.
5. J. A. Goff, B. J. Kraft, L. A. Mayer, S. G. Schock, C. K. Sommerfield, H. C. Olson, S. P. S. Gulick, and S. Nordfjord, "Seabed characterization on the New Jersey middle and outer shelf: Correlatability and spatial variability of seafloor sediment properties", *Mar. Geology*, **209**, pp. 147–172, 2004.
6. T. A. Davies, J. A. Austin Jr., M. B. Lagoe, and D. Milliman, "Late quaternary sedimentation off New Jersey: New results using 3-D seismic profiles and cores", *Mar. Geology*, **108**, pp. 323–343, 1992.
7. A. Turgut, D. Lavoie, D. J. Walter, and W. B. Sawyer, "Measurements of bottom variability during SWAT New Jersey Shelf Experiments", *Impact of Littoral Environmental Variability on Acoustic Predictions and Sonar Performance*. Norwell, MA: Kluwer, pp. 91–98, 2002.
8. J. A. Goff and J. A. Austin, Jr., "Seismic and bathymetric evidence for four different episodes of iceberg scouring on the New Jersey outer shelf: Possible correlation to Heinrich events", *Mar. Geology*, **266**, pp. 244–254, 2009.

Geo-acoustic Inversion Using Shallow Water Ambient Noise and Analysis of Uncertainty

Jianheng Lin[a], Baoyou Yin[a,b] and Xuejuan Yi[a]

[a] *Qingdao Laboratory, Institute of Acoustics, Chinese Academy of Sciences, Qingdao 266023*
[b] *Key Laboratory of Underwater Acoustic Environment, Chinese Academy of Sciences, Beijing 100190*

Abstract. Ambient noise in shallow water undergoes multiple reflections between the sea surface and seabed, thus, the spatial structure of ambient noise field depends strongly on the geo-acoustic characteristics of the seabed. A great number of measurements show that the noise vertical coherence is a stable feature, changing little with sea states, which reflects the invariance of seabed parameters. Therefore, the noise vertical coherence can be used to estimate geo-acoustic parameters. In this paper, geo-acoustic parameters are estimated by the vertical coherence of measured ambient noise. The differential evolution method is applied in the global optimization parameters search. However, geo-acoustic inversion is a strongly nonlinear process, so the inversion solution may be uncertain and non-unique. According to Bayesian theory, the solution is fully characterized by its posterior probability density (PPD), which combines prior information about the forward model with information from measured data. A fast Gibbs sampler approach was used to calculate the multi-dimensional integrals of the PPD in the paper, and the validity of the inversion method based on the vertical coherence of the ambient noise was checked. For these inversions two different seabed models were employed. First, the seabed was assumed to be a homogeneous fluid half-space. Another seabed model added a fluid sediment layer on top of a homogeneous fluid half-space. In the end, substituting these inversion results into the forward model, comparisons were made to estimate the uncertainties of inversion results by the broadband vertical coherence of shallow water noise data.

Keywords: Ambient Sea Noise, Vertical Coherence, Matched Field Inversion, Fast Gibbs Sampler, Uncertainty Anlysis
PACS: 43.30Pc

1. INTRODUCTION

In shallow water, geo-acoustic parameters such as sound speed, density and attenuation coefficients play very important roles in acoustic field predication, matched field localization etc. Seabed parameters can be directly obtained by on-site sample surveys; however, acoustic inversion methods are usually applied in engineering situations, using acoustic propagation, reverberation and ambient sea noise data etc. Geo-acoustic parameter inversion by ambient sea noise uses natural sound such as wind-generated noise and rain noise, and needs no transmitted sound sources, which makes it safe and convenient to use. Especially, one of the benefits of the ambient noise inversion technique is that it can be accomplished using a simple two- element array.

CP1272, *Shallow-Water Acoustics, Second International Shallow-Water Acoustics Conference*
edited by J. Simmen, E. S. Livingston, J. X. Zhou, and F. H. Li

Buckingham[1], Deane[2] and Carbone[3] estimated the compressional and shear wave speeds, and attenuation coefficients of a shallow water seabed from the vertical coherence of ambient noise. They presented inversions from two shallow water sites: the Cortes Bank and the North Celtic Sea. Harrison, Simons and Baldacci[10-13] deduced the bottom reflection properties from ambient noise. They believed that it was a simple technique to extract geo-acoustic parameters from ambient noise measurements. Siderius and Geostoft[14,15] presented a passive method to determine water depth and seabed sub-bottom layering using ambient noise. Arvelo and Jr.[16] quantified the robustness and constraints of estimating bottom loss via simulations using processed ambient noise data collected during ASIAEX 2001. In this paper, we present inversions using ambient noise data in shallow water in China. An analysis of estimated geo-acoustic parameters is also given.

This paper is organized as follows. The method of estimating geo-acoustic parameters using vertical coherence characteristics of ambient noise is introduced in Section 2, including the ambient sea noise model, objective function, optimization algorithm and quantifying uncertainty in ambient noise inversion. In Section 3, seabed parameters are estimated using ambient noise data measured in Chinese shallow water in recent years, and the analyses for uncertainty of the inversion results are made. Conclusions are given finally.

2. GEO-ACOUSTIC PARAMETERS INVERSION BY AMBIENT SEA NOISE

In this paper the noise inversion method is based on the vertical coherence properties of ambient noise fields. The vertical coherence function is the normalized form of noise spectral density, which is a relatively stable characteristic, varying little with sea states. Deane and Buckingham[2] investigated ambient noise data measured at different times and sea states to verify that the vertical coherence of ambient sea noise is a robust and repeatable random process, which can be inverted for seabed parameters.

The inversion method involves four aspects as follows: the forward model, objective function, global optimization algorithm, and uncertainty analysis in geo-acoustic inversion.

2.1 Forward Model

The speed of running a forward model will affect the speed of the whole inversion process, because the forward model will be run repeatedly. In order to improve the inversion speed, we assume a range-independent and azimuthal symmetry environment, and we assume the surface noise sources are a plane of omnidirectional point sources which are Poisson distributed in space and time and which are located just below the sea surface. Carbone[3] gave an expression of the cross-spectral density of ambient noise received by a hydrophone pair in a water column:

$$S_{mn} = 4\pi v Q^2 \int_0^\infty rG(r,\omega,z_{rm})G^*(r,\omega,z_{rn})dr\,, \qquad (1)$$

where Q is the source strength, G is the Green's function, the asterisk denotes complex conjugate, and z_{rm} and z_{rn} are the depths of two receivers ($m, n = 1, 2$).

The noise vertical coherence is simply a normalized version of the cross-spectral density. It is expressed as a following broadband function:

$$C_{model}(\omega) = \frac{S_{12}(\omega)}{\sqrt{S_{11}(\omega)S_{22}(\omega)}}. \quad (2)$$

Fig. 1 shows sensitivities of seabed sound speed, density and attenuation to broadband noise coherence functions. We can conclude that the sound speed is highly sensitive to both the real and the imaginary parts of coherence, while density and attenuation are weakly sensitive to the imaginary part; the density and attenuation are very weakly sensitive to the real part below 500 Hz. Therefore, the real part of the coherence function is used to estimate seabed geo-acoustic parameters, and the frequency band from 400 Hz to 2kHz is chosen.

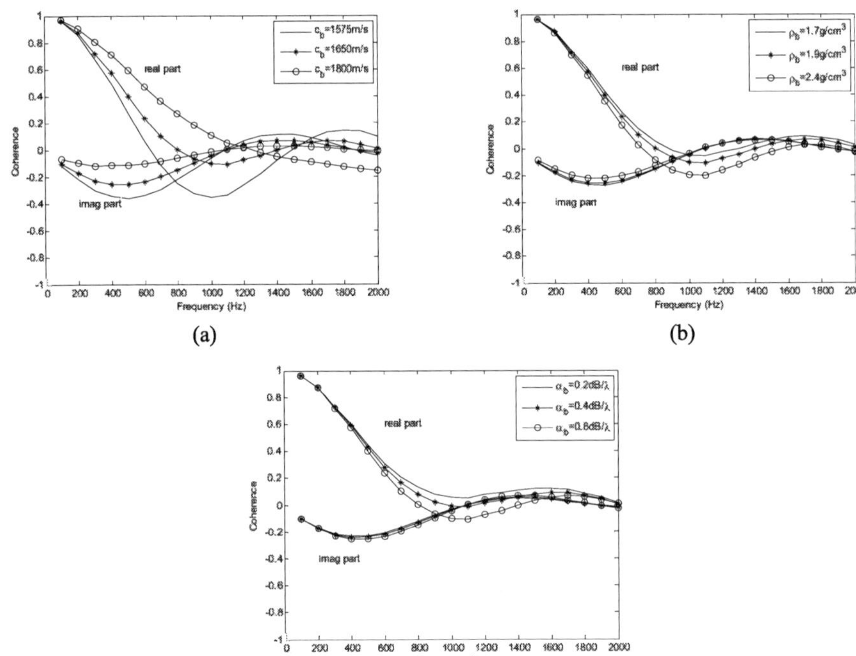

(c)

FIGURE 1. The sensitivity curves of different seabed parameters: (a) sound speed (b) density (c) attenuation

2.2 Objective Function

Dividing measured noise data into K segments and performing an *FFT* transform on this data, the noise cross-spectral density between two phones is given by:

$$\hat{S}_{mn} = \frac{1}{K}\sum_{k=1}^{K} X_m(f,\Delta T_k)X_n^*(f,\Delta T_k). \quad (3)$$

Accordingly, the vertical coherence of the measured noise field is given by:

$$C_{data}(f)=\frac{\hat{S}_{12}(f)}{\sqrt{\hat{S}_{11}(f)\hat{S}_{22}(f)}}. \tag{4}$$

The objective function demonstrates the matched relation between the measured sound field and the modeled replica fields. According to the sensitivities of the three seabed parameters, the objective function may be written as follows:

$$O=\sqrt{\frac{1}{N}\sum_{n=1}^{N}\left[\mathrm{Re}\left(C_{data}(f_n)-C_{model}(f_n)\right)\right]^2}, \tag{5}$$

where N is the number of frequency points, and ***Re*** denotes the real part. Because the range of the real part of the noise vertical coherence is from -1 to 1, the normalized objective function will change from 0 to 1.

2.3 The Differential Evolution (DE) Global Optimization

Differential Evolution[4] is a parallel direct search method which is powerful enough to reliably converge to the true optimum. DE requires only a few control variables which can be drawn from a well-defined numerical interval, so it is easy to operate.

DE utilizes N parameter vectors $x_{i,P}$ (i = 0, 1, 2, …, N-1) as a population for each generation G. N does not change during the minimization process. A uniform probability distribution is supposed in the initial population randomly. For each vector $\boldsymbol{x}_{i,P}$ (i = 0, 1, 2, …, N-1), a trial vector $\boldsymbol{x}_{i,P+1}$ is generated according to Eq. (6):

$$v_{i,P+1}=x_{r1,P}+F(x_{r2,P}-x_{r3,P}), \tag{6}$$

where r_1, r_2, r_3□[0, N-1], are integer and mutually different, and F is a real and constant factor which controls the amplification of the differential variation ($x_{r2,p}$-$x_{r3,p}$).

In order to increase the diversity of the parameter vectors, the vector $\boldsymbol{u}=(u_0, u_1, \ldots, u_{D-1})^T$ with

$$u_j=\begin{cases} v_j & j=\langle n\rangle_D,\langle n+1\rangle_D,\ldots,\langle n+L-1\rangle_D \\ x_{i,P} & j\in[0,D-1] \end{cases} \tag{7}$$

is formed, where the acute brackets $\langle\ \rangle_D$ denote the modulo function with modulus D, and n is a randomly chosen integer from the interval [0, D-1]. The integer L, which denotes the number of parameters, is drawn from the interval [1, D].

In order to decide whether the new vector $\boldsymbol{u}$ shall be a population member of generation P+1, it will be compared with $x_{i,P}$. If vector $\boldsymbol{u}$ yields a smaller objective function value than $x_{i,P}$, $x_{i,P+1}$ is set to u, otherwise the old value is retained.

2.4 Uncertainty Analyses in Geo-acoustic Inversion

Geo-acoustic inversion is an optimization process for multi-parameters, so there may be many combinations with different parameter values. Without some indication about uncertainties, no confidence can be placed in the model parameters obtained by inversion, and the inversion procedure will not be integral, either. A statistical method is valid to analyze the uncertainty of the inversion results. According to the Bayesian theorem, the inversion result can be fully characterized by its posterior probability density, which combines prior information about the forward model ***d*** with information from measured data ***m***. The conditional posterior probability density can be expressed as:

$$P(\mathbf{m} \mid \mathbf{d}) \propto L(\mathbf{d} \mid \mathbf{m}) P(\mathbf{m}), \tag{8}$$

where $L(\mathbf{d}|\mathbf{m})$ is the likelihood function, $P(\mathbf{m})$ is the prior probability distribution. For the assumption of Gaussian errors, the likelihood function is of the form $L(\mathbf{d}|\mathbf{m})$□exp[-$E(\mathbf{m},\mathbf{d})$], where ***E*** represents the error function, expressed as follows:

$$E(\boldsymbol{m}) = -\frac{|\boldsymbol{d} - C(\boldsymbol{m})|^2}{v}, \tag{9}$$

where ***v*** is the variance of the data errors.

Hence, the PPD becomes

$$P(\mathbf{m} \mid \mathbf{d}) \propto \exp[-E(\mathbf{m}, \mathbf{d})] P(\mathbf{m}), \tag{10}$$

when normalized,

$$P(\mathbf{m} \mid \mathbf{d}) = \frac{\exp[-E(\mathbf{m}, \mathbf{d})] P(\mathbf{m})}{\int \exp[-E(\mathbf{m'}, \mathbf{d})] P(\mathbf{m'}) d\mathbf{m'}}, \tag{11}$$

where the domain of integration spans the multi-dimensional model space.

To interpret the multi-dimensional PPD requires calculation of its moments, such as the mean, covariance, and marginal distributions. A fast Gibbs sampler algorithm is introduced to calculate these quantities.

3. INVERSION RESULTS AND UNCERTAINTY

In this section, ambient noise data are used to estimate geo-acoustic parameters based on vertical coherence of surface noise.

3.1 The First Experimental Site

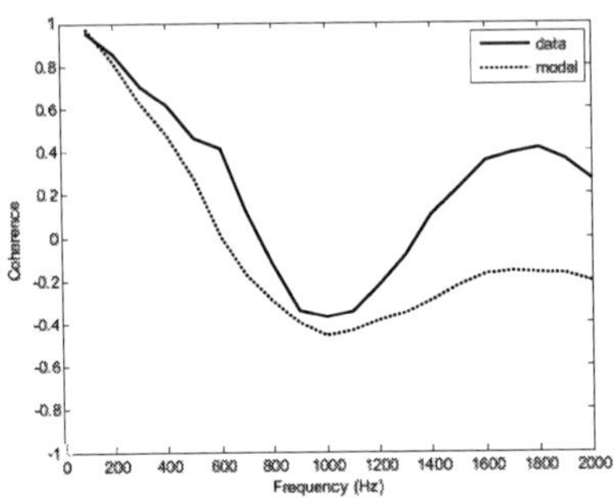

FIGURE 2. Comparison between inversion results and experiment data.

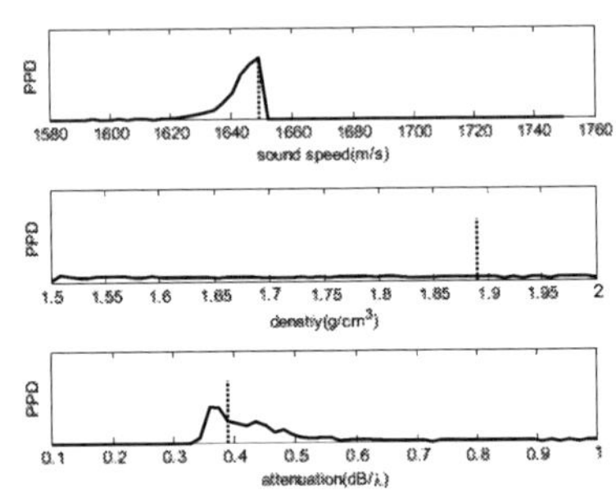

FIGURE 3. Estimated marginal PPDs for the estimated parameters.

The sound speed profile (SSP) of the experimental site is typically downward refractive. The wind speed is only 2.8 m/s during the experiment. A Sea Chart shows the seabed composition at this site is silt. Ignoring the shear effects, the seabed can be simplified as a fluid half-space. Comparison between the noise vertical coherence using the estimated seabed parameters and the measured vertical coherence is shown in Fig. 2. As can be seen, when the frequency is low, the difference between them is small, whereas the difference clearly increases with frequency. The error might be attributed to high frequency scattering from the surface or the bottom.

The searching ranges for these three parameters, the inversion results using ambient noise data and the propagation data of explosive sound obtained at the same experimental site[9] are given in Table 1. Fig. 3 shows the estimated marginal PPDs for this experiment. The solid curve indicates the marginal probability density function, while the dashed one corresponds to the best parameter value. The density is difficult to estimate.

TABLE 1. Inversion results.

	Search ranges	Propagation Inversion results	Noise inversion results
sound speed (m/s)	1580~1750	1600.9	1649
density (g/cm^3)	1.5~2.0	1.79	1.89
attenuation (dB/λ)	0.1~1.0	0.29	0.39

3.2 The Second Experimental Site

The water depth is about 82 m, with a constant sound speed profile in winter. Fig. 4 shows the comparison between measured and modeled noise vertical coherences when the seabed is still simplified as a fluid half-space. The best parameters are as follows: sound speed of 1657.6 m/s, density of 1.7 g/cm^3, and attenuation of 0.7 dB/λ. However, there are significant differences between the two curves.

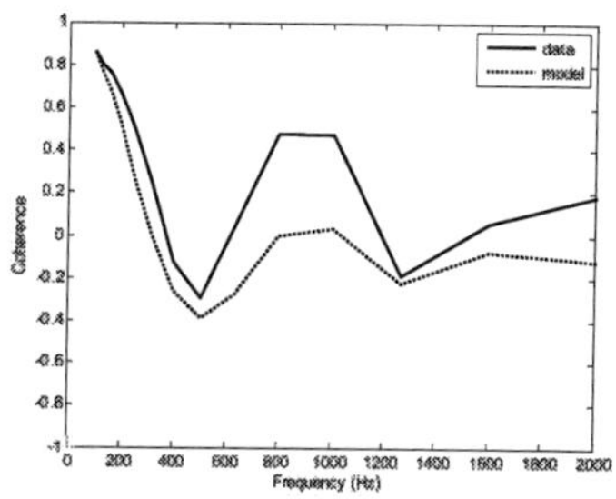

FIGURE 4. Comparison between inversion results and experimental data

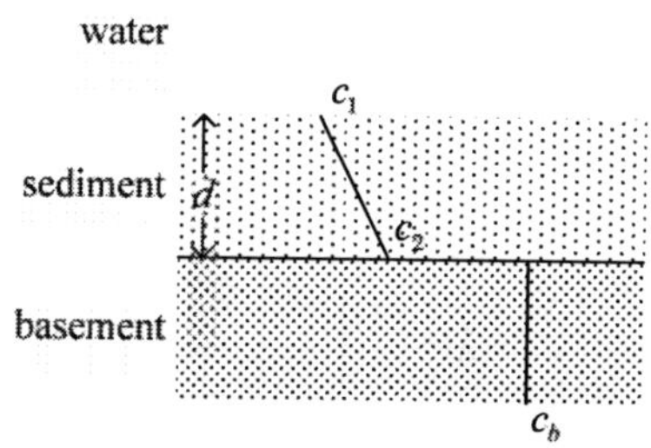

FIGURE 5. The layered seabed model used in inversion

From Fig.4, we find that the position of zero crossing of the modeled noise vertical coherence is different from the measured data. While the layered structure of the seabed may affect the noise vertical coherence[8], the obvious effect of the layered structure is that it will shift the zero crossing of the real component of the coherence to a higher frequency. In order to check this seabed effect, a layered seabed model is introduced as shown in Fig. 5. This model is characterized by a thin layer of fluid sediment over an infinite half-space. Because the sensitivities of density and attenuation to noise vertical coherence is small, we regard both in sediment as equal to the values in basement: $\rho = 1.7$ g/cm^3 , $\alpha = 0.7$ dB/λ.

The searching ranges of the parameters to be inverted and inversion results are given in Tab. 2. The measured noise vertical coherence is compared with calculated values using the ambient noise model as shown in Fig.6. They coincide at lower frequencies. Parameter estimates are plotted in Fig. 7. In fact, the layered structure is not acquired in the experiment because of equipment limitations. So what we have done using a two-layered-seabed is just a try; it should be verified in the future.

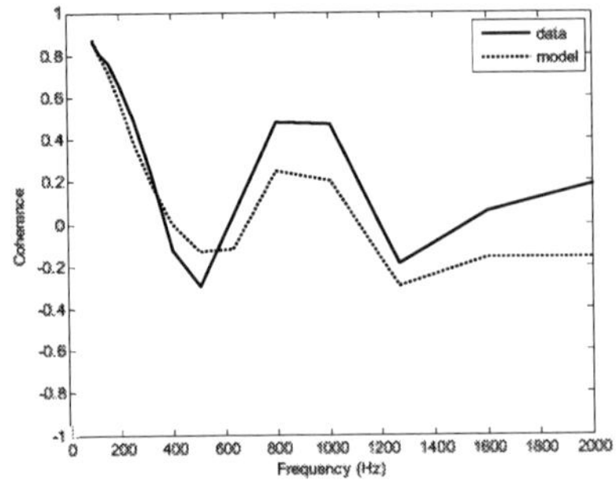

FIGURE 6. Comparison between inversion results and experiment data.

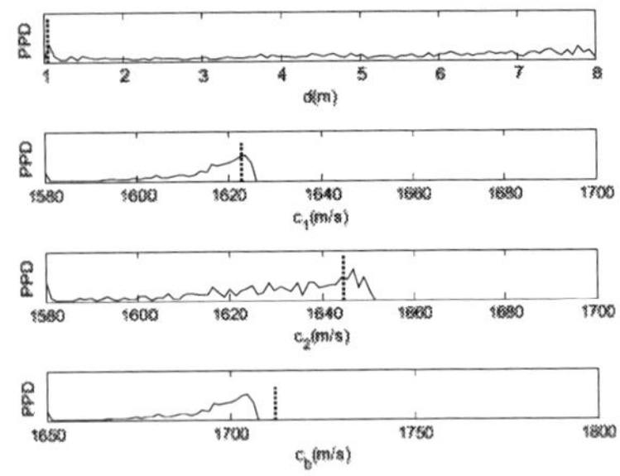

FIGURE 7. Estimated marginal PPDs for the estimated parameters.

TABLE 2. Inversion results.

	Searching ranges	**Noise inversion results**
c_1 (m/s)	1580~1700	1622.7
c_2 (m/s)	1580~1700	1644.5
c_b (m/s)	1650~1800	1712.1
d (m/s)	1~8	1.0583

4. CONCLUSIONS

The sensitivities of different seabed parameters to noise vertical coherence were discussed based on a simple range-independent ambient sea noise model. Then, geo-acoustic parameters were estimated by inversion of the measured noise vertical coherence with a global optimization technique as an automatic search method. In the two experiments in China, we assumed the seabed to be a fluid half-space firstly, and the sound speed, density and attenuation were obtained. In the second experiment, we assumed the seabed included a thin fluid sediment layer on top of a homogenous fluid basement, and got a reasonable coincidence. At the end of each inversion procedure, the estimation of uncertainty of parameters to be inverted was made. We can conclude that:

(1) The inversion results were in good agreement with the measured data in lower frequencies when the seabed was considered as a half-space in the first experiment, while the fit is somewhat less precise at frequencies above 1600 Hz.

(2) The fit in the second experiment was not good if we continued to consider the seabed as a homogenous half-space. However, the results have been improved if assuming a single-sediment model.

(3) The model dependence was a significant factor to affect inversion results because different seabed models might produce different effects on vertical coherence.

(4) Because density and attenuation are less sensitive to noise vertical coherence, improving the inversion precision of the two parameters is a problem to be solved in the future.

ACKNOWLEDGMENTS

The authors would like to acknowledge the Key Laboratory of Underwater Acoustic Environment, Chinese Academy of Sciences providing the ambient noise data. In addition, we thank Carbone and Buckingham for fundamental work in ambient noise model and inversion procedure. We also thank Chapman and Yongmin Jiang for their guidance on signal processing. This work was supported by the National Natural Science Foundation of China No. 10810101032.

REFERENCES

1. M. J. Buckingham. "Theoretical model of ambient noise in a low-loss, shallow water channel", *J. Acoust. Soc. Am.* **67** (4), 1186-1192 (1980).
2. G. B. Deane and M. J. Buckingham. "Vertical coherence of ambient noise in shallow water overlying a fluid seabed", *J. Acoust. Soc. Am.* **102** (6), 3413-3424 (1997).
3. N. M. Carbone, G. B. Deane, and M. J. Buckingham. "Estimating the compressional and shear wave speeds of a shallow water seabed from the vertical coherence of ambient noise in the water column", *J. Acoust. Soc. Am.* **103** (2), 801-813 (1998).
4. R. Storn and K. Price. "Differential Evolution – a simple and efficient adaptive scheme for global optimation over continuous space", Technical Report TR-95-012, ICSI, 1995.
5. S. E. Dosso. "Quantifying uncertainty in geoacoustic inversion. I. A fast Gibbs sampler approach", *J. Acoust. Soc. Am.* **111** (1), Pt. 1, 129-142 (2002).
6. S. E. Dosso. "Quantifying uncertainty in geoacoustic inversion. II. Application to broadband, shallow-water data", *J. Acoust. Soc. Am.* **111** (1), Pt. 1, 143-159 (2002).
7. E. K. Westwood. "A normal mode model for acousto-elastic ocean environments", *J. Acoust. Soc. Am.* **100** (6), 3631-3645 (1996).
8. Baoyou Yin, Jianheng Lin, and Xuejuan Yi, et al. "Sensitivity of Geoacoustic Parameters to Vertical Coherence of Ocean Ambient Noise", *Technical Acoustics*, **28**(5), Pt. 2, 84-86(2009).
9. Licheng Lu and Li Ma. "Geoacoustic inversions from experiment data in shallow water," *Technical Acoustics*, **27**(5), Pt. 2, 56-57(2008).
10. C. H. Harrison and D. G. Simons. "Geoacoustic inversion of ambient noise: A simple method", *J. Acoust. Soc. Am.* **112** (4), 1377-1389 (2002).
11. C. H. Harrison and A. Baldacci. "Bottom reflection parameters deduced from ambient noise: simulation and experiment", in proceeding of Theoretical and Computational Acoustics 2003, 162-177 (2003).
12. D. G. Simons, C. V. Moll, and C. H. Harrison. "Inversion of shallow water ambient noise data by means of differential evolution as a global search method", in A. Caiti, N. R. Chapman, J. P. Hermand and S. M. Jesus eds. "Acoustic Sensing Techniques for the Shallow Water Environment", printed in the Netherlands, 265-249 (2006).
13. C. H. Harrison. "Reflection loss and sub-bottom profiling with ambient noise", in A. Caiti, N. R. Chapman, J. P. Hermand and S. M. Jesus eds. "Acoustic Sensing Techniques for the Shallow Water Envrionment", printed in the Netherlands, 281-288 (2006).
14. M. Siderius, C. H. Harrison, and M. B. Porter. "A passive fathometer technique for imaging seabed layering using ambient noise", *J. Acoust. Soc. Am.* **120** (3), 1315-1323 (2006).
15. P. Gerstoft, W. S. Hodgkiss, M. Siderius, et al. "Passive fathometer processing", *J. Acoust. Soc. Am.* **123** (3), 1397-1305 (2008).
16. J. I.Arvelo and Jr. "Robustness and constraints of ambient noise inversion", *J. Acoust. Soc. Am.* **123** (2), 679-686 (2008).

Bottom Parameters Inversion Research based Tow-Ship Noise

Fuchen Liu Shihong Zhou

Hangzhou Applied Acoustic Research Institute, Hangzhou, China

Abstract. This paper discusses geoacoustic inversion from tow-ship noise data acquired via a horizontal towed array. Self-noise geoacoustic inversion involves the estimation of bottom parameters such as sound speeds, densities and attenuation. This bottom parameter method is designed using a Line-spectrum of tow-ship noise through near-field propagation mode and genetic arithmetic. A global search algorithm, i.e., GA is applied to the optimization problem. The experimental results are particularly encouraging in view of the high level of interference shown to be tolerated from nearby shipping.

Keywords: Geoacoustic inversion, self-noise, tow ship. horizontal linear array,

INTRODUCTION

Nowadays with the development of passive sonar systems, *in-situ* bottom parameter inversion has become more and more important for sonar performance evaluation and prediction, especially in shallow water areas. This has lead to increasing interest in self-noise inversion, wherein plant and hydrodynamic noise generated by the tow ship itself is received on the array and processed to extract estimates of local geoacoustic parameters such as bottom velocity. An introduction to the self-noise inversion problem and associated physics can be found in [1], [2], [3].

This paper examines self-noise geoacoustic inversion from the technique of near-field–matched-field processing (MFP). It uses a Line-spectrum of tow-ship noise, otherwise wide-band noise which comes from the changing position of the tow-ship, to determine the distance from the source to the array, without a single equivalent source center. Solution of the geoacoustic inversion problem itself is posed in a multiparameter optimization framework and solved by an efficient global search based on genetic algorithms (GA).

To validate the robustness of the method, effective bottom parameters are estimated by using experimental data acquired from one experiment in 2008. The results obtained are considerably encouraging.

INVERSION MODEL AND ARITHMETIC

In fig.1, the sea bottom consists of one sediment layer and one basement layer, and c_p , ρ , α , $c_p^{'}$,

CP1272, *Shallow-Water Acoustics, Second International Shallow-Water Acoustics Conference*
edited by J. Simmen, E. S. Livingston, J. X. Zhou, and F. H. Li

ρ', α' represent the compressional velocity, density and attenuation parameters of the sediment and the basement respectively. Otherwise four parameters, depth of water column, depth of sediment, towed depth of HLA and range between the equivalent acoustic center of the ship and the HLA are also crucial to inversion.

Ship-radiated noise exists as a series of inherent line-spectrua, which can be received by a sonar array after near-field propagation. Inversion is shown in the chart below.

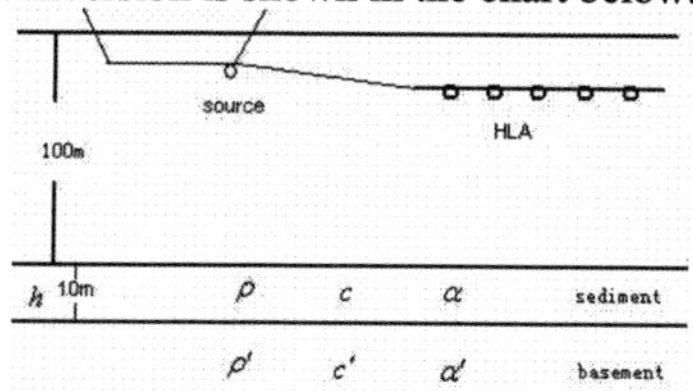

FIG.1 Typical geometry for shallow water towed-array and inversion paramters

Firstly, the inherent line-spectrum is attained by custom beam-forming processing towards the towed ship's tracing azimuth, which can filter interference from any other azimuth. Noticeably, the tracing azimuth from the ship is between ten degrees and twenty degrees because of muti-path propagation from the source signal. The data vector $B(f)$ consists of all hydrophone line-spectra.

Secondly, the match vectors $P(f)$ are obtained through near-field model computing. Here, the fast-field propagation model based wave-number integrator is applicable in near-field acoustic field modeling in Eq.1.

$$P_i(f) \approx \sqrt{\frac{1}{2\pi r}} e^{i[k_r r-(m+\frac{1}{2})\frac{\pi}{2}]} \int_0^{\infty} g(k_r, z)\sqrt{k_r} e^{ik_r r} dk_r \qquad (1)$$

Thirdly, a Bartlett matched processor is chosen because of robust performance in following Eq.2. Genetic arithmetic is applicable to the global search parameter space based crossover heuristic function.

$$MF = \sum_{f_l}^{f_h} \frac{\sum |P \cdot B^H|^2}{\sum |P|^2 \cdot \sum |B|^2} \qquad (2)$$

RESULT OF SEA EXPERIMENTAL DATA

An experiment was conducted in an oceanic waveguide with a transition layer. The water depth was 90m while the acoustic speed decreased from 1540 m/s to 1525 m/s shown in Fig.2. The towed ship signal was composed of many line-spectra. A spectrum line of 86 Hz exists in the low frequency band according to time and frequency analysis (data length : 100s) in Fig.3.

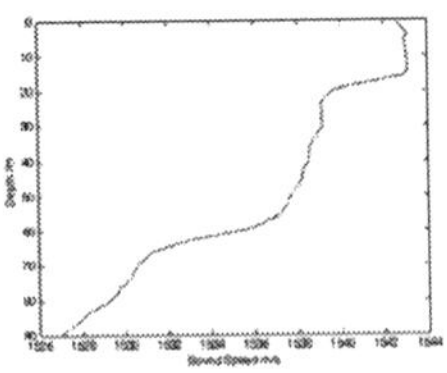

FIGURE 2 Experiment sea ssp

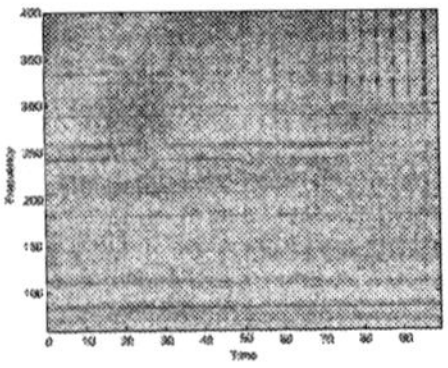

FIGURE 3 Received signal time and frequency analysis

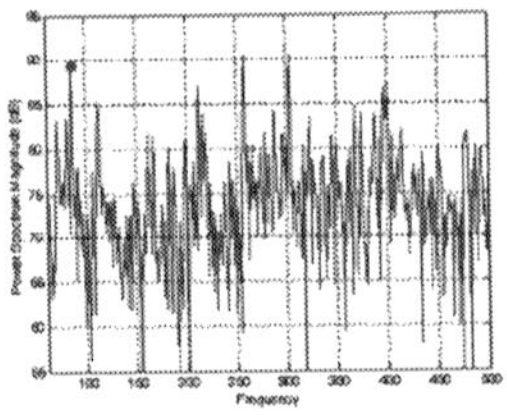

FIGURE 4 power spectrum magnitude of one array cell

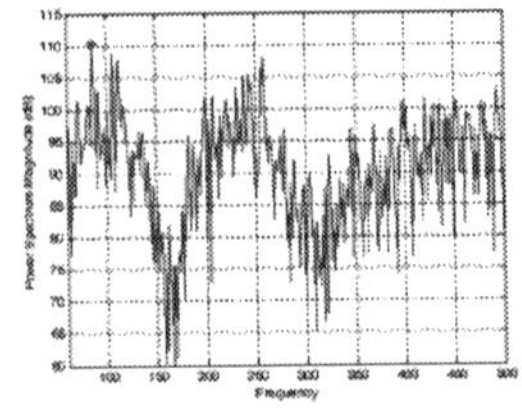

FIGURE 5 power spectrum magnitude of CBF

Fig.4 indicates the spectrum of one hydrophone, whose power spectrum magnitude is 89 dB; in Fig.5 the power spectrum magnitude is 110dB after custom beam-forming towards tracing azimuth. This increment accords with array plus. Thereby, a conclusion can be made that this spectrum line is from the towed ship rather than some other target source.

In genetic arithmetic, the population is set to 60, and the fitness function is a crossover heuristic. After 113 iterations, the Optimization Value correlation coefficient between the matched vector and the data vector is convergent to 0.81 in Fig.7. The optimization convergence values of ten parameters are shown in Table.1. The converge process of the genetic arithmetic is shown in Fig.6.

Basement sound speed	1703.7 m/s	Sediment sound speed	1628.9 m/s
Basement density	1.97 kg/m^3	Sediment density	1.788 kg/m^3
Basement attenuation	0.82 dB/λ	Sediment attenuation	0.53 dB/λ
Depth of Sediment	18.27 m	Depth of source	1.71 m
Range	0.343 km	Depth of HLA	5.93 m

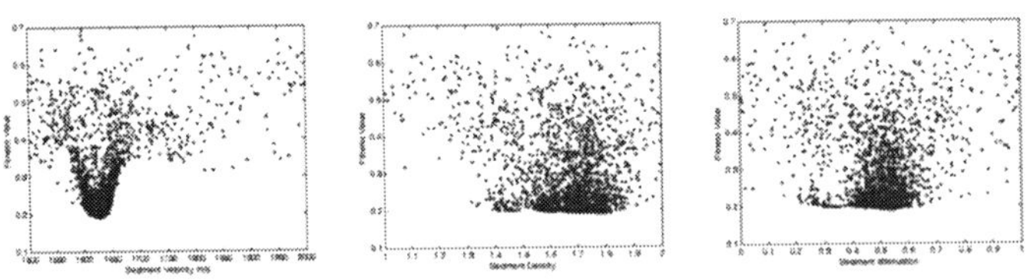

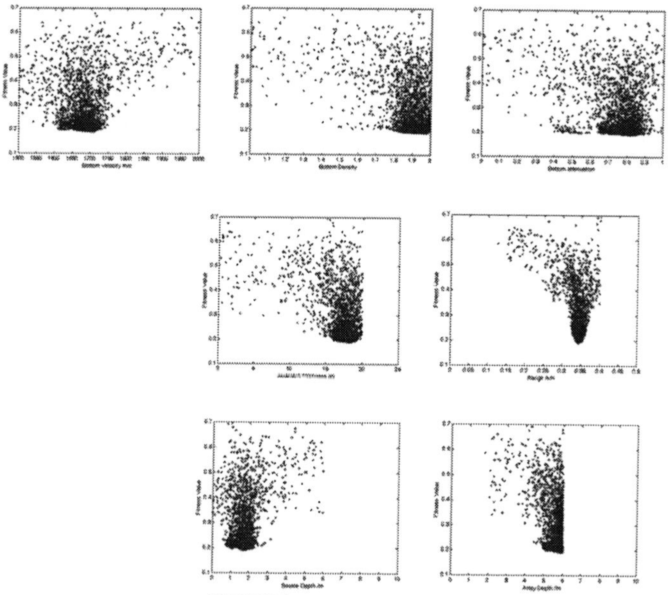

FIGURE 6 parameter converge process of genetic arithmetic

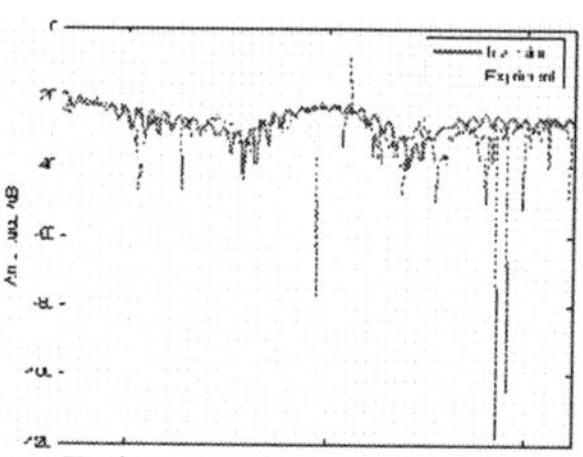

FIGURE 7 Each array cell transmission loss contrast between experiment data and inversion data

CONCLUSION

The following conclusion has been made. Parameters like sound speed of sediment and basement , depth of water , depth of sediment , towed depth of HLA and range between the equivalent acoustic centers of the ship and the HLA all are sensitive and converge rapidly to real values, which are consistent with measured values. In contrast, parameters like density, attenuation parameters of sediment and basement are not sensitive. In a word, this method is effective for inversion of sound speed of basement, which is valuable to sonar performance's evaluation and prediction. In future, we will develop a new optimized matched arithmetic to reduce iterative times, and develop a wide-band algorithm to improve its performance.

REFERENCES

1. D J Battle, P Gerstoft, W S Hodgkiss, et al. Bayesian model selection applied to self-noise geoacoustic inversion[J]. JASA, 2004, 116: 2043- 2056.
2. D J Battle, P Gerstoft, W A Kuperman, et al. Geoacoustic inversion of tow-ship noise via near-field-matched-field processing[J]. IEEE J Ocean Engineering, 2003,OE-28: 454-467.
3. C Park, W Seong, P Gerstoft. Geoacoustic inversion in time domain using ship of opportunity noise recorded on a horizontal towed array[J]. JASA, 2005, 117: 1933-1941.
4. C Park, W Seong, P Gerstoft, et al. Time-domain geoacoustic inversion of high-frequency chirp signal from a simple towed system[J]. IEEE J Ocean Engineering, 2003, 28(3).
5. N R Chapman, R M Dizaji, R L Kirlin. Geoacoustic inversion using broad band ship noise[C]. Proceedings of the 5Pth ECUA, Lyon, 2000: 787-792.
6. M Siderius, P Nielsen, P Gerstoft. Range-dependent seabed characterization by inversion of acoustic data from a towed receiver array[J]. JASA, 2002, 112:1523-1535.

Matched-Field Geoacoustic Inversion By Inverting Ship-Noise Data

Licheng Lu and Li Ma

Key Laboratory of Underwater Acoustic Environment, Chinese Academy of Sciences, Beijing, 100190, China
Institute of Acoustics, Chinese Academy of Sciences, Beijing, 100190, China

Abstract. We report the results of geoacoustic inversion by inverting ship noise data obtained with a vertical array during an East Sea Experiment in April 2006. The inversion process employs a Hybrid optimization and Bayesian inversion techniques based on Matched-Field processing to explore parameters of the seabed such as speed, density, attenuation and so on. The Hybrid optimization, which combines global simulated annealing and downhill simplex, is used to interrogate geoacoustic parameters from multiple data portions as the ship navigates around the vertical array. Bayesian inversion quantifies the uncertainty of geoacoustic parameters using a Gibbs Sampling (GS) approach to estimate posterior probability density (PPD). An isotropy sediment overlay on an infinite basement geoacoustic model is assumed and ship noise tones of approximately 100 Hz and 405 Hz are used in the inversion. Sediment speed and thickness are well determined, as is the basement speed. The density and attenuation are not well determined, mostly because the sound interacted little with the seabed for short transmission distance. Ship noise transmission loss (TL) comparisons modeled at 100 Hz, 405 Hz and 695 Hz are exhibited.

Keywords: Geoacoustic Inversion, Ship Noise, Bayesian Estimation, Transmission Loss.
PACS: 43.30.Pc

I. INTRODUCTION

The problem of estimating seabed geoacoustic parameters from measured ocean acoustic fields at an array of sensors based on matched-field processing has received considerable attention in recent years [1-8]. Matched-field inversion represents a strongly nonlinear problem for which a direct solution is not available. Many attempts have been made to estimate the optimal geoacoustic model by minimizing the error function, which is defined as a measure of misfit between the measured acoustic fields and modeled replicas. Optimization algorithms, such as simulated annealing, genetic algorithms, and hybrid optimizations that combine global and local methods, have been developed for searching the optimal modeled parameters. An important part of geoacoustic inversion is quantifying the uncertainties of the modeled parameters. Appraisal analysis based on Bayesian theory provides formalism for estimating parameters of uncertainties for strongly nonlinear inversion. In Bayesian formulation, the solution is characterized by its posterior probability density (PPD), which can be computed by a Gibbs-Sampling approach.

CP1272, *Shallow-Water Acoustics, Second International Shallow-Water Acoustics Conference*
edited by J. Simmen, E. S. Livingston, J. X. Zhou, and F. H. Li

Much previous research in geoacoustic inversion with measured acoustic data was done with active sources, such as an impulsive source [9] or other special waveforms [7]. As surface ships in shallow water are abundant and radiate broadband noise, it is the purpose of this paper to use ship noise to invert seabed properties. Geoacoustic inversion results of inverting ship noise data received by a vertical line array (VLA) during the East Sea Experiment in April 2006 are reported. The inversion process uses a Hybrid optimization and Bayesian inversion techniques based on Matched-Field processing to explore parameters of the seabed, such as speed, density, attenuation, etc. The remainder of this paper is organized as follows. Section II briefly describes the East Sea Experiment and the relevant environment. Section III reviews the inversion procedures of hybrid optimization and GS approach, and describes the seabed model and other parameters. Section IV presents the results and the comparison of ship noise TL and modeled TL. Section V summarizes and discusses this work.

II. DESCRIPTION OF THE EXPERIMENT

A. Experimental Geometry and Data Spectrum

The East Sea Experiment was carried out in the littoral region in April 2006. The data were recorded from an 18-element ship-hanged VLA at sample rate 6 kHz. The VLA was evenly spaced at a separation of 2 m from 2 m to 36 m below the sea surface, of which the 7th, the 9th and the 11th hydrophones were bad and were abandoned. Fig. 1 shows the ship track and VLA location, where the ship approached to the VLA from the east at about 360 m distance and then went past to the west.

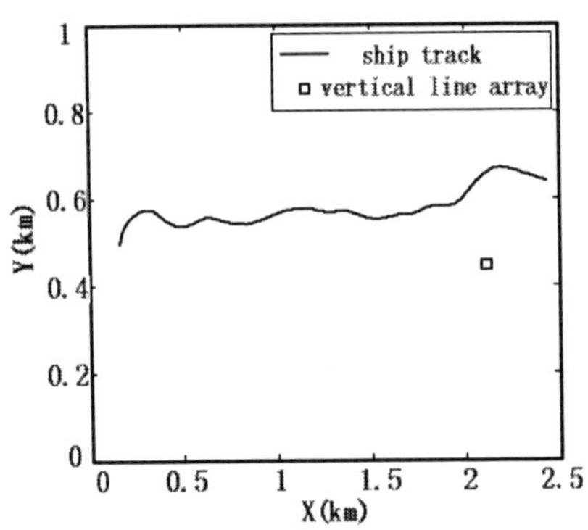

FIGURE 1. The relevant position of ship track and VLA

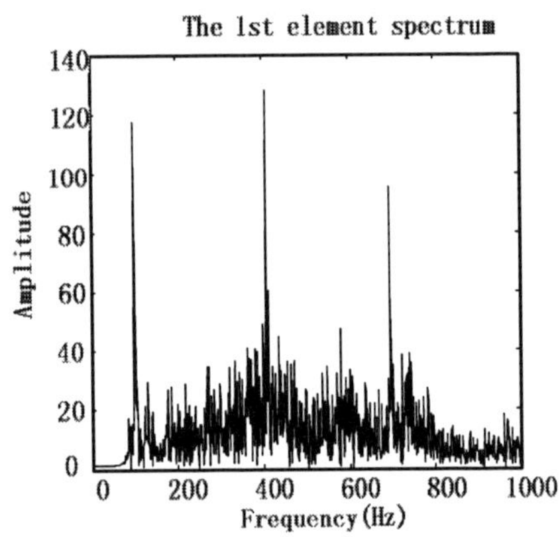

FIGURE 2. The spectrum of ship noise signal

Data analyses of this paper for inversion contain only the first 36 s, in which the ship steered about 180 m at 10 knots speed. Every 1.2 s data segment was transformed to frequency domain with a Kaiser window to get acoustic field information for inversion. One of the data segments spectrum is shown in Fig. 2, from which there are three ship noise tones at about 100 Hz, 405 Hz and 695 Hz. Only the 100 Hz and 405 Hz tones are used for geoacoustic inversion.

B. Water Column Sound Speed Profile

The water column sound speed profiles (SSP) were measured using CTD at the VLA position. Fig. 3 shows the four measurements during the experiment from 06:33am to 13:24pm. The differences of these SSP are no more than 3 m/s and can be negligible. So the polyfit of the SSP used in inversion is the thick line, and it has a slightly positive gradient.

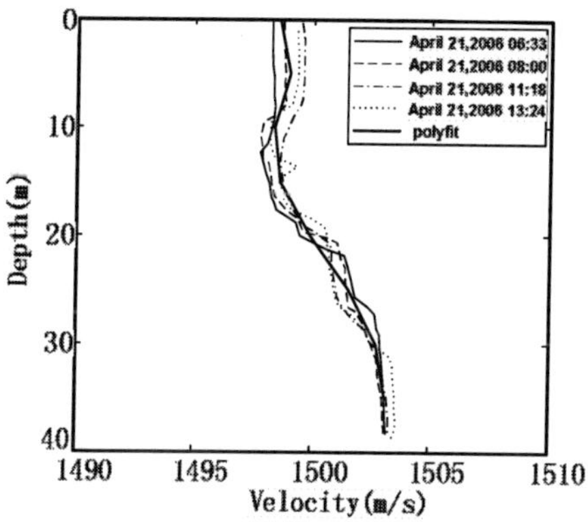

FIGURE 3. The sound speed profiles

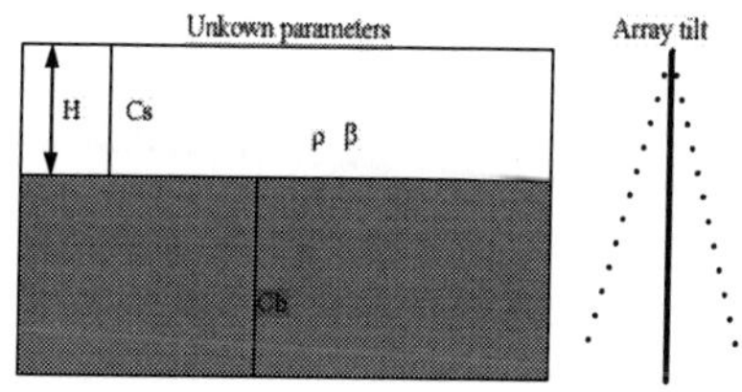

FIGURE 4. Geoacoustic model and array geometry

III. GEOACOUSTIC INVERSION PROCEDURES

A. Acoustic Propagation Model

The water depth in the experiment area is about 38 m-39 m, and the distance between the ship and the VLA is about 220 m-360 m. Although the distance is 5-9 times the water depth, it is also believed that near field computation is required for low frequency sound propagation. The PE (RAM [13]) acoustic model is used here for computing the acoustic modeled field. The number of Padé coefficients was set to two and range and depth mesh size were set to 1/10λ and 1/40λ, respectively, where λ is the acoustic wavelength in meters.

B. Geoacoustic Model and Array Geometry

The seabed of the experiment area is very soft on the bottom surface, so an isotropy sediment overlay on an infinite basement geoacoustic model is assumed. The sediment and basement have the same density and attenuation, as dictated in Fig. 4. There are five unknown parameters in the geoacoustic model in total. The VLA may be tilted and curved from the vertical position for reasons of sea flow. To account for the array relative positions, a rigid and linear VLA shape is assumed and the geometry of the VLA is in term of angle θ departure from vertical (Fig. 4).

C. Hybrid Optimization

The hybrid optimization combines simulated annealing and downhill simplex algorithm (ASSA) [10], which is used to minimize the error function defined as misfit between acoustic data and the modeled field. In matched-field inversion processing, commonly the error function is defined as

$$E(\boldsymbol{m}) = 1 - \sum_f \frac{\left| \boldsymbol{p}_f^\dagger(\boldsymbol{m}) * \boldsymbol{d}_f \right|^2}{\left| \boldsymbol{p}_f(\boldsymbol{m}) \right|^2 \left| \boldsymbol{d}_f \right|^2} \tag{1}$$

$\boldsymbol{p}$ and $\boldsymbol{d}$ represent the modeled field vector and experimental data vector respectively, $\boldsymbol{m}$ represents the unknown parameters, the superscript † denotes the conjugate transpose, and f is the different frequency band.

The simulated annealing employs the Metropolis algorithm, for which randomly generated parameters yielding lower error function than the current parameters are automatically accepted. Those yielding a higher error function by condition of the Boltzmann probability distribution

$$P(\Delta E) = \exp(-\Delta E / T) \tag{2}$$

where ΔE signifies a positive increase in error function at temperature T. In this strongly nonlinear inversion problem, the search method usually converges slowly. Faster convergence in the hybrid algorithm is achieved by employing downhill simplex search, as described in Ref. 10.

The hybrid optimization is used to invert every 1.2 s segment data to get optimal parameters. Table I lists each inversion parameter along with their bounds.

Table I. Inversion parameters boundary

Model parameters		**Search Upper**	**Search Lower**
Description	**Symbol**		
Sediment thickness	H	1	20
Sediment speed(m/s)	C_s	1480	1600
Basement increase speed(m/s)	C_b	1	200
Density(g/cm^3)	ρ	1.2	2.0
Attenuation(dB/λ)	β	0.0	0.5
Array tilt	θ	-5	5
Source depth(m)	SD	0	3
Source range(m)*	R	-20	20
Water depth(m)	WD	36	40

* Source range is the GPS range ± 20m

D. Bayesian Estimation

Following the Bayes' rule, the parameters $\boldsymbol{m}$ posterior probability density (PPD) can be expressed

$$P(\boldsymbol{m} \mid \boldsymbol{d}) \propto P(\boldsymbol{d} \mid \boldsymbol{m}) P(\boldsymbol{m}) \tag{3}$$

The $P(\boldsymbol{m} \mid \boldsymbol{d})$ represents the conditional probability density function (PDF) of $\boldsymbol{m}$ in given $\boldsymbol{d}$, $P(\boldsymbol{d} \mid \boldsymbol{m})$ is the condition PDF of $\boldsymbol{d}$ in given $\boldsymbol{m}$, $P(\boldsymbol{m})$ is the PDF of $\boldsymbol{m}$. When the experimental data $\boldsymbol{d}$ is already obtained, $P(\boldsymbol{d} \mid \boldsymbol{m})$ is interpreted as a function of $\boldsymbol{m}$, known as the likelihood function $L(\boldsymbol{d} \mid \boldsymbol{m})$ and Eq. (3) becomes

$$P(\boldsymbol{m} \mid \boldsymbol{d}) \propto L(\boldsymbol{d} \mid \boldsymbol{m}) P(\boldsymbol{m}) \tag{4}$$

Under the assumption of Gaussian errors, the likelihood function is the form

$$L(\boldsymbol{d} \mid \boldsymbol{m}) \propto \exp\left[-\phi(\boldsymbol{m})\right] \tag{5}$$

where the $\phi(\boldsymbol{m})$ is a measure misfit between the modeled data and experimental data and will be discussed later. The PPD can be written

$$P(\boldsymbol{m} \mid \boldsymbol{d}) = \frac{\exp\left[-\phi(\boldsymbol{m})\right]}{\int \exp\left[-\phi(\boldsymbol{m}')\right] d\boldsymbol{m}'} \tag{6}$$

This assumes that $\boldsymbol{m}$ has uniform distribution. The multidimensional PPD is interpreted in terms of properties defining parameter estimates and uncertainties, such as mean and marginal probability distribution, defined as

$$\langle \boldsymbol{m} \rangle = \int \boldsymbol{m}' P(\boldsymbol{m}' \mid \boldsymbol{d}) d\boldsymbol{m}' \tag{7}$$

$$P(m_i \mid \boldsymbol{d}) = \int \delta(m_i' - m_i) P(\boldsymbol{m}' \mid \boldsymbol{d}) d\boldsymbol{m}' \tag{8}$$

For nonlinear problems, analytic solutions to Eq.(7) and (8) generally do not exist and numerical approaches must be applied. Marginal probability distributions are determined by applying Gibbs sampling [11-12] to sample $\phi(\boldsymbol{m})$.

The measure misfit of $\phi(\boldsymbol{m})$ following Ref. 12 can be written

$$\phi(\boldsymbol{m}) = \sum_f \frac{\left(1 - E_f(\boldsymbol{m})\right) \left|\boldsymbol{d}_f\right|^2}{\sigma_f^2} \tag{9}$$

$E_f(\boldsymbol{m})$ is the formulation of Eq. (1), σ_f^2 is the corresponding estimate of the variance associated with the combination of model mismatch, ocean, and measured noise, which only depends on frequency f and estimates following the Ref. 14 and 12 as uncorrelated with identical variance

$$\sigma_f^2 = E_f(\widehat{\boldsymbol{m}}) \left|\boldsymbol{d}_f\right|^2 / N_e \tag{10}$$

where $\widehat{\boldsymbol{m}}$ is the optimal parameters, N_e is the effective number of acoustic modes. The $\widehat{\boldsymbol{m}}$ of this paper is the mean of 30 data segments inversion results and N_e is 3 and 12 for 100 Hz and 405 Hz, respectively, considering the near field including more acoustic modes.

IV. RESULTS ANALYSIS

A. Optimization Inversion

The Hybrid optimization inversion was applied to the first 36 s data portion. The typical initialization of this algorithm is: starting temperature 0.3, temperature reduction factor 0.99. The results are shown in Fig. 5, from which some features can be inferred:

(1) Sediment thickness H, sediment speed C_s and substrate speed C_b are well determined with small variance. Sediment speed is slightly lower than water

column speed. For near field, sound penetrates the seabed more deeply and interacts with the lower layer, so lower layer parameter is very sensitive.

(2) Density ρ and attenuation β are not well determined, mostly due to the short transmission distance. However the attenuation still concentrates on a relatively small range, perhaps explained by the reflection coefficient of two layers of media, where the attenuation is exhibited in the exponential item.

(3) Source depth SD is not resolvable in the inversion. Water Depth WD and array tilt θ are also well-determined consistent with measurements.

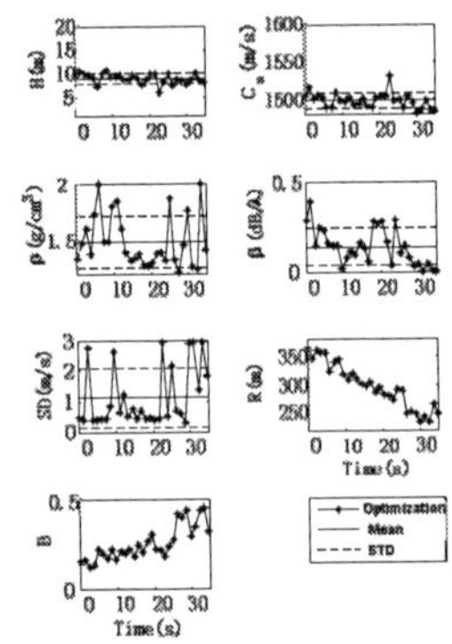

FIGURE 5. The optimal results of thirty data segments

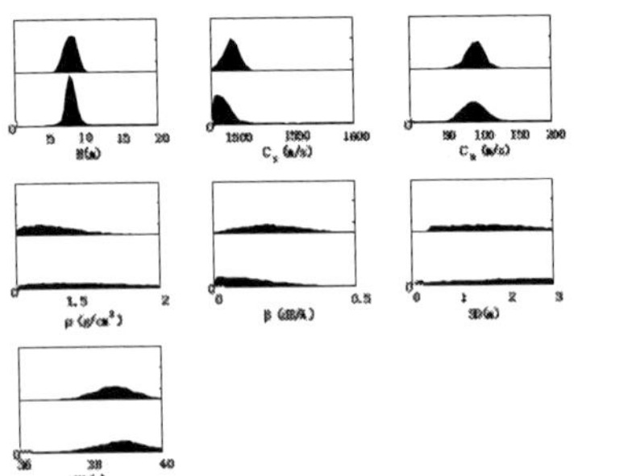

FIGURE 6. Comparison of Bayesian inversion between the 20th and 30th data segments

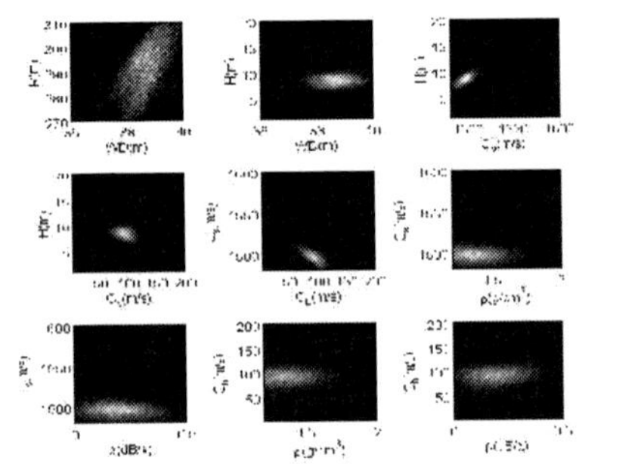

FIGURE 7. Selected 2D marginal PPDs of t he 20th data segment

B. the Bayesian Inversion

This section presents the Bayesian inversion estimates for the 20th and 30th data segments. Based on the results of optimization inversion, the variance σ_f^2 was computed, and marginal probability distributions are sampled using a GS approach. The one-dimension distributions except for array tilt and source range are shown in Fig. 6. The sediment thickness, sediment speed, substrate speed and water depth are consistent for the two data segments. Density and source depth distribution are broader. It's also clear that the sensitive parameters have narrower uncertainty.

Another factor that affects model parameter uncertainties is the parameters coupled. Fig. 7 shows the two-dimension (2D) cross sections for the selected parameters. The 2D dependence of R and WD shows a strong correlation and this can be explained by the waveguide invariant. Sediment thickness and speed exhibit a positive correlation, while sediment speed and substrate speed exhibit a negative correlation.

C. TL Comparison and Localization

The inversion quality can be checked by comparisons of modeled TL and measured TL. Fig. 8 shows the measured TL received on the 1st element of the VLA, whose depth is about 36 m. The calculations of the modeled TL were performed by the RAM propagation program using the means of the inversion parameters as: sediment thickness is 8.93 m, sediment speed is 1497.6 m/s, substrate speed is 1604.4 m/s, density is 1.49 g/cm^3, the attenuation is 0.14 dB/λ and water depth is 38.29 m.

Another validation of inversion results is performing matched-field processing for ship localization. Fig. 9 shows the Bartlett ambiguity surface for the frequencies of 100 Hz, 405 Hz and 695 Hz at the R=1.04 km. It can be seen that the distinguishing target is localized, consistent with the measured position.

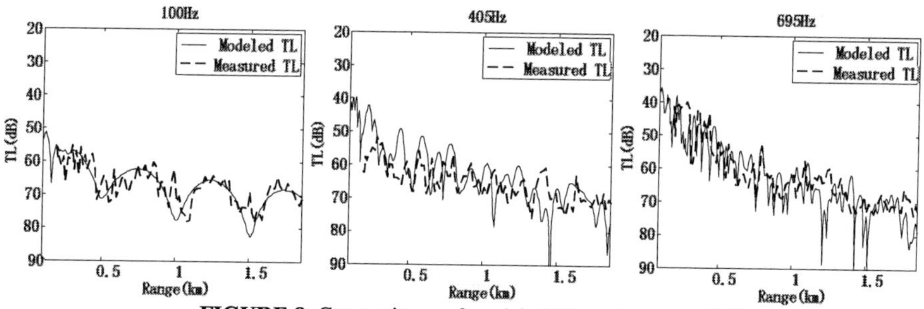

FIGURE 8. Comparisons of modeled TL and measured TL

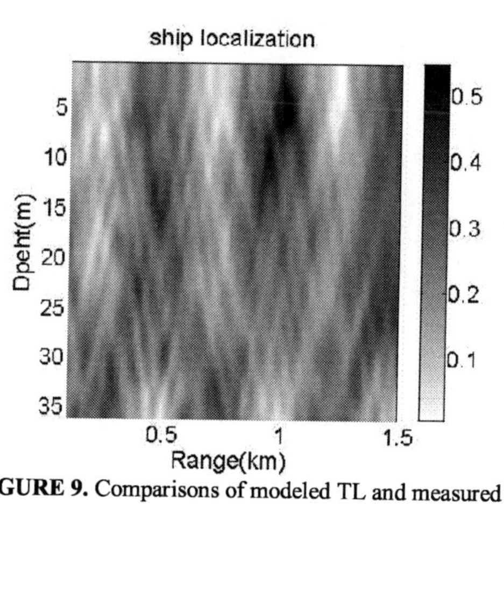

FIGURE 9. Comparisons of modeled TL and measured TL

V. SUMMARY

This paper reports the geoacoustic inversion results based on ship-noise data obtained during an East Sea Experiment in 2006. An isotropy sediment overlay on an infinite basement geoacoustic model is assumed and ship noise tones about 100 Hz and 405 Hz are used in the inversion. Sediment speed and thickness are well determined, as is the basement speed. Ship noise TL comparisons with modeled TL at 100 Hz, 405 Hz and 695 Hz are shown to be consistent. Geoacoustic inversion results have successfully been applied to matched-field processing for ship localization.

ACKNOWLEDGMENT

This work is supported by the National Natural Science Foundation of China (No.10774156). The author would like to thank the East Sea Experiment participants and Jinrong Wu for his experiment data.

REFERENCES

1. M. D. Collins, W. A. Kuperman, and H. Schmidt, "Nonlinear inversion for ocean bottom properties," J. Acoust. Soc. Am. 92, 2770-2883 (1992).
2. S. E. Dosso, M. L. Yeremey, J. M. Ozard, and N. R. Chapman, "Estimation of ocean-bottom properties by matched-field inversion of acoustic data," IEEE J. Ocean. Eng. 18, 232-239 (1993).
3. P. Gerstoft, "Inversion of seismo-acoustic data using generic algorithms and a posterior probability distribution," J. Acoust. Soc. Am, 95, 770-782 (1994).
4. A Tolstoy, "Using matched-field processing to estimate shallow water bottom properties from shot data taken in the Mediterranean Sea," IEEE J. Ocean. Eng. 21, 471-479 (1996).
5. A. Tolstoy, N. R. Chapman, and G. E. Brooke, "Workshop'97: Benchmarking for geoacoustic inversion in shallow water," J. Comput. Acoust. 6, 1-28 (1998).
6. Z. H. Michalopoulou and U. Ghosh-Dastidar, "Tabu for matched-field source localization and geoacoustic inversion," J. Acoust. Soc. Am. 115, 135-145 (2004).
7. Chen-Fen Huang and William S. Hodgkiss, "Matched-Field geoacoustic inversion of low-frequency source tow data from the ASIAEX East China Sea Experiment," IEEE J. Ocean. Eng, 29, 952-963 (2004).
8. Dag Tollefsen, M.J Wilmut, "Estimates of Geoacoutic Model Parameterss from Inversion of Horizontal and Vertical Line Array Data," IEEE J. Ocean. Eng, 30, 592-597 (2005).
9. D. P. Knobles, R. A. Koch, L. A. Thompson, K. C. Focke, and P. E. Eisman, "Broadband sound propagation in shallow water and geoacoustic inversion," J. Acoust. Soc. Am. 113, 205-222 (2003).
10. S. E. Dosso, M. J. Wilmut, and A. S. Lapinski, "An adaptive hybrid algorithm for geoacoustic inversion," IEEE J. Ocean. Eng. 26, 324-336 (2001).
11. S. E. Doss, "Quantifying uncertainty in geoacoustic inversion. I. A fast Gibbs sampler approach," J. Acoust. Soc. Am. 111, 129-142 (2002).
12. S. E. Doss and P. L. Nielson, "Quantifying uncertainty in geoacoustic inversion II. Application to broadband, shallow-water data," J.Acoust. Soc. Am. 111, 143-159 (2002).
13. M. Collins, "A split-step padé solution for the parabolic equation method," J. Acoust. Soc. Am. 93, 1736-1742 (1993).
14. P. Gerstoft and C. F. Mecklenbrauker, "Ocean acoustic inversion with estimation of a posterior probability distribution," J. Acoust. Soc. Am. 104, 808-819 (1998).

Long Range Sediment Tomography in the East China Sea

Gopu R. Potty and James H. Miller

Department of Ocean Engineering, University of Rhode Island, Narragansett, RI 02882, USA.

Abstract. Results of the long range sediment tomography inversions using the broad band acoustic data from the Asian Seas International Acoustic Experiment (ASIAEX) is discussed in this paper. A hybrid inversion scheme is used for the sediment tomography technique which utilizes the modal travel time data. The conventional approach is modified in this study incorporating the model sensitivity to tune the parameter search intervals. Model parameters with high sensitivity are searched at closer intervals and *vice versa*. This results in computational efficiency by concentrating search effort to model subspace which is more sensitive. The results of the tomographic inversions compare well with core measurements and estimates using other techniques.

Keywords: dispersion, normal modes, nonlinear inversion, genetic algorithms.
PACS: 43.30.Pc, 43.30.Bp

INTRODUCTION

This paper discusses the application of a long-range sediment tomography technique [1] to data collected in the East China Sea during the Asian Seas International Acoustics Experiment (ASIAEX). This technique is based on the group speed dispersion data obtained from broadband acoustic sources measured at 30 km from the small (38 g) explosive sound sources (Wide Band Sources (WBS)). Our method uses acoustic normal mode arrival times as the observed data [1]. The modal arrivals are stable and they can be identified when their amplitudes are above a certain threshold. We use advanced time-frequency analysis techniques such as wavelet based methods to identify individual modal arrivals. The weight of the WBS, the source-receiver range, water depth and sediment type are major factors which influence the performance of the tomography technique [1].

Global optimization schemes and hybrid schemes have been extensively used for the inversion of geoacoustic parameters. Most commonly used methods are Genetic Algorithms (GA) [1], Simulated Annealing (SA) and Differential Evolution (DE). A Genetic Algorithm (GA) is used for the global search in the present study. Global search methods are effective at finding solutions in the neighborhood of the optimum solution, but local optimization methods are much better at finding the exact value of the minima since they use gradients or curvature for their search. The variance of the geoacoustic estimates can be reduced by applying a local method as the final step in

CP1272, *Shallow-Water Acoustics, Second International Shallow-Water Acoustics Conference*
edited by J. Simmen, E. S. Livingston, J. X. Zhou, and F. H. Li

the optimization procedure. We use the hybrid approach developed by Potty *et al.* [1] applying a Levenberg - Marquardt method along with the GA.

In the inversions presented in this article, sampling of the model space has been done in an optimum manner reflecting the sensitivity of the model parameters. Model parameters were sampled based on the relative changes in the objective function caused by small perturbations in them i.e., model sensitivity. Figure 1(a) shows the relative sensitivity of model parameters (representing the compressional wave speeds in the sediment) as a function of depth below seafloor for all frequencies. These values were calculated based on the difference in group speed values caused by a perturbation in model parameters at these depths from group speed values calculated for a background profile.

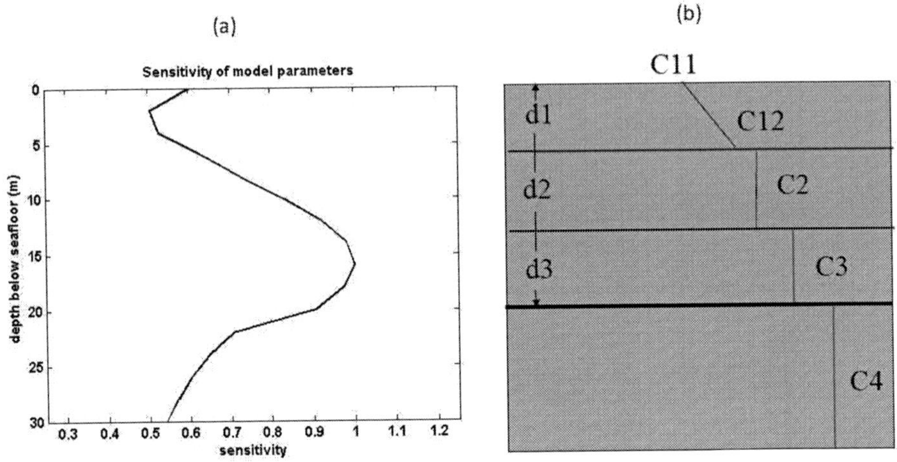

FIGURE 1. Left panel shows the model sensitivity as a function of depth. Model parameters considered are the compressional wave speeds at these depths. The sensitivity is normalized to the maximum value. The right panel shows the geoacoustic model used for the inversion. The focus is on the layers near the water-sediment interface (0-2 m), the next layer with low sensitivity (2 – 5 m), third layer with very high sensitivity (5 – 22 m) and the basement below that. Inversion searches for the optimum values of the thickness of the layers and the compressional wave speeds in the layers (marked C11, C12, C2, C3 and C4 in (b))

We can see that model parameters have very high sensitivity at depths 5 to 22 m. So one can expect to get relatively good estimates of the model parameters at these depths. The sensitivity at depths 2 to 4 m is lower compared to the surficial sediments (0 to 2 m). Also, the sensitivity is low below 22 m. We have used this sensitivity information along with background geoacoustic and seismic data to identify the sediment layers and to set up a model sampling approach. This approach is discussed in more detail in the following section. The geoacoustic model shown in Figure 1(b) includes three layers of sediment over an acoustic basement. These layers can be related to sediments in different layers identified by the seismic data from the area, which is discussed in a later section.

SENSITIVITY BASED MODEL SEARCH

The long range sediment tomography inversions are based on hybrid optimization scheme which combine a Genetic Algorithm (GA) with a local approach such as Levenberg-Marquardt method [1]. Details of this hybrid inversion scheme are provided later in this article. In the conventional inversion scheme the model parameters are sampled with equal importance without considering their resolution. The resolvability of a model parameter is reflected in its resolution length. An inversion scheme will be efficient if of the search effort corresponding to each model parameter is proportional to its resolution length. This can be accomplished if the model subspace corresponding to less sensitive parameters is sampled coarsely whereas the subspace corresponding to highly sensitive parameters is sampled fine. The GA was modified to force it to sample certain parameters with low sensitivity in coarse manner in this study. The selection of the parameters to be sampled coarse and the 'coarseness' of the sampling was based on the sensitivity of the model parameters as shown in Figure 1(a).

Ideally the intensity of sampling of a model parameter should reflect its resolution length. But resolution of model parameters is not often available *a priori* for most of the practical problems. One alternative then is to use the model sensitivity as a proxy to resolution and use it to adjust the model sampling [2,3]. We have used the relative sensitivity shown in Figure 1(a) to identify the model parameters that need only coarse sampling. Based on the sensitivity the sediment compressional wave speeds in the third layer was sampled twice (approximately equal to the inverse of the sensitivity) as fine as the compressional wave speed in the basement. The compressional wave speeds in first and second layers were not adjusted even though they have a lower sensitivity since the layers are very thin and the sensitivity appears to transition from high to low over a short depth. Other model parameters associated with the sediment model, layer thickness, were defined based on the model sensitivity and *a priori* seismic data.

A PRIORI GEOACOUSTIC DATA

The location in the East China Sea where the Asian Seas International Acoustic Experiment (ASIAEX) experiment was carried out is bisected by a boundary between siltier surficial sediments to the west and sandier surficial sediments to the east [4]. The sources of the western silty sediments are the large rivers to the north such as the Yangtze. The source of the eastern sandy sediments is the Kuroshio Current, which transports these sediments from the south [4]. The ASIAEX experiment area straddles a border between □mud-and-sand□ and □sand. The water-gun measurements taken at the experimental location as part of this experiment indicate that the sub-bottom structure consists of a thin (0-3 meters thick) surficial layer of sandy sediment of variable thickness overlying an extensive package of silty marine sediments with relatively uniform acoustic attributes. River channels dissect this unit and channel fill consists of sediment that produces an extremely heterogeneous distribution of acoustic attributes as shown in Figure 2(a,b). Core analysis shows that the surface sediment layer varies in compressional wave speed from a low near 1600 m/s in the west side of the experiment area to 1660 m/s in the east side of the experimental area.

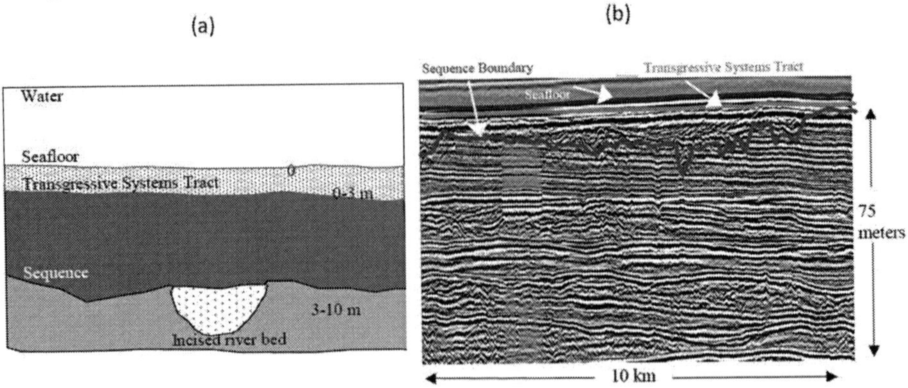

Figure 2. Left panel shows the schematic of the sediment structure in the East China Sea. The transgressive systems tract (TST) is 0 to 3 meters below the seafloor. The sequence boundary is typically 3 to 10 meters below the seafloor. The right panel shows the sub-bottom profile from water-gun data. The seafloor, the transgressive systems tract and the sequence boundary are also marked.

Based on results of the sensitivity study shown in Figure 1(a) and the sediment layering information from the seismic study shown in Figure 2(b), the sediment geoacoustic model as shown in Figure 1 (b) was used to represent the bottom. The sediment consists of 3 layers over an acoustic basement. The thickness of the sediment layers and the compressional wave speeds in these layers were considered unknowns. For the top layer the compressional wave speeds at the top and bottom of the layer were treated as unknowns.

INVERSION

The inversion scheme for long range sediment tomography is based on matching the group speed dispersion characteristics of the broadband signal. In order to apply the inversion approach described above we need to measure the modal arrival times. The broadband data from WBS is analyzed using wavelet based time- frequency analysis techniques [1,4]. The wavelet decomposition aids in the identification of the individual mode arrivals and to measure the modal arrival times. The inversion scheme used in the present study uses a GA for the global search in order to generate 'candidate' models. GA is a very popular search method and has been extensively used in underwater acoustics applications [1], [4]. The objective of GA is to seek a model which gives a globally optimal data misfit within a pre-defined finite dimensional parameter space. The fitness is measured in terms of an objective function, which is minimized in the optimization process. The objective function represents the mismatch between the observed data and the theoretical predictions. The objective function for the present inversion is based on minimizing the difference between group speed values calculated using the observed time of arrivals and the predicted group speeds. This function is given by

$$E(\mathbf{m})=||\mathbf{w}(\mathbf{d}^{obs}-\mathbf{d}^{pred})||^2$$

where E(**m**) is the objective function for the parameter set **m** and **w** is a diagonal weight matrix. The weight associated with the ith data point is obtained as the reciprocal of the standard deviation of the data σ_i, which is determined from the time-frequency representation of the broadband data.

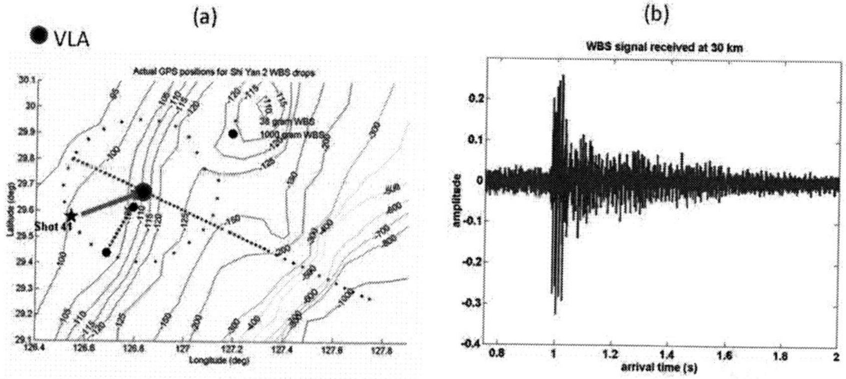

Figure 3. The bathymetry and the location of the source (Shot 41) and receiver (VLA) are shown in the left panel. Acoustic sources were deployed in a circular arrangement around the receive array which was at the center (sources were also deployed along a line which are not considered in this study). The continuous line shown is the acoustic path from the source to the receiver corresponding to the data used in our inversion. The acoustic signal received at the VLA (in one hydrophone) from shot 41 is shown in the right panel. Note that the amplitude is arbitrary in this figure.

The theoretical group speed predictions are calculated using the normal mode approach. Since the global methods are less efficient in finding the exact minimum compared to local methods a hybrid approach is used in this study. Once the global method narrow the search to a region close to the global minimum, a Levenberg-Marquardt method is employed to 'fine tune' the inversion. This local search is initiated at the best model the GA provides. The use of Levenberg Marquardt method in combination with the GA search constitutes the hybrid approach which we employ for the sediment tomography technique.

Based on this sensitivity study and the data from geophysical surveys (Miller *et al.* [3]) the sediment was modeled as indicated in Figure 1(b). This sediment model was used in the inversion scheme to estimate the compressional wave speed profile. The compressional wave speed and thickness of these layers are the unknown parameters. In the first layer the compressional speeds at the top and bottom of the layer are treated as unknowns. In addition, the water depth, source-receiver range and coefficients of Empirical Orthogonal Functions (EOFs) representing the variations in sound speeds in water column were also included as unknowns in the inversion scheme. The EOFs are obtained from the ocean sound speeds calculated from CTD measurements.

The acoustic signal from one of the WBS charges deployed during the Asian Seas International Acoustic Experiment (ASIAEx) received at the VLA at a depth of 78.6

m is shown in Figure 3(b). We can see the direct blast arriving at 1 s and late arrivals coming until approximately 1.6 s. This signal is analyzed for its time-frequency behavior using a wavelet-based analysis. This time-frequency analysis yields the arrival pattern of individual modes at various frequencies as shown in Figure 4(a). It is possible to identify up to seven modes from the wavelet scalogram (Figure 4(a)) of the acoustic signal shown in Figure 3(b). From this diagram the arrival times corresponding to individual modes at frequencies ranging from 20 Hz to 600 Hz are "peak picked". The arrival times shown in the figure are arbitrary. The tonal signature seen at around 360 Hz corresponds to the ship noise. In addition to the arrival time corresponding to a spectral peak, the width at 95 % peak level is also measured. This is used in the objective function as the standard deviation of the data.

RESULTS

The compressional wave speed values estimated from the inversion are shown in Figure 4(b). Inversion is carried out using acoustic signals from a single charge to estimate a depth dependent compressional wave speed profile assuming range independence. The bathymetry along the propagation path from WBS charge 41 to the receiver at the center of the circle is fairly uniform and hence assumed range independent. The arrival times obtained from the time-frequency diagrams were used as data for the inversion scheme. The sediment model for the inversion of depth dependent compressional wave speed consists of three layers of sediment and a basement.

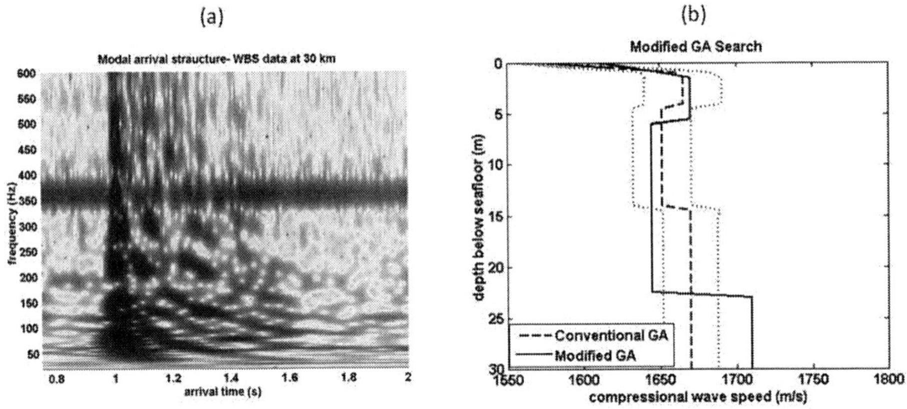

Figure 4. The arrival time structure of the acoustic signal received at 30 km range is shown in the left panel. The arrival times corresponding to acoustic normal modes 1 to 7 are used as the data for the inversion. The sediment model that matches these arrival time data is shown in the right panel. The continuous line is the best model from the inversion using optimum model search. The dashed line is the output from the conventional search. The two dotted lines on either side of this are line corresponding to one standard deviation.

Figure 4(b) shows the range averaged compressional wave speeds along the acoustic propagation path (continuous line from shot 41 to center of the circle in

Figure 3(a)). The top sediment layer if 2 m thick and the compressional wave speed in this layer increases from 1560 m/s to 1670 m/s. This compares well with previous inversions [5] and gravity core data [4]. The second layer is 5 m thick and the compressional wave speed in this layer is 1670 m/s. The layer thickness and compressional wave speed in this layer compare well with the conventional inversion which is also shown in Figure 4(b). The third layer 16 m thick and the compressional wave speed in this layer is lower at 1645 m/s. These values and the compressional wave speed in the basement are different from the conventional inversion. We don't have any deep core measurements near the propagation path for comparison and hence it is difficult to make a comparison. Computationally the modified search performed better than the conventional one and converged to the optimum value with less number of iterations.

CONCLUSIONS

Compressional wave speeds were estimated using long range sediment tomography technique. Data from the Asian Seas International Acoustic Experiment (ASIAEX) were inverted using an inversion scheme based on matching the travel time dispersion of the acoustic normal modes. The model parameter search was modified by dividing the model parameters into subspaces. Model parameters with high sensitivity were sampled at a finer scale compared to parameters with low sensitivity. The results compare well with previous inversions and core data.

ACKNOWLEDGMENTS

This study was sponsored by the Office of Naval Research (USA), Ocean Acoustic Division.

REFERENCES

1. G. R. Potty, J. H. Miller, J. F. Lynch and K. B. Smith, ``Tomographic inversion for sediment parameters in shallow water," J. Acoust. Soc. Am., 108(3), 973-986, (2000).
2. G. Nolet and R. Montelli, "Optimal parametrization of tomographic models," *Geophys. J. Int.,*. **161**, 365–372, (2005).
3. G. R. Potty, J. H. Miller, C. J. Lazauski, and C-S. Chen, "Optimal search strategies for nonlinear, multi-parameter geoacoustic inverse problems," J. Acoust. Soc. Am. 113, 2190 (2003).
4. J. H. Miller, L. Bartek, G. R. Potty, D. J. Tang, J. Na and Y. Qi, " Sediments in the East China Sea," IEEE J. Oceanic. Eng., 29(4), 940-951, (2004).
5. G. R. Potty, J. H. Miller, P. H. Dahl, and C. J. Lazauski, "Geoacoustic inversion results from the ASIAEX East China Sea Experiment," IEEE J. .Eng., 29(4), 1000-1010, (2004).

Acoustic remote sensing of the seabed using ambient noise

Martin Siderius

Department of Electrical and Computer Engineering, Portland State University, Portland OR, U.S.A., 97201 siderius@pdx.edu

Abstract. In this paper, results will be described that use measured ambient noise data to characterize the seabed. Knowledge of seabed properties such as the layering and the compressional sound speed and attenuation are important for sonar performance prediction (e.g. transmission loss) to estimate the effectiveness of various sonar systems. Recently developed techniques will be described in which seabed layers are imaged using the ocean ambient noise field as the sound source. This so called passive fathometer technique exploits the naturally occurring acoustic sounds generated on the sea-surface, primarily from breaking waves. The method is based on the cross-correlation of noise from the ocean surface with its echo from the seabed which recovers travel times to significant seabed reflectors. To derive a complete picture of the geo-acoustic properties of the seabed, a second method that uses the intensity of the vertically beamformed noise data to infer bottom reflection loss will also be used. The combination of these two ambient noise approaches provides for a detailed description of the seabed geo-acoustic properties.

Keywords: Ambient Noise, Seabed layering, Adaptive Beamforming
PACS: PACS numbers: 43.30Nb, 43.30Wi, 43.30Pc, 43.30Gv

INTRODUCTION

There are many methods for characterizing the properties of the seabed using acoustics, but here we will consider using background ambient noise as the sound source for remote sensing the seabed. Using the sea-surface breaking waves as the sound source allows us to simplify the measurement geometry (i.e., no sound projectors). The ambient noise field also contains energy over a broad band of frequencies and since it is completely passive these measurements can be made anywhere without concerns about environmental impact. Two ambient noise techniques that have recently been developed will be described here. The first method uses the coherent ambient noise field to estimate the seabed layering [1] and the second uses the intensity of vertically beamformed noise to determine the magnitude of the bottom reflection loss [2]. Both techniques are important for determining a complete geoacoustic description of the seabed that can be used for transmission loss (TL) prediction. The passive fathometer requires beamforming and two versions have been formulated. The first uses conventional beamforming and the second uses adaptive methods such as the Minimum Variance Distortionless Response (MVDR). The beamforming is an important component of the processing to allow practical averaging times. A brief description of all of these methods is presented here and additional details can be found in the cited references.

CP1272, *Shallow-Water Acoustics, Second International Shallow-Water Acoustics Conference*
edited by J. Simmen, E. S. Livingston, J. X. Zhou, and F. H. Li

CONVENTIONAL PASSIVE FATHOMETER PROCESSING

The cross-correlation between the noise from the ocean surface and its echo from the seabed produces a time trace with peaks representing the water sediment interface as well as sub-bottom layers. One of the key differences between typical noise correlation approaches and the passive fathometer method is the use of a vertical hydrophone array. The array is used to emphasize the noise directly above rather than from horizontal directions. The details of the conventional passive fathometer are described here and in the next section the extension for adaptive processing.

In the original formulation [1], conventional beamforming was used and is briefly summarized here. The hydrophone data for each channel at angular frequency ω, are written as a column vector $\mathbf{p} = [p_1, p_2 ... p_M]$ for the M hydrophones. Each entry is determined through a Fourier Transform of an ambient noise time series measured on each channel, $p_m(\omega) = \mathscr{F}\{p_m(t)\}$. The number of points in the FFT processing will be referred to as the snapshot size.

With conventional beamforming, the weight for the mth hydrophone steered at angle θ for plane waves arriving at grazing angle θ between the hydrophones separated by distance d is written

$$w_m = e^{-imkd\sin\theta} . \tag{1}$$

The array is referenced to the shallowest hydrophone which is element $m = 0$. The wavenumber is $k = \omega/c$ and c is the sound speed in the water (around 1500 m/s).

Writing the steering weights as a column vector, $\mathbf{w} = [w_0, w_1 ... w_{M-1}]$, the beam, $b(\theta)$, is written,

$$b(\theta) = \mathbf{w}^\dagger \mathbf{p}, \tag{2}$$

where $\dagger$ represents the conjugate transpose operation. The conventional instantaneous beam power for a given direction θ is computed by the following,

$$B(\theta) = \mathbf{w}^\dagger \mathbf{p}(\mathbf{w}^\dagger \mathbf{p})^\dagger = \mathbf{w}^\dagger \mathbf{p}\mathbf{p}^\dagger \mathbf{w} = \mathbf{w}^\dagger \mathbf{K}\mathbf{w}, \tag{3}$$

where the single snapshot cross-spectral density matrix (CSDM) $\mathbf{K}$ is identified as $\mathbf{p}\mathbf{p}^\dagger$. Forming the CSDM allows for multiple snapshots to be averaged before beamforming.

An example of conventional beamforming ocean noise can be illustrated using data from the MAPEX2000 experiment [3] (see Fig. 1). These data were collected by the NATO Undersea Research Centre in the Strait of Sicily between Italy and Malta in November 2000. The array had 64 hydrophones of different separations and here, the 0.5 m equally spaced 32 hydrophones of the array were used for analysis. The array was positioned in about 130 m water depth and moored to the bottom. Three minutes of data were averaged to form the CSDM used for beamforming. Figure 1 shows the beamformed output normalized for a maximum of 0 dB. The positive angles correspond to the beams steered towards the surface (up-looking beams) which capture signals traveling downward from the surface. The negative angles are beams steered down towards the seabed (down-looking beams) and capture signals traveling upward from the seabed. In Fig. 1, the higher intensity noise is evident on the up-looking beams since the down-looking beams see the surface through a seabed reflection and therefore experience losses. Also note the high intensity noise coming from near horizontal. These

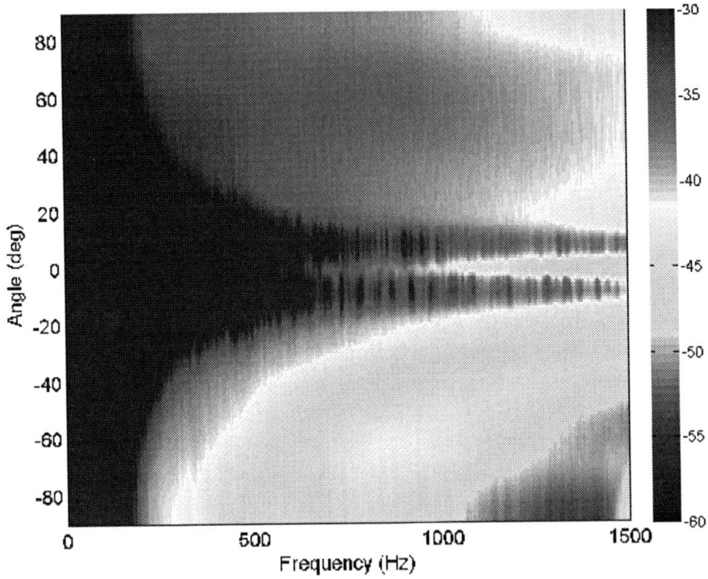

FIGURE 1. Conventional beamforming applied to 3 minutes of ocean noise data from the MAPEX2000 experiment.

data consist of a variety of noise sources including breaking waves and distant shipping. For passive fathometer processing, the interest is in correlating the up-looking beam with the down-looking beam rather than forming the beam power at each grazing angle. To steer an up-looking beam directly upward towards the surface, $\theta = +90^\circ$ and to steer directly towards the seabed, $\theta = -90^\circ$. The steering weights are $w_m = e^{\mp(imkd)}$ and the up-looking weight will be denoted $\mathbf{w}_+$ and the down-looking as $\mathbf{w}_-$. Therefore,

$$b_{up} = \mathbf{w}_+^\dagger \mathbf{p}, \tag{4}$$

and the downward beam,

$$b_{dn} = \mathbf{w}_-^\dagger \mathbf{p}. \tag{5}$$

Expressed in the frequency domain, the correlation of these two beams is,

$$C = (\mathbf{w}_-^\dagger \mathbf{p})(\mathbf{w}_+^\dagger \mathbf{p})^\dagger = \mathbf{w}_-^\dagger \mathbf{p}\mathbf{p}^\dagger \mathbf{w}_+ = \mathbf{w}_-^\dagger \mathbf{K}\mathbf{w}_+. \tag{6}$$

But, for conventional processing, the down-looking beam is just the conjugate of the up-looking beam, $\mathbf{w}_- = \mathbf{w}_+^*$. The conventional beamforming correlation then can be written simply using $\mathbf{w} = \mathbf{w}_-$,

$$C = \mathbf{w}^\dagger \mathbf{K}\mathbf{w}^*. \tag{7}$$

The conventional passive fathometer in (7) differs only from the conventional beamformer output given by (3) in that the $\mathbf{w}$ to right is conjugated. As a result, B is positive real and C is a complex number. The corresponding phase of C contains the information on the reflection arrival times. As with conventional beamforming, the CSDM given in Eq. (7) can be formed over as many snapshots of data, **p**, as needed to obtain the desired averaging. Averaging time may vary depending on factors such as array motion, noise level and number of elements in the array.

ADAPTIVE PASSIVE FATHOMETER PROCESSING

As evident in Fig. 1, there is significant noise intensity at grazing angles near horizontal. This high intensity noise away from the $+90°$ and $-90°$ directions (i.e., away from endfire) degrade the passive fathometer output due to leakage through sidelobes. Sidelobe leakage is a common problem in beamforming and one of the mitigation strategies is to use adaptive methods such as Minimum Variance Distortionless Response (MVDR) [4, 5]. To adaptively beamform using MVDR, the steering weights $\tilde{\mathbf{w}}$ are computed according to,

$$\tilde{\mathbf{w}} = \frac{\mathbf{K}^{-1}\mathbf{w}}{\mathbf{w}^\dagger\mathbf{K}^{-1}\mathbf{w}}. \tag{8}$$

Using (7) and continuing to assume $\tilde{\mathbf{w}}_- = \tilde{\mathbf{w}}_+^*$, the MVDR correlation at frequency ω is,

$$C_A = \tilde{\mathbf{w}}^\dagger\mathbf{K}\tilde{\mathbf{w}}^*. \tag{9}$$

Note that for the typical application of MVDR beamforming (as opposed to cross-correlating beams), the beamformer output can be written as was done in (3) but with adaptive weights,

$$B_A(\theta) = \tilde{\mathbf{w}}^\dagger\mathbf{K}\tilde{\mathbf{w}}. \tag{10}$$

And, using (8), the beamformer output reduces to a much simpler expression that uses the conventional weights,

$$B_A(\theta) = \left[\frac{\mathbf{K}^{-1}\mathbf{w}}{\mathbf{w}^\dagger\mathbf{K}^{-1}\mathbf{w}}\right]^\dagger \mathbf{K}\left[\frac{\mathbf{K}^{-1}\mathbf{w}}{\mathbf{w}^\dagger\mathbf{K}^{-1}\mathbf{w}}\right] = \left(\mathbf{w}^\dagger\mathbf{K}^{-1}\mathbf{w}\right)^{-1} \tag{11}$$

MVDR is sometimes implemented with this expression since it is less computationally involved than Eqs. (8) and (10).

For adaptive processing, the weights for up- and down-looking beams are not necessarily conjugates of each other (i.e., $\tilde{\mathbf{w}}_- \neq \tilde{\mathbf{w}}_+^*$) and the correct expression for the correlation is

$$C_A = \tilde{\mathbf{w}}_-^\dagger\mathbf{K}\tilde{\mathbf{w}}_+. \tag{12}$$

For implementation, we derive the up- and down-looking beam adaptive weights separately using (8) and then compute the correlation using (12). The time-series passive fathometer response is simply the inverse Fourier transform of C or, $c(t) = \mathscr{F}^{-1}\{C(\omega)\}$ (or the inverse Fourier transform of C_A). For a 1-D medium with a single point source the Green's function is proportional to the noise cross-correlation $c(t)$[6]. This model resembles the main wave propagation for the fathometer and thus the reflection response is proportional to $c(t)$.

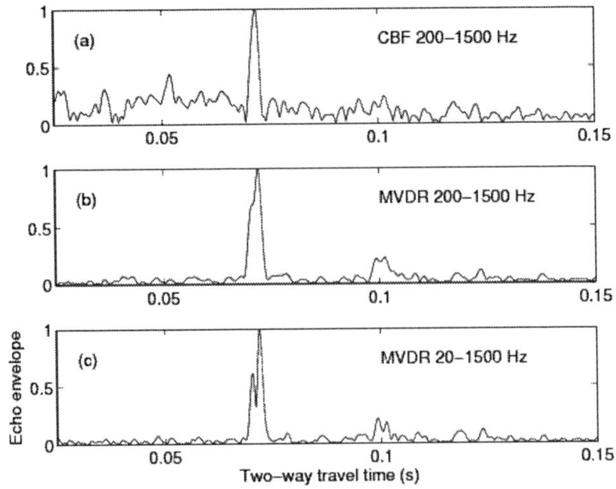

FIGURE 2. Example of conventional beamforming (a) and adaptive (MVDR) beamforming (b) on the same cross-spectral density matrix of data. The envelope of $c(t)$ is shown using a 200–1500 Hz band. Panel (c) shows MVDR results in the 20–1500 Hz band.

An example of the difference between conventional and MVDR adaptive beamforming is shown in Fig. 2. Both results use the same CSDM averaged over 3 minutes (same CSDM as used in Fig. 1). The envelope of the time domain expression, $c(t)$, is shown in the panel (a) for conventional beamforming in the band 200–1500 Hz processed using (6). Panel (b) shows the MVDR results in the same band using (9). Panel (c) of Fig. 2 shows MVDR over a larger bandwidth of 20–1500 Hz. The MVDR better rejects the horizontal noise and therefore it is possible to include lower frequencies which penetrate farther into the seabed and with the increased bandwidth better time resolution is possible.

The adaptive passive fathometer results can also be compared with standard seismic measurements (using active sonar) that were taken at the experiment site. Figure 3 shows the passive fathometer time trace (far right of figure) along with a corresponding seismic profile taken in the same location. The fixed array was located at the south end of the seismic track; the approximate array location is near range 11 km in the figure. The adaptive passive fathometer time trace gives an indication of some interfering layers near the water-sediment interface which is consistent with the joining of two layers in the seismic profile. Also, the passive fathometer shows two sub-bottom layers approximately 20 m below the water-sediment interface. Recall that in Fig. 2 the conventional processing did not resolve layers near the water-sediment interface and the deeper layers are not identifiable.

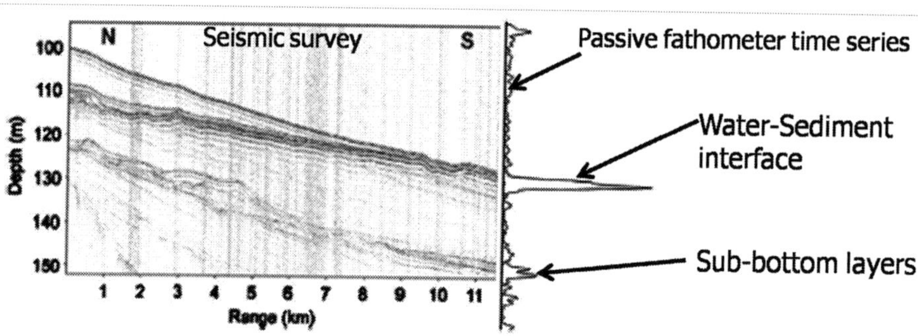

FIGURE 3. Left side shows the sub-bottom profile for the MAPEX2000bis site (sub-bottom profile from Ref. [7]) along with the adaptive passive fathometer result shown on the right. The approximate position of the vertical array is at 11 km in range.

GEO-ACOUSTIC INVERSION

With the adaptive passive fathometer, a second ambient noise technique is used here and referred to as the R-θ method. This uses the intensity of vertically beamformed noise to determine an approximation to the magnitude-squared bottom reflection loss (i.e. power reflection loss) [2]. The reflection intensity (R-θ) and the adaptive passive fathometer returns are the observables that can be used to determine a complete geoacoustic description of the seabed through inverse methods.

The geoacoustic inversion approach first parameterize the seabed into layers based on the adaptive passive fathometer returns. Once these layers are determined they are fixed. Next, a random guess is made for the compressional sound speed, density and attenuation for each layer. Using numerical methods we construct simulated observables just like the measured observables using the seabed properties from the random guess. A cost function is defined to quantify the agreement between the measurement and the simulation. A search is performed over all possible combinations of seabed properties to determine the set that produces the best match with the data. A genetic algorithm is used to direct the search and make the process more efficient. The final result is a fairly detailed description of the seabed consisting of several layers and each with a geoacoustic description. An important point is that generating the simulated observables is efficient since only a numerical bottom loss calculation is needed along with signal processing to account for beamforming effects; no actual sound propagation modeling is required.

We also compared the ambient noise inverted seabed properties with a matched field inversion approach. Matched field geo-acoustic inversion (MFI) is a model based technique that has been applied successfully in characterizing the seabed for the most important parameters for propagation prediction. This is a remote sensing method that uses down-range acoustic measurements to infer properties of the seabed. Computer simulations are used to model the down-range acoustic response to different seabed types, and efficient search algorithms (e.g. genetic algorithms) are applied to find the

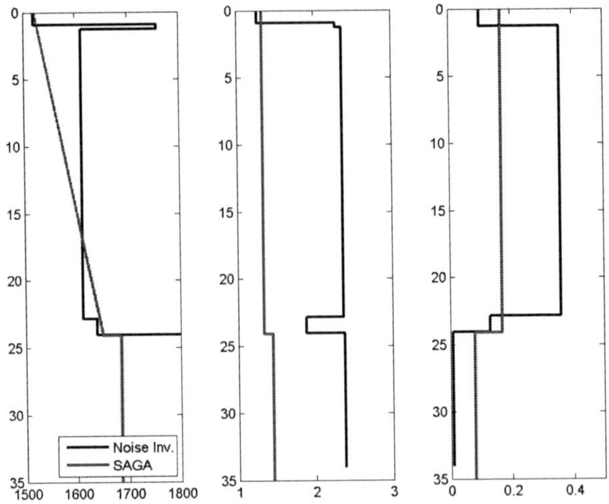

FIGURE 4. Comparisons between inverting ambient noise data (black line) and using a set of tonal transmissions with matched field inversion approach using SAGA (red). The most sensitive parameter is the sound speed shown in the left panel. The middle panel is the comparison between density estimates and the right panel for attenuation.

environment giving an optimal match between modeled and measured data. This may sound similar to the ambient noise approach but with some big differences 1) this requires a sound source with a range separation from the array and 2) the inversion involves propagation modeling 3) the resulting estimates are a type of integrated or effective seabed over the source-array range 4) the sound field interacts with a possibly rough surface that may be difficult to model. The matched field inversion used the sound source transmission at 3.5 km received on the vertical array and a set of tonal transmissions at 200, 275, 350, 425, 500 and 575 Hz. In an attempt to keep the results unbiased the SAGA MFI software was used [8]. SAGA is more or less a turnkey package for data of this type (i.e., tonals recorded on a vertical array of hydrophones). The results comparing the two inversions are shown in Fig. 4. The results show reasonable agreement especially near the water-sediment interface. In particular, the seabed sound speed, which is by far the most sensitive parameter agree well. The MFI technique uses a single layer over a halfspace description of the seabed and this is simpler than that from the passive fathometer processing. Still, the deep reflector at about 22 m depth is nearly the same for both inversion approaches. Part of the discrepancy between the two inversion approaches for the density and attenuation at the larger depths is likely due to decreased sensitivity of these measurements to those parameters.

CONCLUSION

Using a fixed receiver array, the ambient noise field was measured and estimates for the sediment properties were made. The beamformed noise intensity pattern was used to estimate the magnitude of the reflection loss curve (R-θ method) while the coherent noise field is used to determine the layering (adaptive passive fathometer method). The adaptive passive fathometer was also used to correct for the position of the array relative to the seabed. This type of seabed characterization can be used for predicting transmission loss and for other environmentally sensitive signal processing such as matched field processing.

ACKNOWLEDGMENTS

The author would like to gratefully acknowledge support for this research by the Office of Naval Research Ocean Acoustics Program.

REFERENCES

1. M. Siderius, C. H. Harrison, and M. B. Porter, *J. Acoust. Soc. Am.* **120**, 1315–1323 (2006).
2. C. H. Harrison, and D. G.Simons, *J. Acoust. Soc. Am.* **112**, 1377–1389 (2002).
3. M. Siderius, P. L. Nielsen, and P. Gerstoft, *J. Acoust. Soc. Am.* **112**, 1523–1535 (2002).
4. H. L. Van Trees, *Detection, Estimation, and Modulation Theory: Part IV: Optimum Array Processing*, John Wiley & Sons, Inc., New York, New York, 2002.
5. H. L. Van Trees, *Underwater Acoustic System Analysis*, Prentice-Hall Inc., Englewood Cliffs, New Jersey, 1984.
6. P. Gerstoft, W. S. Hodgkiss, M. Siderius, C.-F. Huang, and C. H. Harrison, *J. Acoust. Soc. Am.* **123**, 1297–1305 (2008).
7. C. W. Holland, *IEEE Journal of Oceanic Engineering* **27**, 454–470 (2002).
8. P. Gerstoft, SAGA Users guide 5.4, an inversion software package, Tech. rep., SACLANT Undersea Research Centre, La Spezia, Italy, SM–333 (1997).

Seabed Geoacoustic Inversion from Long-Range Broadband Sound Propagation in the Yellow Sea

Lin Wan[a], Ji-Xun Zhou[a, b] and Peter Rogers[a]

a. School of Mechanical Engineering, Georgia Institute of Technology, Atlanta, GA, 30332, USA
b. State Key Laboratory of Acoustics, Institute of Acoustics, Chinese Academy of Sciences, China
E-mail:lin.wan@gatech.edu

Abstract. In the summer of 1996, a joint China-U.S. underwater acoustics experiment was conducted near the geographic center of the Yellow Sea with a water depth of 75 m. Hundreds of broadband explosive sources were deployed. The seabed sound speed in this experiment has been inverted using two methods based on data-derived mode depth functions and modal arrival time differences. The seabed attenuation has been estimated from three different approaches using modal attenuation coefficients, modal amplitude ratios, and transmission loss measurements [Wan, Zhou, and Rogers, *J. Acoust. Soc. Am.*, **128**(2), 2010]. The present paper analyzes the vertical coherence of the sound propagation as a function of range and frequency. The seabed sound attenuation is inferred by minimizing the difference between the theoretical and measured vertical coherence. The seabed attenuation, inverted from long-range broadband sound propagation with different methods over a frequency range of 80-1000 Hz, can be expressed as $\alpha_b = (0.34 \pm 0.02) f^{1.87\pm0.04} (dB/m \cdot kHz)$.

Keywords: Seabed geo-acoustic inversion, shallow water acoustics, spatial coherence.
PACS: 43.30.Ma, 43.30.Bp, 43.30.Re

INTRODUCTION

The geo-acoustic properties of the seabed play an important role in shallow water sound propagation. Many inversion techniques have been developed to estimate low-frequency seabed sound speed and attenuation as a function of frequency. These broadband inversion techniques involve dispersion analysis of broadband signals [1,2], data-derived mode depth functions [3], measurements of modal attenuation coefficients [4,5], measurements of modal amplitude ratios [6-8], transmission loss (TL) measurements [9,10], and vertical coherence of reverberation and propagation [11].

The seabed sound speed and attenuation in the Yellow Sea '96 experiment (YS '96), inverted from long range broadband acoustic measurements with different methods in a frequency range of 80-1000 Hz, have been reported in [12]. The seabed sound speed has been inverted using two inversion schemes. The first scheme is based on extracting the normal mode depth functions from the cross-spectral density matrix (CSDM). In the second scheme, the modal arrival time differences of broadband signals are used to obtain the seabed sound speed. Wan *et al.* [12] have used the

CP1272, *Shallow-Water Acoustics, Second International Shallow-Water Acoustics Conference*
edited by J. Simmen, E. S. Livingston, J. X. Zhou, and F. H. Li

inverted seabed sound speed as a constraint condition to estimate the seabed sound attenuation by three different approaches. The first approach involves measurement of individual normal mode attenuation coefficients. The second approach is based on measured normal mode amplitude ratios. In the third approach, the seabed sound attenuation is inverted by finding the best match between the measured and modeled broadband TLs.

In this paper, those inversion methods in [12] are briefly introduced and the inversion results are summarized. Then, the vertical coherence of sound propagation in YS '96 is analyzed as a function of range and frequency. The sound attenuation in the seabed is inverted by matching the calculated vertical coherence with the experimental data.

AT-SEA EXPERIMENT

The Yellow Sea '96 experiment, the first joint China-U.S. shallow water acoustics experiment, was conducted near the geographic center of the Yellow Sea. The experimental site has a very flat bottom, and the water depth is 75 m. From August 22, 1996 to August 25, 1996, about 300 broadband explosive sources (TNTs) were deployed at varying distances from the source ship (research vessel *Shi Yan 2*). The measurements were made by a 32-element hydrophone vertical line array with an element spacing of 2 m. The vertical line array was deployed from the research vessel *Shi Yan* 3, which held stationary. The depth of each hydrophone is shown by dot in figure 1. The averaged water column sound speed profile (SSP) in figure 1 was determined from the conductivity-temperature-depth data measured hourly at the location of *Shi Yan* 3 during the experiment. The averaged SSP shows a strong and sharp thermocline. The nominal depth of the source detonations was 50 m, which was below the thermocline. Just prior to the Yellow Sea '96 experiment, scientists from the South China Sea Institute of Oceanology took bottom core data and reported the density of the seabed to be 1.85g/cm^3.

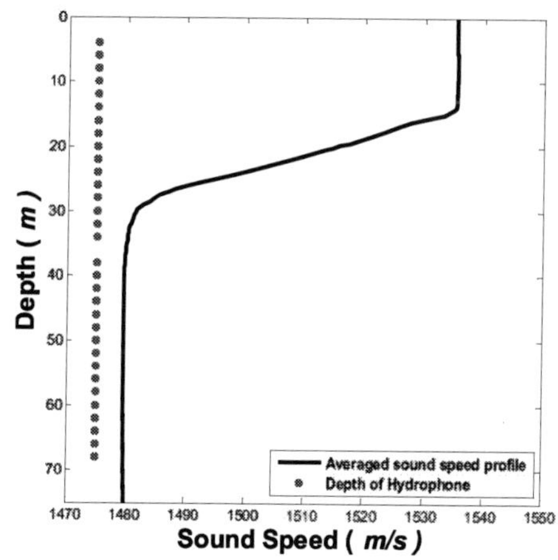

FIGURE 1. Water column SSP and depth of hydrophone

SEABED SOUND SPEED INVERSION BY DATA-DERIVED MODE DEPTH FUNCTIONS

The eigenvectors of the CSDM measured on a vertical line array correspond to the mode shapes [3]. In YS'96, the data-derived mode depth function is obtained using singular value decomposition of the CSDM, which is constructed using measured pressure spectra over multiple-frequency bins in a narrow band at a fixed source-receiver distance. The theoretical mode depth function is calculated by a shooting method [3] based on a depth-separated finite difference wave equation. The seabed sound speed corresponding to the calculated mode that best matches the data-derived mode is $1593 \pm 15m/s$ at 100 Hz.

SEABED SOUND SPEED INVERSION USING MEASURED MODAL ARRIVAL TIMES

The acoustic propagation of a broadband signal exhibits dispersion characteristics in shallow water. The group velocities differ for different frequencies and modes. In YS'96, the experimental modal arrival time differences corresponding to various modes and frequencies are extracted from time-frequency analysis of experimental signals measured at sufficiently long range. The theoretical modal arrival times are calculated using normal mode code, such as KRAKEN [13]. The inverted seabed sound speed is obtained by adjusting the seabed sound speed inputs to KRAKEN such that the difference between modeled and measured group velocity dispersion curves is minimized. The inverted results over a frequency range from 150Hz to 400 Hz are shown in figure 2(a).

The average value of inverted seabed sound speed from the CSDM-based inversion and the dispersion-based inversion is 1588 m/s. This value is used as a constraint condition in the seabed sound attenuation inversion using modal attenuation coefficients, modal amplitude ratios, transmission loss data, and vertical coherence measurement.

SEABED ATTENUATION INVERTED FROM MEASURED MODAL ATTENUATION COEFFICIENTS

The normal-mode attenuation coefficients depend on the seabed attenuation [14]. The seabed attenuation can thus be inverted by minimizing the difference between the theoretical modal attenuation coefficients and the measured results. In YS'96, the experimental modal attenuation coefficients of the first three modes over a frequency range of 100Hz to 400Hz are extracted from a plot of measured modal amplitude versus ranges. The inverted seabed attenuations are shown in figure 2(b) by crosses.

SEABED ATTENUATION INVERTED FROM MEASURED MODAL AMPLITUDE RATIOS

The modal amplitude ratios are sensitive to the seabed attenuation. Therefore, the seabed attenuation can be estimated by finding the best match between the theoretical modal amplitude ratios and the measured results. In YS'96, the experimental modal amplitude ratios of the 2nd, 3rd and 4th mode relative to the 1st mode over a frequency range of 160Hz to 400Hz are extracted to determine the seabed attenuation. The inverted seabed attenuations are shown in figure 2(b) by downward-pointing triangles.

SEABED ATTENUATION INVERSION USING TL DATA

The seabed attenuation can also be determined by comparing TL data with predictions along the propagation direction. In YS'96, the seabed attenuation is inverted by matching the modeled and measured TL as a function of frequency and range when receivers are located both above and below the thermocline. The inverted seabed attenuations as a function of frequency from 80 Hz to 1000 Hz are shown in figure 2(b) by upward-pointing triangles.

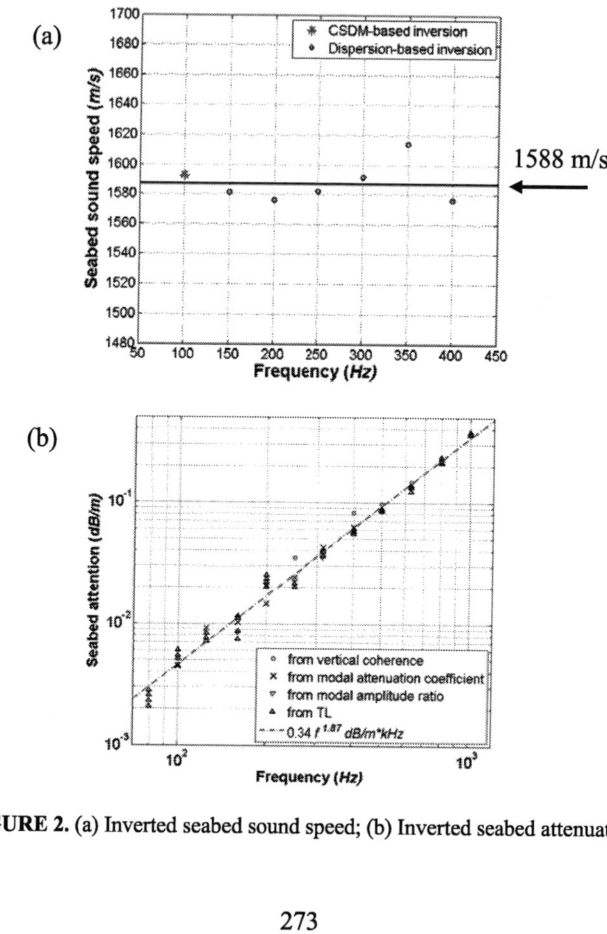

FIGURE 2. (a) Inverted seabed sound speed; (b) Inverted seabed attenuation

SEABED GEO-ACOUSTIC INVERSION USING VERTICAL COHERENCE MEASUREMENTS

The spatial coherence of the sound field in shallow water is sensitive to the seabed parameters. Zhou *et al.* have derived a mathematical model to calculate reverberation vertical coherence in shallow water and have used reverberation vertical coherence data to invert the seabed sound speed and attenuation [11]. Vertical coherence of sound propagation is defined in this paper as the normalized cross correlation between the sound pressures received by two hydrophones separated from the vertical line array.

The mathematical expression for experimental vertical coherence can be express as

$$\rho(\Delta T) = \int_{t}^{t+\Delta T} p_1(t)p_2(t)dt \Bigg/ \sqrt{\int_{t}^{t+\Delta T} p_1^{\,2}(t)dt \times \int_{t}^{t+\Delta T} p_2^{\,2}(t)dt} \tag{1}$$

where, ΔT is the integration time.

In this paper, the experimental vertical coherence is evaluated for the stated center frequency with a 1/3 octave bandwidth using an integration time of 683 ms. The measured veridical coherence at each distance is averaged over data from three pairs of hydrophones, located at depths from 42 m to 52 m. The separation between a pair of hydrophones is 4 m.

In terms of normal modes, the mathematical expression for the theoretical vertical coherence can be expressed as [11]

$$\rho(\Delta z, z, z_s, r) = \frac{\sum_n \overline{|\Psi_n(z_s)|^2}\,\overline{|\Psi_n(z)|^2}\cos[k(z)\Delta z \sin\theta_n(z)]\exp(-2\beta_n r)/k_n}{\sum_n \overline{|\Psi_n(z_s)|^2}\,\overline{|\Psi_n(z)|^2}\exp(-2\beta_n r)/k_n} \tag{2}$$

where, Δz is the vertical separation of a pair of hydrophones, z_s is the source depth, z is the receiver depth, r is the source-receiver distance, Ψ_n is the eigenfunction of the nth normal mode, k_n is the longitudinal wave number, β_n is the modal attenuation coefficient of the nth normal mode, and θ_n is the mode angle. k_n and the mode angle are related by: $\cos\theta_n(z) = \dfrac{k_n}{k(z)}$ and $k(z) = \dfrac{\omega}{c(z)}$.

According to WKB approximation,

$$\overline{|\Psi_n(z)|^2} = \frac{2}{S_n \tan\theta_n(z)} \tag{3}$$

where, S_n is the cycle distance of the nth normal mode.

If a receiver is near a turning point, $\tan\theta_n(z) \approx 0$, Eq. (3) can not be used. Instead, Zhang and Jin [15] proposed the following expression

$$\overline{|\Psi_n(z)|^2} = \frac{1}{S_n[k^2(z)D(z) + k^2(z) - k_n^2]^{1/2}} \tag{4}$$

where,

$$D(z) = 0.875 \left| \frac{1}{\pi f} \frac{dc(z)}{dz} \right|^{2/3} \tag{5}$$

The theoretical vertical coherence can be calculated using KRAKEN, Eq. (2), Eq. (3), Eq. (4), and Eq. (5). A half-space bottom model with three-parameters (sound speed, density and attenuation) is used. The density (1.85 g/cm^3) is taken from the core analysis. The seabed sound speed input to KRAKEN, 1588 m/s, is an average value obtained from the CSDM-based inversion and the dispersion-based inversion. The seabed sound attenuation is estimated by adjusting the sound attenuation inputs to KRAKEN such that the difference between modeled and measured vertical coherence is minimized. The following cost function is defined to calculate the difference.

$$E(\alpha_b) = \sqrt{\sum_{i=1}^{N_r} |\hat{\rho}(i) - \rho(i)|^2 \Big/ N_r} \tag{6}$$

where, N_r is the number of distances, $\hat{\rho}(i)$ is the measured vertical coherence at the ith distance, $\rho(i)$ is the theoretical coherence at the ith distance.

The vertical coherence from 7 km to 19 km is used for inversion. The inversion results, obtained from 250 Hz to 800 Hz by minimizing the cost function (Eq. 6), are listed in Table 1 and shown in figure 2(b) by circles.

TABLE 1. Inverted seabed sound attenuation as a function of frequency

f **(Hz)**	**250 3**	**15 4**	**00**	**500 6**	**30**	**800**
α_b (dB/m)	0.03479	0.03782	0.08157	0.09824	0.14715	0.22826

Figures 3 and 4 show the theoretical and measured vertical coherence as a function of range at 400 Hz and 800 Hz, respectively. The theoretical predictions based on the inverted seabed sound attenuation are in good agreement with the measured results.

(a)

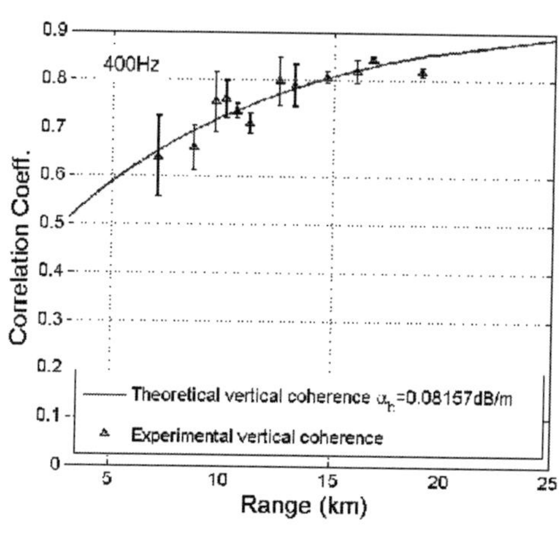

FIGURE 3. Data-model comparison at 400 Hz. Error bars show the standard errors.

(b)

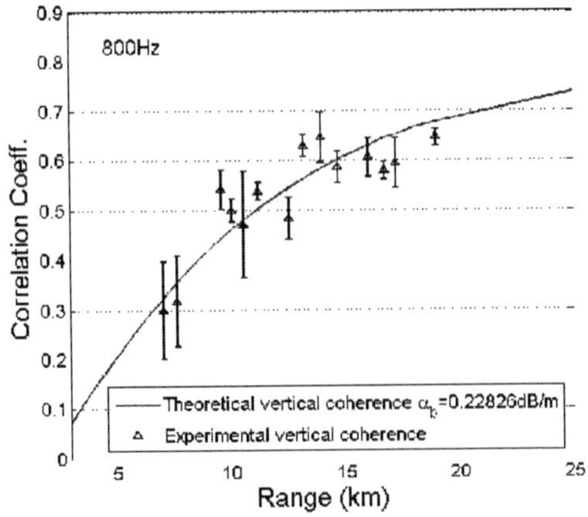

FIGURE 4. Data-model comparison at 800 Hz. Error bars show the standard errors.

SUMMARY

The seabed sound speed and attenuation as a function of frequency at the YS '96 experimental site are inverted by different broadband inversion techniques, which involve dispersion analysis of broadband signals, mode depth functions, modal attenuation coefficients, modal amplitude ratios, transmission loss, and vertical coherence of sound propagation.

The resultant averaged sound speed is 1588 m/s in the seabed. The ratio of the seabed sound speed to the sound speed in the water column near the seabed is 1.073. The inverted sound speed in the seabed at the YS'96 site are close to the geo-acoustic inversion results in the Yellow Sea by Zhou [5][8], Li *et al.*[16], Rogers *et al.*[17].

A power law form of frequency dependence for the seabed attenuation in a given frequency band is widely used in the ocean acoustics community:

$$\alpha_b = \alpha_0 f^n (dB/m \cdot kHz) \tag{7}$$

where α_0 and n are empirically derived constants, f is frequency in kilohertz. The constants (α_0 and n) and their uncertainties can be uniquely determined using the power law fitting method. The inverted seabed sound attenuation shown in figure 2(b) exhibits nonlinear frequency dependence [$\alpha_b = (0.34 \pm 0.02) f^{1.87\pm0.04} (dB/m \cdot kHz)$], which is similar to the low-frequency seabed attenuation data obtained from 20 locations in different coastal zones around the world by Zhou *et al.* [18].

ACKNOWLEDGMENTS

The authors thank Dr. Ellen Livingston and Dr. Benjamin Reeder for their support and helpful comments. This work was sponsored by the Office of Naval Research (ONR), including an ONR Acoustics Graduate Traineeship Award to the first author. The YS '96 experiments were partially funded by the Chinese Academy of Sciences.

The efforts of the ship manager, captains and crews of *Shi Yan 3* and *Shi Yan 2* are gratefully acknowledged.

REFERENCES

1. G. R. Potty, J. H. Miller, J. F. Lynch and K. F. Smith, "Tomographic inversion for sediment parameters in shallow water", *J. Acoust. Soc. Am.* **108**, 973-986, 2000.
2. Z. L. Li and R. H. Zhang, "Geoacoustic inversion based on dispersion characteristic of normal modes in shallow water", *Chin. Phys. Lett.* **24**, 471-474, 2007.
3. P. Hursky, W. S. Hodgkiss and W. A. Kuperman, "Matched field processing with data-derived modes," *J. Acoust. Soc. Am.* **109**, 1355-1366, 2001.
4. F. Ingenito, "Measurement of mode attenuation coefficients in shallow water," *J. Acoust. Soc. Am.* **53**, 858-863, 1973.
5. J. X. Zhou, "Normal mode measurements and remote sensing of sea-bottom sound velocity and attenuation in shallow water," *J. Acoust. Soc. Am.* **78**, 1003-1009, 1985.
6. G. R. Potty, J. H. Miller and J. F. Lynch, "Inversion for sediment geoacoustic properties at the New England Bight", *J. Acoust. Soc. Am.* **114**, 1874-1887, 2003.
7. C. T. Tindle, "Attenuation parameters from normal mode measurements", *J. Acoust. Soc. Am.* **71**, 1145-1148, 1982.
8. J. X. Zhou, X. Z. Zhang, P. H. Rogers and J. Jarzynski, "Geoacoustic parameters in a stratified sea bottom from shallow-water acoustic propagation," *J. Acoust. Soc. Am.* **82**, 2068-2074, 1987.
9. Z. H. Peng, J. X. Zhou, P. H. Rogers and R. H. Zhang, "Sea-bed acoustic parameters from dispersion analysis and transmission loss in the East China Sea", *IEEE J. Ocean. Eng.* **29**, 1038-1045, 2004.
10. I. Rozenfeld, W. M. Carey, P. G. Cable, and W. L. Siegmann, "Modeling and analysis of sound transmission in the Strait of Korea," *IEEE J. Ocean. Eng.* **26**, 809-819, 2001.
11. J. X. Zhou, X. Z. Zhang, P. H. Rogers, J. A. Simmen, P. H. Dahl, G. L. Jin and Z. H. Peng, "Reverberation vertical coherence and sea-bottom geoacoustic inversion in shallow water," *IEEE J. Ocean. Eng.* **29**, 988-999, 2004.
12. L. Wan, J. X. Zhou, and P. H. Rogers, "Low frequency sound speed and attenuation in sandy seabottom from long-range broadband acoustic measurement", *J. Acoust. Soc. Am.*, **128**, 2010.
13. M. B. Porter, "The KRAKEN Normal Mode Program," Naval Research Lab., Washington D. C., NRL/MR/5120-92-6920, 1992.
14. R. A. Koch, C. Penland, P. J. Vidmar and K. E. Hawker, "On the calculation of normal mode group velocity and attenuation," *J. Acoust. Soc. Am.* **73**, 820-825, 1983.
15. R. H. Zhang and G. L. Jin, "Normal-mode theory of the average reverberation intensity in shallow water," *J. Sound Vib.,* 119, 215-223, 1987.
16. F. Li and R. H. Zhang, "Bottom sound speed and attenuation inverted by using pulsed waveform and transmission loss," *Acta Acoustica,* **25**, 297-302, 2000.
17. P. H. Rogers, J. X. Zhou, X. Z. Zhang, and F. Li, "Seabottom acoustic parameters from inversion of Yellow Sea experimental data," in *Experimental Acoustic Inversion Methods for Exploration of the Shallow Water Environment,* edited by A. Caiti, J-P Hermand, S. M. Jesus, and M. B. Porter, Kluwer Academic, Dordrecht, 2000, pp. 219-234.
18. J. X. Zhou, X. Z. Zhang, and D. P. Knobles, "Low-frequency geoacoustic model for the effective properties of sandy seabottoms," *J. Acoust. Soc. Am.* **125**, 2847-2866, 2009.

An inversion method of geoacoustic properties based on towed tilted line array in shallow water

Kunde Yang, Yuanliang Ma, Xuegang Li and Haibin Qiu

Institute of Acoustic Engineering, Northwestern Polytechnical University, Xi'an, 710072, P. R. China

Abstract. This paper studied a method for extracting seabed reflection losses based on a towed line array for large scale sea areas with high resolution. The experimental system consisted of a towed tilted line array and three acoustic sources to cover a large scope of grazing angles (from 10° to 70°). The simulation results in shallow water show that different arrays have different experimental configurations in order to cover the same scope of grazing angles. The vertical line array must operate with several acoustic sources and requires a stationary platform of the array. The towed horizontal line array needs only one acoustic source but requires a large aperture line array. The towed tilted line array combines the merits of the vertical line array and the horizontal line array, which can invert the seabed acoustic parameters in a moving and high resolution way, and requires only a smaller aperture line array. The error resulting from the distortion of the array shape was analyzed for the inversion method.

Key words: Underwater acoustic field prediction, Seabed acoustic parameters, Seabed reflection loss, Towed tilted line array

PACS: 43.30. +m, 92.10.VZ

I. INTRODUCTION

A shallow water seabed is an interface of the ocean waveguide. Consequently, its topography, structure and physical characteristics are very important for acoustic propagation. Here the performance of a sonar system is determined by the seabed absorption, scattering and reflection. A fast inversion scheme for the seabed acoustic parameters is very important for the underwater acoustic field prediction[1], matched field localization[2] or underwater acoustic communication[3].

Several methods are already used to estimate the acoustic parameters of the seabed, such as matched field inversion[4-7], mode amplitude inversion[8] and shallow bottom profile inversion[9]. Each method has its own advantages and disadvantages. Matched field inversion is usually used in long range acoustic field. The average

CP1272, *Shallow-Water Acoustics, Second International Shallow-Water Acoustics Conference*
edited by J. Simmen, E. S. Livingston, J. X. Zhou, and F. H. Li

effect of the changing environment of the water and seabed can be obtained. It has small sensitivity to the density and attenuation coefficient and needs to invert many parameters. Schock[9] presented a method to estimate seabed acoustic parameters based on the complex Biot model and the bottom profile vertical reflection data. However it needed to estimate[13] parameters in the Biot model. Holland and Dettmer[10-11] extracted the seabed reflection loss curve through a single fixed hydrophone and an impulsive broadband moving source. This method can estimate the sound velocity and density of the seabed surface in a better way.

The methods above usually use a vertical line array to receive the acoustic data, these are inconvenient to invert the seabed parameters in a moving way. Recently the matched field inversion method based on the towed horizontal line array[12] was studied; however this requires a large aperture line array. This paper presents a fast inversion scheme to obtain the seabed acoustic parameters; it consists of a towed tilted line array and three acoustic sources, and covers a large scope of grazing angles. Then we can invert many other important parameters such as density, sound velocity, seabed layer thickness and seabed attenuation coefficient using the method in Yang's paper[13]. Its obvious advantage is to invert the seabed acoustic parameters in a moving, high resolution way by using a smaller aperture line array.

II. CALCULATION OF SEABED REFLECTION LOSS

This section provides a calculation method to determine the seabed reflection loss. The seabed reflection wave's sound pressure amplitude $Pr1$ satisfies the following equation,

$$20\log Pr_1 = 20\log Ps - RL - 20\log r_1 - \alpha \cdot r_1 \times 10^{-3} \quad (1)$$

where RL is the seabed reflection loss in dB, Ps is the sound pressure amplitude of the acoustic source level, r_1 (m) is the propagation distance of the seabed reflection wave, absorption coefficient $\alpha = \frac{0.102 f^2}{1+f^2} + \frac{40.7 f^2}{4100+f^2}$ (dB/km), f (kHz) is the frequency of the transmitted signals.

The direct arrival wave's sound pressure amplitude Pr_2 satisfies equation (2),

$$20\log Pr_2 = 20\log Ps - 20\log r_2 - \alpha \cdot r_2 \times 10^{-3} \quad (2)$$

where r_2 (m) is the propagation distance of the direct arrival wave. If the acoustic source has the same directivity both in the seabed reflection direction and the direct arrival direction, equation (3) can be derived by using Eq. (1) and Eq. (2). Otherwise, the differences of the directivity should be corrected.

$$RL = 20\log\frac{Pr_2}{Pr_1} - 20\log\frac{r_1}{r_2} - \alpha(r_1 - r_2)\times 10^{-3} \quad (3)$$

The seabed reflection loss curve with the frequency, sound velocity, density and grazing angle changes is shown in Fig. 1. The seabed medium in Fig. 1 (a) is divided into 3 layers, from top to bottom, the sound velocity is followed by 1550 m/s, 1650 m/s and 1800 m/s, the seabed attenuation coefficient is followed by 0.25 dB/λ, 0.5 dB/λ and 0.75 dB/λ, and the seabed density is followed by 1.5 g/cm3, 2.0 g/cm3 and 2.1 g/cm3. The seabed in Fig. 1 (b) is the single-layer uniform medium, the seabed velocity is 1600 m/s, and the attenuation coefficient is 0.5 dB/λ. The seabed in Fig. 1(c) is the single-layer uniform medium, the seabed density is 2.0 g/cm3, and the attenuation coefficient is 0.5 dB/λ.

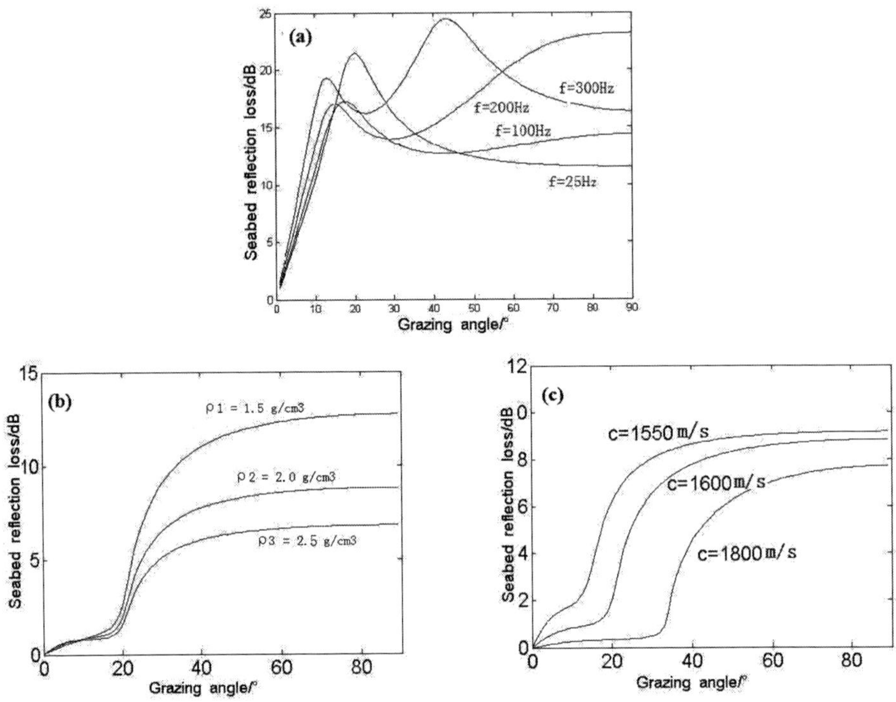

FIGURE 1. The seabed reflection loss curves, sound velocity c =1500m/s, water density ρ=1.0g/cm3. (a) Frequency sensitivity curve. (b) Density sensitivity curve. (c) Sound velocity sensitivity curve.

Ⅲ. SYSTEM SIMULATION AND DATA PROCESSING

The simulation conditions are as follows. The curve shown in Fig. 2 is the sound velocity profile, the depth of the shallow water is 100m, the acoustic source frequency is 2000Hz, the seabed sound velocity is 1610m/s, the seabed density is 1.80g/cm3, and the seabed attenuation coefficient is 0.3 dB/λ.

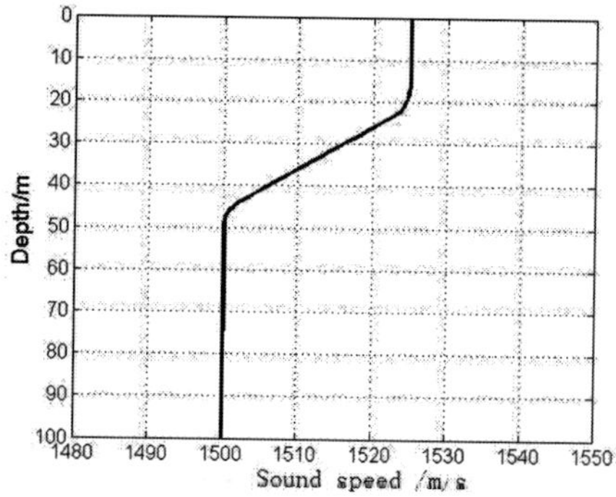

FIGURE 2. The sound velocity profile in shallow water

The system considered here (Fig. 3) consists of three broadband sources S1, S2, S3 and a towed tilted line array R. The towed line array consists of 21 hydrophones. The distance between the adjacent hydrophones is 4m. The length of the line array in the water is 95m, which is less than the depth of the shallow water. The tilt angle of the towed line array remains about 30 degrees. The sources are towed by 3 streamers and transmit at uniform time intervals.

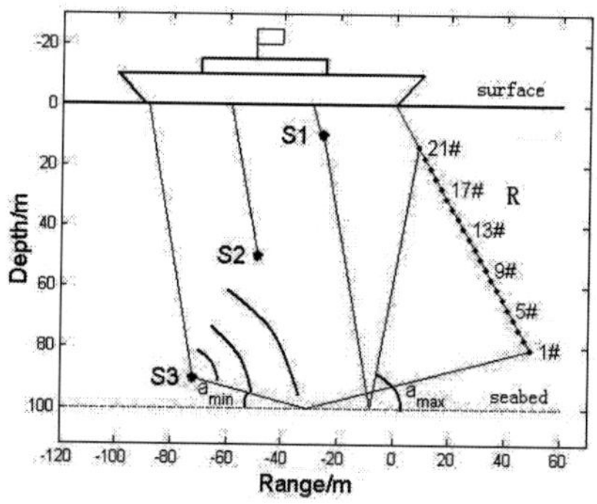

FIGURE 3. The system configuration (a ship towing 3 acoustic sources and a tilted line array).

The ship should cruise at uniform speed to keep the gesture of the system unchanged. In order to avoid interferences with each other, the three sources transmit signals in a looped way (S1—S2—S3—S1…), so that there is only one source working at a time. It is assumed that the ship velocity is 5m/s, each source works for one second, and the seabed parameters have small changes in short distance.

As the effect of the multi-path interference, some multi-path signals may not be distinguished when they arrive at the same time. The following method is used to obtain useful information. As shown in Fig. 4, we consider five arrival signals including d (the direction arrival wave), s (the surface reflection wave), b (the bottom reflection wave), sb (the surface and bottom reflection wave) and bs (the bottom and surface reflection wave), respectively.

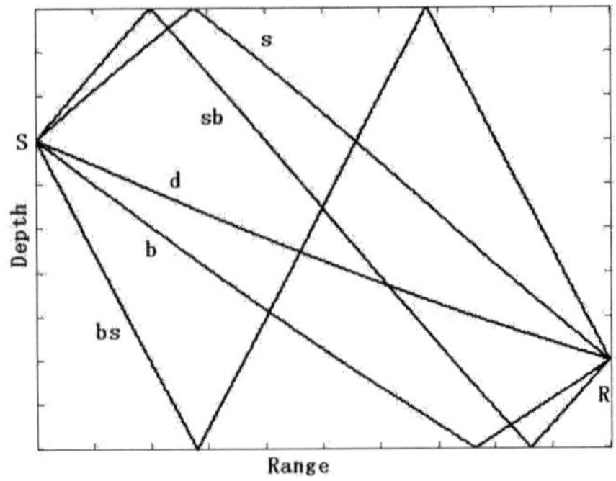

FIGURE 4. The multi-path signal configuration

Multi-path time delayed graphics are obtained by processing the data of the multi-path arrival signals. According to Eq. (3), we use d (direction arrival wave) and b (bottom reflection wave) to compute the seabed reflection loss (*RL*). The hydrophones whose receiving signals (b or d) interfere with other signals should be found out and eliminated for extracting the reflection loss. As shown in Fig. 5(a), arrival signal b interferes with s in 1# and 2# hydrophones, so we have to discard data in these hydrophones. Also 11#, 12# hydrophones in Fig. 5(b) and 1#, 2#, 3#, 20#, 21# hydrophone in Fig. 5(c) have the same problems. The color bar indicates the intensity of the arrival signal.

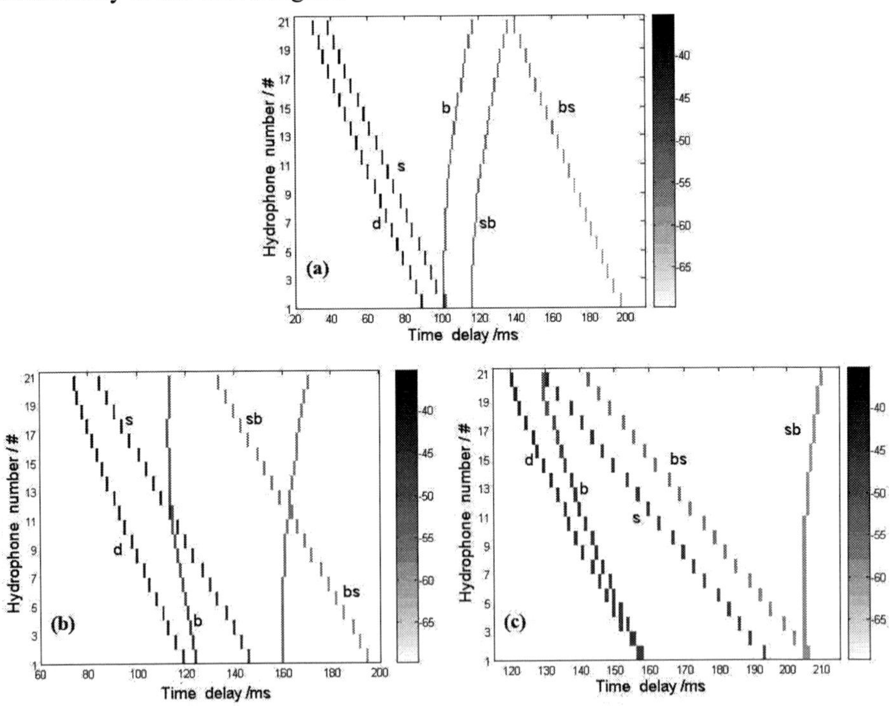

FIGURE 5. Multi-path arrival time delay. (a) The source S1, (b) The source S2, (c) The source S3.

Eq. (3) is applied to calculate the seabed reflection loss curve as shown in Fig. 6. The curve covers the grazing angles from 11° to 73°. The small circles in Fig. 6 are useless points, which correspond to the nine useless data points above mentioned. The towed tilted line array can invert the seabed acoustic parameters in a moving and high resolution way; it requires only a smaller aperture line array. After obtaining the seabed reflection loss curve, we can invert other seabed parameters, such as density, sound velocity, attenuation coefficient and so on.

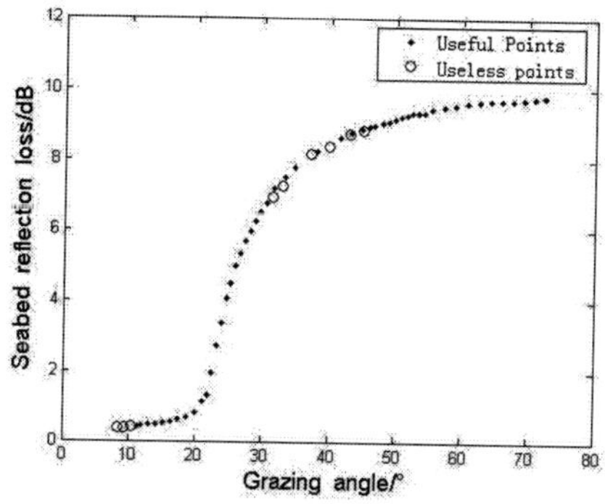

FIGURE 6. The seabed reflection loss curve

Ⅳ. ERROR ANALYSIS

The gesture of the towed tilted line array under the water has a great impact on the simulation results. It is very necessary to record the gesture of the line array accurately by using some depth sensors and tilted sensors. Then the actual shape of the line array can be calculated[14]. It provides a necessary error compensation for the system configuration.

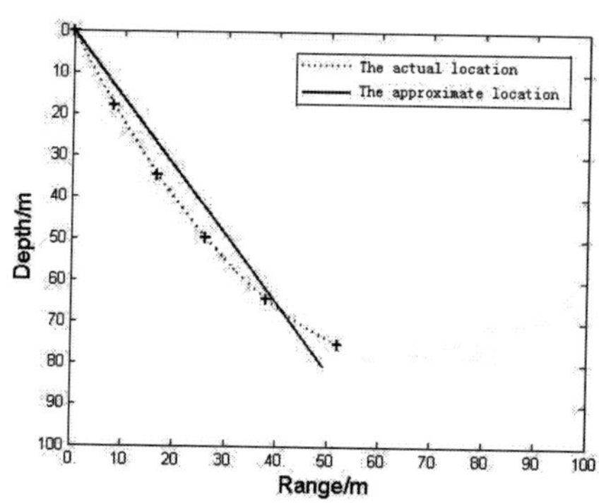

FIGURE 7. The actual location and the approximate location of the towed tilted line array

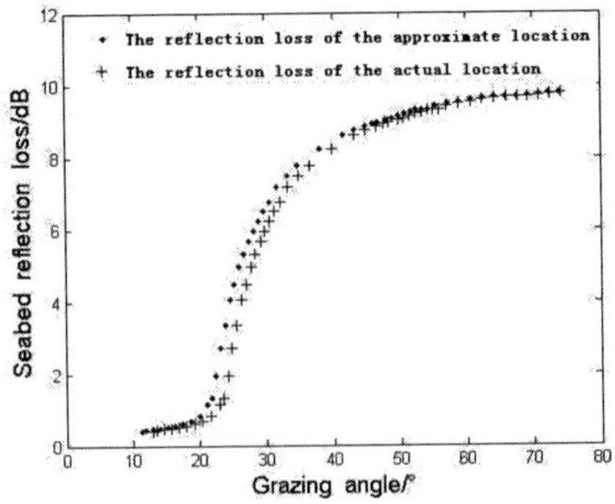

FIGURE 8. Seabed reflection loss corresponding to different shape of the line array

As shown in Fig. 7, the solid curve is the approximate location of the towed tilted line array, and the dotted curve is the actual location of the line array. Having obtained the actual location of the line array, a more accurate seabed reflection loss curve is calculated. The black dots curve in Fig.8 denotes the reflection loss of the approximate location, and the plus curve denotes the actual location of the line array. Then a more accurate curve for seabed reflection loss can be obtained, which covers the grazing angles from 13° to 73°. When the density of the line array is suitable, the acoustic module of the line array is similar to a straight line. The towed fish can also be used to control the gesture of the line array.

V. DISCUSSION AND SUMMARY

A fast inversion scheme for the seabed acoustic parameters has been discussed,. It consists of a towed tilted line array and three acoustic sources to cover a large scope of grazing angles. We have also simulated the seabed reflection loss by using a towed horizontal line array and a vertical line array. The simulation results show that different arrays have different experimental configurations. In order to cover the same scope of grazing angles, the vertical line array must operate with several acoustic sources and require a stationary platform of the array. The towed horizontal line array needs only one acoustic source but requires a large aperture line array. The towed tilted line array combines the merits of the vertical line array and the horizontal line array, which can invert the seabed acoustic parameters in a moving and high resolution way and requires only a smaller aperture line array. The error resulting from the distortion of the array shape is analyzed for the inversion method.

As the ship velocity and the sea environment are changing with time, it may be hard to keep the system steady while the system platform is moving. But we can use some high-precision sensors to control and record the system gesture, and then make error analysis in the data processing. This paper simulates the system and calculates the seabed reflection loss based on the Ray model. Future work involves doing experiments that can confirm the conclusion of the simulation.

ACKNOWLEDGMENTS

Project supported by the National Natural Science Foundation of China (Grant No. 10774119), the program for New Century Excellent Talents in University (Grant No. NCET-08-0455), the Natural Science Foundation of Shaanxi Province (Grant No. SJ08F07), the NPU Foundation for Innovation, and the NPU Foundation for Fundamental Research.

REFERENCES

1. W. Luo and H Schmidt, *J. Acoust. Soc. Am*, 2009, 125 pp52.
2. E. K. Skarsoulis and G. S. Piperakis, *J. Acoust. Soc. Am.* 2009, 125 pp1384.
3. J. W. Yin, J. Y. Hui, L. X. Guo, *Acta Physica Sinica*, 2008, 57 pp1753.
4. P. Gerstoft and C. F. Mecklenbrauker, *J. Acoust. Soc. Am.* 1998, 104 pp808.
5. R. Chapman, S. Chin-Bing, D. King and R. B. EVANS, *IEEE J. Oceanic Eng.* 2003, 28 pp320.
6. K. D. Yang, R. Chapman, Y. L. Ma, *J. Acoust. Soc. Am*, 2007,121 pp833.
7. K. D. Yang, Y. L. Ma, C. Sun and J. H. Miller and G. R. Potty, *IEEE J. Oceanic Eng*, 2004, 29 pp964.
8. Z. L. Li, J. Yan, F. H. Li and L. H. Guo, *Acta Acustica*, 2002, 27 pp487.
9. S. G. Schock, 2004 *IEEE J. Oceanic Eng.* 29 pp1200.
10. C. W. Holland, 2003, 114 pp1861.
11. J. Dettmer, S. E. Dosso and C. W. Holland, *J. Acoust. Soc. Am*, 2007,122 pp3327.
12. M. R. Fallat, L. P. Nielsen, S. E. Dosso and M. Siderius, *IEEE J. Oceanic Eng*, 2005, 30 pp198.
13. K. D. Yang, Y. L. Ma, *Acta Physica Sinica*, 2009, 58 pp1798.
14. Q. H. Li, Design principle of digital sonar, Anhui: Anhui Education Press, 2002, pp426

Sound Speed Profiles Inversion in Shallow Water Using a Parallel Genetic Algorithm

Yanxin Yu [a, b, c], Zhenglin Li [a, b]

[a]*State Key Laboratory of Acoustics, Institute of Acoustics, Chinese Academy of Sciences, Beijing 100190, China,*
[b]*Haikou laboratory, Institute of Acoustics, Chinese Academy of Sciences, Haikou 570105, China,*
[c]*Graduate School, Chinese Academy of Sciences, Beijing 100190, China*

Abstract. The sound speed profile and bottom parameters play important roles in shallow water sound propagation. And much attention has been paid to how to obtain them. In concurrence with in-situ measurements, many inversion methods, such as the matched field inversion, have been put forward to invert the sound speed profile from acoustic signals. However, the matched field inversion may cost too much time in replica field calculations. The feasibility and robustness of an acoustic tomography scheme with matched field processing in shallow water is studied here. The sound speed profile is described by the empirical orthogonal functions (EOFs). The acoustic signals from a vertical linear array in ASIAEX2001 East China Sea are analyzed to invert sound speed profiles combined with the estimated EOFs. The parallel genetic algorithm is adopted as the optimization algorithm to increase the inversion speed. The results show that the inverted sound speed profiles are in good agreement with CTD measurements. Moreover, *a posteriori* probability analysis is used to verify the validity of the inverted results.

Keywords: matched-field processing, sound speed profile, parallel genetic algorithm
PACS: 43.30.Pc, 43.30.Wi

I. INTRODUCTION

As the starting point for inferring oceanographic features in the water column, it is necessary to recover the sound speed as a function of space. Compared with that in deep water, the sound speed profile (SSP) in shallow water is much more complicated. In concurrence with in-situ measurements, many inversion methods, for example matched-field inversion[1, 2], have been put forward to invert SSPs from acoustic signals. The empirical orthogonal functions are used to describe the SSP to reduce the total number of inverted parameters[3] in SSP inversion. However, this method may cost too much time in replica field calculation for matched-field processing (MFP). In an effort to obtain quasi-realtime inversion results, parallel computation is used in this paper.

The paper is organized as follows. Section II presents the general framework of MFP procedure; using *a posteriori* probability distribution to evaluate the inversion results is also discussed. In Section III, a brief description of the experiment is given, and the results of the inversion are presented. Summary and conclusion are provided in Section IV.

CP1272, *Shallow-Water Acoustics, Second International Shallow-Water Acoustics Conference*
edited by J. Simmen, E. S. Livingston, J. X. Zhou, and F. H. Li

II. SOUND SPEED PROFILE INVERSION ALGORITHM

Figure 1 shows the realization of MFP. First the SSPs are reconstructed through the mean SSP and EOFs. Then the replica acoustic field is calculated and correlated with the measured acoustic field to obtain the cost function. If the end condition is not satisfied, the unknown parameters are updated by the genetic algorithm (GA). Otherwise, the inverted coefficients of EOFs are given.

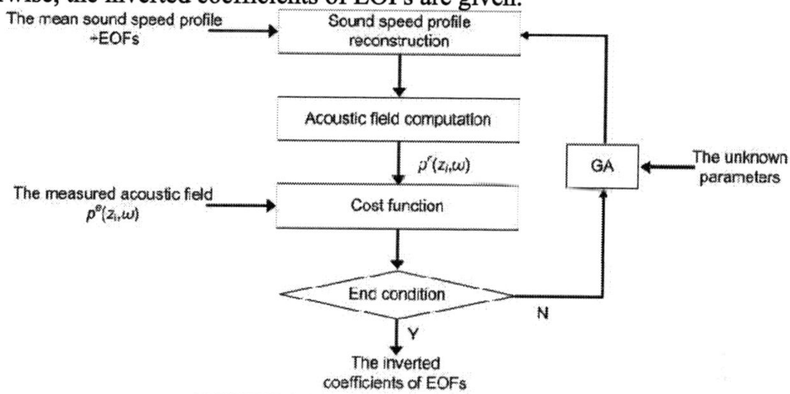

FIGURE 1. The flowchart of MFP inversion.

A. Sound Speed Profile Representation with EOFs

The sound speed structure in the water column, which is determined by the water temperature, salinity and the pressure, strongly influences the sound propagation in shallow water. And the sound speed in the surface layer is mainly determined by the temperature structure. In general, it varies temporally and spatially. However, it is assumed that SSPs do not change with range in this paper. In other words, the matched-field inversion aims at obtaining the equivalent sound speed profile in the water column, taking no account of the fluctuation caused by internal waves and tide, etc.

Parameter inversion is a multi-dimensional optimization problem, which means searching the matched solution in the unknown parameter space. In order to reduce the search space, the environment should be characterized in as few parameters as possible. EOFs are efficient basis functions for an expansion of sound speed profiles. In general, to get a highly accurate representation of a profile, only the first few orders of EOFs are needed.

Experimental data taken from the Asian Sea International Acoustic Experiment (ASIAEX) in the East China Sea[4] were analyzed to confirm that the sound speed profiles recovered by the first four orders of EOFs could be highly accurate. Fifty-four SSPs measured by CTD during the experiments are shown in Figure 2(a). The thick solid line represents the mean SSP. It can be seen that the sound speed fluctuates greatly at the depth between 0 to 70 meters. The first 4 orders of EOFs are shown in Figure 2(b).

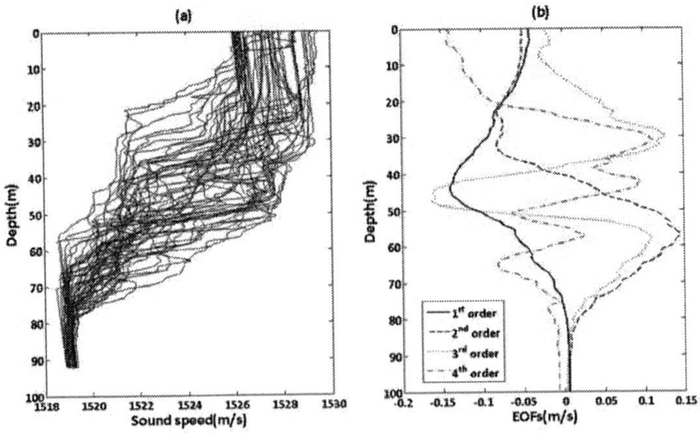

FIGURE 2. (a) The measured SSPs and the mean SSP; (b) The first four-order EOFs.

Figure 3 shows one of the actual SSPs measured by CTD and the corresponding SSP represented by the first four orders of EOFs. It can be seen that they are well matched.

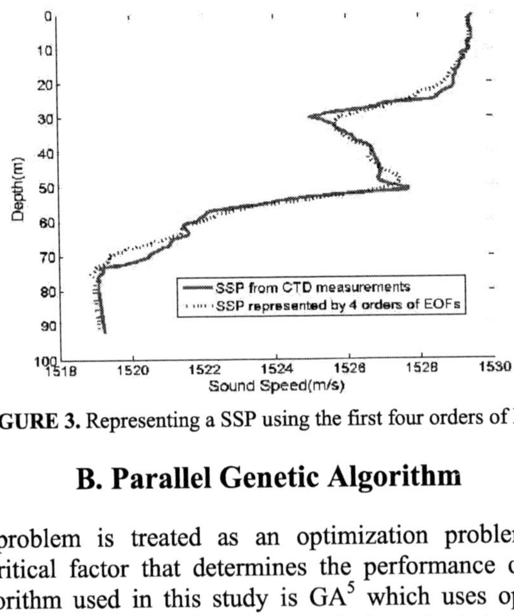

FIGURE 3. Representing a SSP using the first four orders of EOFs.

B. Parallel Genetic Algorithm

The inverse problem is treated as an optimization problem, and the search technique is a critical factor that determines the performance of the scheme. The optimization algorithm used in this study is GA[5] which uses operators that mimic evolution to improve a collection of candidate solutions. It starts with a population of individuals that are codified to represent solutions to the inversion problem. The cost function of each solution is computed based on the difference of the measured data and the computed data.

GA has inherent parallelism, because the fitness evaluation, usually the most computationally demanding step, can be carried out independently for each individual in the population. A simple technique to parallelize GA is to divide the task of evaluating the population among several processors.

It is convenient to parallelize the GA as a master-slave program. The flowchart of the parallel genetic algorithm (PGA) is shown in Figure 4. The parallel algorithm works with the Message Passing Interface (MPI) on a PC cluster.

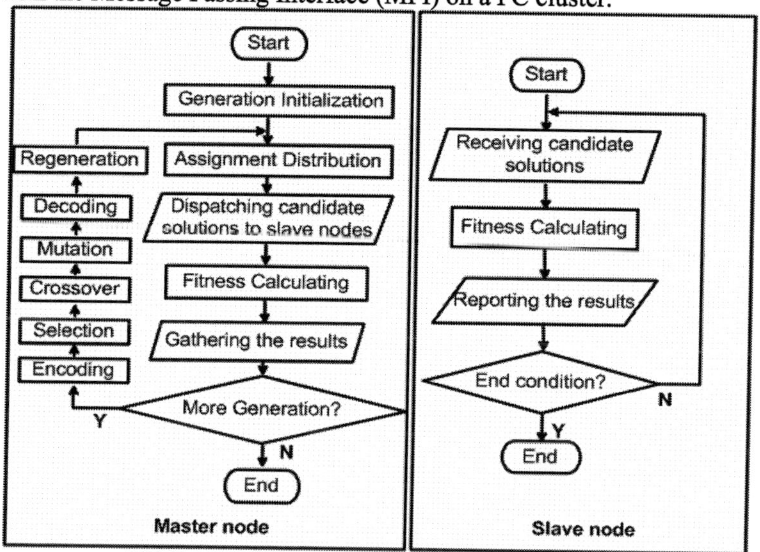

FIGURE 4. The flowchart of parallel genetic algorithm.

C. Multi-frequency Matched Field Processing

Matched-field processing has been widely used in the underwater acoustic inverse problem. It makes use of spatial and frequency characteristics of the acoustic field in the water column. The cost function used in this paper is the multi-frequency incoherent Bartlett processor [6]:

$$E = \frac{1}{M_f} \sum_{i=1}^{M_f} \frac{\left| \sum_{j=1}^{N} p_j^*(f_i) q_j(f_i) \right|}{\sqrt{\left[\sum_{j=1}^{N} |p_j(f_i)|^2 \right] \left[\sum_{j=1}^{N} |q_j(f_i)|^2 \right]}}. \quad (1)$$

Here, $p_j(f_i)$ is the experimental complex pressure field at f_i received by the jth hydrophone, while $q_j(f_i)$ is the modeled complex pressure field. The superscript represents the complex conjugation. N is the hydrophone number of vertical linear array (VLA), M_f is the total number of frequency bins, and i is the frequency bin index.

D. Results Evaluation of *a Posteriori* Distribution

An *a posteriori* probability distribution of the unknown parameters $\sigma(m_k)$ combines *a priori* information $\rho(m_k)$ about the model parameters with information from an observed data set $\{m_k\}$, and is used to estimate the uncertainties of the inverted results [7, 8].

During the inversion process, the samples of the search spaces $\{m_k\}$ and the corresponding cost function $\phi\,(m_k)$ are saved, and these observations are used to estimate the *a posteriori* probability. For the N_{obs} observations, the *a posteriori* probability for the kth model is given by:

$$\hat{\sigma}(m_k) = \frac{l(m_k)\rho(m_k)}{\sum_{j=1}^{N_{obs}} l(m_j)\rho(m_j)} = \frac{\exp(-[\phi(m_k) - \phi(m_0)]/T)}{\sum_{j=1}^{N_{obs}} \exp(-[\phi(m_j) - \phi(m_0)]/T)}, \tag{2}$$

where ϕ is the cost function and m_0 is the best sampled parameter vector in the evolution process. T is the "temperature" as defined in the simulated annealing (SA). Experience shows that a good value for T is the average of the 50 best cost functions obtained in the optimization minus the best value of the cost function[8].

III. EXPERIMENTAL DATA ANALYSIS

A. Experiment Description

The wide-band source (WBS) signals measured in the Asian Sea International Acoustic Experiment (ASIAEX) in East China Sea (ECS) were used to invert the sound speed. The data used in this paper were recorded by a 32-hydrophone vertical linear array (VLA) deployed from the receiver ship *Shi Yan 3* at the location of (29°40.47'N, 126°49.21'E). The VLA spanned water depths from 4.6 to 90.5 meters, and the received signals were recorded by a 32-channel recorder with a sample frequency of 6.0 kHz for each channel. The WBS sources were deployed from *Shi Yan 2,* and the signals received from about 6.8 km to 12.3 km were used for the MFP inversion in this paper.

Assume that the sound speed profiles and geoacoustic parameters were range-independent in the experimental region. And a simplified half infinite liquid bottom model is used, as shown in Figure 5. With the empirical orthogonal functions, the first 4 orders of EOF coefficients could be used to recover an SSP. Considering the uncertainties and errors existing in the measured values of some sensitive parameters, 4 additional parameters, that is, source depth, source-receiver range, water depth, and bottom sound speed, were also inverted.

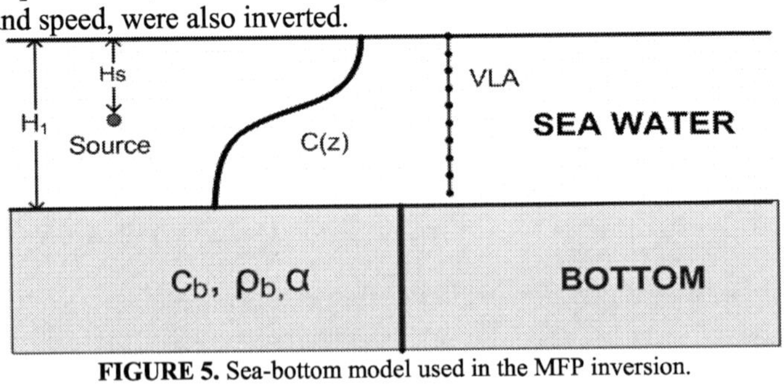

FIGURE 5. Sea-bottom model used in the MFP inversion.

B. Experimental Results

There are 8 parameters to be inverted, and the search bounds for each parameter are shown in Table 1. The frequency band used in the MFP inversion is from 100 Hz to 200 Hz with a frequency step of 2 Hz. The inverted SSPs are shown in Figure 6, where 4 sets of acoustic data were used and WBS@10:* are the inverted SSPs using the signals of WBS deployed at 10:11, 10:17, 10:23, and 10:29 in the morning of June 3, 2001 respectively. CTD@08:41 and CTD@11:46 are the real SSP measured by CTD at 8:41 and 11:46 respectively. Figure 6 shows that the inverted SSPs are in good agreement with the CTD measurements. The inverted results corresponding to WBS@10:29 are also presented in Table 1.

TABLE 1. Parameter search bound and some inverted results for WBS@10:29.

Parameter	Lower Bound	Upper Bound	Best Value
EOF coefficient of 1st order	-35	35	7.02
EOF coefficient of 2nd order	-30	30	-6.11
EOF coefficient of 3rd order	-25	25	-3.02
EOF coefficient of 4th order	-20	20	-3.13
Source depth (m)	45	55	54.42
Water depth (m)	102	107	105.36
Bottom sound speed (m/s)	1580	1640	1598.1
Range (m)	R_{GPS} -200	R_{GPS} +200	12496

R_{GPS} stands for the range between the receiver ship and the source ship measured by GPS.

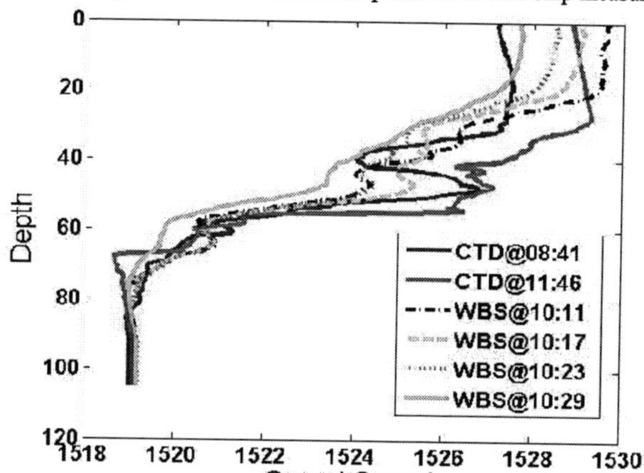

FIGURE 6. The measured and inverted SSPs.

C. *A Posteriori* Statistics

The multidimensional integration was carried out to obtain the one-dimensional (1-D) marginal *a posteriori* probability densities for each parameter. The inverted result, corresponding to WBS@10:29 in Figure 6, was given as an example to estimate the uncertainties. Figure 7 shows the corresponding marginal probabilities. The down-arrows indicate the best values during the evolution process. It is observed that most parameters are well determined, and for the source-receiver range, water depth and the 4th order EOF coefficient there are some ambiguities in the values.

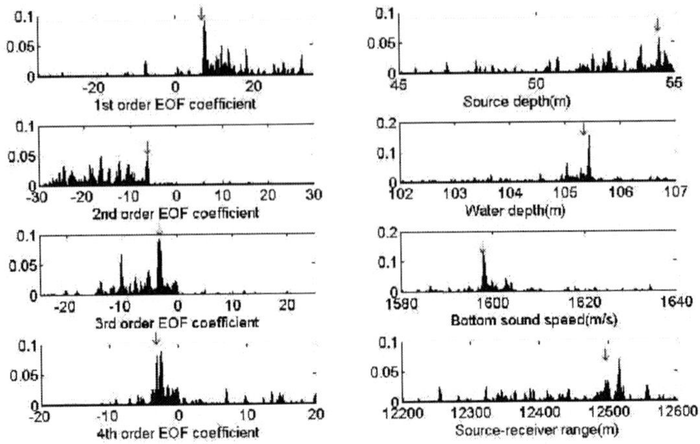

FIGURE 7. The estimated *a posteriori* distribution for an inversion result.

D. Parallel Performance

Speedup refers to how much a parallel algorithm is faster than a corresponding sequential algorithm in a parallel computing. By speedup S_p for p processors we mean the ratio of the execution time of the sequential algorithm T_1 to that of the parallel algorithm with p processors T_p.

The following tests are carried out on a PC cluster. For the same environment and the same experimental data, the MFP inversion using PGA is performed on variant processors. The execution time and corresponding speedup of the results are shown in Table 2 and Figure 8. The dotted line in Figure 8 represents linear speedup, that is, S_p equals the number of processors; and the stars indicate the actual speedup when the MFP inversion is performed on variant processors. It can be seen that the parallel algorithm runs with nearly linear speedup, which means the algorithm exhibits good scalability. In the future, if a PC cluster with 80 processors is used, the inversion can be done within 5 minutes.

TABLE 2. The execution time on variant processors.

Number of Processors	1	2	4	6	8
Execution Time/minutes	303	150	79	58	41

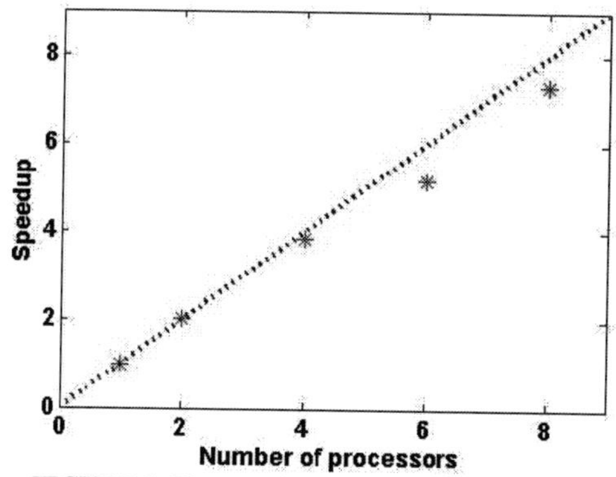

FIGURE 8. The speedup vs. number of processors.

IV. SUMMARY AND CONCLUSION

The empirical orthogonal functions are used to describe the sound speed profiles, and the WBS signals received in ASIAEX ECS 2001 are used to invert the SSPs with the matched-field processing. The parallel genetic algorithm is adopted as the optimization approach. The results show that the SSPs inverted from acoustic signals are in good agreement with CTD measurements and the parallel genetic algorithm has approximately linear speedup for the MFP inversion.

ACKNOWLEDGMENTS

The authors would like to thank the ASIAEX ECS 2001 experimental team.

REFERENCES

1. A .Tolstoy et al., *J. Acoust. Soc. Am.*, **89**(3): 1119-1127 (1991).
2. M. I. Taroudakis and M. G. Markaki, *J. Acoust. Soc. Am.*, **102**(2): 885-895 (1997).
3. L. R. LeBlanc and F. H. Middleton, *J. Acoust. Soc. Am.*, **67**(6): 2055-2062 (1980).
4. Z. L. Li et al., *IEEE J. Ocean. Eng.*, **29**(4): 973-979 (2004).
5. W. Banzhaf et al., "Genetic Programming Fundamentals" in *Genetic Programming - An Introduction*, edited by E. Beller, San Francisco, CA: Morgan Kaufmann, 1998, pp.105-236.
6. E. K.Westwood, *J. Acoust. Soc. Am.*, **91**(5): 2777-2789 (1992).
7. P.Gerstoft, *J. Acoust. Soc. Am.*, **95**(2):770-782 (1993).
8. P. Gerstoft and C. F.Mecklenbräuker, *J. Acoust. Soc. Am.*, **104**(2): 808-819 (1998).

Study on Geoacoustics Interface Wave Scattering Characteristic

Xinyi Guo and Li Ma

Institute of Acoustics, Chinese Academy of Sciences, Beijing China 100190

Abstract. This paper introduces the propagation characterization of interface waves in elastic half space. In a layered medium, frequency dispersion exists and propagation models of interface waves can be researched using mode analysis. For a typical layered structure model, the propagation modes of interface waves can be calculated and the Green's function can be obtained using propagation modes. In practical problems, there are buried objects in the medium. An interface wave scattering integral equation is derived, and scattering interface wave fields can be simulated by computer. For checking this arithmetic, the scattering displacement field can be calculated using the first mode of an interface wave. The convergence rapidity of the method has been proved to be very high by some practical examples, and the scattering strength can be obtained simultaneously. Sometimes, there are multiple objects buried in the medium, and the interface wave scattering integral equation can be spread into calculation of a multi-objects scattering problem by the integral equation of the scattering elastic wave field associated with multi-inhomogeneities. In this paper, the high mode scattering of an interface wave can be calculated also. In order to achieve the detection buried objects using interface wave, there is an experiment for verifying this method. In this experiment, the interface wave signal extraction, propagation velocity, propagation attenuation, and several other aspects are analyzed. The time difference between direct wave and echo wave in the received signal can be obtained according to the geometric relationship of buried objects, source and geophone array locations. The echo signal exists in the received signal.

Keywords: Interface wave; Scattering wave; Buried objects detection

PACS: 43.30.Ft

INTRODUCTION

A geo-acoustic interface wave is a kind of guided-wave transmission along the interface of two different media. At present, there are some papers reporting a

CP1272, *Shallow-Water Acoustics, Second International Shallow-Water Acoustics Conference*
edited by J. Simmen, E. S. Livingston, J. X. Zhou, and F. H. Li

detecting method using interface waves[1]. Based on interface wave propagation characteristics, this paper establishes an interface wave scattering model, and obtains an interface wave scattering sound field for buried objects. We present an experiment to verify theoretical and experimental results.

EQUATION OF INTERFACE WAVE SCATTERING

Elastic wave scattering can be expressed as integral

$$u_q(\mathbf{x}) = \int \left(-\partial_{j'} G_{qi}(\mathbf{x},\mathbf{x}')\right)\left(\Delta C_{ijkl}(\mathbf{x}')\partial_{l'} + \Delta\rho(\mathbf{x}')\omega^2\delta_{ik}\right)u_k(\mathbf{x}')d\mathbf{x}' + \int G_{qi}(\mathbf{x},\mathbf{x}')f_i(\mathbf{x}')d\mathbf{x}' \tag{1}$$

In general, it is difficult to calculate three-dimensional interface wave scattering using Eq. 1. We need to transform this equation according to characteristics of interface wave propagation. For scattering, there is a general view that the total wave field is equal to incident wave plus scattered wave. Scattered wave field series expansion of displacement can be expressed

$$\mathbf{u}(\mathbf{x}) = \mathbf{u}^0(\mathbf{x}) + \sum_{p=1}^{N} \mathbf{u}^{(p)}(\mathbf{x}) \tag{2}$$

where $\mathbf{u}^{(p)}(\mathbf{x})$ is the p-th scattering displacement field. Inserting Eq. 2 to Eq. 1 we can obtain Solution for each scattering.

$$u_q^{(p)}(\mathbf{x}) = \int \left(-\partial_{j'} G_{qi}(\mathbf{x},\mathbf{x}')\right)\left(\Delta C_{ijkl}(\mathbf{x}')\partial_{l'} + \Delta\rho(\mathbf{x}')\omega^2\delta_{ik}\right)u_k^{(p-1)}(\mathbf{x}')d\mathbf{x}' \tag{3}$$

Scattering series 0-order is the last term of Eq. 1. It indicates the source wave field in the reference medium.

$$u_q^0(\mathbf{x}) = \int G_{qi}(\mathbf{x},\mathbf{x}')f_i(\mathbf{x}')d\mathbf{x}' \tag{4}$$

Only one iteration for Equation (3) can obtain the solution of the Born approximation. This iterative series is convergent.

In theory, Green's function includes all kinds of waves in a reference medium. This paper discusses interface wave scatter, and we only consider the interface wave item in Green's function. If interface waves propagate in a horizontal uniform medium, the Green's function is equal to the sum of all propagation modes. For isotropic, or horizontal isotropic elastic media, the fundamental solution of interface waves in elastic half space can be expressed [2]

$$G_{qi}(\mathbf{x},\mathbf{x}')=\sum_m Z_q^m(z)\,g_i^m(x,y;\mathbf{x}') \tag{5}$$

where, $Z_q^m(z)$ is expressed

$$Z_q^m(z)=\begin{pmatrix} 1/k_m\left(-V^m(z)\partial/\partial x+W^m(z)\partial/\partial y\right) \\ 1/k_m\left(-W^m(z)\partial/\partial x-V^m(z)\partial/\partial y\right) \\ U^m(z) \end{pmatrix} \tag{6}$$

displacement potential $g_i^m(x,y;\mathbf{x}')$ can be expressed as

$$g_i^m(x,y;\mathbf{x}')=\frac{-i}{8c_m u_m J_1^m}Z_i^m(z')H_0^{(2)}(k_m R) \tag{7}$$

U^m, V^m and W^m is the three-dimensional composition of the Eigenfunction in mode m. U^m and V^m represent the vertical composition and horizontal composition of a Rayleigh wave Eigenfunction respectively; and W^m represents the horizontal composition of a Love wave Eigenfunction. k_m, c_m and u_m are wave number, phase velocity and group velocity of m mode interface wave. R is the horizontal range of position $\mathbf{x}$ and $\mathbf{x}'$, $H_0^{(2)}$ is the Hankel function of 2nd kind, order 0, J_1^m is normalized energy integral

$$J_1^m = \frac{1}{2}\int_0^\infty dz\rho^0\left[\left(U^m\right)^2 + \left(V^m\right)^2 + \left(W^m\right)^2\right] \tag{8}$$

Eq. 5 is a general expression synthesizing all modes of Rayleigh wave and Love wave in an elastic wave field. For a homogeneous isotropic half-space, there is no Love wave, so W^m is equal to zero. If the energy of an interface wave concentrates on low mode, m is equal to 1 approximatively. With Eq. 5 substituted into Eq. 3, we can obtain the interface wave scattering field. In Figure 1, the solid line is the theoretical calculation results of interface wave scattering target strength.

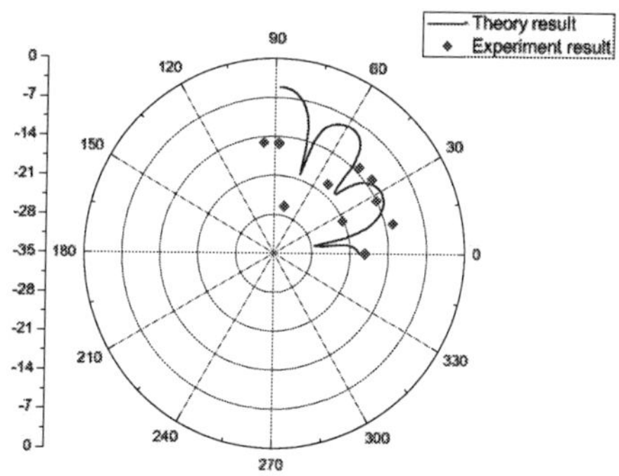

FIGURE 1. Theory and experiment values of interface wave target strength (dB)

EXPERIMENT

In this experiment, we use the vector polarization filtering method from the three-component geophone signal to extract interface wave components. As known, interface wave propagation is reverse elliptical motion; the major axis is perpendicular to the direction of wave propagation, the minor axis is parallel to the wave propagation direction. So Medium particle vibration exists polarization in horizontal and vertical direction. Three-component geophones receive both horizontal and vertical signals at the same time. Interface wave components can be extracted from the signal using Polarization filtering. The specific process is as follows:

V and R are the vertical and radial real signal respectively. The real signal can change into a complex signal by Hilbert transform

$$\mathbf{V} = Ve^{j(\omega t+\phi v)}, \quad \mathbf{R} = \mathrm{Re}^{j(\omega t+\phi r)} \tag{9}$$

Vertical and radial components of interface signals exists phase difference. Complex power is formed for V and R

$$\mathbf{P} = \mathbf{R}^{*}\mathbf{V} \tag{10}$$

For the interface wave, the imaginary part of complex power is not equal to zero, and another kind of wave is equal to zero.

$$\mathrm{Im}(\mathbf{P}) = \mathrm{Im}\left(\mathbf{R}^{*}\mathbf{V}\right) \tag{11}$$

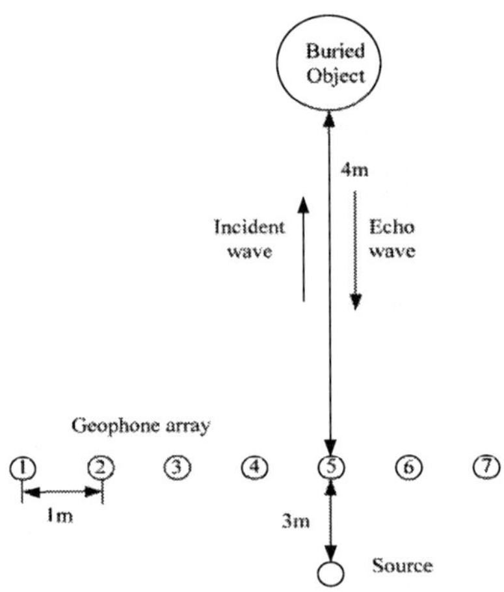

FIGURE 2. Geophones and source location

Figure 2 is an Experimental schematic. In Figure 3, the solid line is the existence of buried object condition, and the dashed line is the non-existence of object condition by compared the imaginary part of complex power. From Figure 2 and Figure 3, we can obtain the time difference according to geometric relationship and signal waveform. Here, interface wave velocity is 110 m/s. The target strength can be obtained from scattered wave energy, and is shown as splashes in Figure 1.

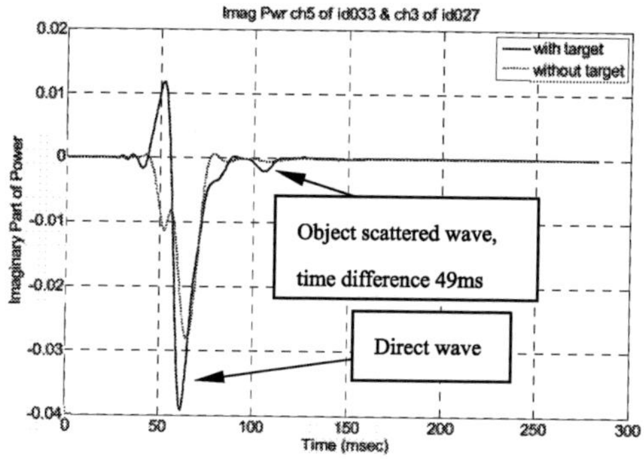

FIGURE 3. Geophone imaginary part of complex power with and without target

CONCLUSION

Scattered wave fields of interface waves can be obtained by theoretical and experimental methods. Results show that the experimental data and theoretical calculation fit well in the angle between an incident wave and an echo less than 60°, and in the 60°~90° angle, the experimental data are lower than the theoretical value by approximately 5~8 dB. This is a result of experimental conditions and accuracy.

REFERENCE

1. Eric S., Wilson P. S., Bacon F. W. et al., *J. Acoust. Soc. Am.,* **103**, 2333-2343 (1998).
2. Aki K., Richards P. G., *Quantitative Seismology: Theory and Method*, San Francisco, W. H. Freeman and Company, 1980.

Coherent reverberation model based on adiabatic normal mode theory in a range dependent shallow water environment

Zhenglin Li[1, 2] Renhe Zhang[1] Fenghua Li[1]

1 State Key Laboratory of Acoustics, Institute of Acoustics, Chinese Academy of Sciences, Beijing 100190, China,
2 Haikou Laboratory, Institute of Acoustics, Chinese Academy of Sciences, Haikou 570105, China

Abstract. Ocean reverberation in shallow water is often the predominant background interference in active sonar applications. It is still an open problem in underwater acoustics. In recent years, an oscillation phenomenon of the reverberation intensity, due to the interference of the normal modes, has been observed in many experiments. A coherent reverberation theory has been developed and used to explain this oscillation phenomenon [F. Li et al, Journal of Sound and Vibration, 252(3), 457-468, 2002]. However, the published coherent reverberation theory is for the range independent environment. Following the derivations by F. Li and Ellis [D. D. Ellis, J. Acoust. Soc. Am., 97(5), 2804-2814, 1995], a general reverberation model based on the adiabatic normal mode theory in a range dependent shallow water environment is presented. From this theory the coherent or incoherent reverberation field caused by sediment inhomogeneity and surface roughness can be predicted. Observations of reverberation from the 2001 Asian Sea International Acoustic Experiment (ASIAEX) in the East China Sea are used to test the model. Model/data comparison shows that the coherent reverberation model can predict the experimental oscillation phenomenon of reverberation intensity and the vertical correlation of reverberation very well.

Keywords: Coherent reverberation model, oscillation phenomenon, range dependent environment
PACS: 43.30.Gv, 43.30.Bp

I. INTRODUCTION

In shallow water active sonar applications, the predominant background interference is often bottom reverberation. Several reverberation models based on ray [1], parabolic equation (PE) [2], or normal mode [3, 4] theories have been developed. Those models have generally treated the propagation from a source to a scatterer and from that scatterer to the receiver. Moreover, much research about vertical coherence of shallow water reverberation has also been done [5-9]. However, two Yellow Sea Experiments observed an oscillation phenomenon of bottom reverberation loss, which is due to the interference of the normal modes and cannot be explained by the incoherent theory of reverberation. A coherent reverberation theory has been developed to explain the oscillation phenomenon [10, 11]. The reverberation model

CP1272, *Shallow-Water Acoustics, Second International Shallow-Water Acoustics Conference*
edited by J. Simmen, E. S. Livingston, J. X. Zhou, and F. H. Li

based on the normal mode theory, combined with ray-mode analogies, is one of the most practical models. However, the reverberation model is valid only for a range independent environment.

The objective of this paper is to develop a general reverberation model based on adiabatic normal mode theory, which is valid for the range dependent shallow water environment. This model can be used to predict the coherent or incoherent reverberation field, as well as the reverberation intensity and reverberation spatial correlation. The reverberation experiment, as a part of the 2001 Asian Sea International Acoustic Experiment (ASIAEX2001) in the East China Sea (ECS), is conducted in a relatively sloping environment. The experiment data are used to test the coherent reverberation model. The following parts of this paper are arranged as follows. Section II presents the coherent model of reverberation. In section III, numerical and experimental results are presented. Summary and conclusion are given in Section IV.

II. THE COHERENT REVERBERATION MODEL BASED ON ADIABATIC NORMAL MODE THEORY

The basic idea of the coherent reverberation model is to follow the derivations by Li[10] and Ellis[4], where scattered fields from different normal modes are summed coherently.

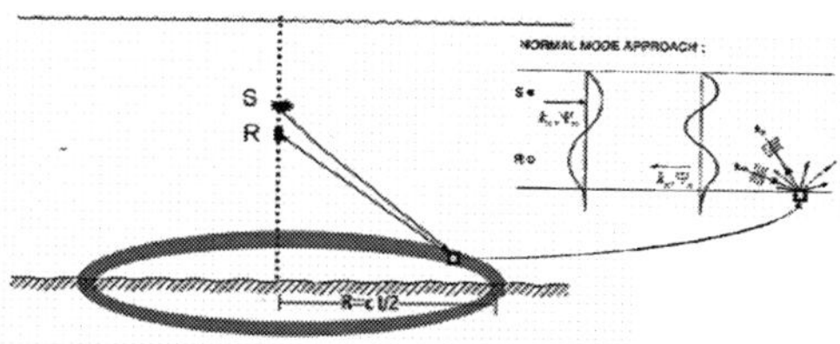

FIGURE 1. Schematic illustrating reverberation scattered from the bottom.

The reverberation field at the receiver $(z_r, 0)$ is often calculated by the summation of the sound pressure scattered from the bottom scattering elements over the area insonified by the incident pulse

$$p(t) = \sum_i p_i^{scatt} , \tag{1}$$

where p_i^{scatt} is the sound pressure scattered from the ith bottom scattering element in a bottom circular ring with the radius of $ct/2$ and width of $c\tau/2$, τ being the signal duration.

The propagation field in a range dependent environment can be expressed by the adiabatic normal mode theory [12]:

$$p_{ref=1m}(r,z) \approx \frac{i\sqrt{2\pi}}{\rho(z_s)} e^{-i\pi/4} \sum_m \Psi_m(z_s)\Psi_m(r,z) \frac{e^{i\int_0^r k_m(r)dr}}{\sqrt{\int_0^r k_m(r)dr}}, \tag{2}$$

where z_s is the source depth, $k_m(r)$ represent the local eigenvalues, $\Psi_m(r,z)$ are the eigenfunctions of local modes. Near an interface, the mode function $\Psi_m(r,z)$ can be written as

$$\Psi_m(r,z) = A_m(r,z)\left[e^{-i\varphi_m(r,z)} + R\,e^{i\varphi_m(r,z)}\right], \tag{3}$$

where $A_m(r,z)$ is the amplitude of mode function at local range and satisfies

$$4|A_m(r,z)|^2 = [\Psi_m(r,z)]^2 + \gamma_m^{-2}(\partial\Psi_m(r,z)/\partial z)^2. \tag{4}$$

$\varphi_m(r,z)$ is the vertical phase

$$\varphi_m(r,z) = \begin{cases} \tan^{-1}[\gamma_m\Psi_m(r,z)/(\partial\Psi_m(r,z)/\partial z)] + m\pi & m \text{ is odd} \\ \tan^{-1}[\gamma_m\Psi_m(r,z)/(\partial\Psi_m(r,z)/\partial z)] + (m+1)\pi & m \text{ is even} \end{cases}, \tag{5}$$

and $R = |R|e^{i\alpha}$ is the reflective coefficient at the interface.

The incident field due to mode m on the boundary at z_b is

$$p_m^{inc}(r,z_b) = \frac{i\sqrt{2\pi}}{\rho(z_s)} e^{-i\pi/4} \Psi_m(z_s) A_m(r,z_b) \frac{e^{i\left[\int_0^r k_m(r)dr - \varphi_m(r,z_b)\right]}}{\sqrt{\int_0^r k_m(r)dr}}. \tag{6}$$

Using the reciprocity principle, the scattered field propagating in mode n from a scattering element on the boundary at z_b back to the receiver at z_r is

$$p_n^{scatt}(r,z_b) = \frac{i\sqrt{2\pi}}{\rho(z_s)} e^{-i\pi/4} \Psi_n(z_r) A_n(r,z_b) \frac{e^{i\left[\int_0^r k_n(r)dr - \varphi_n(r,z_b)\right]}}{\sqrt{\int_0^r k_n(r)dr}}. \tag{7}$$

The reverberation field can be obtained by coherently summing the scattered fields from different modes

$$p_{rec}(r,z_r,z_s) = \mathbf{P}^{\mathbf{inc}} \cdot \mathbf{G} \cdot \mathbf{P}^{\mathbf{scat}}, \tag{8}$$

where the element of the scattering coefficient matrix $\mathbf{G}$ corresponding to being incident from mode m and scattering to mode n can be expressed as

$$g_{mn} = |g_{mn}| e^{i\xi_{mn}} . \quad (9)$$

The coherent reverberation loss (*RL*) can be expressed in decibels as

$$RL(r,z) \approx -10\log[|p_{rec}(r,z_r,z_s)|^2 \pi rc] . \quad (10)$$

The smoothed RL can be obtained by averaging in a frequency band

$$\overline{RL}(r,z) = -10\log\left[\frac{\pi rc}{\Delta f}\int_{f_0-\Delta f/2}^{f_0+\Delta f/2} |p_{rec}(r,z_r,z_s,f)|^2 df\right] . \quad (11)$$

where f_0 is the central frequency, and Δf is the frequency band. The reverberation vertical correlation for two receivers at different depths is defined as

$$RCorr(r,z_1,z_2) = \operatorname{Re}\left(\sum_f p_{rec}(r,z_1,z_s) p_{rec}^*(r,z_2,z_s) \Big/ \sqrt{\sum_f |p_{rec}(r,z_1,z_s)|^2 \sum_f |p_{rec}(r,z_2,z_s)|^2} \right) . \quad (12)$$

In the coherent reverberation model, for a range dependent environment, Eq. (8) can be generalized by replacing the incident field and scattered field in Eq. (6) and Eq. (7) with equation

$$p_m(r,z_b) = \frac{i\sqrt{2\pi}}{\rho(z_s)} e^{-i\pi/4} \Psi_m(z_s) A_m(z_b) \frac{e^{i[k_m r - \varphi_m(z_b)]}}{\sqrt{k_m r}} . \quad (13)$$

Meanwhile, the incoherent reverberation model can be derived by incoherently summing the scattered fields from different modes,

$$p_{rec}(r,z_r,z_s) = \sqrt{|\mathbf{P}^{\mathbf{inc}}|^2 \cdot |\mathbf{G}|^2 \cdot |\mathbf{P}^{\mathbf{scat}}|^2} . \quad (14)$$

III. NUMERICAL AND EXPERIMENTAL RESULTS

A. Numerical Results

The first numerical example is the calculation of bottom backscattering in a typical shallow-water environment. The environment, as shown in Fig. 2, is a Pekeris waveguide with water depth 100 m, sound speed 1500 m/s, bottom sound speed 1800 m/s, density 1.8 g/cm^3, and bottom absorption coefficient 0.35 dB/λ. The bottom scattering function **G** follows the Mackenzie-Lambert rule [13]:

$$g_{mn} = \mu \sin\theta_m \sin\theta_n , \qquad (15)$$

where $10\log\mu = -27\text{dB}$.

Figure 3 shows the reverberation loss at 100 Hz versus time by the reverberation model. The source depth and receiver depth are 50 m and 70 m respectively. The coherent reverberation gives a more or less random distribution compared to the incoherent values. This is due to the transmission loss fluctuations and to the absence of averaging in the azimuth, range or frequency domains. As an example, Fig. 4 gives the reverberation loss versus time averaged in a 1/3 Oct. frequency band at a central frequency of 100Hz from Eq. (11). Comparing this with Fig.3, it can be seen that the random distribution of coherent reverberation loss is relatively small. To investigate the effects of a sloped bottom on the reverberation fields, the sloped bottom, shown as the solid line in Fig. 5, is assumed in calculation. Figure 6 compares the averaged reverberation in the 1/3 Oct. band for the sloped bottom model and the flat bottom model. The reverberation loss with a sloped bottom shown as the thick solid line and thick dashed line in Fig. 6 is larger than the results calculated by the flat bottom model. This is mainly because the backscattering angles grow smaller during the scattering process near the slope bottom boundary.

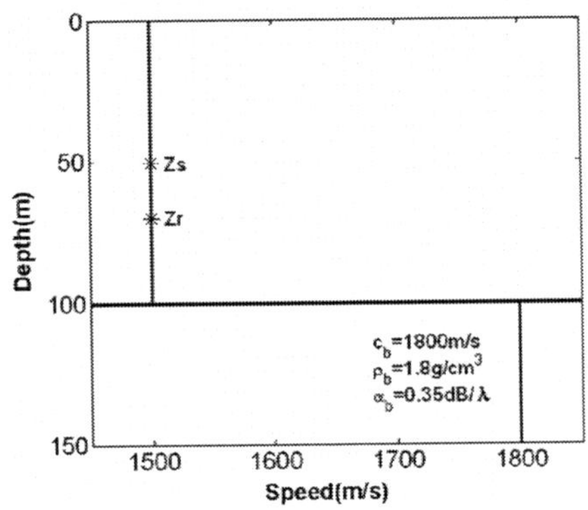

FIGURE 2. Ocean environment.

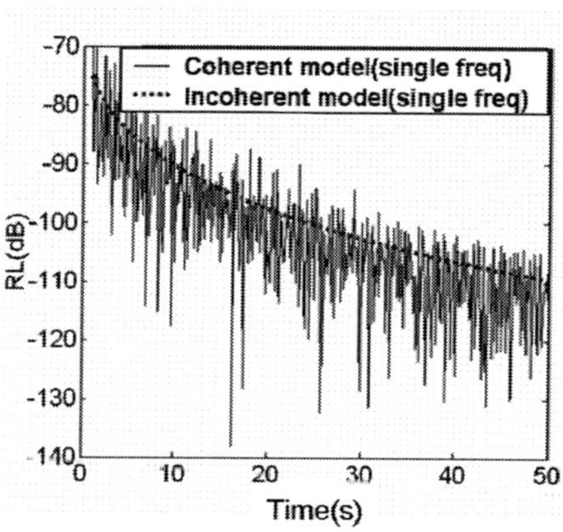

FIGURE 3. Reverberation loss versus time at frequency 100Hz calculated by using the coherent and incoherent model.

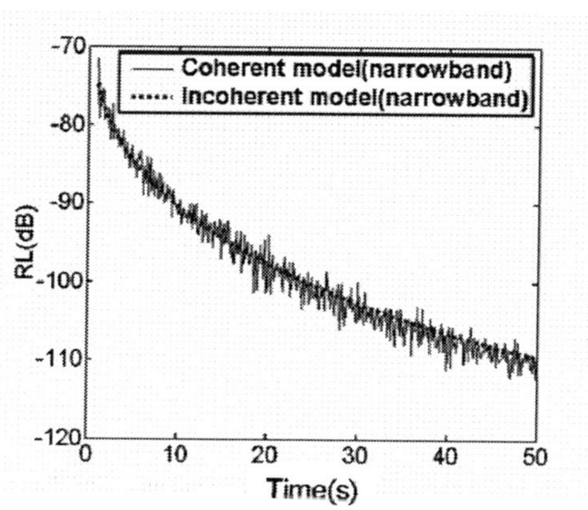

FIGURE 4. Averaged reverberation loss in 1/3 Oct. band at central frequency 100Hz.

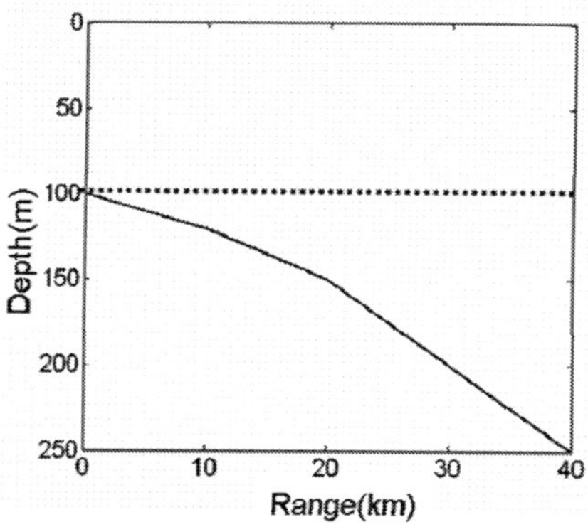

FIGURE 5. Slope/flat bottom model.

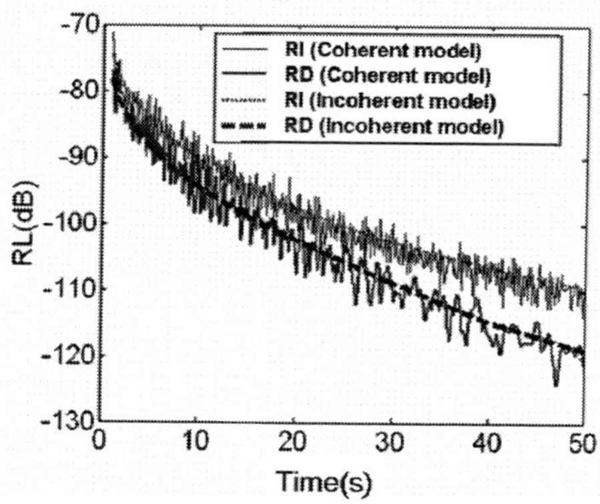

FIGURE 6. Comparison of averaged reverberation in 1/3 Oct. band for slope bottom model (RD) and flat bottom model (RI).

A. Experimental Results

As a part of ASIAEX, reverberation measurements were performed on June 3, 2001 at a site in the East China Sea, centered at 29°40.67' N and E126°49.39' E. The reverberation data were collected by using explosive sources with 1000 gram TNT charges [9]. The water depth at the ASIAEX site is 104 m. The sea bottom in the reverberation analysis area (radius = 15 km) has a relatively small slope and changes from the minimal depth of 98 m to the maximal depth of 110 m. The reverberation data were collected by a 32-element vertical line array (VLA), and fed to a 32-channel digital recorder. The signal sampling frequency was 6000 Hz. The hydrophone array, which was made by Institute of Acoustics, Chinese Academy of Sciences, was suspended from a Chinese research vessel, *Shi Yan 3*. The sound velocity profile (SVP) measured by CTD at a time that was close to the period of reverberation measurements is shown in Fig. 7. Figure 8 shows a time series for 1000 Hz, 1/3 octave mono-static reverberation.

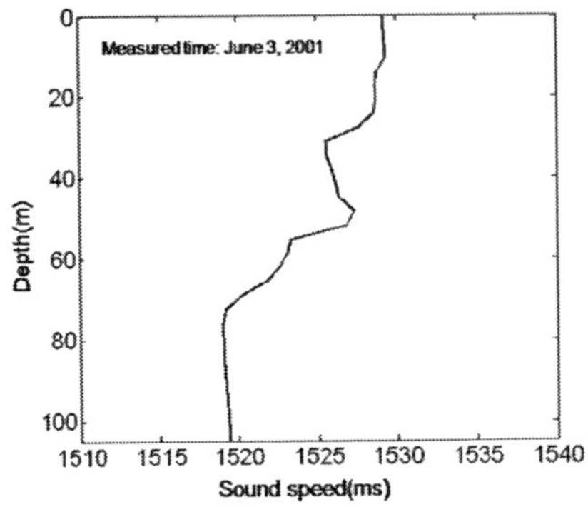

FIGURE 7. The sound velocity profile during the period of reverberation measurements

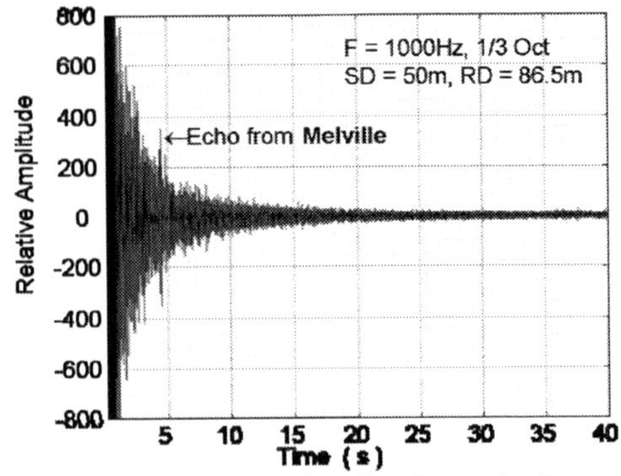

FIGURE 8. A reverberation time series from ASIAEX (1000 Hz, 1/3 Octave).

The reverberation loss and vertical correlation were analyzed. As an example, Fig. 9 shows the experimental reverberation loss in the 1/3 octave at 1000Hz from the hydrophone at the depth of 80.5 m. Meanwhile, the numerical result calculated by the coherent reverberation model is also shown in Fig.9 as the dotted line. In calculation, the seabed is assumed to be a half infinite liquid model, and the bottom density, sound speed, and attenuation are 1.86 g/cm^3, 1610 m/s[14], and 0.15 dB/m[15], respectively. It can be seen from Fig. 9 that both the decreasing trend and the random fluctuation of the numerical result agree with the experimental data very well. The vertical correlation of reverberation versus time for two receivers with space 2 m are shown in Fig. 10, where the solid line is the experiment data and the dotted line is the numerical result from Eq. (12). It is shown that both the experiment data and the numerical result of vertical correlation have significant fluctuation, and have the same increasing tread from -0.4 to 0.2. There is a small difference in the fluctuation, which is caused by no effects of random phase in the scattering function in numerical result.

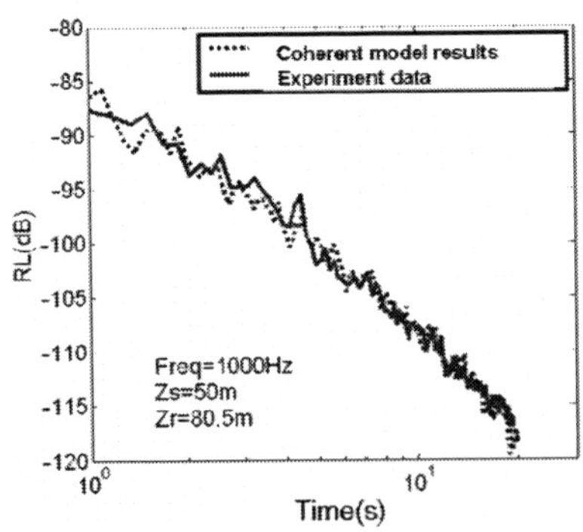

FIGURE 9. Model/data comparison of reverberation intensity

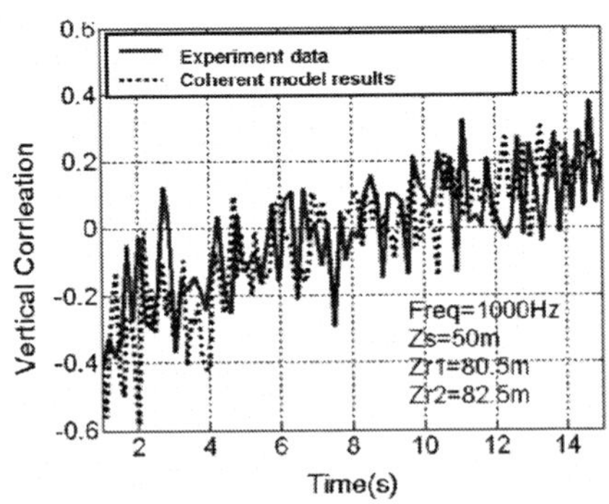

FIGURE 10. Model/data comparison of reverberation vertical correlation.

IV. SUMMARY AND CONCLUSION

A general reverberation model based on the adiabatic normal mode theory in a range dependent shallow water environment is presented. From this, the coherent or incoherent reverberation field, as well as the reverberation intensity or reverberation vertical correlation can be predicted. Observations of reverberation from the ASIAEX 2001 ECS are used to test the model. The model/data comparison shows that the coherent reverberation model can predict reverberation intensity and vertical correlation very well. The reverberation model can also be used to simulate the reverberation field for an active sonar simulator.

ACKNOWLEDGMENTS

The authors would like to thank the ASIAEX ECS 2001 experimental team; many thanks go to Professor Jixun Zhou and Dr. Jeffrey A. Donnell, Georgia Institute of Technology, for polishing the paper. The work was supported by the Knowledge Innovation Program of the Chinese Academy of Sciences Grant No. KZCX1-YW-12-2 and the National Natural Science Foundation of China under Grant No. 10974218.

REFERENCES

1. C. Wu, *Acta Acustica*, 4: 114 - 119 (1979).
2. M. D. Collins, G. J. Orris and W. A. Kuperman, *Ocean Reverberation*, Netherlands: Kluwer Academic Publishers, 1993, pp. 119-224.
3. R. Zhang and G. Jin, *Journal of Sound and Vibration*, 119(2): 215-223 (1987).
4. D. D. Ellis, *J. Acoust. Soc. Am.*, 97(5): 2804-2814 (1995).
5. R. J. Urick and G. R. Lund, *J. Acoust. Soc. Am.*, 47: 342-349 (1970).
6. J. X. Zhou, *Acta Oceanologia Sinica*, 1(2): 212-217 (1979). [in Chinese]
7. F. Li and J. Liu, *Acta Acustica*, 28(6): 494-503 (2003). [in Chinese]
8. J. Liu, F. Li and Z. Peng, *Chin. Phys. Lett.*, 20(2): 2188-2191 (2003).
9. J. X. Zhou, X. Z. Zhang, el al, *IEEE J. Oceanic Engineering*, 29(4): 988-999 (2004).
10. F. Li, J. Liu, Z. Li et al., *Journal of Sound and Vibration*, 252(3): 457-468 (2002).
11. F. Li, J. Liu, Z. Li et al., *Science in China Ser. G*, 48(4): 413-421, (2005).
12. F. B. Jenson, W. A. Kuperman, M. B. Porter et al., *Computational Ocean Acoustics*, New York: AIP Press, 1994, pp. 271-337.
13. K. V. Mackenzie, *J. Acoust. Soc. Am.*, 33: 1498-1504 (1961).
14. Z. Li, R. Zhang, J. Yan et al, *IEEE J. Oceanic Engineering*, 29(4): 973-979 (2004).
15. P. H. Dahl, R. Zhang, J. H. Miller el al, *IEEE J. Oceanic Engineering*, 29(4): 920-928 (2004).

Low Frequency Volume Reverberation Measurements in Turbid Seawater

Yongwei LIU [a], Qi LI, Dejiang SHANG, Daheng YANG, Chao ZHANG, Rui TANG, Dajing SHANG and Mengying CHEN

[a] *National Laboratory of Underwater Acoustic Technology, Harbin Engineering University, Harbin, China,150001.*

Abstract. Shallow coastal waters are characterized by high levels of suspended sediment particles relative to the open ocean. This kind of seawater is also characterized as turbid seawater. An experimental investigation on volume reverberation has been done in the sea area outside the Yangtze River Estuary, where the seawater contains a lot of suspended sediment particles. Sound scattering data at 8 sites in the frequency range from 10 to 40 kHz, in steps of 1 kHz, has been obtained. The results demonstrate that volume backscattering intensity in turbid seawater may be expected to fall in the range from -60 to -100 dB *re* m^{-1}. Because sound absorption caused by suspended sediment particles in turbid seawater is much greater than that in clear seawater, the volume backscattering intensity does not change with the pulse length of the transmitted signal, even when the pulse length is less than 0.5 ms. Salinity in the sea area has little effect on volume backscattering intensity. However, suspended sediment particles may have a great effect on it. Therefore, if the concentration of suspended sediment particles changes from 31 mg/L to 47 mg/L, the volume backscattering intensity may be changed by 30 dB.

Keywords: Turbid Seawater; Volume Reverberation; Suspended Sediment Particles.
PACS: 43.30Ft, 43.30Gv, 43.30Re

INTRODUCTION

There are many causes of acoustic scatter in the ocean, such as marine organisms, sediment particles, bubbles, the water mass induced by local temperature inhomogeneity, and so on. In addition, the rough sea surface and sea bottom not only reflect sound wave, but also scatter sound wave. All these scatters comprise inhomogeneity in the ocean, and make the physical property of the medium discontinuous. The process of sound scattering initiates when sound wave is projected into this medium. Some part of the incident sound energy propagates continually along the origin direction; the other part of this sound energy scatters omnidirectionally, then, the scattered sound field generates. The latter is one of the reasons why sound energy attenuates along the propagation direction. The sound scattering along the anti-incident direction is also characterized as volume reverberation. This is the main background interference of active sonar. Meanwhile, volume reverberation is one of main factors influencing the detection of a target near the sea bottom. Past research has mainly been done on the properties of the deep scattering layer, sea surface reverberation and sea bottom reverberation [1-7]. However, research on volume

CP1272, *Shallow-Water Acoustics, Second International Shallow-Water Acoustics Conference*
edited by J. Simmen, E. S. Livingston, J. X. Zhou, and F. H. Li

reverberation in shallow coastal waters is limited. If the property of volume reverberation is investigated, it may be helpful in the design of sonar and the detection of suspended or submerged targets in shallow coastal waters.

VOLUME REVERBERATION

Volume reverberation follows the transmitted signal closely and is made up of superposed sound scattering from mass irregular scatters in the ocean. The parameter used here to describe the ability of the volume to produce sound reverberation is its backscattering intensity $S_{s,v}$. This is defined as the ratio in decibels of the backscattered sound intensity, I_{scat}, from a unit volume of the medium, to the incident intensity, I_{inc}, if the former is measured in the far-field of the scattering volume and referred back to a unit distance from its acoustic center. Then, the volume backscattering intensity will be expressed as

$$S_{s,v} = 10\log_{10}\frac{I_{scat}}{I_{inc}} \tag{1}$$

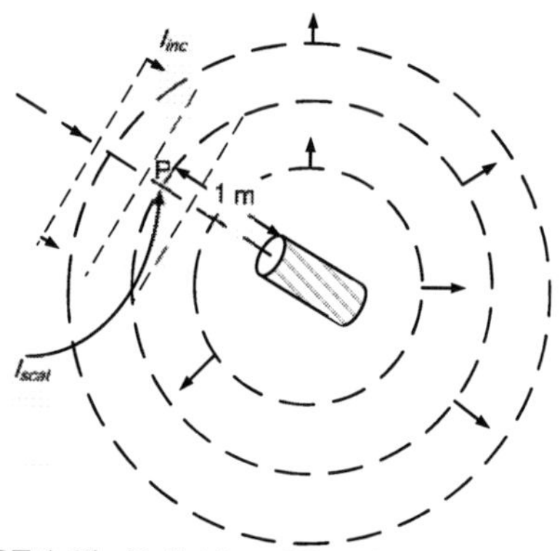

FIGURE 1. The Definition of Sound Scattering Intensity

In FIGURE 1, the direction of sound scattering points to the sound source. The quantity $S_{s,v}$ is also characterized as the volume backscattering intensity. The reference point is at *P*, which is 1 m from the unit volume. Volume reverberation is a very complex process, which is influenced by many factors. The assumption is as follows. First, the propagation of sound scattering is linear; the propagation loss can be expressed as spherical attenuation and sound absorption. Second, at any instant, the distribution of the scatters in a volume is homogeneous. Third, the density of the scatters is so great that there are a lot of scatters in each element of the volume. Fourth, the duration of the pulse is so short that the propagation effect in the element can be ignored. Fifth, there is no multiple scattering, namely, the reverberation generated by the reverberation can be ignored [8].

SEA TEST

In July 2008, low frequency volume reverberation measurements were obtained in the sea area outside the Yangtze River estuary. At the experiment sites, the transducer and hydrophone were put at a suitable depth, according to sea conditions. The signal generator transmitted pulses of a single frequency. Then, the signal was amplified by a power amplifier and sent to the transducer. After 1 ms, a data acquisition system amplified and recorded the volume reverberation signal, which was collected by the hydrophone. The cross section of temperature, salinity and turbidity was collected by the CTD and OSB. The content of the oxygen and the concentration of suspended sediment particles at each experimental site were collected by other instruments before or after the acquisition of underwater acoustic data.

EXPERIMENTAL RESULTS

At one experiment site, the collected volume reverberation signal is shown in FIGURE 2. The frequency of the signal is 24 kHz, and the duration of the pulse is 0.4 ms.

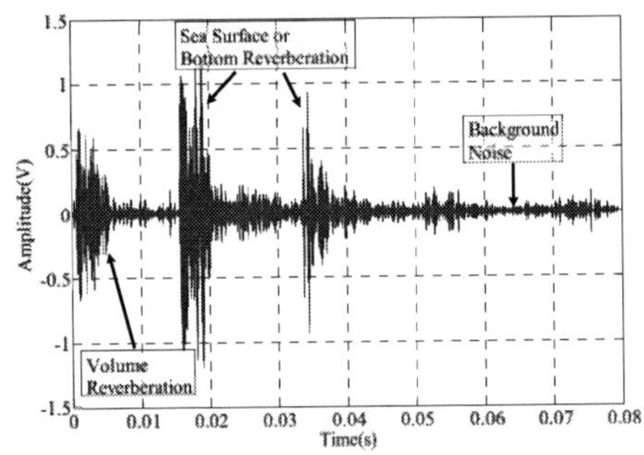

FIGURE 2. The Volume Reverberation Signal of 24 kHz

The volume backscattering intensity with the temperature, salinity, oxygen content, turbidity, the concentration of suspended sediment particles and frequency at eight experimental sites is shown in FIGURES 3 and 4.

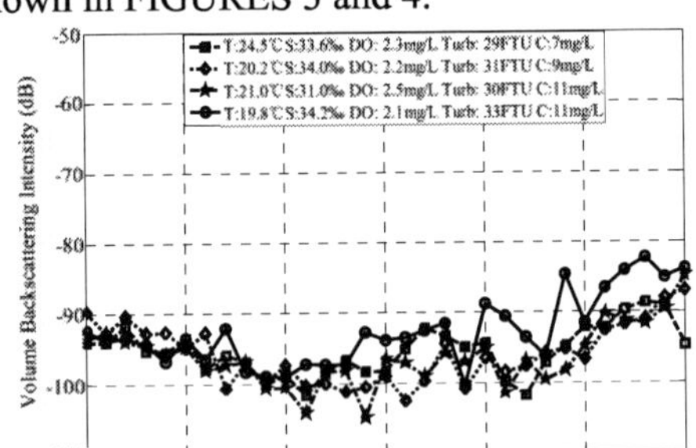

FIGURE 3. The Volume Backscattering Intensity at Different Experimental Sites.

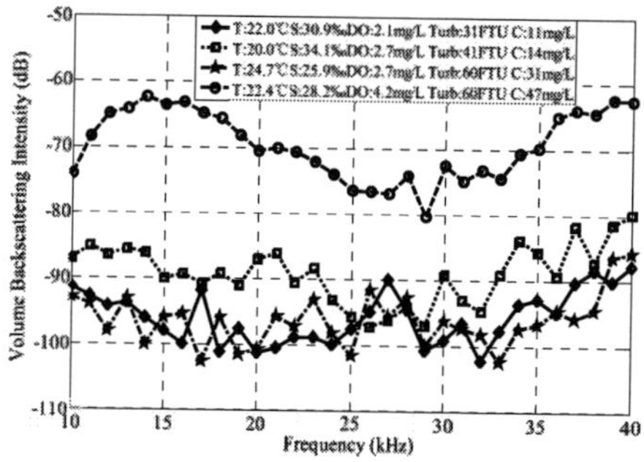

FIGURE 4. The Volume Backscattering Intensity at Different Experimental Sites.

The volume backscattering intensity at different sites may be expected to fall in the range from -60 to -100 dB *re* m^{-1}. Here, *T* denotes temperature; *S* denotes salinity; *DO* denotes the content of the oxygen; *Turb* denotes turbidity; *C* denotes the concentration of suspended sediment particles. At the experiment sites, volume backscattering intensity has no obvious dependence on the frequency. Generally speaking, the main scatters inducing volume reverberation are the plankton with high distribution density, besides the fish with swimming bladders. However, most plankton, especially zooplankton, has the phenomenon of diurnal vertical migration. They ascend to the surface at night, and descend when the dawn comes. This phenomenon has been obviously seen on many pelagic crustaceans. Other crustaceans, such as medusans, siphonosomes, ctenophores, chaetognaths, pteropods, and cephalopods, also have the habit of diurnal vertical migration. The light intensity is one of most important zoology factors, which induces zooplankton's vertical migration. In order to avoid the effect of zooplankton on the measurement, the experiments have been done at the noon. At that time, the light intensity is high and zooplankton usually descends to a deep position. Moreover, the phytoplankton does not generate volume reverberation. The dissolved oxygen in the seawater has the most important effect on the survival of the fish. The oxygen content at most experimental sites is less than 4.0 mg/L, which indicates that the sites belong to anaerobic environments, and fish can not live in these environments. In other words, it indicates that the swimming bladders of the fish are not the main source of volume reverberation. The bubbles induced by the winds, usually exist several meters below the surface. Nevertheless, the transducers at the experimental sites are 10 m below the sea surface. The chosen scattering layer is a spherical one and only 2 m from the transducer when the volume backscattering intensity is being calculated. Therefore, the bubbles induced by the winds are not the main reason for generating volume reverberation. The salinity usually does not induce volume reverberation. Although volume reverberation caused by a single suspended sediment particle is small, a lot of suspended sediment particles will make up a 'snow' structure. With the temperature change of ambient seawater, all these cause the sound parameters of the seawater to be non-homogeneous [9-12]. When sound wave propagates in the seawater, which contains a lot of suspended sediment particles, the intensive sound scattering process generates. The suspended sediment particles in the seawater make sound parameters non-homogeneous. Volume backscattering intensity may be changed by 30 dB due to the concentration change of suspended sediment particles.

The volume backscattering intensity with different pulse length is shown in FIGURES 5 and 6. In these figures, five cycles and ten cycles denote the number of sine waves filled up in the CW pulse.

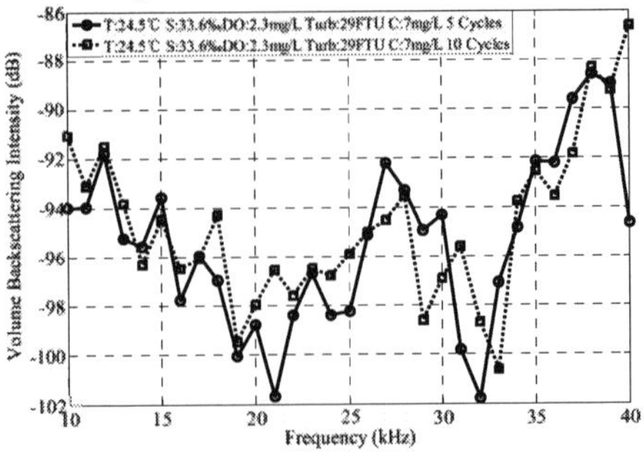

FIGURE 5. The Volume Backscattering Intensity of Different Signal Length.

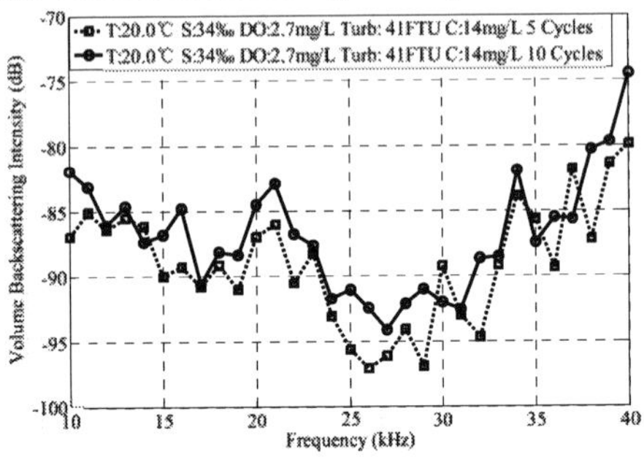

FIGURE 6. The Volume Backscattering Intensity of Different Signal Length.

If the duration is less than 10 ms, the volume backscattering intensity is linear to the length of the pulse [13]. However, the volume backscattering intensity is not linear to the length of the pulse in FIGURES 5 and 6. Because sound absorption in turbid seawater is bigger than that in the water, when the length of the pulse is increasing, the effect of sound absorption in the scattering layer must be considered. The intensive effect of sound absorption in the scattering layer will make the volume backscattering intensity independent of the length of the pulse. Then, the volume backscattering intensity is saturated.

ACKNOWLEDGMENTS

The author would like to thank Professor Ping CAI, Shoubin KE, Fengyang CHI, Chao ZHONG and Yan XIAO for their assistance with data processing.

REFERENCES

1. Barry A.Gold, *J. Acoust. Soc. Am.* **40**, 689-696(1966)
2. Tokuo Yamamoto, *J. Acoust. Soc. Am.* **99**, 866-879(1996)
3. Peter H.Wiebe, Timothy K. Stanton, Mark C. Benfield, et al., *IEEE J. Ocea. Eng.* **22**, 445-463(1997)
4. Shi Zhong, Zhang Shuying, Hamilton L J., *ShengXueXueBao/Acta Acustica.* **23**,221-228(1998)
5. Timothy C. Gallaudet, Christian P.de Moustier, *J. Acoust. Soc. Am.* **112**,489-503(2002)

6. Timothy C. Gallaudet, Christian P. de Moustier, *J. Acoust. Soc. Am.* **114**,707-725(2003)
7. V.E.Sklyarov, A. V. Berezutskii, *Proc. of SPIE.* **5397**,191-198(2004)
8. Liu Bosheng, Lei Jiayi, *The Principles of Underwater Acoustics*, Harbin: Harbin Engineering University Press, 1993, pp.187-195.
9. XiaMen Marine Lives College, *Planktology in the Ocean*, Beijing: Agriculture Press, 1981, pp. 97-218.
10. Yang Dongfang, Gao Zhenhui, *The ecology in the JiaoZhou Bay and Yangtze River Estuary*, Beijing: Ocean Press, 2007, pp. 200-206,341-347.
11. Bao Qinghua, Chai Xinyu, *Acta Oceanologica Sinica.* **17**, 59-64(1995)
12. Zhang Huaijing, Zhai Shikui, Fan Dejiang, et al., *Marine Geology & Quaternary Geology.* **27**, 1-10(2007)
13. John J.Keane, *J. Acoust. Soc. Am.* **43**, 566-570(1968)

Shallow Water Reverberation Model Based on Bottom Reflection Parameters

Jinrong Wu, Erchang Shang and Tianfu Gao

Key Laboratory of Underwater Acoustic Environment
Institute of Acoustics, Chinese Academy of Sciences, Beijing, China, 100190

Abstract. The modal shallow water reverberation theory and the energy-flux shallow water reverberation theory were combined to develop a new model of waveguide reverberation based on bottom reflection parameters P and Q. There are only three environmental parameters (P, Q and backscattering coefficient μ) in the new reverberation model. The parameters have clear physical meaning and the new reverberation model can be tested by the experiment data easily. The new shallow water reverberation model was compared with the modal reverberation model and the Yellow Sea reverberation experiment data. The results show that the new model can explain shallow water reverberation in most cases.

Keywords: Energy-flux, reverberation, reflection parameters
PACS: 43.30.Gv

I INTRODUCTION

Reverberation is the main interference to active detection, and for this reason reverberation has been of interest for a long time, especially in the case of shallow water waveguides. There have appeared many shallow water reverberation models, and these models can be separated into two groups: phenomenological models and physical based models. Most of the shallow water reverberation models are phenomenological, such as normal mode model[1-5], energy-flux model[6,7], spectral representation model[8], PE model[9] and so on. In these models, the key component of the bottom back-scattering process has been treated with help either by introducing an empirical law (e.g. Lambert's law) or using some kind of assumed model with parameters to be determined [10,11]. These methods are practical and heuristic, but not rigorous. And the physical picture of the phenomenological reverberation models is not clear. The other group of reverberation models is physical based models, sometimes called full-wave reverberation models. These models deal with two-way sound propagation and backscattering at the same time. The analytic expressions of backscattering sound fields were deduced from wave equations using the Born approximation due to the rough interface of the bottom[12] or volume inhomogeneities in the bottom[13], or both of the two mechanisms[14,15]. These models are rigorous in theory and provide clear physical pictures for reverberation, but most of these models are untested against shallow water reverberation data because some environmental parameters in the model are difficult to obtain.

CP1272, *Shallow-Water Acoustics, Second International Shallow-Water Acoustics Conference*
edited by J. Simmen, E. S. Livingston, J. X. Zhou, and F. H. Li

In this paper, a new model of waveguide reverberation based on bottom reflection parameters was presented. The new model has clear physical meaning and can be tested. It's simple and only has three environmental parameters, which can be measured or inversed easily. This paper is organized as follows: the modal reverberation model based on the Perturbation theory is introduced in Section II. In Section III, the shallow water reverberation model based on P and Q is presented. Numerical simulation cases are presented in section IV. The summary and conclusions are addressed in section V.

II MODAL REVERBERATION MODEL

The scenario of the shallow-water reverberation problem that we are going to deal with is depicted in Fig.1. For simplicity, only rough interface induced reverberation in Pekeris waveguide was considered here. The sound speed and density in water is c_0 and ρ_0 seperately.The constant water depth is H. The water-sediment rough interface has a fluctuated height η, with $<\eta> = 0$. In the liquid halfspace, the sound speed and density is c_b and ρ_b. In the water column, R_0 represents the source point, R represents the field point and R_1 represents the scattering point.

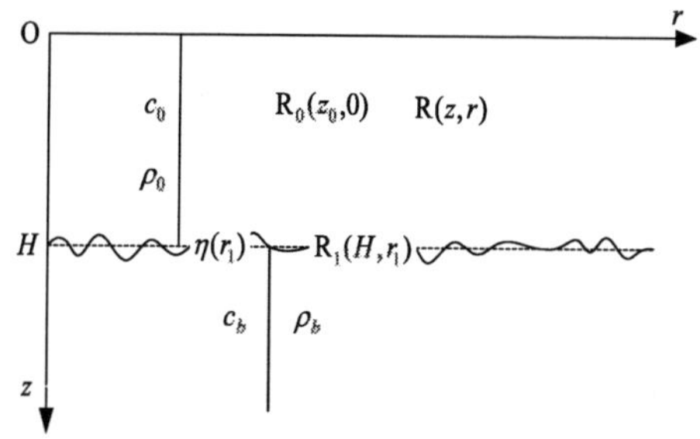

FIGURE 1. Shallow water reverberation scenario

By using the Bass perturbation theory, the continuities of the field at the rough surface are replaced by the inhomogeneous boundary conditions at the smooth interface $z = H$, as follows [12,15,16]

$$\frac{\partial u_1(\mathrm{R},\mathrm{R}_1)}{\partial z} - \frac{\partial u_2(\mathrm{R},\mathrm{R}_1)}{\partial z} = V(\mathrm{R}_1)G^i(\mathrm{R}_1,\mathrm{R}_0) \tag{1}$$

$$\rho_0 u_1(\mathrm{R},\mathrm{R}_1) - \rho_b u_2(\mathrm{R},\mathrm{R}_1) = p(\mathrm{R}_1)G^i(\mathrm{R}_1,\mathrm{R}_0) \tag{2}$$

where

$$V(\mathrm{R}_1) = (k_0^2 - k_b^2/\alpha)\eta(\mathrm{R}_1) + (1-1/\alpha)\nabla_\perp \bullet (\eta\nabla_\perp) \tag{3}$$

$$p(\mathrm{R}_1) = (\rho_b - \rho_0)\eta(\mathrm{R}_1)\frac{\partial}{\partial z} \tag{4}$$

where $k = \omega/c_0$, $\alpha = \rho_b/\rho_0$, and u_2 , u_1 are the scattered field under and above the interface respectively. The initial incident field (with $\eta = 0$) of a unit point source at R_0, G^i (R_1,R_0) can be written as a summation of proper modes (we assumed that $c_b > c_0$) :

$$G^i(\mathrm{R}_0,\mathrm{R}_1)=i\pi\sum_m^M \varphi_m(z_0)\varphi_m(H)H_0^{(1)}(K_m r_1) \quad (5)$$

where $H_0^{(1)}$ is the first kind zero order Hankel function, $\varphi_m(z)$ is the normalized mode function, K_m is the complex modal wave number : $K_m = k_m + i\beta_m$, and r_c is the center range of the scattering area (we assumed that the scattering area has limited size).

According to Green's theorem, under the Born approximation, the scattered field of a point source in the water is given by :

$$u_1(\mathrm{R}_0,\mathrm{R})=\int d\mathrm{R}_1\{G(\mathrm{R},\mathrm{R}_1)V(\mathrm{R}_1)G^i(\mathrm{R}_1,\mathrm{R}_0)-\frac{1}{\rho_w}\frac{\partial G(\mathrm{R},\mathrm{R}_1)}{\partial z}p(\mathrm{R}_1)G^i(\mathrm{R}_1,\mathrm{R}_0)\} \quad (6)$$

Substituting eq.(3), eq.(4) and eq.(5) into eq.(6), we get

$$u_1(\mathrm{R}_0,\mathrm{R})\cong\frac{2\pi i}{k_0 r_c}\sum_m^M\sum_n^M \varphi_m(z_0)\varphi_n(z)\exp[-(\beta_m+\beta_n)r_c]S_{mn}^R K^R(k_m,k_n) \quad (7)$$

where

$$S_{mn}^R \equiv \varphi_m(H)C_{mn}^R\varphi_n(H) \quad (8)$$

$$C_{mn}^R=[k_0^2-k_b^2/\alpha+(1-\alpha^{-1})k_m k_n+(1-\alpha)(\alpha)^{-2}\gamma_m\gamma_n]=k_0^2 c_{mn}^R \quad (9)$$

$$c_{mn}^R=\{1-(c_0/c_b)^2/\alpha+(1-\alpha^{-1})\cos\theta_m\cos\theta_n+(1-\alpha)(\alpha^{-2}) \\ \times[\cos^2\theta_m-(c_0/c_b)^2]^{1/2}[\cos^2\theta_n-(c_0/c_b)^2]^{1/2}\}\cong \text{const.}=c^R \quad (10)$$

$$\gamma_m=(k_m^2-k_b^2)^{1/2}\,,\qquad k_m=k_0\cos\theta_m \quad (11)$$

$$K^R(k_m,k_n)\equiv\int dr_1\eta(r_1)\exp[i(k_m+k_n)r_1] \quad (12)$$

The scattering field given by eq.(7) is for CW wave, the scattering field of a pulse signal $s(t)$ with a spectrum $s(\omega)$ can be obtained by Fourier transform :

$$u_1(\mathrm{R}_0,\mathrm{R};t)=\int d\omega[s(\omega)u_1(\mathrm{R}_0,\mathrm{R};\omega)\exp(i\omega t)] \quad (13)$$

Substituting eq.(7) into eq.(13), and making a narrow-band approximation,

$$k_m(\omega)\cong k_m(\omega_0)+(\omega-\omega_0)\frac{\partial k_m(\omega_0)}{\partial\omega} \quad (14)$$

Then, we get

$$u_1(\mathrm{R}_0,\mathrm{R};t)=\frac{2\pi i}{k_0 r_c}\sum_m^M\sum_n^M s(t-t_{mn})\varphi_m(z_0)\varphi_n(z)\exp[-(\beta_m+\beta_n)r_c]S_{mn}^R K^R(k_m,k_n) \\ \cong\frac{2\pi i}{k_0 r_c}s(t-t_c)\sum_m^M\sum_n^M\varphi_m(z_0)\varphi_n(z)\exp[-(\beta_m+\beta_n)r_c]S_{mn}^R K^R(k_m,k_n) \quad (15)$$

where

$$t_{mn}=[\frac{\partial k_m}{\partial\omega}+\frac{\partial k_n}{\partial\omega}]r_c=[\frac{1}{u_m}+\frac{1}{u_n}]r_c\approx\frac{2r_c}{c_0}=t_c \quad (16)$$

In eq.(14) we neglect the pulse spreading caused by the $(\partial^2/\partial\omega^2)k_m$ term. The modification of the pulse amplitude spreading should be considered for very long range and very wide band pulses. Such a case has been discussed in Ref.[4,13].

We take the statistical average of the incoherent summation of $<|u_1(\mathrm{R}_0, \mathrm{R}; t)|^2_{\mathrm{inc}}>$ as the intensity of the reverberation - $I_R(\mathrm{R}_0, \mathrm{R}; t)$, and consider the insonified area as a ring with area: $A = 2\pi r_c \Delta r$, and $\Delta r \cong c_0 \tau_0/2$, where τ_0 is the duration of the signal $s(t)$, then from eq.(15), we get

$$I_R(\mathrm{R}_0,\mathrm{R};t) = (\frac{2\pi}{k_0 r_c})^2 s^2(t-t_c)(2\pi r_c)\sum_m^M\sum_n^M \varphi_m^2(z_0)\varphi_n^2(z)\exp[-2(\beta_m+\beta_n)r_c][S_{mn}^R]^2\Gamma(k_m,k_n) \tag{17}$$

where

$$\Gamma(k_m,k_n) = \int_{r-\Delta r/2}^{r+\Delta r/2} dr' \int_{r-\Delta r/2}^{r+\Delta r/2} dr'' <\eta(r')\eta(r'')> \exp[i(k_m+k_n)r' - i(k_m+k_n)r''] \tag{18}$$

To evaluate the integration in eq.(18), set $r'' = r' + x$, and assume Δr is much larger than the correlation scale L, we get

$$\Gamma(k_m,k_n) \cong (\Delta r)\sigma_\eta^2 \int_0^\infty dx R^\eta(x)\exp(i2k_0x) = \frac{c_0\tau_0}{2}\sigma^2 P^\eta(2k_0) \tag{19}$$

where $R^\eta(x)$ and σ_η^2 are the correlation function and MS value of the fluctuation η respectively. P^η is the power spectrum of the fluctuation η. Here, we consider a widely accepted spectrum of interface roughness – the Goff-Jordan spectrum[17] :

$$P^\eta(2k_0) = \pi L[1+(2k_0L)^2]^{-3/2} \tag{20}$$

where parameter L is the horizontal correlation scale, in general, L ranges 10-20 m. Substituting eq.(19) into eq.(17), we get

$$I_R(\mathrm{R}_0,\mathrm{R};t) = E_0(\frac{2\pi}{k_0 r_c})^2(\pi r_c c_0)\sum_m^M\sum_n^M \varphi_m^2(z_0)\varphi_n^2(z)\exp[-2(\beta_m+\beta_n)r_c]\Theta_{mn}^R \tag{21}$$

where $E_0 = s^2(t-t_c)\tau_0$ is the energy of the initial pulse signal, and the MBSM is given by

$$\Theta_{mn}^R = \sigma_\eta^2 P^\eta(2k_0)[S_{mn}^R]^2 \tag{22}$$

Consider that the bottom reflection phase shift ϕ_b at small grazing angle can be approximated as follows

$$\phi_b \cong -\pi + \mathrm{P}\theta \tag{23}$$

where P is a parameter. For a liquid half space bottom, parameter P can be related to the physical properties of the sediment as follows[18] :

$$\mathrm{P} \cong 2(\rho_b/\rho_0)[1-(c_0/c_b)^2]^{-1/2} \tag{24}$$

Under the approximated bottom reflection phase shift given by eq.(23), the mode function can be written as [15,18]

$$\varphi_m(H) = (2/H_{eff})^{1/2}\sin[m\pi(H/H_{eff})] \tag{25}$$

where

$$H_{eff} = H + \Delta H = H + \mathrm{P}/(2k_0) \tag{26}$$

The corresponding modal dispersion equation is

$$H_{eff}(k_0^2 - k_m^2)^{1/2} = m\pi \tag{27}$$

or

$$\sin\theta_m = m\pi/(k_0 H_{eff}) \tag{28}$$

Substituting eq.(26) and eq.(28) into eq.(25), we get the angular dependency of $\varphi_m(H)$:

$$\varphi_m(H) \approx (-1)^{m+1}(2/H_{eff})^{1/2}\sin(\theta_m \mathrm{P}/2) \tag{29}$$

Finally, substituting eq.(29) into eq.(8), we can obtain the scattering angular dependency due to interface roughness given by S_{mn}^R as:

$$S_{mn}^R = (-1)^{m+n} k_0^2 c^R A_m A_n \sin(\theta_m \mathrm{P}/2)\sin(\theta_n \mathrm{P}/2) \tag{30}$$

where A_m is the normalized amplitude of the mode function, for Pekeris waveguide,

$$A_m = (2/H_{eff})^{1/2} \tag{31}$$

Substituting eq.(30) into eq.(22), we get the MBSM due to interface roughness as:

$$\Theta_{mn}^R = \sigma_\eta^2 P^\eta(2k_0)[k_0^2 c^R A_m A_n \sin(\theta_m \mathrm{P}/2)\sin(\theta_n \mathrm{P}/2)]^2 \tag{32}$$

The MBSM defined in this way has the factor of modal amplitude $A_m A_n$. It is well known that the $A_m A_n$ is waveguide dependent (see eq.(26) and eq.(31)). By extracting this factor out, we can alternatively redefine a more intrinsic MBSM as follows:

$$[\Theta_{mn}^R]^{\mathrm{int}} = \sigma_\eta^2 P^\eta(2k_0)[k_0^2 c^R]^2 \sin^2(\theta_m \mathrm{P}/2)\sin^2(\theta_n \mathrm{P}/2) \tag{33}$$

The MBSM given by eq.(33) only depends on bottom properties which can affect the feature of scattering and does not depend on the type of waveguide. And eq.(21) can be rewritten as[15]

$$I_R = E_0\left(\frac{2\pi}{k_0 r_c}\right)^2(\pi r_c c_0)\sum_m^M\sum_n^M \varphi_m^2(z_0)\varphi_n^2(z)A_m^2 A_n^2 \exp[-2(\beta_m+\beta_n)r][\Theta_{mn}^R]^{\mathrm{int}} \tag{34}$$

III REVERBERATION MODEL BASED ON P AND Q

In Section II, the shallow-water reverberation modal model was deduced based on the perturbation theory. In this section, the shallow-water reverberation modal model will be rewritten in the sketch of an energy-flux model[6,7]. Then the shallow water reverberation model based on bottom reflection parameters P and Q will be presented.

By using the WKB approximation, and supposing there is no turning point in the water column, then the eigenfunction can be written as[18]

$$\varphi_m(z) = 2(k_m)^{1/2}(D_m+\delta_m)^{-1/2}[k_0^2(z)-k_m^2]^{-1/4}\sin\left[\int_0^z (k_0^2(z)-k_m^2)^{1/2}dz\right] \tag{35}$$

where D_m is the modal cycle distance and δ_m is the displacement due to bottom reflection phase shift

$$D_m = 2\int_0^H (k_m)^{1/2}[k_0^2(z)-k_m^2]^{-1/2}dz \tag{36}$$

$$\delta_m = (\partial/\partial k_m)\phi_b \tag{37}$$

Vertical wavenumber k_{zm} is

$$k_{zm} = \sqrt{k_0^2(z)-k_m^2} \tag{38}$$

and, neglecting the displacement δ_m, eq.(35) can be written as

$$\varphi_m(z) \approx 2\sqrt{\frac{k_m}{D_m k_{zm}(z)}}\sin\left[\int_0^z k_{zm}(z)dz\right] \tag{39}$$

In Pekeris waveguide, eq.(39) becomes

$$\varphi_m(z) = 2\sqrt{\frac{k_m}{D_m k_{zm}}} \sin(k_{zm} \cdot z) \tag{40}$$

The average energy of the mode was defined as

$$I_m = \frac{1}{H}\int_0^H \varphi_m^2(z)dz \tag{41}$$

Substituting eq.(40) into eq.(41), then

$$I_m = \frac{2k_m}{D_m \cdot k_{zm}} - \frac{k_m}{H \cdot D_m \cdot k_{zm}^2} \sin(2k_{zm}H) \tag{42}$$

In Pekeris waveguide, the modal cycle distance can be written as

$$D_m = 2\int_0^H \frac{dz}{\tan\theta_m(z)} = \frac{2H}{\tan\theta_m} \tag{43}$$

Eq.(43) was taken into eq.(42), then

$$I_m = \frac{1}{H} - \frac{\sin(2k_{zm}H)}{2H^2 k_{zm}} \approx \frac{1}{H} \tag{44}$$

For Pekeris waveguide

$$A_m = (2/H_{eff})^{1/2} \tag{45}$$

$$H_{eff} = H + \Delta H = H + \mathrm{P}/(2k_0) \approx H \tag{46}$$

The attenuation term,

$$\exp(-2\beta_m r) = \exp(2\ln|V(\theta)|\, r/D_m) \tag{47}$$

The intensity reflection coefficient is approximated as

$$V(\theta) = \exp(-\alpha\theta/2) \tag{48}$$

Based on small angle approximation, $\tan\theta \approx \theta$, and using eq.(43) and eq.(48), eq.(47) can be re-expressed as

$$\exp(-2\beta_m r) = \exp(-\alpha r\theta_m^2/2H) = \exp(-\frac{\alpha r\theta_m^2}{2H}) \tag{49}$$

By using the dispersion equation, the relation between θ_m and m is approximately given by [16].

$$\theta_m \approx \frac{m\pi}{k_0 H}, \quad d\theta_m \approx \frac{\pi}{k_0 H} dm, \quad dm \approx \frac{k_0 H}{\pi} d\theta_m \tag{50}$$

Using eq.(44),(45),(49),(50), the averaged reverberation intensity of the normal mode reverberation in shallow water, eq.(34) can be written as

$$I_R = 16 \cdot I_0 \frac{\pi c\tau_0}{H^2 r}\int_0^{\theta_c}\int_0^{\theta_c} \exp(-\alpha r\theta_i^2/2H)[\Theta_{mn}]^{\text{int}} \exp(-\alpha r\theta_o^2/2H)d\theta_i d\theta_o \tag{51}$$

where $I_0 = E_0/\tau_0 = s^2(t - t_c)$.

In the energy flux reverberation model, the reverberation can be written as[7]

$$I = I_0 \frac{\pi c\tau_0}{H^2 r}\int_0^{\theta_c}\int_0^{\theta_c} \exp(-\alpha r\theta_i^2/2H) M_b \exp(-\alpha r\theta_o^2/2H)d\theta_i d\theta_o \tag{52}$$

Comparing eq.(51) and eq.(52), we can find that the two reverberation theories are equal, provided:

$$16 \cdot [\Theta_{mn}]^{\text{int}} = M_b \quad (53)$$

This result is the same as the result in reference [16]. And eq.(52) and eq.(53) are the new energy-flux model of waveguide reverberation based on Perturbation theory.

Shang[18] has proposed that in a small grazing angle area, the bottom-reflection coefficient $V(\theta)$ can be simplified by two parameters P and Q

$$\ln V(\theta) \approx -\mathrm{Q}\theta \quad (54)$$

$$\arg V(\theta) \approx -\pi + \mathrm{P}\theta \quad (55)$$

In Pekeris waveguide, P was difined in eq.(24)

$$\mathrm{Q} \approx 2[(\rho_b/\rho_0)(c_0/c_b)2\eta] / [1-(c_0/c_b)^2]^{3/2} \quad (56)$$

$$\eta = \frac{Kc_b}{2\pi(8686)} \quad (57)$$

where K is the attenuation coefficient of sediment(dB/kHz-m), the unit of sound speed in sediment c_b is m/s.

In the normal mode theory, eigenvalue k_m and eigenfunction $\varphi_m(z)$ are affected by parameter P. And mode attenuation β_m is affected mainly by parameter Q . That is to say, P controls the phase information of the sound field, and Q controls the energy of the sound field.

Considering eq.(48), eq.(52), eq.(53), eq.(54), the shallow water reverberation model based on P and Q can be written as,

$$I(r) = I_0 \frac{\pi c\tau}{H^2 r} \int_0^{\theta_c} \int_0^{\theta_c} \exp[-\mathrm{Q}(f) r\theta_i^2 / H] M_b(\theta_i, \theta_o) \exp[-\mathrm{Q}(f) r\theta_o^2 / H] d\theta_i d\theta_o \quad (58)$$

$$M_b(\theta_i, \theta_o) = \mu(f)\sin^2[\theta_i \mathrm{P}(f)/2]\sin^2[\theta_o \mathrm{P}(f)/2] \quad (59)$$

where

$$\mu(f) = 16 \cdot \sigma_\eta^2 P^\eta (2k_0)[k_0^2 c^R]^2 \quad (60)$$

Ⅳ NUMERICAL SIMULATION ON A PEKERIS WAVEGUIDE

In the numerical simulation, the environmental parameters were listed in Figure 2: a Pekeris waveguide with water depth $H = 50$ m, sound speed in water $c_0 = 1500$ m/s, sound speed in sediment $c_b = 1623$ m/s, the density in sediment $\rho_b = 1.77$ g/cm^3, and the attenuation in sediment $\alpha_b = 0.673$ dB/kHz-m. For rough interface, consider the RMS value of roughness $\sigma = 0.1$ m, the space correlation scale $L = 10$ m.

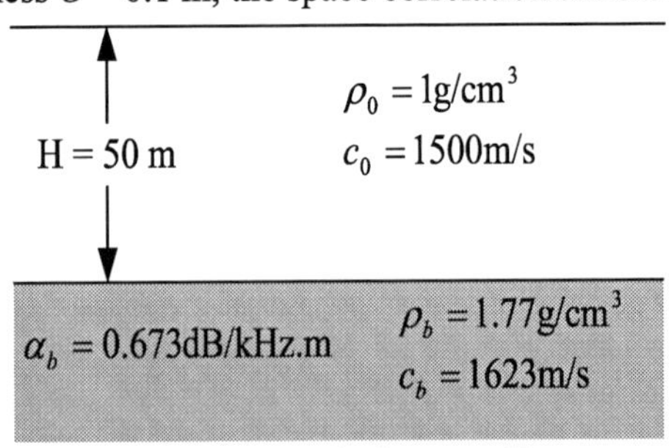

FIGURE 2. Pekeris waveguide

According to the PQ definition in reference [18], P = 9.27 and Q = 0.88 for the Pekeris waveguide shown in Figure 2. The reverberation data was calculated based on the theories listed in Section II and Section III. Figure 3 and Figure 4 are the comparison between a modal reverberation model and an energy-flux reverberation model corresponding to fc = 300 Hz and fc = 900 Hz respectively. In the comparison, source depth and the receiver depth are both 25 m. The two theories agree well in most part of the water, except near the boundary; for example, source depth and receiver depth are both 5 m, see Figure 5. However, the two theories agree well in most parts of the water column.

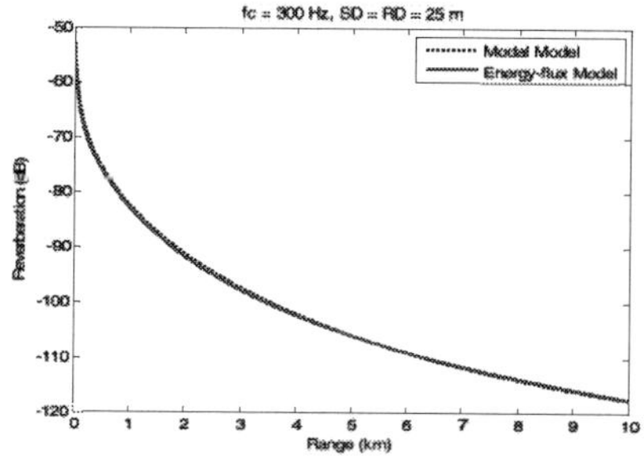

FIGURE 3. Comparison between modal model and energy-flux model(fc = 300 Hz)

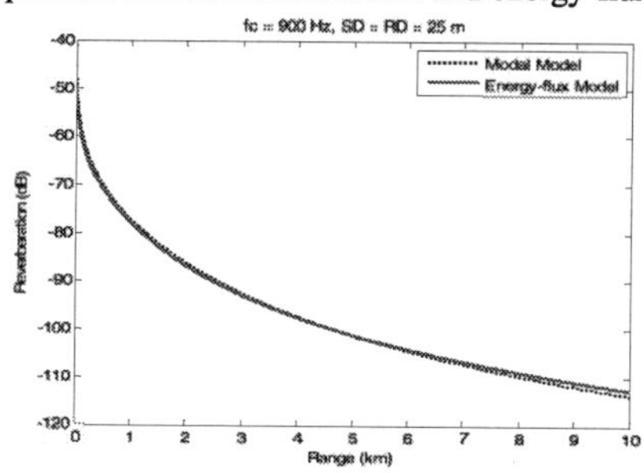

FIGURE 4. Comparison between modal model and energy-flux model(fc = 900 Hz)

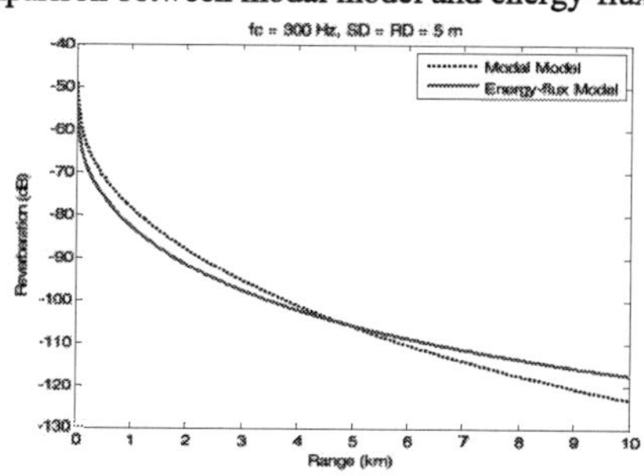

FIGURE 5. Comparison between modal model and energy-flux model near boundary

V CONCLUSION AND DISCUSSION

A new shallow water reverberation model based on bottom reflection parameters was given in this paper. The new model has only three environmental parameters (P,

Q and μ). P illustrates the angular dependency of backscattering; Q controls the reverberation intensity decaying; and μ shows the backscattering strength in the model. The three parameters can be inverted easily. However, the new model only considers average reverberation intensity, and has neglected many factors such as source depth, receiver depth, and so on. And further work should be done on this new model in the future. For example, it should be extended to non-Pekeris waveguides.

ACKNOWLEDGMENTS

This work was supported by the National Natural Science Foundation of China under No. 10604062 and No. 10874201.

REFERENCES

1. H.P. Bucker and H.E. Morris, "Normal-mode reverberation in channels or ducts," J. Acoust. Soc. Am. 44, pp.827-828 (1968).
2. R.H. Zhang and G.L. Jin, "Normal mode theory of average reverberation intensity in shallow-water," J. Sound and Vib, 119, pp.215-223 (1987).
3. D.D. Ellis, "A shallow-water normal mode reverberation model," J. Acoust. Soc. Am.,97, pp.2804-2814 (1995).
4. K. LePage, "Bottom reverberation in shallow-water: coherent properties as a function of bandwidth, waveguide characteristics and scatterer distribution," J. Acoust. Soc. Am., 106, pp.3240-3254 (1999).
5. V.A.Grigor'er, V.M.Kuz'kin, and B.G.Petnikov, "Low-frequency bottom reverberation in shallow-water ocean regions," Acoustical Physics, Vol.50 (1), pp.37–45 (2004).
6. J.X. Zhou, D. Guan , E.C. Shang and E. Luo, "Long-range reverberation and bottom scattering in shallow-water, Chinese J. Acoustics, 1, pp. 54-63 (1982).
7. C.H. Harrison, "Closed-form solutions for reverberation and signal excess with mode-stripping and Lambert's law," J. Acoust.Soc.Am., 114, pp.2744-2756 (2003).
8. B.H. Tracy and H. Schmidt, "Seismo-Acoustic field statistics in shallow water," IEEE, JOE, 22, pp.317-331 (1997).
9. G.Bouchage, K.D.LePage, "A shallow-water reverberation PE model," Proceedings of the Sixth European Conference on Underwater Acoustics, ECUA'2002, pp.125-130, (2002).
10. C,W. Holland, "Constrained comparison of ocean waveguide reverberation theory and observation," J. Acoust. Soc. Am.,120, pp.1922-1931 (2006).
11. G. L. Jin, R. H. Zhang and X. F. Qiu, "Characteristics of shallow water reverberation and inversion for bottom properties", Proceedings of SWAC, Ed. Zhang and Zhou, pp. 303-308 (1997).
12. T.F. Gao, "Relation between waveguide and non-waveguide scattering from a rough interface," Acta Acust. 14, pp.126-132 (1989).
13. D.J. Tang, "Shallow-water reverberation due to sediment volume inhomogeneities," in Shallow-water Acoustics, Ed. by R.H. Zhang and J.X. Zhou, pp.323-328 (1997).
14. A. N. Ivakin, "A unified approach to volume and roughness scattering", J. Acoust. Soc. Am. 103, 827-837 (1998).
15. E.C.Shang, T.F.Gao and J.R.Wu, "A shallow-water reverberation model based on Perturbation theory," IEEE, Journal of Oceanic Engineering, Vol.33(4), pp.451-461 (2008).
16. E.C. Shang, "Some new challenges in shallow water acoustics," in Progress in Underwater Acoustics, Ed. by H.M.Merklinger, Plenum, New York (1987).
17. J.A. Goff and T. Jordan, "Stochastic modeling of seafloor morphology : Inversion of sea beam data for second order statistics,"JGR, 93, pp. 13589-13608 (1993).
18. T.C. Wang and E.C. Shang, Underwater Acoustics (in Chinese), Scientific Publisher, Beijing China (1981).

T-matrix formulation of scattering by an obstacle near a planar sediment boundary and application to detection using iterative time reversal

Yingzi Ying and Li Ma

Institute of Acoustics, Chinese Academy of Sciences, 21 Beisihuanxi Road, Beijing 100190, China
yingyz05@mails.gucas.ac.cn

Abstract. In this paper, the transition matrix (abbreviated to T-matrix) formulation for scattering is adapted to the case of an obstacle immersed in a fluid half-space and in proximity to a planar interface. The formulation indicates the multiple scattering effects between the obstacle and interface, which may result in super-resonant phenomena. We investigate the application of single-channel iterative time reversal to mitigate the difficulty presented by interface reverberation by exploiting the target resonances.

Keywords: transition matrix, multiple scattering, iterative time reversal
PACS: 43.20.Fn, 43.60.Tj

INTRODUCTION

The transition (T-) matrix method, which is an efficient computational scheme for analyzing the scattering of waves for difficult geometries, is originally introduced by Waterman [1] in acoustics and electro-magnetics [2], and then is extended to the elastic scattering circumstance by Varatharajulu and Pao [3]. The matrix maps the vector of the expansion coefficients of the incident field to that of the scattered field. Upon selecting the complete orthonormal basis set, the T-matrix is only related to the geometries and the material parameters. Here the scheme is adapted to the case of acoustic scattering by an elastic obstacle immersed in a fluid half-space and in proximity to a planar interface.

The presence of interface reverberation hinders the detection of a target resting on the sea floor. The iterative time reversal process, in which the time reversed echo is adopted as the renewed interrogation pulse for the next iteration, will lead echoes to converge to a narrowband signal that corresponds to a scattering object's dominant resonance mode and thus mitigate the difficulty by exploiting the resonances. An at-sea experiment in bottom target detection using single channel iterative time reversal has been performed in the Yellow Sea, China, and the results illustrate the feasibility of the method.

CP1272, *Shallow-Water Acoustics, Second International Shallow-Water Acoustics Conference*
edited by J. Simmen, E. S. Livingston, J. X. Zhou, and F. H. Li

T-MATRIX FOR SCATTERING OF AN OBSTACLE NEAR A PLANAR INTERFACE

A review of the scheme for single obstacle scattering in free space

We first take the case of stationary acoustic scattering by a single elastic obstacle D in an inviscid fluid, and the scheme of the T- matrix formulation is as follows:

1. Give integral representations of fields both inside and outside the obstacle.
2. Select basis functions and expand the incoming and scattered fields, as well as various Green's tensors.
3. Apply the boundary conditions and remaining unknown surface fields.

The exterior pressure field and interior displacement field can be expressed by the Helmholtz-Kirchhoff integral equation representations with null-field hypothesis [4, 5]

$$\int_{\partial D} [p_+ \nabla' g - (\nabla' p_+) g] \cdot \mathrm{d}s' = \begin{cases} p - p^{\mathrm{in}} = p^{\mathrm{sc}}, & r \in \mathbb{R}^3 \setminus \bar{D}, \\ -p^{\mathrm{in}}, & r \in D, \end{cases} \tag{1}$$

$$-\frac{k_{\mathrm{s}}}{\rho\omega^2} \int_{\partial D} [u_- \cdot (n \cdot \Sigma) - t_- \cdot G] \, \mathrm{d}s' = \begin{cases} u, & r \in D, \\ 0, & r \in \mathbb{R}^3 \setminus \bar{D}, \end{cases} \tag{2}$$

where g is the scalar Green's function, and G and Σ are the Green's displacement dyadic and Green's stress triadic for an unbounded region, respectively.

The solid-fluid boundary conditions for the elastic object immersed in the fluid are the continuity of the normal components of displacement and surface traction, and the inner tangential of surface traction vanishes:

$$n \cdot u_- = n \cdot u_+, \; n \cdot t_- = n \cdot t_+, \; t_{-\tan} = 0, \quad \text{on } \partial D, \tag{3}$$

where n is the outward pointing normal to the surface, and a plus and minus sign denote the limiting value taken from the outside and inside, respectively.

The spherical scalar partial wave basis for pressure expansion in the fluid

$$\phi_{\sigma mn}(r) = Y_{\sigma mn}(\theta, \phi) h_n(k_0 r), \tag{4}$$

and the spherical vector partial wave basis for particle displacement expansion in the elastic body

$$\psi_{1\sigma mn}(r) = [n(n+1)]^{-1/2} \nabla \times [r Y_{\sigma mn}(\theta, \phi) h_n(kr)], \tag{5}$$

$$\psi_{2\sigma mn}(r) = (1/k_{\mathrm{s}}) \nabla \times \psi_{1\sigma mn}(r), \tag{6}$$

$$\psi_{3\sigma mn}(r) = (k_{\mathrm{p}}/k_{\mathrm{s}})^{3/2} (1/k_{\mathrm{p}}) \nabla [Y_{\sigma mn}(\theta, \phi) h_n(kr)], \tag{7}$$

can be selected as basis functions with a differential and normalized notation. Here h_n is the outgoing spherical Hankel function, and

$$Y_{\sigma mn}(\theta, \phi) = \left[\varepsilon_m \frac{2n+1}{4\pi} \frac{(n-m)!}{(n+m)!} \right]^{1/2} \mathrm{P}_n^m(\cos\theta) \begin{cases} \cos m\phi, & \sigma \text{ is even} \\ \sin m\phi, & \sigma \text{ is odd} \end{cases} \tag{8}$$

is the normalized real spherical harmonics, where ε_m is the Neumann factor that $\varepsilon_0 = 1$ and $\varepsilon_m = 2$ for $m > 0$, and P_n^m is the associated Legendre polynomial.

We shall suppress indices to simplify the notations:

$$\phi_{\sigma mn} \equiv \phi_n, \tag{9}$$

$$\psi_{\tau\sigma mn} \equiv \psi_n, \tag{10}$$

and the sets $\{\phi_n\}$ and $\{\psi_n\}$ are complete [1, 2]. Both the incident and secondary scattered pressure fields outside the obstacle, and the displacement field inside the obstacle can now be expanded as

$$p^{\mathrm{in}} = \sum_n a_n \operatorname{Re} \phi_n, \tag{11}$$

$$p^{\mathrm{sc}} = \sum_n f_n \phi_n, \tag{12}$$

$$u = \sum_n b_n \operatorname{Re} \psi_n. \tag{13}$$

The operator Re denotes the regular part of the bases that take spherical Bessel rather than Hankel Functions to avoid singular value at the origin. Our goal is to figure out the T-matrix that maps the vector of the expansion coefficients of the incident field to that of the scattered field

$$T: a \mapsto f. \tag{14}$$

We expand the scalar Green's function g in the spherical scalar wave functions

$$g(r,r') = ik_0 \sum_n \phi_n(r \vee r') \operatorname{Re} \phi_n(r \wedge r'), \tag{15}$$

and the Green's dyadic G can be expanded in the spherical vector wave functions

$$G(r,r') = i \sum_n \psi_n(r \vee r') \operatorname{Re} \psi_n(r \wedge r'). \tag{16}$$

By inserting the expansions of incident pressure p^{in}, scattered pressure p^{sc} and the Green's function g into the pressure integral representation in Eq. (1), after identifying coefficients we get

$$a_n = -i\lambda_0 \int_{\partial D} \left[p_+ \nabla' \phi_n - (\nabla' p_+)\phi_n\right] \cdot \mathrm{d}s', \tag{17}$$

$$f_n = -i\lambda_0 \int_{\partial D} \left[p_+ \nabla' \operatorname{Re} \phi_n - (\nabla' p_+) \operatorname{Re} \phi_n\right] \cdot \mathrm{d}s'. \tag{18}$$

Similarly, by using the expansions for displacement u, the Green's displacement dyadic G, and the Green's stress triadic Σ into the displacement integral representation in Eq. (2), then

$$b_n = -i \frac{k_{\mathrm{s}}}{\rho \omega^2} \int_{\partial D} \left[u_- \cdot t(\psi_n) - t_- \cdot \psi_n\right] \mathrm{d}s', \tag{19}$$

$$0 = -i \frac{k_{\mathrm{s}}}{\rho \omega^2} \int_{\partial D} \left[u_- \cdot t(\operatorname{Re} \psi_n) - t_- \cdot \operatorname{Re} \psi_n\right] \mathrm{d}s'. \tag{20}$$

Since the displacement field is not the quantity of interest, Eq. (19) will not be used. The interior surface displacement field and the exterior surface pressure field can be expanded as [6],

$$u_- = \sum_n \alpha_n \operatorname{Re}\psi_n + \beta_n \psi_n, \tag{21}$$

$$p_+ = \sum_n \gamma_n \operatorname{Re}\phi_n. \tag{22}$$

Using the surface field expansions and the boundary conditions in Eq. (17), Eq. (18) and Eq. (20), we get

$$\begin{cases} f & = i\operatorname{Re} Q^1\alpha + i\operatorname{Re} Q^1\beta, \\ a & = -iQ^1\alpha - iQ^1\beta, \\ 0 & = Q^2\gamma + Q^3\alpha. \end{cases} \tag{23}$$

By solving the set of the equations, we get

$$f = \underbrace{-\operatorname{Re} QQ^{-1}}_{T} a, \tag{24}$$

where $Q = Q^1(Q^3)^{-1}Q^2$, and the entries of the matrices are

$$Q^1_{nn'} = \frac{k_0^2}{\rho_0\omega^2}\int_{\partial D}[n\cdot\operatorname{Re}\psi_{n'}\lambda_0\nabla\cdot\nabla\phi_n - n\cdot t(\operatorname{Re}\psi_{n'})n\cdot\nabla\phi_n]\,\mathrm{d}s', \tag{25}$$

$$Q^2_{nn'} = \frac{k_s^3}{\rho\omega^2}\int_{\partial D}\lambda_0 n\cdot\nabla(\operatorname{Re}\phi_{n'})n\cdot t(\operatorname{Re}\psi_n)\mathrm{d}s', \tag{26}$$

$$Q^3_{nn'} = \frac{k_s^3}{\rho\omega^2}\int_{\partial D}[\operatorname{Re}\psi_{n'\tan}\cdot t(\operatorname{Re}\psi_n) - n\cdot t(\operatorname{Re}\psi_{n'})n\cdot\operatorname{Re}\psi_n]\,\mathrm{d}s', \tag{27}$$

respectively.

To sum up, the T-matrix computers the expansion coefficients of the scattered field given that of the incident field. Upon selecting the complete orthonormal basis sets, the T-matrix is only related to the geometry and material parameters.

Adapted to the case near a planar interface

For an elastic object resting on the planar sediment interface, as depicted in Fig. (1), the integral representation Eq. (1) can be adapted by changing the integration domain:

$$\int_{\partial D_1\cup\partial D_2}[p_+\nabla' g - (\nabla' p_+)g]\cdot\mathrm{d}s' = \begin{cases} p - p^{\mathrm{in}} = p^{\mathrm{sc}}, & r\in\mathbb{R}^3\setminus(\bar{D}_1\cup\bar{D}_2), \\ -p^{\mathrm{in}}, & r\in D_1\cup D_2, \end{cases} \tag{28}$$

We assign the position of the transceiver as the origin of the whole system, and the scattered field can be expanded as follows:

$$p^{\mathrm{sc}} = \sum_n f_n \operatorname{Re}\phi_n. \tag{29}$$

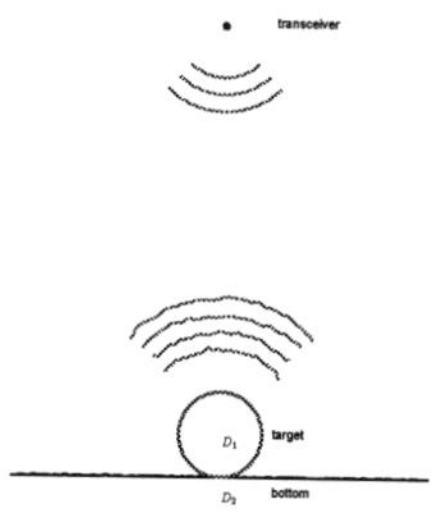

FIGURE 1. Geometry of the at-sea experiment. The blue asterisk denotes the transceiver, and the elastic object is labeled as D_1, while D_2 is the sea bottom.

Here we define the translation properties for the scalar spherical wave function

$$\phi_n(r+r') = \sum_{n'} \sigma_{nn'}(r')\phi_{n'}(r), \tag{30}$$

$$\operatorname{Re}\phi_n(r+r') = \sum_{n'} R_{nn'}(r') \operatorname{Re}\phi_{n'}(r). \tag{31}$$

By inserting the expansions of the incident pressure, scalar Green's function and the translation properties into Eq. (28), and identifying the coefficients of the regular functions, we get

$$R^{-1}(r_0^i)a = a^i - \sum_{j\neq i} \sigma(-r_0^i + r_0^j) f^j, \quad \forall i,j = 1,2, \tag{32}$$

where r_0^i, a^i and f^i are the positions of origins at the system coordinate, expansion coefficients of incident and scattered fields both for elastic obstacle ($i=1$) and sediment ($i=2$), respectively. By using the translation of the coefficients of the scattered field

$$f = \sum_{i=1,2} R(r_0^i) f^i, \tag{33}$$

together with Eq. (32), we get the total T-matrix for the elastic object resting on the interface

$$T = \sum_{i,j=1,2} R(r_0^i)T^i \left[\delta_{mn} - \sum_{n\neq m} \sigma\left(-r_0^m + r_0^n\right) T^n\right]_{n,m=1,2} R^{-1}(r_0^j), \tag{34}$$

where T^i is the T-matrix of the sediment or the obstacle, that is $f^i = T^i a^i$, for $i = 1,2$.

To sum up, the total T-matrix for scattering of an obstacle near a planar interface depends on the geometry and type of target, boundary-target separation, sediment type, and the orientation of the target.

DETECTION USING ITERATIVE TIME REVERSAL

The iterative time reversal is like acoustic ping-pong, because it can

1. Illuminate the target with a broadband interrogation pulse.

2. Measure and window backscattered wave and apply a bandpass filter.
3. Reverse the data stream in time and re-transmit it.
4. Repeat the above procedures iteratively.

We use the iterative time-reversal mirror to automatically generate a matched filter for the spectral response of the target scattering, rather than spatial focusing [7].

For an omnidirectional point source located at r_s emanating unit amplitude monochromatic signal, the incident pressure field is

$$p^{\mathrm{in}}(r)=\frac{\exp(\mathrm{i}k_0|r-r_\mathrm{s}|)}{|r-r_\mathrm{s}|},\quad r\in\mathbb{R}^3\setminus\{r_\mathrm{s}\},\ \text{and}\ r_\mathrm{s}\in\mathbb{R}^3\setminus(\bar{D}_1\cup\bar{D}_2),\tag{35}$$

then the expansion coefficients of the incident pressure are

$$a_n=\left(\operatorname{Re}\phi_n,p^{in}\right)=\int_{[0,2\pi]\times[0,\pi]}\operatorname{Re}\phi_n p^{in}\mathrm{d}\Omega.\tag{36}$$

The inner product is defined as the integration evaluated by the solid angle in the full space.

The single-channel in-out LTI system transfer function equals to the scattered pressure field received at the same position, that is

$$\begin{aligned}h&=p^{sc}\\&=f^{\dagger}\phi\\&=(Ta)^{\dagger}\phi\\&=\left(T\left(\operatorname{Re}\phi,p^{in}\right)\right)^{\dagger}\phi\end{aligned}\tag{37}$$

The scattering of a target near a planar waveguide boundary is a multiple scattering process, which will result in the super-resonant phenomenon [8], and approach greatest return near normal incidence.

Assume e_0 is the initial interrogation pulse, then received signal is e_0h, and time reversed version of the first iteration is $e_0^*h^*$, and of the second is e_0hh^*, and so on. So the renewed input signal for the next emission after n^{th} iteration is

$$e_n(\omega)=\begin{cases}\left(e_0^*|h|^{n-1}h^*\right)(\omega), & \text{if } n \text{ is odd},\\ \left(e_0|h|^n\right)(\omega), & \text{if } n \text{ is even}.\end{cases}\tag{38}$$

The amplitude is $|e_n|=|e_0||h|^n$ for both cases, and its normalized form will converge to

$$\frac{|e_n(\omega)|}{\max\{|e_n(\omega)|\}}\to\mathbf{1}_{\{\omega:\,\omega=\omega_0\}}(\omega)\ \text{as}\ n\to\infty,\tag{39}$$

where the indicator function of set A is $\mathbf{1}_A(\omega)=1$ if $\omega\in A$, and $\mathbf{1}_A(\omega)=0$ if $\omega\in A^{\mathrm{c}}$. The converged-upon frequency ω_0 corresponds to a maximum of the amplitude of LTI system transfer function,

$$\omega_0:=\underset{\omega\in[\omega_1,\omega_2]}{\arg\max}\,|h(\omega)|\tag{40}$$

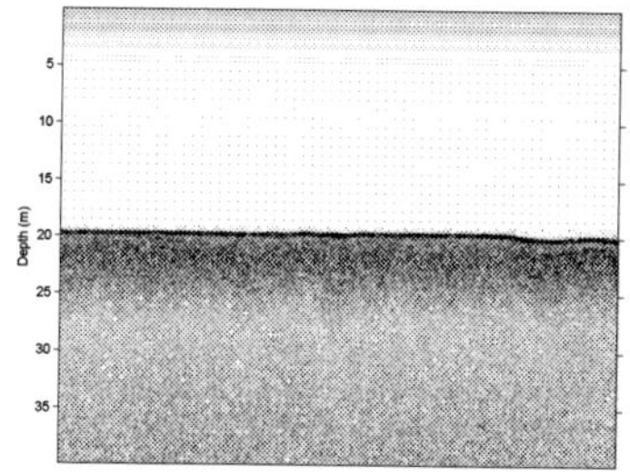

FIGURE 2. Sub-bottom profile of the experimental sea area.

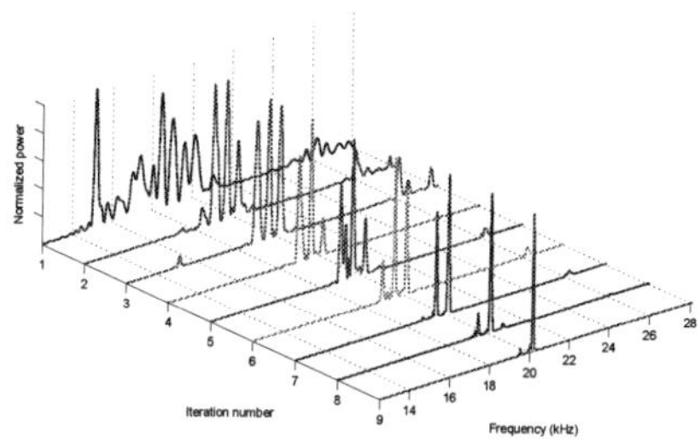

FIGURE 3. Echo frequency components of the first nine iterations.

The frequency interval $[\omega_1, \omega_2]$ is limited by the bandpass filter, and we always select it in the band of a relatively flat domain of transceiver frequency response, which may also be calibrated.

An at-sea experiment in bottom target detection using single-channel iterative time reversal has been performed in the Yellow Sea, China. The cylindrical-like shell 53 cm in external diameter and 260 cm long filled with concrete is resting on the sea floor, and Fig. (2) is the sub-bottom profile.

A 2-cycle CW of 15 kHz center frequency is emanated as the initial interrogation pulse. Fig. (3) illustrates the normalized spectra of the echoes of nine iterations. At the beginning of the process, the echo frequency content is expanded, and as the iteration number increases, the echoes will gradually converge to a single-frequency signal. The response peak of the transceiver is avoided by adjusting the bandpass filter in the flat response interval.

SUMMARY

The obstacle-boundary multiple scattering will change the appearance of the target scattering strength (form function). The iterative process involving a single-channel time reversal mirror will automatically turn the frequency of the interrogation signal to a best one that matches the target scattering. In the absence of the target, the same procedure will not lead to resonance convergence, as the flat sea floor has no dominant sharp frequency response. We judge the existence of the target with this acoustic signature.

ACKNOWLEDGMENTS

The work was supported by the Innovation Foundation of Chinese Academy of Sciences through grant CXJJ-260.

REFERENCES

1. P. C. Waterman, *J. Acoust. Soc. Am.* **45**, 1417–1429 (1969).
2. P. C. Waterman, *Phys. Rev. D* **3**, 825–839 (1971).
3. V. Varatharajulu, and Y.-H. Pao, *J. Acoust. Soc. Am.* **60**, 556–566 (1976).
4. B. B. Baker, and E. T. Copson, *The Mathematical Theory of Huygens' Principle*, Chelsea Pub Co., 3 Sub edition (April 1988).
5. Y.-H. Pao, and V. Varatharajulu, *J. Acoust. Soc. Am.* **59**, 1361–1371 (1976).
6. C. Müller, *Symposia Mathematica* **XVIII**, 353 (1976).
7. C. Prada, J.-L. Thomas, and M. Fink, *J. Acoust. Soc. Am.* **97**, 62–71 (1995).
8. G. C. Bishop, and J. Smith, *J. Acoust. Soc. Am.* **105**, 130–143 (1999).

Adaptive Bubble Pulse Cancellation and Its Applications

Feng-Xiang GE, Yan ZHANG, ZhengLin LI and Renhe ZHANG

State Key Laboratory of Acoustics, Institute of Acoustics, Chinese Academy of Sciences, Beijing 100190

Abstract. Explosive sources are widely used as sound sources in underwater acoustics experiments. Unfortunately, the resulting direct pulse is often corrupted by bubble pulses, which has an undesired influence on the observation and its applications. Canceling these bubble pulses will greatly enhance the quality of the observation and will be helpful for its applications. To cancel the bubble pulse, an adaptive filter, updated from an adaptive least squares algorithm in order to model the physical transmission from the explosive source to the receiver, is proposed. The auto-correlation functions of both the original field trial data and the filtered results show the bubble pulses in the observation can be effectively cancelled. The normal modes of signal propagation can be reconstructed from the filtered results with an adaptive time-frequency representation. The estimated time delay of different normal modes is further applied in geo-acoustic inversion.

Keywords: Bubble Pulse, Adaptive Filtering, Geo-acoustic Inversion.
PACS: 43.60.Mn, 43.60.Pt

INTRODUCTION

Explosive sources are now widely used as sound sources in underwater acoustics experiments [1]-[7], due to their many attractive features, e.g. high source-level, wide bandwidth, etc. Unfortunately, the resulting direct pulse is often corrupted by bubble pulses, which can have an undesirable influence on the observation and its applications. Canceling these bubble pulses (see [2]-[4] and references therein) will greatly enhance the quality of the observation and be helpful for its applications.

In this paper, an adaptive bubble pulse cancellation is proposed, where the adaptive filter is used for bubble pulse cancellation. The adaptive filter is typically updated by using an adaptive least squares algorithm to model the physical transmission from the explosive source to the receiver. The correlation functions on both the original field trial data and the filtered results show the bubble pulses in the observation can be effectively cancelled. The normal modes of the filtered results are reconstructed in its time-frequency representation, which also shows that the bubble pulse is successfully cancelled from the observation.

Bubble pulse cancellation is important as a technical approach for the study of ocean channels and geo-acoustic inversion by using explosive signals. In this paper, the filtered result is further used for a geo-acoustic inversion based on the dispersion

CP1272, *Shallow-Water Acoustics, Second International Shallow-Water Acoustics Conference*
edited by J. Simmen, E. S. Livingston, J. X. Zhou, and F. H. Li

characteristic of normal modes. It is shown the geo-acoustic parameters can be successfully inverted with the experimental data.

The paper is organized as follows. Section II describes the observation and the involved bubble pulses in underwater acoustics experiments. In Section III, we present the solution to cancel the bubble pulses. Results show that the bubble pulses corrupted in the observation can be effectively cancelled. An application in geo-acoustic inversion is addressed in Section IV. Conclusions are reported in Section V.

PROBLEM FORMULATION

Explosive sources are widely used as sound sources in underwater acoustics experiments. The direct pulse in the observation is often corrupted by bubble pulses, which has an undesirable influence on the observation and its applications. These are illustrated in FIGURE 1 and FIGURE 2. FIGURE 1 shows an observation taken close to an explosive source, where the direct pulse and bubble pulses are distinctly separated in the time domain. However, with the increase of propagation distance, the direct pulse in the observation will be corrupted by bubble pulse due to dispersion. A time-frequency representation of observation far away from the explosive source is illustrated in FIGURE 2.

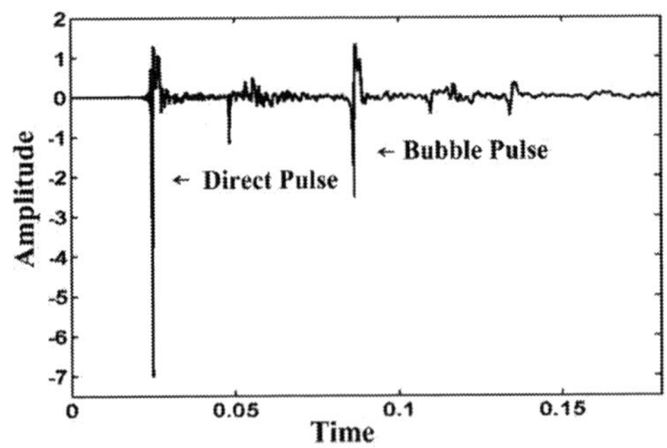

FIGURE 1. Observation taken close to an explosive source

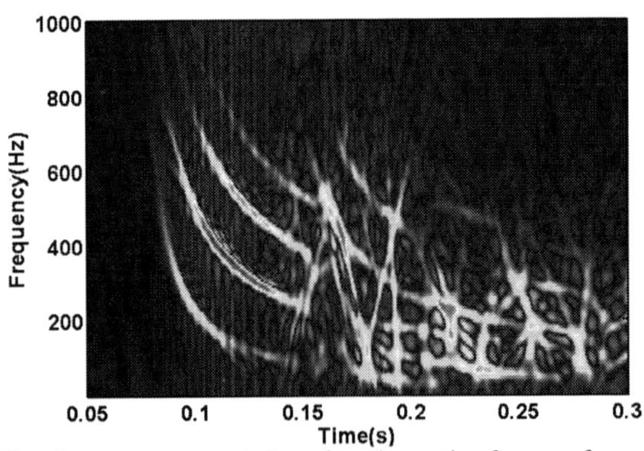

FIGURE 2. Time-frequency representation of an observation far away from an explosive source

Generally, the observation $r(t)$ can be described as:

$$r(t) = x(t) + \sum_{i=1}^{N} \lambda_i \cdot x(t-\tau_i) + n(t), \quad i = 1,2,\cdots N, \tag{1}$$

where $x(t)$ is the direct pulse, λ_i and τ_i are the random multi-path amplitudes and time delays corresponding to different bubble pulses, which are the scatter characteristics and propagation fading, $n(t)$ is the additive white Gaussian noise (AWGN).

In this paper, we hope to cancel the bubble pulse and get the following estimates:

$$\hat{r}(t) = x(t) + \sum_{i=1}^{N} \tilde{\lambda}_i \cdot x(t-\tau_i) + \tilde{n}(t), \quad i = 1,2,\cdots N, \tag{2}$$

where direct pulse $x(t)$ is held and the other bubble pulses are suppressed or even cancelled clearly, in other words, $\tilde{\lambda}_i << \lambda_i$. Thus, it will be of benefit to later applications.

ADAPTIVE BUBBLE PULSE CANCELLATION

A. Adaptive Filtering

Before we present the detailed solution to cancel the bubble pulses adaptively, let's have a brief introduction to adaptive filtering [8], which is a key component in the proposed solution. A flowchart of adaptive filtering is shown in FIGURE 3.

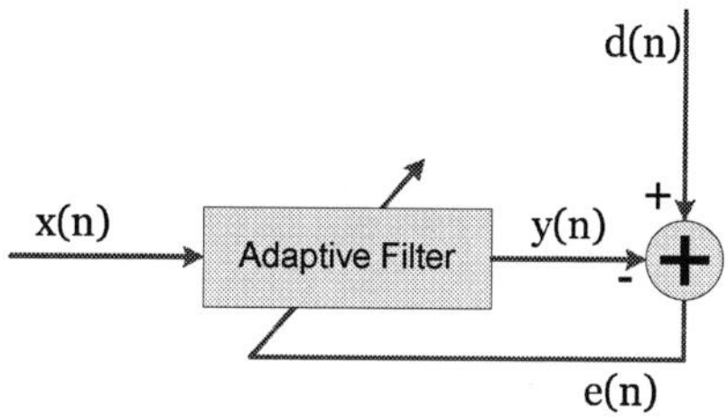

FIGURE 3. Flowchart of adaptive filtering

In FIGURE 3, $\mathbf{x}(n)$ is the input signal vector, $d(n)$ is the reference signal and $e(n)$ is the error signal. Then we have:

$$\begin{cases} e(n) = d(n) - y(n) \\ y(n) = \mathbf{w}^H(n) \cdot \mathbf{x}(n) \\ \mathbf{w}(n+1) = f\{\mathbf{w}, \mathbf{x}, e(n), \boldsymbol{\mu}\} \end{cases}, \tag{3}$$

where $y(n)$ is the output signal and $\mathbf{w}(n)$ is the tap-weight vector of the adaptive filter. The tap-weight vector is updated by using current and previous information, and the function $f\{\mathbf{w}, \mathbf{x}, e(n), \boldsymbol{\mu}\}$ is related to the different adaptive algorithms, such as

LMS, RLS, non-linear least squares, etc. $\boldsymbol{\mu}$ is a step-size parameter vector to adjust the adaptive filtering and H denotes Hermitian transpose.

Adaptive filtering has two important elements. First is a fundamental assumption. That is, the observation may have many components, but always can be divided into two parts. One is of interest and other is to be filtered or cancelled. The reference signal must correlated with one of these two parts; consequently, it is uncorrelated with the other part. The second element is to fix the reference signal, which is the other necessary input of adaptive filtering and always is the desired signal. Its choice always depends on the application. When these two elements are ready, we can get the error signal and use it as feedback to adjust the adaptive filter. Based on different criteria, we can drive the different algorithms and the corresponding structure of adaptive filter.

B. Adaptive Bubble Pulse Cancellation

The observation is shown in FIGURE 4, where the upper plot is the observation from a set of field trial data and the lower is its auto-correlation function estimate. We can see there is another distinct side-lobe, whose peak is at τ_1, besides the main-lobe; this means there is a strong bubble pulse whose time delay is the time difference between these two peaks. Certainly, there are also many bubble pulses in the observation, which are just a bit weaker compared with the strong bubble pulse.

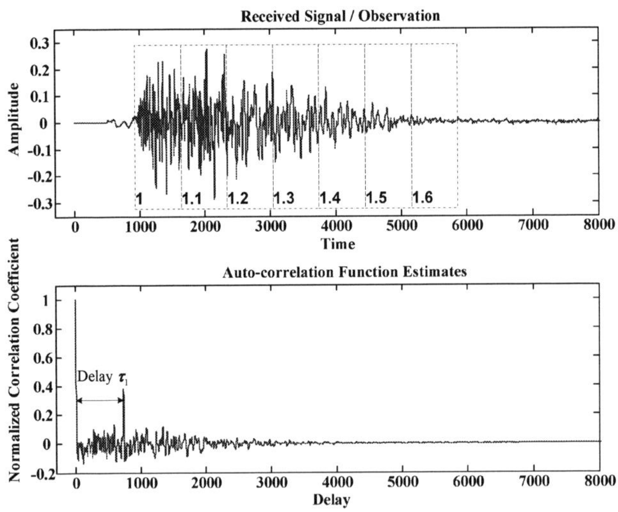

FIGURE 4. Observation and its auto-correlation function estimates

In FIGURE 4, we can notice there are some important features in the auto-correlation function estimates. First, the main-lobe is very narrow, which means the correlation time of the direct pulse and the bubble pulse is short, then we will avoid canceling the direct pulse itself when we can cancel the bubble pulses. Second the time delay between the main-lobe and the strongest side-lobe is larger, which means the reference signal can be taken from the observation before the first bubble pulse.

Always the bubble pulse is uncorrelated with noise. Thus all the necessary elements for adaptive filtering are ready and it can be used for bubble pulse cancellation.

The implementation of the adaptive bubble pulse cancellation is shown in FIGURE 4. First we have a rough estimate of the delay between the direct pulse and the strongest bubble pulse. With the estimated time delay, we can take the segment before the strong bubble pulse in the observation as the reference signal, e.g. the segment in the block 1 in FIGURE 4. Generally, the segment in block 1 is a part of the direct pulse and its duration is a bit less than τ_1 to exclude the strong bubble pulse. With the reference signal, adaptive filtering can be carried out one by one in the following segments, e.g. from the block 1.1 to 1.6. Replicas of bubble pulse segments are expected to be generated in these segments (e.g. block 1.1. to 1.6) of the observation and can be subtracted from the observation at that time of observation. The process will be repeated to cover the whole observation.

In the above adaptive filtering, just a segment of direct pulse is used for a reference signal and thus only the bubble pulse segments which are highly correlated to the selected segment of the direct pulse can be cancelled. In order to further cancel the bubble pulses and improve the quality of the observation, adaptive filtering is sequentially applied to the above filtered results for a new round of adaptive filtering, where the reference signal is taken from the filtered results and then used to generate the replica of the other segments of bubble pulses. Certainly, it depends on the different applications and their requirements.

C. Experimental Results

The results of adaptive bubble pulse cancellation are shown in FIGURE 5, where the upper plot is the filtered observation and the lower plot is its auto-correlation function estimate. It is shown that the strongest bubble pulse is highly suppressed. Furthermore, the other bubble pulses are also suppressed to some extent.

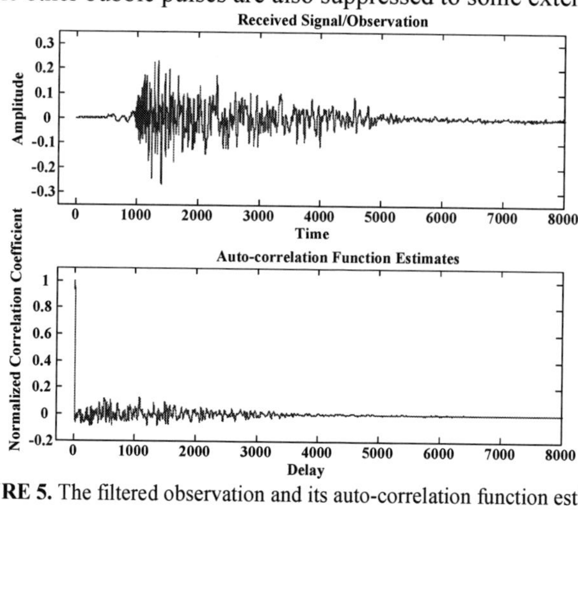

FIGURE 5. The filtered observation and its auto-correlation function estimates

The time-frequency representation of the original observation and the filtered results are shown in FIGURE 6. Comparing FIGURE 6 with FIGURE 2, we can see a much clearer time-frequency representation. It is shown that the bubble pulses in the observation are suppressed effectively.

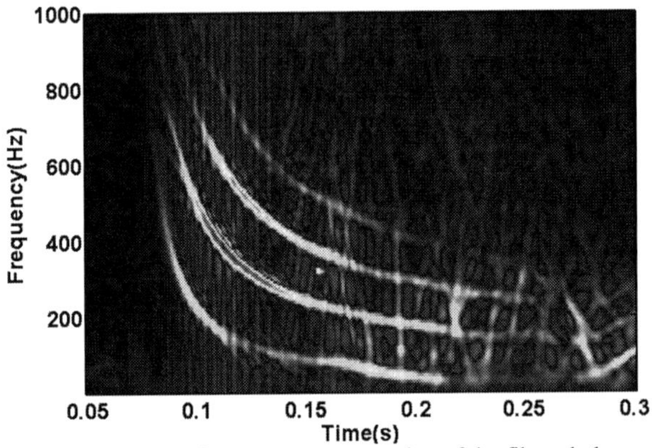

FIGURE 6. Time-frequency representation of the filtered observation

APPLICATION IN GEO-ACOUSTIC INVERSION

As an application of the adaptive bubble pulse cancellation, the filtered results are used for geo-acoustic inversion based on the dispersion characteristic of normal modes [6, 7]. The shallow-water model is shown in FIGURE 7, which consists of a sediment layer and a basement layer.

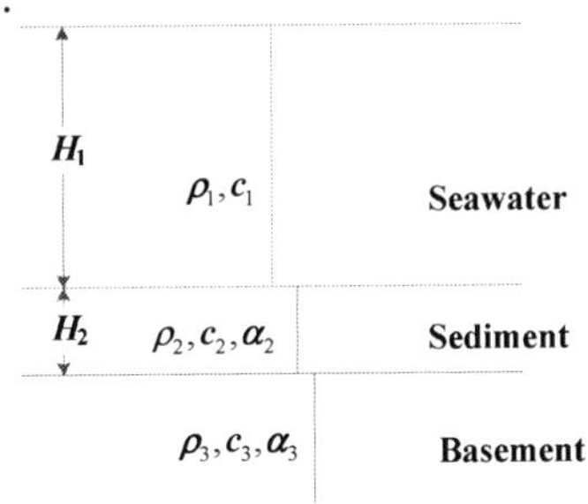

FIGURE 7. Shallow-water model and related parameters

The cost function based on the group delays of normal modes is given by

$$\hat{\boldsymbol{\Omega}} = \underset{(\Omega)}{\arg\min}\left\{\sum_{n,m}\sum_{f}\psi\left(\left[\Delta T_n^{f_i,f_j}, \Delta\widehat{T}_n^{f_i,f_j}\right],\left[\Delta T_{n,m}^{f}, \Delta\widehat{T}_{n,m}^{f}\right]\right)\right\}, \tag{4}$$

where $\Delta\widehat{T}_n^{f_i,f_j}$ is the difference of estimated group delays between the nth mode at frequencies f_i and f_j, and $\Delta\widehat{T}_{n,m}^{f}$ is the difference of estimated group delays between the nth and mth modes at frequency f, $\Delta\widehat{T}_n^{f_i,f_j}$ and $\Delta\widehat{T}_{n,m}^{f}$ can be taken from the time-frequency representation in FIGURE 6. $\Delta T_n^{f_i,f_j}$ and $\Delta T_{n,m}^{f}$ are the corresponding theoretical values and can be expressed analytically by a parameter vector ($\mathbf{\Omega} = (\boldsymbol{H}, \rho, c, \alpha)$), where $\boldsymbol{H}$ is the depth, ρ is the density, $\boldsymbol{c}$ is the sound speed, and $\boldsymbol{\alpha}$ is the absorption coefficient at different layers.

By using the inverted parameters ($\hat{\mathbf{\Omega}}$), the normal mode arrival time T_n^f can be calculated as follows,

$$T_n^f = \left(r/v_n^f - r/v_1^{f_r}\right) + \hat{T}_1^{f_r}, \tag{5}$$

where v_n^f is the group velocity of the nth mode at frequency f and given distance r, f_r is a reference frequency, $v_1^{f_r}$ is the group velocity of the 1st mode at frequency f_r and given distance r, and $\hat{T}_1^{f_r}$ is the estimated arrival time of the nth mode at frequency f_r.

The bottom acoustic parameters are inverted based on the group delays of normal modes; the calculated group delay of the normal modes based on the above inverted parameters is superimposed here as a comparison. By using the inversion scheme described above, it is shown the experimental normal mode arrival times are well matched to the numerical results on the inverted parameters.

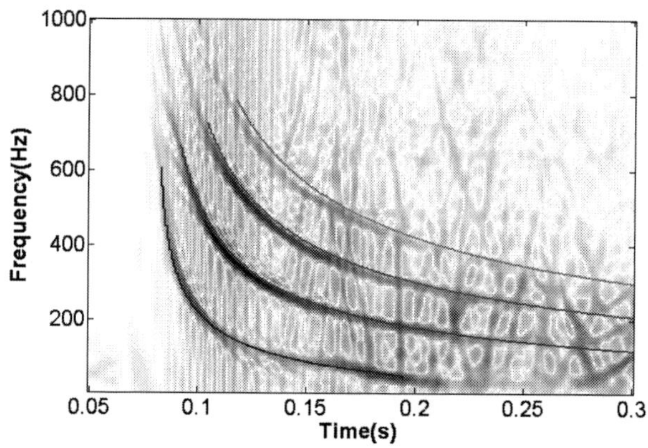

FIGURE 8. Comparison of the experimental and numerical results on the inverted parameters

CONCLUSION

Explosive sources are widely used as sound sources in underwater acoustics experiments. However, the bubble pulses will produce an undesired influence on the observation and its applications. A solution based on adaptive filtering to cancel bubble pulses is proposed, where the reference signals are obtained from the observation and the filtered results in a novel way. The final results on the field trial data show that the bubble pulses in the observation can be effectively cancelled. The filtered results are then applied to the geo-acoustic inversion based on the dispersion characteristic of normal modes. It is shown the geo-acoustic parameters can be successfully inverted.

REFERENCES

[1]. R. H. Cole, *Underwater Explosions*, New Jersey: Princeton University Press, 5-11 (1948).

[2]. Y. Zhang, F.H. Li, Z.L. Li, and R.H. Zhang, "The Method Of Bubble Pulse Removal From Explosive Signals And Its Application," *Acta Acustica*, 34(2), 124-130 (2009).

[3]. J. M. Hovem, "Deconvolution for Removing the Effects of the Bubble Pulses of Explosive Charges," *J. Acoust. Soc. Am.*, 47(1), 281–284(1970).

[4]. M. R. Lackoff and L. R. Leblanc, "Frequency-Domain Seismic Deconvolution Filtering," *J. Acoust. Soc. Am.*, 57(1), 151–159 (1975).

[5]. G. R. Potty, J. H. Miller, P. H. Dahl, and C. J. Lazauski, "Geoacoustic Inversion Results from the ASIAEX East China Sea Experiment," *IEEE J. Ocean. Eng.*, 29(4), 1000–1010 (2000).

[6]. Z. L. Li, R. H. Zhang, "Geoacoustic Inversion based on Dispersion Characteristic of Normal Modes in Shallow Water," *CHIN. Phys. Lett.*, 24(2), 471-474 (2007)

[7]. D. M. Zhang, Z. L. Li and R. H. Zhang, "Inversion For The Bottom Geoacoustic Parameters Based on Adaptive Time-Frequency Analysis," *ACTA Acoust.*, 30(5), 415-419 (2005).

[8]. S. Haykin, *Adaptive Filter Theory* (3rd Edition), Prentice Hall, Inc. (1996).

Signal recovery technique based on a physical method of underwater acoustics

Xinyi Guo, Guoqing Wu and Li Ma

Institute of Acoustics, Chinese Academy of Sciences, Beijing China 100190

Abstract. In the underwater sound channel we often use an array to receive signals from distant sources. The received signals are often mixed with environmental interference. In the complex acoustic environment, received signals are distorted greatly and elongated in time. In many practical applications such as sound communications, sound remote sensing and active sonar signals, we hope to obtain the original signal's waveform. In general theory, the received signals are the convolution of emission signals and Green's function of environment. In unknown Green's function of environment, simply relying on the array to record the information to determine the sound source signal wave propagation features and the environment is not enough. However, in certain circumstances, based on a physics method of underwater acoustics, the spread of recovery technology is successful.

Keywords: Signal recovery; Unknown environment; Estimate Green's function

PACS: 43.30.Wi

INTRODUCTION

When transmitting in the ocean acoustic wave guide, a sound signal produces distortion and a waveform is spread and elongated in the time domain. This paper introduces a method using a vertical hydrophones array to recover the original signal from the distortion signal with unknown environment parameters.

CP1272, *Shallow-Water Acoustics, Second International Shallow-Water Acoustics Conference*
edited by J. Simmen, E. S. Livingston, J. X. Zhou, and F. H. Li

THEORY OF ARTIFICIAL TIME REVERSAL DECONVOLUTION

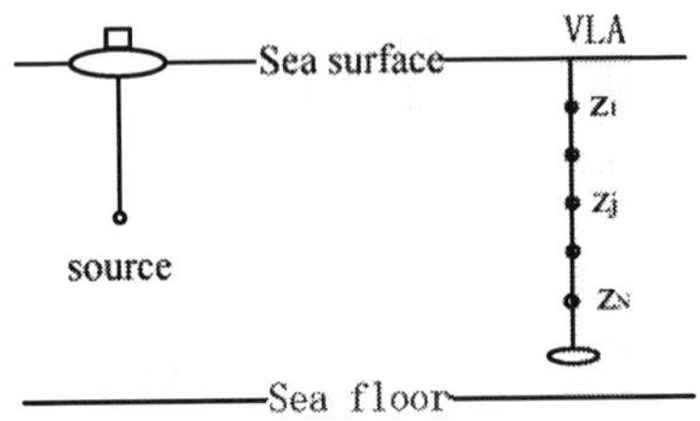

FIGURE 1. Sound source and receiver vertical array geometry location

Figure.1 shows that there is a point source transmitting an acoustic signal in the ocean, and the vertical line array (VLA) is receiving the signal. The transmitted signal $s(t)$ makes a Fourier transform, firstly[1]

$$\tilde{S}(\omega)=\frac{1}{2\pi}\int_{-\infty}^{+\infty}s(t)e^{i\omega t}dt=|\tilde{S}(\omega)|e^{i\theta_s(\omega)} \tag{1}$$

where t is time, ω is transient circular frequency, $\theta(\omega)$ is the phase of signal which is the function of frequency. Acoustic signals propagate through the ocean acoustic channel, and are received by VLA which contains N hydrophones. The j-th hydrophone's depth coordinate is Z_j, and distance coordinate is r_j, received signal is $P_j(t)$, which has a Fourier transform

$$\tilde{P}_j(\omega)=G(r_j,r_s,\omega)\tilde{S}(\omega) \tag{2}$$

where $G(r_j,r_s,\omega)$ is the Green's function between source location and VLA location in ω frequency. The purpose of this method is recovery of the original signal $s(t)$ from the received signal in the condition of uncertain $G(r_j,r_s,\omega)$. ATR(artificial time reversal) is a deconvolution technique which estimates a substitute $G(r_j, r_s, \omega)$ from received signal $\tilde{P}_j(\omega)$, using this substituted Green's function to reconstitute $s(t)$. First, we constitute a normalized data vector from received signal $\tilde{P}_j(\omega)$

$$D_j(\omega) = \frac{\tilde{P}_j(\omega)}{\sqrt{\Sigma_{j=1}^{N} |\tilde{P}_j(\omega)|^2}} = \frac{G(r_j, r_s, \omega) e^{i\theta_s(\omega)}}{\sqrt{\Sigma_{j=1}^{N} |G(r_j, r_s, \omega)|^2}} \tag{3}$$

where the second item is obtained from Eq.2. In Eq.3, the normalization is a nonlinear operation. The normalization in Eq.3 is nearly flat in frequency and introduces no phase shift. In Eq.3 $|\tilde{S}(\omega)|$ can be removed from $\tilde{P}_j(\omega)$, the phase $\theta_s(\omega)$ remains in $D_j(\omega)$. Here, we are removing the signal phase using a phase correction factor $\exp(-i\alpha\omega)$

$$\alpha(\omega) = \arg\{\sum_{j=1}^{N} W_j G(r_j, r_s, \omega)\} + \theta_s(\omega) \tag{4}$$

where W_j are the weights coefficient, which are depth-dependent. Eq.3 can be changed into $\hat{D}_j(\omega)$, which does not include $\theta_s(\omega)$

$$\hat{D}_j(\omega) = D_j(\omega) \exp\{-i\alpha(\omega)\} = \frac{G(r_j, r_s, \omega)}{\sqrt{\sum_{j=1}^{N} |G(r_j, r_s, \omega)|^2}} \exp\left\{ i \arg\left[\sum_{j=1}^{N} W_j G(r_j, r_s, \omega) \right] \right\} \tag{5}$$

The weights selection can be based on normal mode shapes in the ocean environment. The recovery signal can be written as

$$\widetilde{S}_{BP}(\omega)=\sum_{j=1}^{N}\hat{D}_j^*(\omega)\widetilde{P}_j(\omega) \tag{6}$$

COMPUTER SIMULATION OF BANDPASS SIGNAL RECOVERY

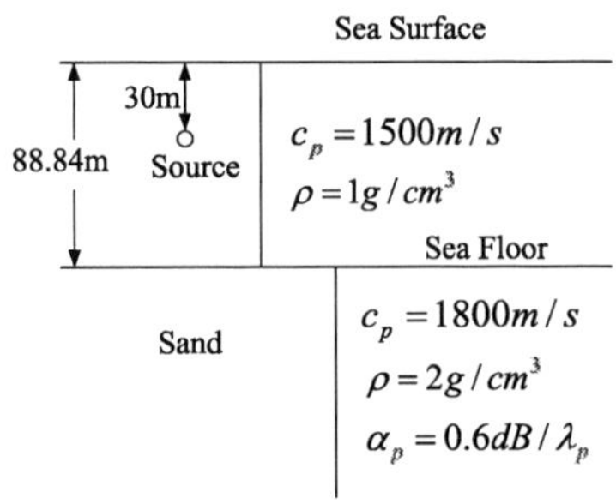

FIGURE 2. Layered environment model

There is a three-layer environment that includes air, seawater and sea bottom shown as Figure 2. Sound speed in seawater is 1500 m/s. In the sand of the sea bottom, the sound speed is 1800 m/s, density is 2 g/cm^3, and attenuation $\alpha_p(dB/\lambda_p)$ in the bottom is 0.6. Ocean depth is 88.84 m, acoustic source depth is 30 m. The VLA includes 32 hydrophones, the first element located at 7 m, and the last one located at 69 m. The range between source and receiver is 47 km. The signal pulse length is 0.1 second. The blanks in front of pulse are 0.1 second and behind pulse are 0.8 second. The centre frequency of signal is 650 Hz, bandwidth is 20%. The pulse's form is a Gaussian random signal, and sample frequency is 6 kHz. Figure.3 (a) shows a simulation result. The upper figure is the original signal; the middle figure is the received multipath signal in 47 km. The lower figure is the correlation coefficient of the original signal and the received signal. The maximum correlation coefficient is 0.3224. Figure.3 (a) shows that multipath signals produce serious distortions in long range propagation. And the correlation between transmitted signal and received signal is low.

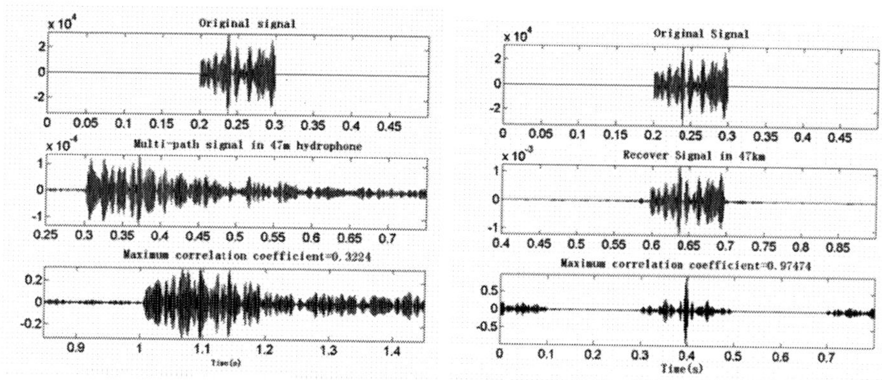

(a) Received distortion waveform and correlation (b) Recovery signal waveform and correlation

FIGURE 3. Simulation of signal recovery in 20dB SNR

The recovery signals technique using method in section 2 are shown in Figure 3 (b). The upper figure is the original signal; the middle figure is the recovery signal in 47 km. The lower figure is the maximum correlation coefficient of 0.97474. The result looks very good. This simulation is in condition of 20 dB signal noise ratio.

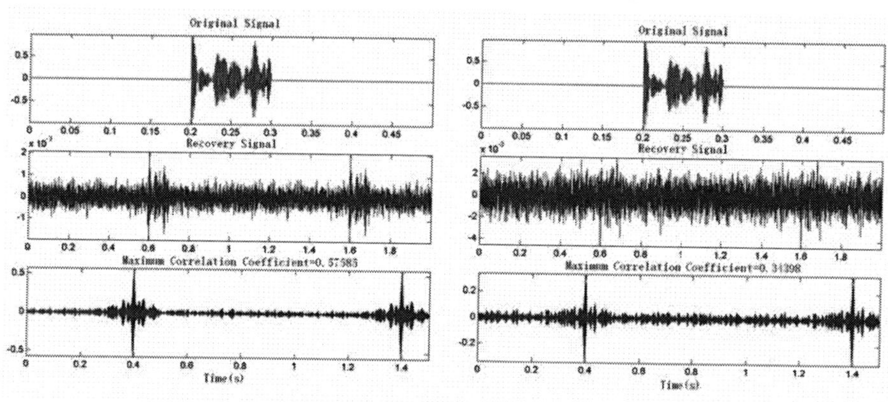

(a) SNR is 0 dB (b) SNR is -10 dB

FIGURE 4. Simulation of signal recovery in low SNR

Figure.4 shows two results in different SNR. Analysis similar to the above, Fig.4 (a) is in 0 dB SNR, the correlation coefficient is 0.57585. Figure.4 (b) is in -10 dB SNR, the correlation coefficient is 0.34398. These results show that the SNR has an obvious effect on the property of signal recovery technique.

CONCLUSION

In this paper, based on signal recovery technology, and obtaining some modeling results, we demonstrate that by using this technique we can obtain recovery signals which approximate the original signals' waveform.

REFERENCE

1. Karim G. Sabra and David R. Dowling, *J. Acoust. Soc. Am.* **116**, 262-271 (2004).
2. Tohyama, M., Tsunehiko, K., and Koike, T. *Fundamentals of Acoustic Signal Processing,* New York, Academic, 1998, pp. 274–314.
3. Hodgkiss, W. S., Song, H. C., Kuperman, W. A., Akal, T., Ferla, C., and Jackson, D. R., *J. Acoust. Soc. Am.* **105**, 1597-1604 (1999).
4. Hursky, P., Hodgkiss, W. S., and Kuperman, W. A., *J. Acoust. Soc. Am.* **109**, 1355-1366 (2001).

Studies of the Characteristics of a Densely-Coupled Array of Underwater Acoustic Transmitting Transducers

Zhengyao He and Yuanliang Ma

Institute of Acoustic Engineering, Northwestern Polytechnical University, Xi'an 710072, China
E-mail:hezhengyao@nwpu.edu.cn

Abstract. The characteristics of a densely-coupled array of underwater acoustic transmitting transducers are studied. At first, the electro-acoustic characteristics such as the admittance, the resonant frequency and the transmitting voltage response, of a low frequency barrel-stave flextensional transducer and a densely-coupled compact array composed of three identical transducers uniformly distributed on a circle with spacing much less than half wavelength, are measured by experiments. Then, the radiation impedances of a single transducer and of transducers in the compact array are calculated by the boundary element model together with the finite element model. Based on the above results, the transducer's equivalent circuit model parameters are calculated in different cases, which include a single transducer in air and in water, and a densely-coupled array of three transducers parallel connected in water. The characteristics of the transducers and array are analyzed by the equivalent circuit model that was obtained. The research results show that when the transducers make up a densely-coupled compact array, the resonant frequency decreases and the transmitting bandwidth broadens. It is also shown that the mutual interactions among elements are significant for the compact array. The mutual radiation resistance between two transducers is close to the self-radiation resistance of the transducers. The vibration velocities of the transducers in the compact array are nearly 1/3 as those of a single transducer, and the radiation acoustic power and transmitting voltage response of the array are nearly the same as those of a single transducer, when the driving voltages of the array and single transducer are unchanged. Furthermore, the transmitting source level of the 3-element compact array is 8.9dB higher than that of the single transducer if the vibration velocities of the transducers in the array are the same as those of the single transducer. The proposed technique can be used to design densely-coupled compact arrays of transmitting transducers.

Keywords: densely-coupled array barrel-stave flextensional transducer finite element model boundary element model equivalent circuit model
PACS: 43.30.Yj, 43.20.Rz

1 INTRODUCTION

The development of advanced active sonar requires low-frequency, high-power, wide bandwidth and compact underwater acoustic arrays of transmitting transducers. In a compact array, the element spacing of the array is much less than half of a wavelength, and the mutual coupling effect among elements is significant. The performance of the compact array will be seriously affected by the mutual interactions.

CP1272, *Shallow-Water Acoustics, Second International Shallow-Water Acoustics Conference*
edited by J. Simmen, E. S. Livingston, J. X. Zhou, and F. H. Li

The barrel-stave flextensional transducer has low weight and volume, its structure and shape are suitable for constituting compact transducer arrays, and its resonant frequency can be very low, thus it is widely used around the world. The detailed characteristics of the barrel-stave flextensional transducer are as follows: low resonant frequency, omnidirectional, high power, high efficiency and low weight and volume, and its limit working depth underwater is deep because its concave shell can increase the prestress when the working depth increases [1,2].

The finite element method can be used to model and analyze the transducers. The resonant frequency, surface vibration displacement and vibration velocity distribution of the transducers can be calculated. The admittance curves, frequency response and radiation directivity curves can all be obtained [3-6]. When the surface vibration velocity distribution of the transducer has been obtained, the boundary element method can be used to calculate the radiation characteristics of the transducer and the array [7-10]. Then, the equivalent circuit model can be used to model and calculate the densely-coupled array of underwater acoustic transmitting transducers.

In this paper, the finite element method and the commercial software ANSYS are used to model the barrel-stave transducer, and the surface vibration displacement and velocity distribution are obtained. Then, these are used as boundary condition and imported into the boundary element calculation software SYSNOISE. The boundary element method is used to calculate the acoustic radiation characteristics of the barrel-stave flextensional transducer and array. Finally, based on the results of theoretical calculation and experimental measurements, the equivalent circuit model is used to analyze the characteristics of a densely-coupled array of barrel-stave flextensional transducers.

2 ACOUSTIC RADIATION CALCULATION OF BARREL-STAVE TRANSDUCER AND ARRAY

Because the barrel-stave flextensional transducer is symmetrical, only the top half of the transducer is modeled and meshed in the three-dimensional boundary element program. Fig. 1 shows the three-dimensional boundary element model and mesh of the top half of the transducer. The transverse direction is the X-axis direction, the longitudinal direction is the Z-axis direction and the origin point O is the center of the transducer. The diameter of the most concave part of the concave vibration shell of the transducer is 65mm. The height of the concave shell is 125mm. The diameter of the two end mass blocks on the top and bottom of the transducer is 90mm, and the height is 40mm. The harmonic response analysis of the transducer is conducted using the finite element method in the ANSYS software. The surface vibration displacement distribution when the transducer vibrates underwater is obtained. The result is then imported into SYSNOISE software exerting on the boundary element model of the transducer. Thus, the surface vibration velocity distribution of transducer is obtained by the SYSNOISE software. Further, the boundary element method is used to calculate the acoustic characteristics of the transducer. In the calculation, the driving voltage of the transducer is 1000V, and the analysis frequency is 1400Hz. Fig. 2 shows the radiated sound pressure amplitude of the 1/4 circle which is 1m from the center of the transducer in the vertical plane (Y=0). The X-axis direction is 0°, and the

Z-axis direction is 90°. The range of the calculated sound pressure level is from 190.44dB to 190.40dB. It is shown that the vertical directivity of the transducer is uniform and the fluctuant amount is within 0.04dB. Fig.3 shows the radiated sound field distribution of the transducer in the vertical plane (Y=0). It is shown that the sound pressure is maximal on the most concave part of the transducer, the sound pressure of the field points decreases when their distance to the origin point increases, and the far-field sound pressure decreases like a spherical wave. The maximal sound pressure level on the surface of the transducer is 218.6dB. The calculated source level of the transducer is 190.44dB, which is the radiated sound pressure level at the distance of 1m from the origin point in the X-axis direction.

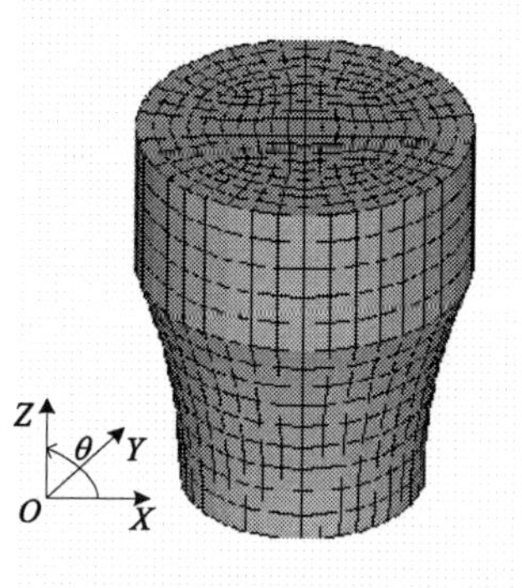

FIGURE 1. Boundary element model and mesh of top half of the transducer.
(Leftside is the sketch map of coordinate.
The origin point is the center of the transducer.)

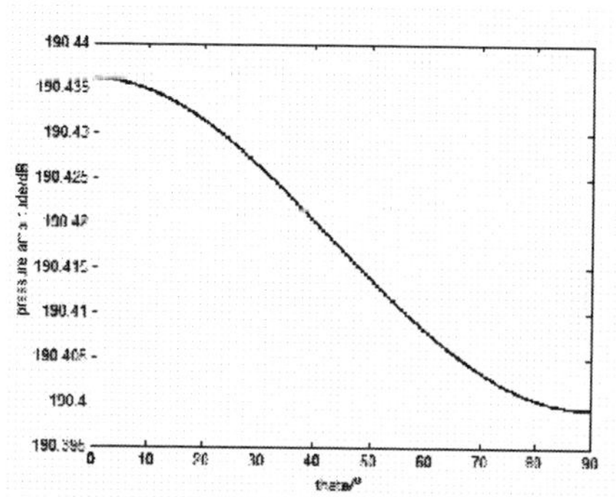

FIGURE 2. Radiated sound pressure amplitude of the 1/4 circle which is 1m from the center of the transducer in the vertical plane.

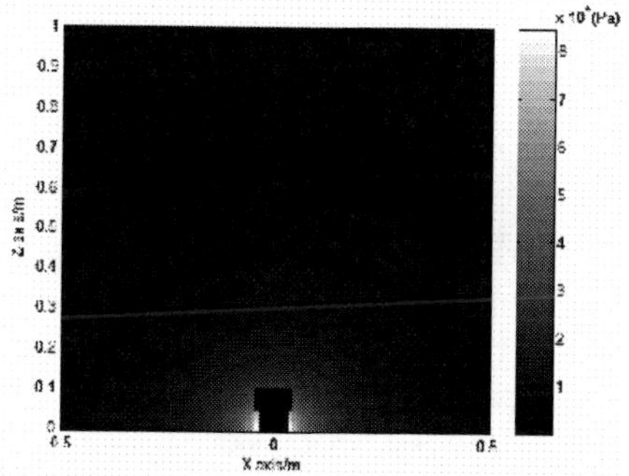

FIGURE 3. Radiated sound field distribution of the transducer in the vertical plane.

The three-dimensional boundary element model and mesh of the top half of the compact transducer array which is composed of three identical barrel-stave flextensional transducers is shown in Fig. 4. The three transducers are the same as the single transducer mentioned above. The maximal diameter of one transducer is 90mm. The three transducers are uniformly distributed on a circle. The diameter of the circumscribed circle of the array is 250mm. The transverse direction is the X-axis direction, the longitudinal direction is the Z-axis direction and the origin point O is the center of the circle which is passing the centers of the three transducers. The same surface vibration velocity distribution as mentioned above for the single transducer case is exerted on the corresponding nodes of the three transducers, then the acoustic radiation characteristics are calculated using the boundary element method in SYSNOSIE software. The analysis frequency is also 1400Hz. Fig. 5 shows the radiated sound pressure amplitude of the 1/3 circle which is 1m from the origin point in the horizontal plane (Z=0). Because the three transducers are uniformly distributed on a circle, only the sound pressure in the range of 0° to 120° directions is calculated. The X-axis direction is 0°, and the Y-axis direction is 90°. The range of the calculated sound pressure level is from 199.15dB to 199.18dB. It is shown that the horizontal directivity of the array is uniform and the fluctuant amount is within 0.03dB. Fig. 6 shows the radiated sound pressure amplitude of the 1/2 circle which is 1m from the origin point in the vertical plane (Y=0). The X-axis direction is 0°, and the Z-axis direction is 90°. The range of the calculated sound pressure level is from 199.15dB to 199.61dB. It is shown that the vertical directivity of the array is also uniform and the fluctuant amount is within 0.46dB. Fig.7 shows the radiated sound field distribution of the compact array in the Y=0 plane. It can be shown that the radiated sound pressure level in the middle of the three transducers is maximal, which is higher than that on the surface of the transducers, and the far-field radiated sound pressure of the array decreases like spherical wave. The maximal sound pressure level in the middle of the array is 224.5dB, and the maximal sound pressure level on the surface of the transducers is 220.7dB. The former is 3.8dB higher than the latter. The calculated source level of the array is 199.15dB, which is the radiated sound pressure level at the distance of 1m from the origin point to the center of one transducer direction in the Z=0 plane. It is shown that the transmitting source level of the array composed of three transducers is 8.7dB higher than that of single transducer. This is because that the spacing between each pair of transducers is much less than half of the wavelength, and the acoustic mutual interactions among transducers are significant. The mutual radiation resistance between every two transducers is close to the self-radiation resistance of the transducers. The radiation power of the transducer and array is proportional to the

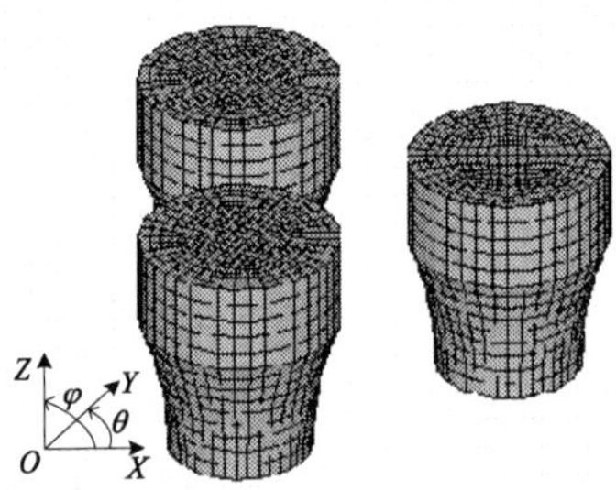

FIGURE 4. Boundary element model and mesh of top half of the compact transducer array. (Leftside is the sketch map of coordinate. The origin point is the center of the circle passing the centers of three transducers.)

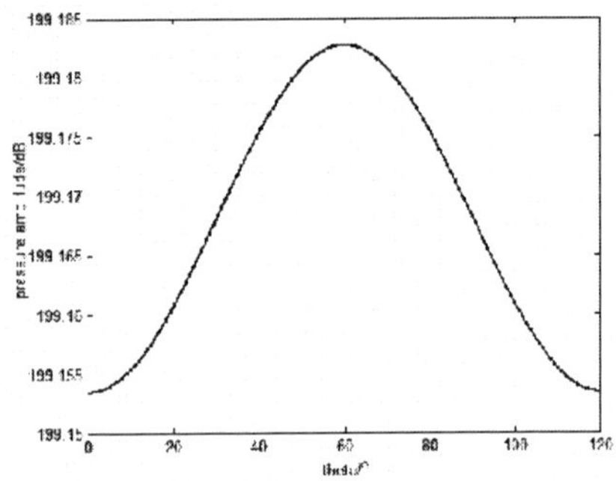

FIGURE 5. Radiated sound pressure amplitude of the 1/3 circle which is 1m from the origin point in the horizontal plane.

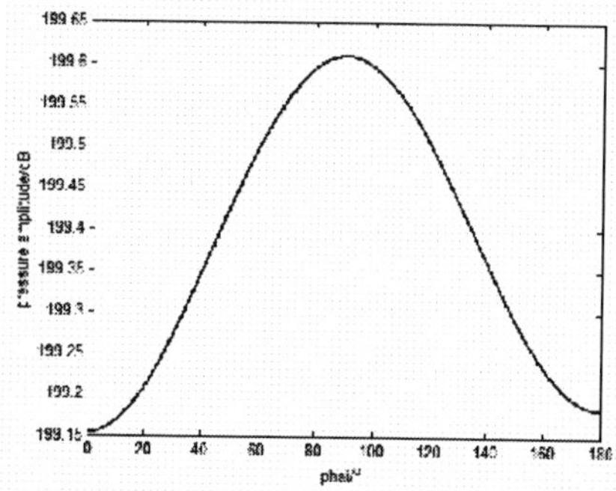

FIGURE 6. Radiated sound pressure amplitude of the 1/2 circle which is 1m from the origin point in the vertical plane.

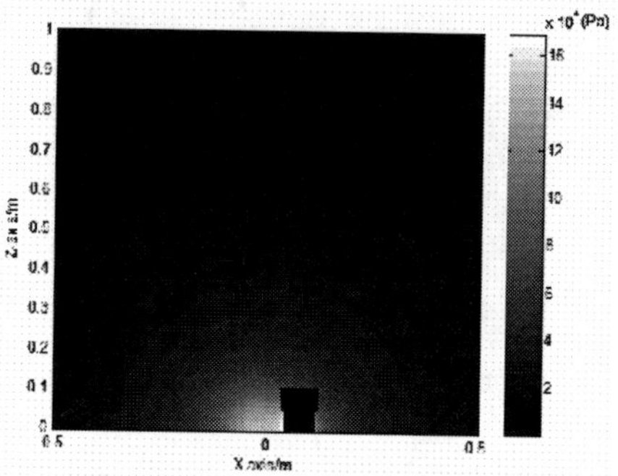

FIGURE 7. Radiated sound field distribution of the transducer array in the Y=0 plane.

square of their volume velocity. When the surface vibration velocity of the transducers in the array is kept at the same value that of a single transducer, the volume velocity of the array becomes 3 times that of the single transducer. Therefore, the radiation power of the array becomes nearly 9 times that of the single transducer.

3 CHARACTERISTICS ANALYZING OF TRANSDUCER AND ARRAY BY EQUIVALENT CIRCUIT MODEL

The electro-acoustic characteristics such as the admittance, resonant frequency and transmitting voltage response, of the barrel-stave transducer and compact array composed of three transducers as mentioned above, are measured by experiments. Table 1 shows the measured admittance, resonant frequency and transmitting voltage response of a single transducer and a compact array of three transducers parallel connected at the frequency of 1.4kHz . From the table, it is shown that the resonant frequency of the compact array decreases compared with that of single transducer. The conductance of the compact array is close to that of the single transducer and the susceptance is approximately 3 times that of the single transducer. And the transmitting voltage response of the compact array is nearly the same as that of the single transducer. These characteristics are due to the mutual interactions among elements in the compact array.

TABLE 1. Measured admittance, resonant frequency and transmitting voltage response of single transducer and compact array at the frequency of 1.4kHz.

	conductance (uS)	**susceptance** (uS)	**resonant frequency** (kHz)	**transmitting voltage response** (dB)
No.1 transducer	120.95	384.08	1.43	129.9
No.2 transducer	146.35	376.174	1.408	130.06
No.3 transducer	123.229	374.55	1.42	129.5
Compact array	129.485	987.678	1.219	129.8

The equivalent circuit model can be used to analyze the characteristics of the compact array. Fig. 8 shows the equivalent circuit of a single transducer in air. Fig. 9 shows the equivalent circuit of a single transducer in water. Fig.10. shows the equivalent circuit of each transducer when three transducers of the compact array are parallel connected in water. The self and mutual radiation impedances of the transducers are calculated by the boundary element model together with the finite element model. Based on the experimental and calculated results, the equivalent circuit model parameters of the transducer in different cases as shown in Figs. 8~10 are calculated. The characteristics of the transducers and array are analyzed by the equivalent circuit models obtained. Figs. 11 and 12 respectively show the admittance of a single transducer in air and in water. The solid line represents the results calculated by the equivalent circuit models, and the dotted line represents the results measured by experiments. Fig. 13 shows the admittance of the compact array of three transducers parallel connected in water. The solid line represents the results calculated by the equivalent circuit models, and the asterisk represents the result measured by experiments. From Figs. 11~13, it is shown that the admittance results calculated by the equivalent circuit models are consistent with those results measured by experiments. The calculated results show that when the transducers make up a densely-coupled compact array, the mutual interactions among elements are significant for the compact array, the resonant frequency decreases, and the transmitting bandwidth broadens. The resonant frequency of a single transducer calculated by the equivalent circuit model is 1.42kHz, and the bandwidth is 0.26kHz. The resonant frequency of the compact array calculated by the equivalent circuit model is 1.24kHz, and the bandwidth is 0.47kHz. It is also shown that the vibration velocities of the transducers in the compact array are nearly 1/3 as those of a single transducer, and the radiation acoustic power and transmitting voltage response of the array are nearly the same as those of a single transducer, when the driving voltages of the array and the single transducer are kept unchanged. The calculated results also show that when the driving voltage increases, leading to the vibration velocities of the transducers in the array that are the same as those of a single transducer, the values for radiation acoustic power and transmitting source level of the 3-element compact array are 8.9dB higher than those of the single transducer. This result, calculated by the equivalent circuit model, is consistent with the result calculated by the boundary element model together with the finite element model as shown in section 2.

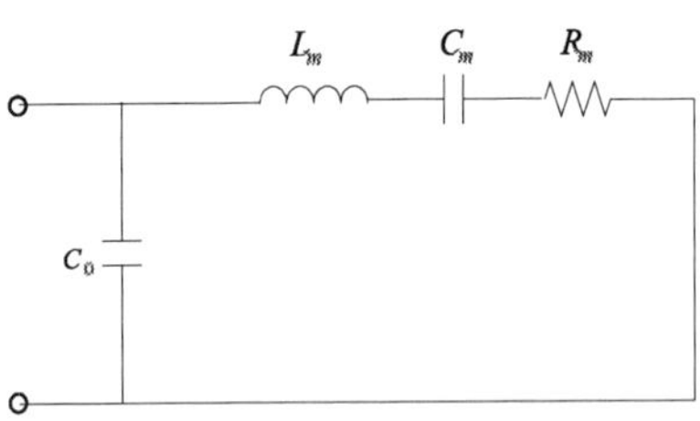

FIGURE 8. Equivalent circuit of single transducer in air.

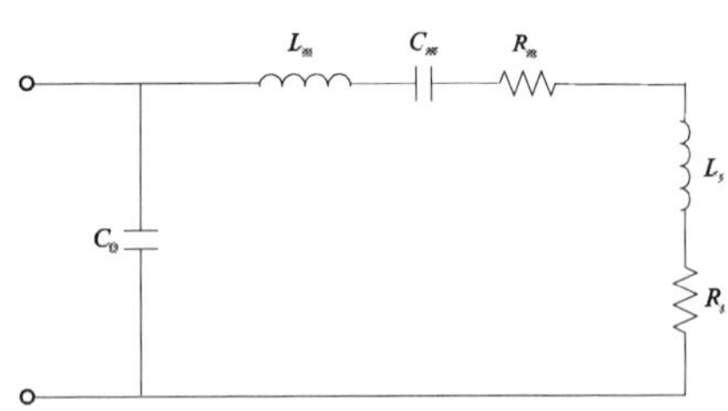

FIGURE 9. Equivalent circuit of single transducer in water.

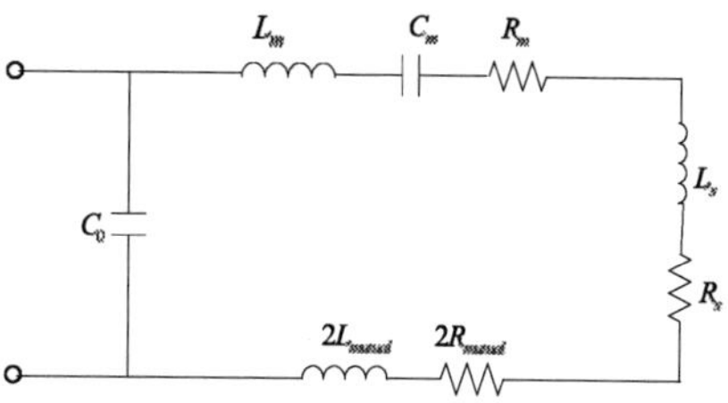

FIGURE 10. Equivalent circuit of each transducer when three transducers of the compact array are parallel connected in water.

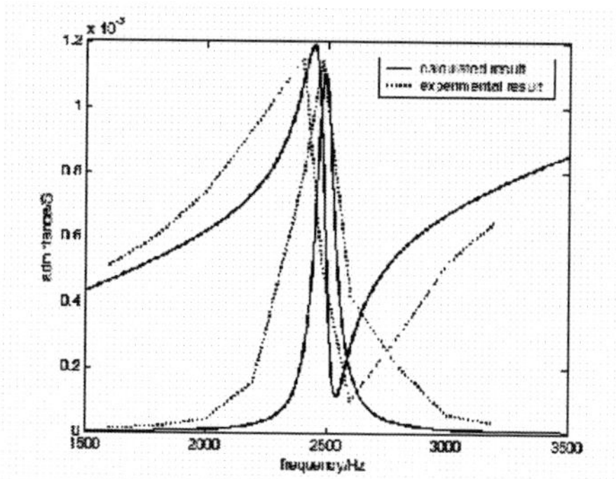

FIGURE 11. Admittance of single transducer in air.

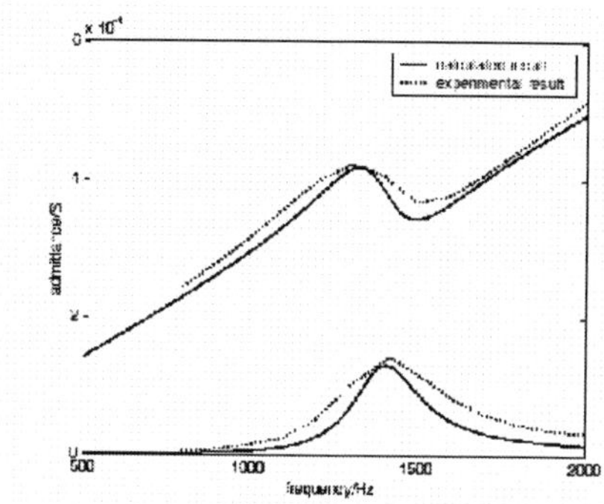

FIGURE 12. Admittance of single transducer in water.

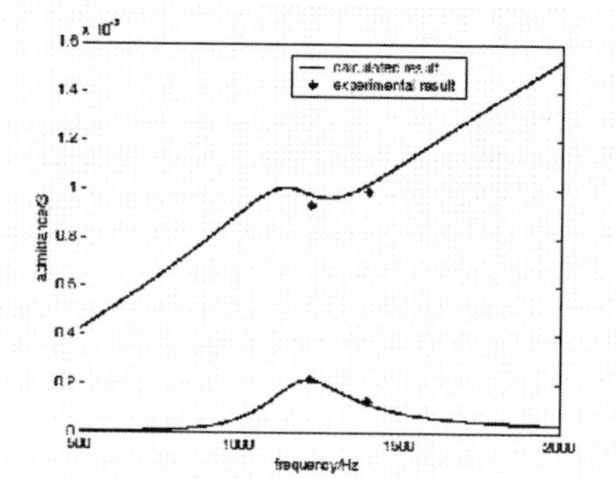

FIGURE 13. Admittance of the compact array of three transducers parallel connected in water.

4 CONCLUSIONS

In this paper, the characteristics of a densely-coupled array of underwater acoustic transmitting transducers are studied using the finite element method, boundary element method, and equivalent circuit model together with experimental results. The compact array is composed of three identical barrel-stave flextensional transducers. It is shown that the mutual interactions among the elements are significant for the compact array. The mutual radiation resistance between any pair of transducers in the array is close to the self-radiation resistance of the transducers. When the same driving voltage is exerted, the vibration velocities of the transducers in the compact array decrease. The radiation acoustic power and transmitting voltage response of the array are nearly the same as those of the single transducer. When the vibration velocities of the transducers in the array are the same as those of the single transducer, the transmitting source level of the 3-element compact array is 8.9dB higher than that of the single transducer.

ACKNOWLEDGMENTS

This work was supported by the National Natural Science Foundation of China (10734030 and 60901076), the NPU Foundation for Fundamental Research (NPU-FFR-JC200805), the China Postdoctoral Science Foundation special funded project (200902604), and the China Postdoctoral Science Foundation funded project (2008 0431248).

REFERENCES

1. Jones D.F., Abboud N.N., "Performance analysis of a low-frequency barrel-stave flextensional projector", in Proceeding of ONR Transducer Materials and Transducers Workshop,1-4 (1996).
2. Jones D.F., Christopher D.A., "A broadband omni-directional barrel-stave flextensional transducer", J. Acoust.Soc.Am.,106, L13-L17 (1999).
3. Hamonic B.,Debus J. C., "Analysis of a radiating thin-shell sonar transducer using the finite-element method", J.Acoust.Soc.Am.,86,1245-1253 (1989).
4. Jarng Soon Suck, "Comparison of barrel-stave sonar transducer simulations between a coupled FE-BEM and ATILA", IEEE Sensors Journal, 3,439-446 (2003).
5. Allik H. , Webman K. M., "Vibrational response of sonar transducers using piezoelectric finite elements" , J.Acoust.Soc.Am.,56,1782-1791 (1974).
6. Smith R.R., Hunt J.T., "Barach D.. Finite element ana-lysis of acoustically radiating structures with applica-tions to sonar transducers.", J.Acoust.Soc. Am.,54, 1277-1288 (1973).
7. Harry A. Schenck, "Improved integral formulation for acoustic radiation problems", J.Acoust.Soc.Am., 44, 41-58 (1968).
8. Cunefare K.A., Koopmann G., Brod K., "A boundary element method for acoustic radiation valid for all wave numbers", J.Acoust.Soc.Am., 85,39-48 (1989).
9. Christan Audoly, "Some aspects of acoustic interact-tions in sonar transducer arrays", J.Acoust.Soc.Am., 89,1428-1433 (1991).
10. He Zhengyao, Ma Yuanliang, "Calculation of baffle effect and mutual interaction between elements for an underwater acoustic conformal array with application to the optimization of projecting beampattern", Chinese Science Bulletin, 52, 2584-2591 (2007).

Statistical Estimation of Source Location in Presence of Geoacoustic Inversion Uncertainty

Chen-Fen Huang* and Peter Gerstoft and William S. Hodgkiss†

*Institute of Oceanography, National Taiwan University, Taiwan 10617
†Marine Physical Laboratory, Scripps Institution of Oceanography, La Jolla, CA 92093-0238

Abstract. A statistical estimation of source location incorporating uncertainty in ocean environmental model parameters is derived using a Bayesian approach. From a previous geoacoustic inversion, a posterior probability distribution (PPD) of the environmental parameters that reflects uncertainty in the ocean environment is obtained. This geoacoustic uncertainty then is mapped into uncertainty in the acoustic pressure field and is propagated through the Bartlett matched-field processor for source localization. Using data from the ASIAEX 2001 East China Sea experiment, the estimated source location and variability over time are compared with the known source positions.

INTRODUCTION

This paper presents a statistical estimation of source location in the presence of ocean acoustic parameter environmental uncertainty. Many Bayesian approaches for increasing robustness to ocean environmental uncertainty have been developed[1, 2, 3, 4, 5, 6, 7]. Averaging over a uniform prior distribution of environmental parameters was proposed earlier in [1, 2] as a method of improving robustness of source localization. Dosso and Wilmut[5, 6] proposed a Bayesian focalization to quantify source localization with environmental uncertainty. In their Bayesian focalization approach all source positions and environmental parameters are inverted at once. For practical problems involving thousands of source position, tracking is an efficient alternative. Kalman and particle filtering [4, 7] can be used efficiently to track the source and geoacoustic parameters.

Estimation of uncertainty in environmental parameters for a single range is well developed using Markov chain Monte Carlo (MCMC) Metropolis-Hasting sampling [8, 9, 10]. The processor derived in this paper translates the uncertainty in environmental parameters, which for example can be from a previous geoacoustic inversion, to the uncertainty in replica fields for matched-field source localization. The resulting processor is a weighted sum of Bartlett processors over the PPD of the environmental parameters. We demonstrate the approach using the source tow data from the ASIAEX 2001 experiment[8, 11]. Using the geoacoustic inversion results for source localization is well known[12], the focus here is on the accompanying source position uncertainty.

THEORY

Figure 1 summarizes the statistical estimation of source position using environmental information obtained from a previous geoacoustic inversion.

CP1272, *Shallow-Water Acoustics, Second International Shallow-Water Acoustics Conference*
edited by J. Simmen, E. S. Livingston, J. X. Zhou, and F. H. Li

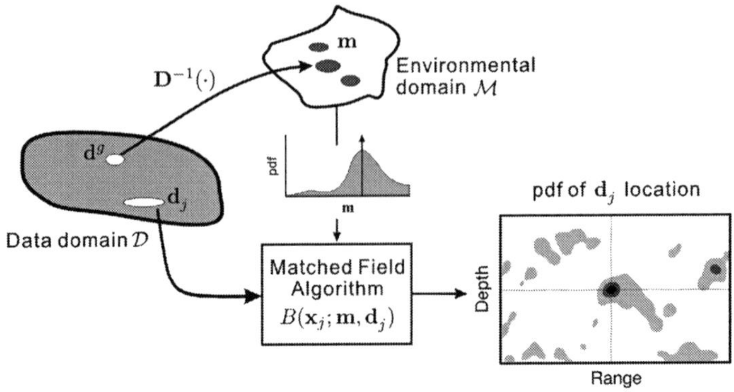

FIGURE 1. Measured data for source localization $\mathbf{d}_j$ ($\in \mathscr{D}$) is received at time j. From previous measured data $\mathbf{d}^g$ ($\in \mathscr{D}$) environmental parameters $\mathbf{m}$ ($\in \mathscr{M}$) are estimated via geoacoustic inversion along with a posterior probability distribution (PPD) of $\mathbf{m}$ that reflects its uncertainty. These environmental parameters then are mapped through a matched-field processor for statistically estimating the source location from $\mathbf{d}_j$.

Classical matched-field processor

The classical matched-field processor consists of systematically placing a single frequency test point source at each point of a search grid, computing the acoustic field (replica) at all the elements of the array and then correlating this modeled field with the observed data from an unknown source. In absence of data error and error in estimating the environmental parameter vector $\mathbf{m}$, the correlation is maximum when the test source location $\mathbf{x}$ is collocated with the true source,

$$\begin{aligned} B(\mathbf{x}_j;\widehat{\mathbf{m}},\mathbf{d}_j) &= |\mathbf{w}(\mathbf{x}_j;\widehat{\mathbf{m}})^H\mathbf{d}_j|^2 \\ &= \mathbf{w}(\mathbf{x}_j;\widehat{\mathbf{m}})^H\mathbf{R}_j\mathbf{w}(\mathbf{x}_j;\widehat{\mathbf{m}}) \end{aligned} \quad (1)$$

where $\mathbf{d}_j$ denotes the observed data at time index j, $\widehat{\mathbf{m}}$ an estimate of the parameter vector; $\mathbf{w}(\mathbf{x};\widehat{\mathbf{m}}) = \mathbf{d}(\mathbf{x};\widehat{\mathbf{m}})/|\mathbf{d}(\mathbf{x};\widehat{\mathbf{m}})|$ is the normalized replica vector with the field solution $\mathbf{d}(\mathbf{x};\mathbf{m})$ computed from a acoustic model using the environmental parameter vector $\mathbf{m}$. When multiple snapshots are available, we assume that the signal source term can vary across snapshots, whereas the error power is the same for all snapshots. Then the observed data are efficiently expressed as a data covariance matrix $\mathbf{R}_j = \langle \mathbf{d}_j\mathbf{d}_j^H \rangle$.[13]

Posterior probability distribution of source position

The uncertainty in environmental parameters is translated to the uncertainty in replica fields for matched-field source localization[8]. The PPD of the source position at time j

given the observation $\mathbf{d}_j$ is found via Bayes' rule,

$$\mathrm{p}(\mathbf{x}_j|\mathbf{d}_j) = \frac{\mathrm{p}(\mathbf{d}_j|\mathbf{x}_j)\mathrm{p}(\mathbf{x}_j)}{\mathrm{p}(\mathbf{d}_j)}, \tag{2}$$

where $\mathrm{p}(\mathbf{x}_j)$ is the PDF of $\mathbf{x}_j$ before observing the data. $\mathrm{p}(\mathbf{d}_j|\mathbf{x}_j)$ is the likelihood function of the source position $\mathbf{x}_j$ at time step j given the data $\mathbf{d}_j$. This likelihood function can be derived from the PPD of $\mathrm{p}(\mathbf{d}_j,\mathbf{m}|\mathbf{x}_j)$, by averaging over the environmental parameter vector $\mathbf{m}$:

$$\begin{aligned}\mathrm{p}(\mathbf{d}_j|\mathbf{x}_j) &= \int \mathrm{p}(\mathbf{d}_j,\mathbf{m}|\mathbf{x}_j)\mathrm{d}\mathbf{m} \\ &= \int \mathrm{p}(\mathbf{d}_j|\mathbf{m},\mathbf{x}_j)\mathrm{p}(\mathbf{m}|\mathbf{x}_j)\mathrm{d}\mathbf{m},\end{aligned} \tag{3}$$

where $\mathbf{m}$ is independent of the source position $\mathbf{x}_j$ and the information about $\mathbf{m}$ was obtained from a previous geoacoustic inversion process using data $\mathbf{d}^g$. Therefore,

$$\mathrm{p}(\mathbf{m}|\mathbf{x}_j) - \mathrm{p}(\mathbf{m}) = \mathrm{p}(\mathbf{m}|\mathbf{d}^g). \tag{4}$$

The likelihood of the environmental parameters $\mathbf{m}$ and the source position $\mathbf{x}_j$ given $\mathbf{d}_j$ is determined from the assumption of data error statistics. The data error consists of both ambient noise as well as error in modeling the environment[8]. For simplicity, the error terms are assumed independent and identically distributed (IID) complex Gaussian random variables with variance ν:

$$\mathrm{p}(\mathbf{d}_j|\mathbf{m},\mathbf{x}_j,\nu) = \frac{1}{\pi^N \nu^N}\exp\left(-\frac{\phi(\mathbf{x}_j;\mathbf{m},\mathbf{d}_j)}{\nu}\right) \tag{5}$$

where $\phi(\mathbf{x}_j;\mathbf{m},\mathbf{d}_j) = \|\mathbf{d}_j\|^2 - B(\mathbf{x}_j;\mathbf{m},\mathbf{d}_j)$ is the Bartlett objective function for the environmental parameter vector $\mathbf{m}$. Note, that in Eq. (5) the complex source term has been estimated by a maximum likelihood approach[13] as opposed to integrating it out[1]. The uncertainty in the error variance ν is accounted for by integrating the complex Gaussian PDF over the nuisance parameter ν assuming a non-informative prior for ν (proportional to $1/\nu$) giving the likelihood function[11]:

$$\mathrm{p}(\mathbf{d}_j|\mathbf{m},\mathbf{x}_j) \propto \frac{1}{\phi(\mathbf{x}_j;\mathbf{m},\mathbf{d}_j)^N} \tag{6}$$

where N is number of the hydrophones. Substituting Eqs. (6), (4) and (3) into (2) yields the PPD of $\mathbf{x}_j$ as:

$$\begin{aligned}\mathrm{p}(\mathbf{x}_j|\mathbf{d}_j) &= \frac{\mathrm{p}(\mathbf{x}_j)}{\mathrm{p}(\mathbf{d}_j)}\int \mathrm{p}(\mathbf{d}_j|\mathbf{m},\mathbf{x}_j)\mathrm{p}(\mathbf{m}|\mathbf{d}^g)\,\mathrm{d}\mathbf{m} \\ &\propto \mathrm{p}(\mathbf{x}_j)\sum_{k=1}^{K}\frac{1}{\phi(\mathbf{x}_j;\mathbf{m}_k,\mathbf{d}_j)^N}.\end{aligned} \tag{7}$$

where $\phi(\mathbf{x}_j;\mathbf{m}_k,\mathbf{d}_j)$ is the Bartlett objective function for an environmental parameter vector $\mathbf{m}_k$ sampled by a Markov chain Monte Carlo (MCMC) algorithm. The statistical estimation of source position is based on Eq. (7).

RESULTS

Data on a vertical line array from the ASIAEX 2001 East China Sea experiment in 106.5-m water depth are used to illustrate the approach[8]. First, an inversion was carried out at source-receiver range of 1.7 km, indicated by a circle of Fig. 1 (a). An environment with 13 unknown parameters with their search bounds indicated in Fig. 2 (b) was estimated, including geometrical, geoacoustic, and ocean sound speed EOF coefficients[8]. The inversion was based on a MCMC sampling giving samples $\mathbf{m}_k$ from which the PPD $\mathrm{p}(\mathbf{m}|\mathbf{d}^g)$ could be estimated and subsequently used as prior information for source localization with new data $\mathbf{d}_j$. Based on Eq. (7), the samples $\mathbf{m}_k$ corresponding to a particular environment then are combined with the source position data $\mathbf{d}_j$ via the objective function $\phi(\mathbf{x}_j;\mathbf{m}_k,\mathbf{d}_j)$ for each possible location $\mathbf{x}_j$. The prior distribution $\mathrm{p}(\mathbf{x}_j)$ was assumed uniform with the depth-range search interval. Summing over all samples gives the PPD of source location $\mathrm{p}(\mathbf{x}_j|\mathbf{d}_j)$.

To estimate statistically the source position given the environmental information obtained from the inversion, a measured data vector $\mathbf{d}_j$ is used where the source range was 2.5 km. The source signal was a CW tone at 195 Hz. The signal-to-noise ratio (SNR) was estimated to be 16.3 dB using the ratio of the mean power received in the signal frequency FFT bin and the (assumed to be noise) power received outside this bin. The observed data cross spectral density matrix $\mathbf{R}$ is normalized by the largest eigenvalue of the matrix, so that the maximum signal power is one. For simulation purposes, the desired SNR is generated by varying the noise variance ν_n according to the array SNR definition[8]:

$$\mathrm{SNR} = 10\log\frac{1}{\nu_\mathrm{n}}, \text{ where } \nu_\mathrm{n} = \frac{E\left[\|\mathbf{n}\|^2\right]}{N}. \tag{8}$$

and $\mathbf{n}$ is the noise in the data (the array SNR should not be confused with the element SNR used in e.g. Ref. [5]). Since noise only can be added to the data, we only can lower the SNR from the maximum of 16.3 dB.

The parameter uncertainties are mapped through to the statistical estimation of source localization using Eq. (7). The replica vector for a hypothesized source position $\mathbf{x}$ and for each environmental parameter vector $\mathbf{m}_k$, $\mathbf{w}(\mathbf{x};\mathbf{m}_k)$, is computed using the normal mode propagation model SNAP[14]. Monte Carlo subsampling[8] of the geoacoustic inversion samples is used to evaluate Eq. (7) with $K = 900$. The range-depth search grid spacing is 8 m in range from 1500 to 3500 m and 1.4 m in depth from 1 to 99 m.

Figure 3(a) shows the effect of the array SNR on the PPD of source range and depth at a source frequency of 195 Hz and range 2.5 km. The intersection of the lines indicates the peak of the PPD, the darkest region indicates the 50% highest posterior density (HPD) in which the source falls given the observed data $\mathbf{d}_j$ and the environmental uncertainty. The range-depth PPD for the various SNRs (Figs. 3(a-1 to a-3)), show that the number of competing peaks increases as SNR decreases. Note that the range-depth PPD for low SNR depends on the noise realization. The benefit here is not only do we have the maximum *a posteriori* (MAP) estimate of source position but also the level of uncertainty.

The effect of SNR on the source position uncertainties is displayed for three source ranges in Fig. 3(b). The estimated SNRs in the observed data for ranges of 1.9, 2.5 and

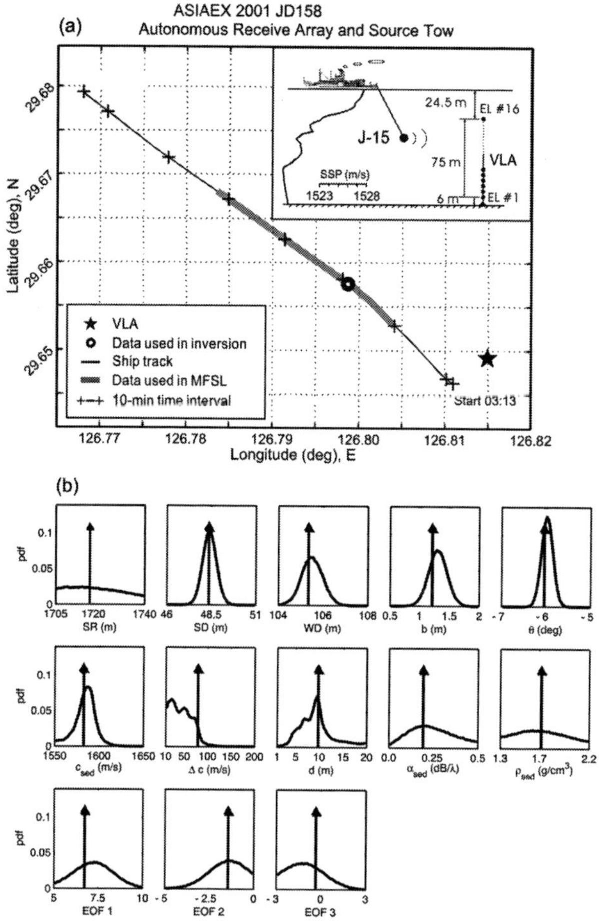

FIGURE 2. (a) Track of R/V *Melville* during the ASIAEX 2001 East China Sea experiment. The experimental geometry is shown in the upper right-hand corner of the figure. (b) 1D marginal posterior probability densities and maximum *a posteriori* (MAP; indicated by arrows) of the model parameters using the measured data obtained at approximately 1.7 km from the source.

3.15 km are 16.9, 16.3 and 15.2 dB, respectively (estimated using adjacent bins, see above). In Fig. 3(b-1) and 3(b-2) the 90% credibility interval (CI) extracted from the 1D marginal distribution is shown. At lower SNR, the noise is dominating the uncertainty of source position. For high SNR, the source position uncertainty will not approach zero since uncertainty in the environmental parameters will dominate. From the PPD surfaces in Fig. 3(a) we can integrate a region around the source position (an acceptable region of 20 m in depth and 200 m in range) obtaining the probability of correct localization PCL[2] (this also has been termed integrated probability, IP[5]). The PCL increases with SNR and tends to decrease with range.

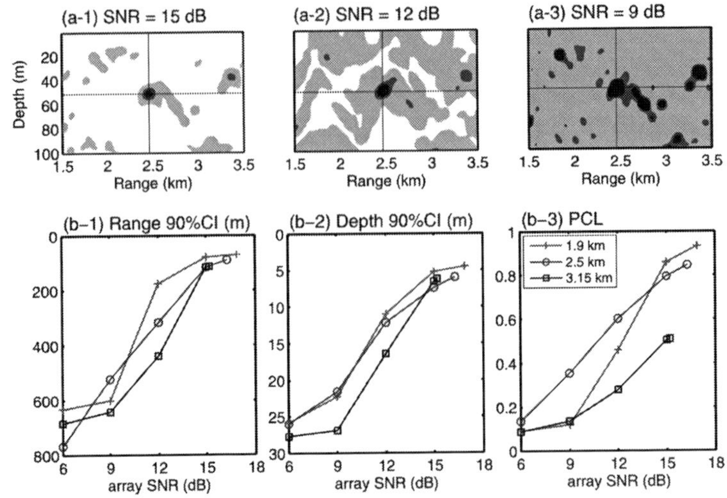

FIGURE 3. (a) Realizations of PPD of source range and depth for 195 Hz at 2.5 km range for three SNRs. Each panel shows the maximum *a posteriori* (MAP) source position (intersection of thin lines) and the highest posterior density (HPD) of source position [with 50% (dark), 75% and 99% (light)]. (b) Source position uncertainty versus SNR for ranges 1.9 (+), 2.5 (○), and 3.15 km (□): (b-1) and (b-2) 90% credibility interval (CI) and (b-3) probability of correct localization (PCL) in a region around the assumed source position.

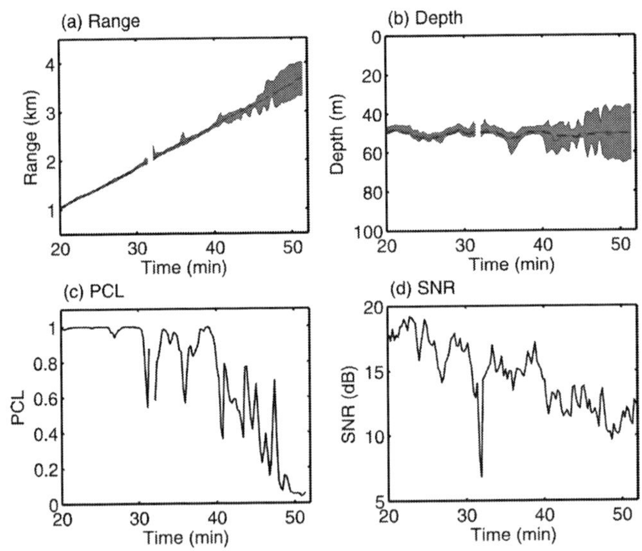

FIGURE 4. MAP estimate of source range (a) and depth (b) (solid) and their 90% CI (gray area) from time 20 to 52 min. The dashed line indicates the range measured by DGPS (a) and the depth by a depth sensor (b). (c) PCL and (d) SNR estimated from the data.

As part of the experiment, a source tow from 1–3.7 km was performed. Using the estimated geoacoustic parameters and their uncertainty as well as the methodology above the MAP estimate of source position and the 90% CI is presented in Fig. 4(a–b). Furthermore, Fig. 4(c–d) shows that as the SNR decreases so does the PCL.

SUMMARY

The source position estimation in the presence of ocean environment uncertainty is investigated using the ASIAEX 2001 source tow data. A simulation shows that the noise deteriorates the localization performance at lower SNR, while the uncertainty in the ocean environment dominates at higher SNR. The continuous localization using the source tow data indicates the localization uncertainty increases with range and 1/SNR.

ACKNOWLEDGMENTS

This work was supported by National Science Council of Taiwan through contract NSC97-2221-E-019-056 and Office of Naval Research under Grant N00014-05-1-0264.

REFERENCES

1. A. M. Richardson, and L. W. Nolte, *J. Acoust. Soc. Am.* **89**, 2280–2284 (1991).
2. J. A. SHOREY, L. W. NOLTE, and J. L. KROLIK, *Journal of Computational Acoustics* **2**, 285–314 (1994).
3. J. A. Shorey, and L. W. Nolte, *J. Acoust. Soc. Am.* **103**, 355–361 (1998).
4. S. L. Tantum, and L. W. Nolte, *J. Acoust. Soc. Am.* **103**, 362–373 (1998).
5. S. E. Dosso, and M. J. Wilmut, *J. Acoust. Soc. Am.* **121**, 2567–2574 (2007).
6. S. E. Dosso, and M. J. Wilmut, *J. Acoust. Soc. Am.* **124**, 82–97 (2008).
7. C. Yardim, P. Gerstoft, and W. S. Hodgkiss, *The Journal of the Acoustical Society of America* **125**, 746–760 (2009).
8. C.-F. Huang, P. Gerstoft, and W. S. Hodgkiss, *J. Acoust. Soc. Am.* **120**, 1932–1941 (2006).
9. Y.-M. Jiang, N. R. Chapman, and M. Badiey, *J. Acoust. Soc. Am.* **121**, 1879–1894 (2007).
10. J. Dettmer, S. E. Dosso, and C. W. Holland, *J. Acoust. Soc. Am.* **125**, 706–716 (2009).
11. C.-F. Huang, P. Gerstoft, and W. S. Hodgkiss, *J. Acoust. Soc. Am.* **119**, 197–207 (2006).
12. D. F. Gingras, and P. Gerstoft, *J. Acoust. Soc. Am.* **97**, 3589–3598 (1995).
13. P. Gerstoft, and C. F. Mecklenbräuker, *J. Acoust. Soc. Am.* **104**, 808–819 (1998).
14. F. B. Jensen, and M. C. Ferla, *SNAP: the SACLANTCEN normal-mode acoustic propagation model*, SACLANT Undersea Research Centre, SM–121, La Spezia, Italy (1979).

An Improved Passive Phase Conjugation Array Communication Algorithm

Ning Jia, Zhongyuan Guo, Jianchun Huang and Geng Chen

Key Laboratory of Underwater Acoustic Environment, Chinese Academy of sciences, Beijing, 100190, China
Institute of Acoustics, Chinese Academy of sciences, Beijing, 100190, China

Abstract. The time-varying, dispersive, multipath underwater acoustic channel is a challenging environment for reliable coherent communications. A method proposed recently to cope with intersymbol interference (ISI) is Passive-Phase-Conjugation (PPC) cascaded with Decision-Feedback Equalization (DFE). Based on the theory of signal propagation in a waveguide, PPC can mitigate channel fading and improve the signal-to-noise ratio (SNR) by using a receiver array. At the same time the residual ISI will be removed by DFE. This method will lead to explosive divergence when the channel is changed by a large amount, because PPC estimates channels inaccurately. An improved algorithm is introduced in this paper to estimate the channel during all the communication process; as a result, the change of the channel can be found in time and the PPC could use more accurate channel estimated. Using simulated and at-sea data, we demonstrate that this algorithm can improve the stability of original algorithm in changed channels.

Keywords: Passive-Phase Conjugation, Decision-Feedback Equalizer
PACS: 43.30 43.60

I. INTRODUCTION

The PPC cascaded with DFE method has been proposed recently to cope with ISI in underwater acoustic communication [1]. In this system, PPC can mitigate channel fading and improve the signal-to-noise ratio (SNR) by using a receiver array, and DFE can remove residual ISI. This method has been proven to provide good performance in shallow water communication [2][3]. The residual ISI depends upon the number of array elements, the complexity of the channel, and their spatial distribution. .

The organization of this paper is as follows. The first section introduces the method of PPC cascaded with DFE and shows some sea experimental results we have obtained using this method. The next section discusses the efficiency of the method

CP1272, *Shallow-Water Acoustics, Second International Shallow-Water Acoustics Conference*
edited by J. Simmen, E. S. Livingston, J. X. Zhou, and F. H. Li

and presents a modified approach. The performance analysis of the proposed method is shown at last.

II. PPC CASCADED WITH DFE APPROACH

The PPC communication system is based on the time-reversal process. It uses a single source and receiver array as shown in Fig.1. The receiver array is used to correlate the channel response and the communication sequence. Before the data is transmitted, a channel probing signal (as shown in Fig.2) is used to measure the channel impulse response for every element of the receiver array. Then the channel impulse response of every element is correlated with the array data received after the probing signal. By using correlation processing the effect of a match filter could be obtained.

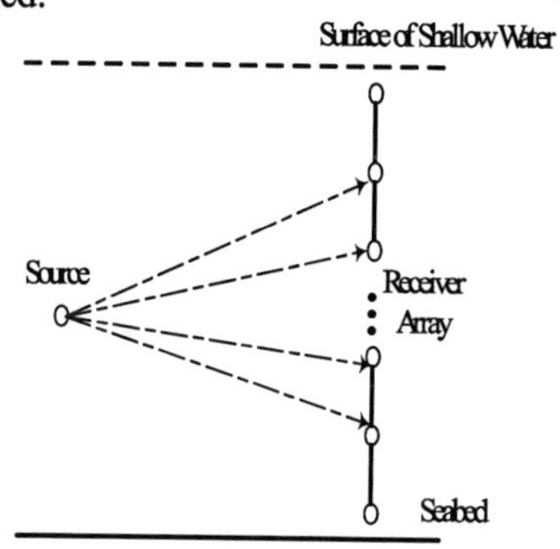

FIGURE 1. System of communication

FIGURE 2. Frame of signal block

Fig.1 shows the communication model used in this paper. A transducer is used to transmit sound signals and a receiver array is deployed almost the whole depth of the water. Fig.2 shows the frame structure of the data; the probe signal before the data signal is used to measure the temporal communication channel.

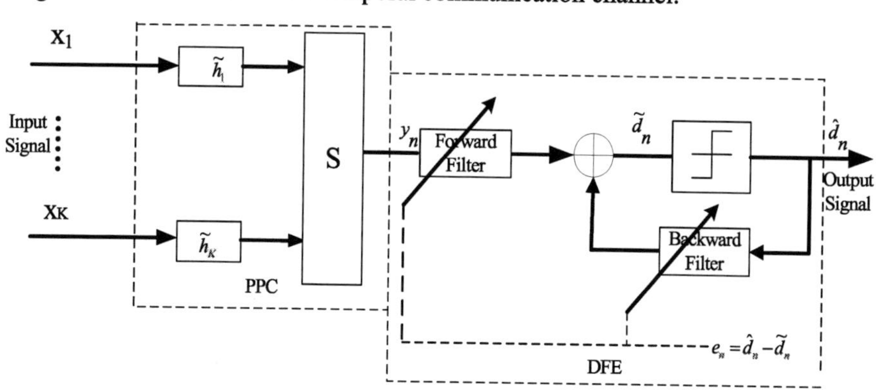

FIGURE 3. PPC combined with DFE method

Fig.3 shows processing structure of the PPC combined with the DFE method. When a known signal $s(t)$ is transmitted from the source in a waveguide, the kth element of the receiver array receives a signal as $x_k(t) = s(t) \otimes h_k(t) + \xi_k(t)$ where $h_k(t)$ is the channel impulse response of the waveguide, $\xi_k(t)$ is the additive noise and $\otimes$ denotes convolution. The PPC method output signal can be written as

$$
\begin{aligned}
y(t) &= \sum_k x_k(t) \otimes \hat{h}_k(-t) \\
&= \sum_k (s(t) \otimes h_k(t) + \xi_k(t)) \otimes \hat{h}_k(-t) \\
&= s(t) \otimes [\sum_k (h_k(t) \otimes \hat{h}_k(-t))] + \sum_k \xi_k(t) \otimes \hat{h}_k(-t) \\
&\equiv s(t) \otimes Q(t) + \varsigma(t)
\end{aligned} \quad (1)
$$

where the function $Q(t)$ is denoted by q-function and is the summation of the auto-correlation of each channel impulse response. And the output signal of DFE can be written as

$$\hat{d}_n = F[\sum_{i=1}^{L} y_{n+i} Wf_i + \sum_{i=1}^{M} \hat{d}_{n-i} Wb_i] \quad (2)$$

where the "F" denotes the decision function, Wf_i and Wb_i are the tap coefficients for the feed-forward and feedback filters, respectively, and $\hat{d}_{n-i}$ denotes a decision symbol of the estimated symbol d_{n-i}.

A sea experiment has been conducted. The water's depth was 37 m, the number of array elements was 32 and these were spaced at 2 m intervals between two elements. The communication code modulation was 16-QAM and the data rate was 14kbps. The communication carrier frequency was 7kHz with 5kHz bandwidth. At a distance of 6km, the result of sea experiment is shown in Fig.4.

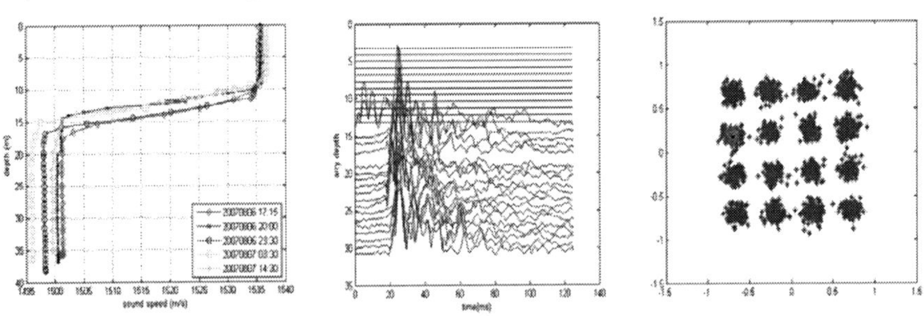

(a) (b) (c)

FIGURE 4. Result of sea experiment 1. (a) Sound-speed profile measured during the experiment. (b) Channel impulse response measured during the experiment. (c) Performance of the method.

Fig.5 shows another experiment which was conducted at the East Sea of China in 39-m-deep water; the modulation was QPSK at a data rate of 4kbps, the carrier frequency was 6kHz with 3kHz bandwidth, the distance between source and receiver was 3km.

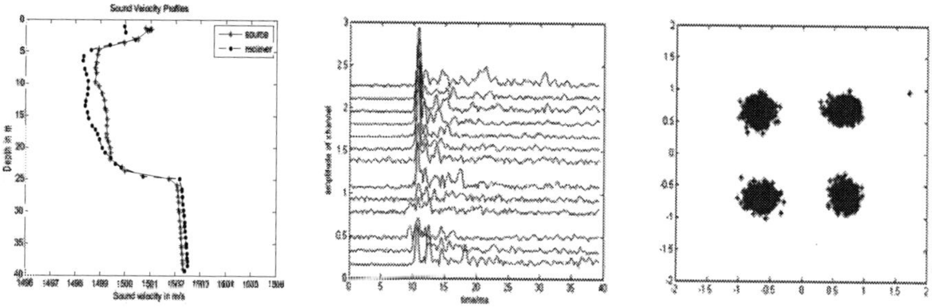

(a) (b) (c)

FIGURE 5. Result of sea experiment 2. (a) Sound-speed profile measured during the experiment. (b) Channel impulse response measured during the experiment. (c) Performance of the method.

The sea experiments show that the method has good performance in shallow water communication; however, the channel response estimation influences the effect of the method especially in cases of a moving source. Fig.6 shows an instance of tremendous channel change found in the sea experiments. The common length of the communication frame is about several seconds, and a shallow water communication channel could change tremendously during the time of one frame. If we used only the channel response estimated by the probe signal at frame head, the additional ISI could appear.

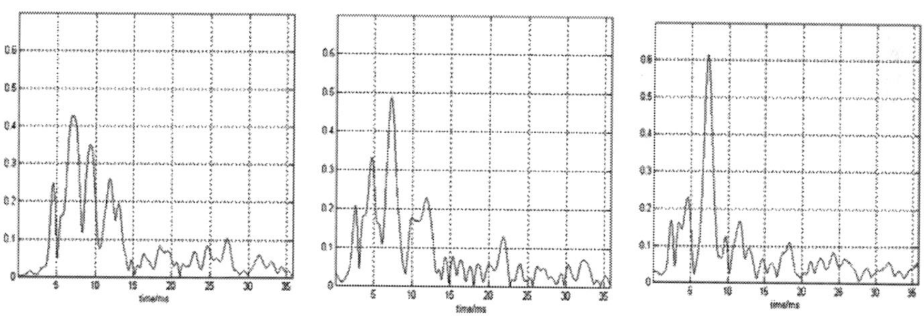

(a) (b) (c)

FIGURE 6. The change of a channel during 2 seconds

III. THE MODIFIED APPROACH

For the purpose of reducing the additional ISI, a modified method was proposed. The new method can estimate and supervise from time to time during all the communication process; the PPC processing should begin in action when the change of the channel is over the threshold we set. The modified structure is shown is Fig.7.

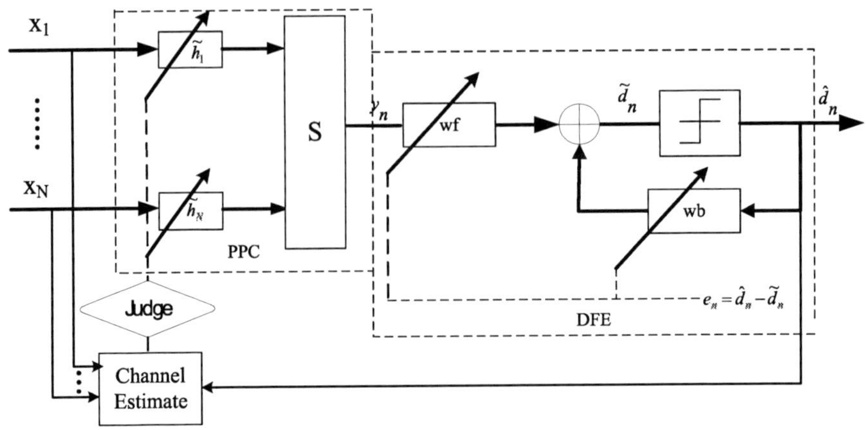

FIGURE 7. The modified approach structure

To verify the efficiency of the modified method, a computer simulation and a sea experiment were conducted. The result of computer simulation is shown in Fig.8.

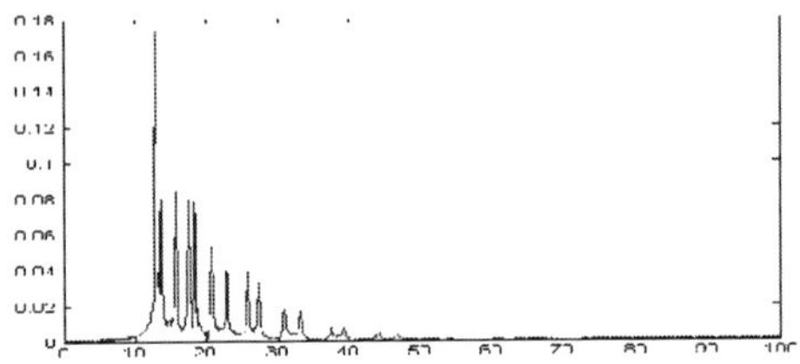

(a)

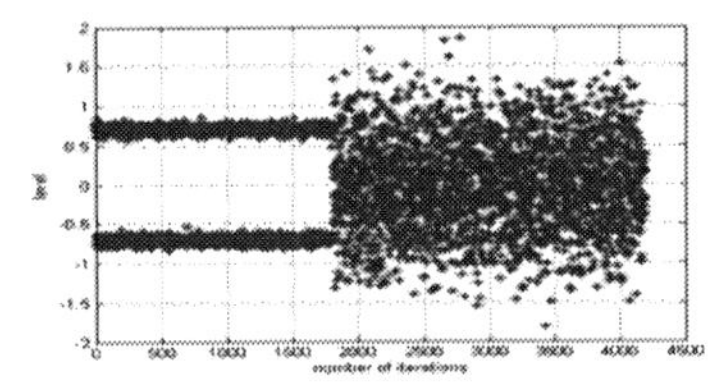

(c)

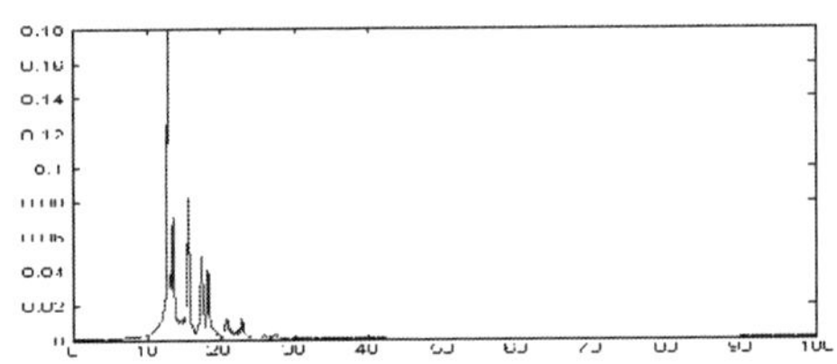

(b)

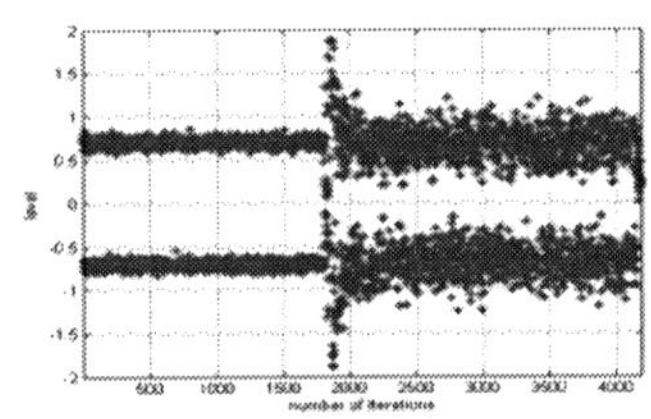

(d)

FIGURE 8. Performance of the modified approach in simulation. (a) Channel response at the beginning of the communication frame. (b)Channel response at the middle of the communication frame. (c) Performance of the PPC cascaded DFE method. (d) Performance of the modified method.

Fig.8 (a) and (b) show the change of the channel during one communication data frame time. Comparing (c) and (d), better performance with the modified method has obviously been obtained.

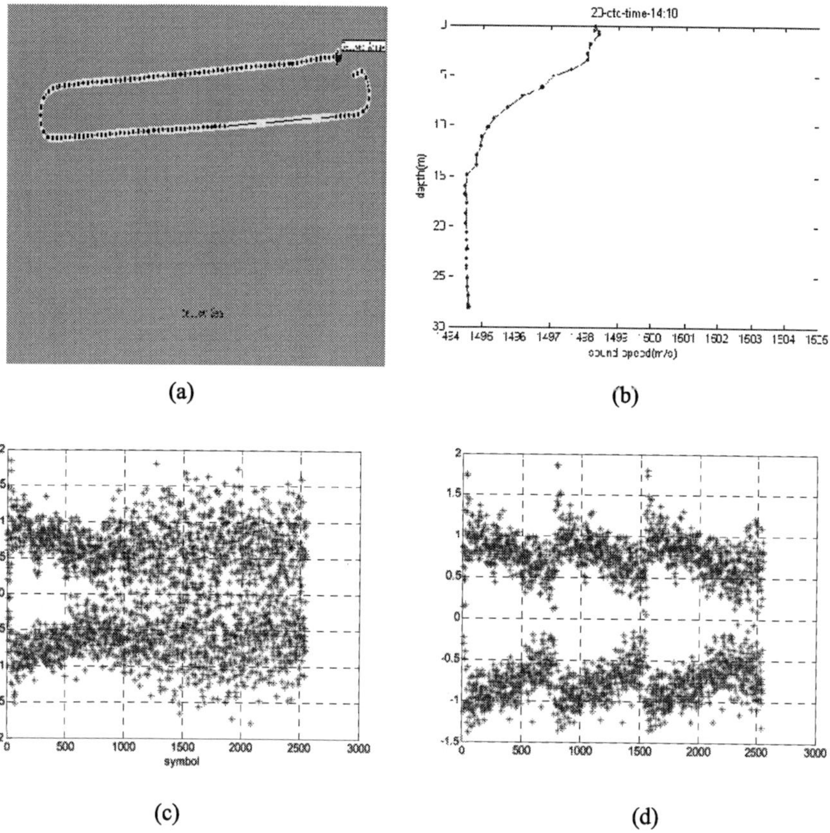

FIGURE 9. Result of sea experiment 3. (a)A vertical receive array was placed at the point of blue flags in the map, and a source moved along the yellow line, the max distance between source and receiver was about 10km. (b)Sound-speed profiles measured in the source area. (c)Performance using PPC cascaded with DFE method. (d) Performance using the modified method.

PPC cascaded with the DFE method does not work robustly because of the time-varying characteristic of the underwater communication channel especially in the case of a moving source. In this case the modified method gave better performance.

IV. CONCLUSION

In this paper a modified method based on PPC cascaded with DFE was proposed, and estimation of the channel response in real time was used to reduce the residual ISI. The computer simulation and sea experiment results show the efficiency of the modified method. As is well known in the case of a moving sound source, the channel response changed from point to point; therefore in order to follow the change of the channel response, a proper algorithm is very important to use. A good algorithm could speed the estimation of the channel response in real time.

REFERENCES

1. H.C.Song,W.S.Hodgkiss,W.A.Kuperman,M.Stevenson,and T.Akal."Improvement of Time-Reversal Communnications Using Adaptive Channel Equalizers," *IEEE J. Ocean. Eng., Vol.31, no.2, pp.487-496, Apr.2006.*
2. G.F.Edelmann,H.C.Song,S.Kim,W.S.Hodgkiss,and T.Akal."Underwater Acoustic Communications Using Time Reversal", *IEEE J. Ocean. Eng., Vol.30, no.4, pp.852 -864, Oct.2005.*
3. T.C.Yang."Differences Between Passive-Phase Conjugation and Decision-Feedback Equalizer for Underwater Acoustic Communications", *IEEE J. Ocean. Eng., Vol.29, no.2, pp.472 -487,Apr.2004.*

Joint Time-Reversal Processing and Spatial Diversity for Distributed Target Detection in Shallow Water

Chunxiao Li[a,b], Xiang Pan[b] and Mingfei Guo[c]

[a]*Zhijiang College of Zhejiang University of Technology, Hangzhou, 310024, P. R. China*
[b]*Department of Information Science and Electronic Engineering, Zhejiang University, Hangzhou 310027, P. R. China*
[c]*MOE Key Laboratory of Mechanical Manufacture and Automation, Zhejiang University of Technology, Hangzhou, P. R. China*

Abstract. Due to the high resolution of time reversal processing in a shallow water waveguide, the target can not be treated as a point-like scatterer, but should be modeled as a distributed target composed of elemental scatterers. In this paper, the problem of using multiple focusing on the target to improve the detection performance is investigated. The signal processing method consists of both coherent time-reversal processing and incoherent accumulation of the output of time-reversal processing from each elemental scatterer. The influence of target scintillations on detection using only time reversal processing and the detection performance of joint processing are investigated by a waveguide experiment. The experimental results show that even though the target has a regular aspect, the detection performances differ significantly when different parts of the target are ensonified due to the wave propagation effects in a shallow water waveguide. The effectiveness of incoherent processing that takes advantage of spatial diversity to improve detection performance is also evaluated.

Keywords: time reversal, spatial diversity, distributed target, detection

PACS: 43.60.Tj, 43.30.Vh, 43.60.Fg

1. INTRODUCTION

It is important for target detection to make good use of the characteristic features of wave propagation in a shallow water waveguide. However, the traditional detectors usually view certain features as nuisance parameters that degrade the sonar performance because they are based on some critical assumptions. The assumption that the acoustic signal in shallow water is plane wave propagation is employed by most traditional detectors, but since the ocean is inhomogeneous and sound is channeled in a waveguide bounded above by the ocean surface and below by the ocean bottom, this assumption is inaccurate. Another assumption is that the target is a point-like scatterer, while in the practical application for sonar, target scintillations lead to target fading that can cause a large degradation in the system's performance. The amount of energy reflected from the target toward the receiver is a function of the target aspect with respect to the transmitter/receiver pair. In addition, due to the effect

CP1272, *Shallow-Water Acoustics, Second International Shallow-Water Acoustics Conference*
edited by J. Simmen, E. S. Livingston, J. X. Zhou, and F. H. Li

of waveguide propagation, it is also influenced by the target position in the waveguide. For example, the energies are very different for two targets at the same range but different depths, even with equal strength.

Great efforts have been made to solve these problems. Based on plane wave propagation, the transmitter array of sensors for active sonar is typically used to steer a beam in a certain direction, and then plane wave beamforming and matched filter are used at the receiver side to determine the presence of the target. Therefore, it is thought that channel multipath is detrimental and should be minimized. Time reversal (TR) processing provides an alternative solution to this problem [1,2,3]. TR transmission can realize spatial focusing and temporal compression at the position of the target by using right wave front matched to the propagation medium. It has been shown that TR processing can make good use of multipath propagation [2,3]. Similarly, most, if not all, sonar processing ignores target scintillations and considers them as a niusance. A new concept of multiple-input multiple-output (MIMO) radar has been introduced in recent years to apply target spatial diversity to improve detection performance [4,5,6].

In free space, it is well known that the resolution of an array depends on the physical extent of the array aperture. But in an ocean waveguide, the spatial complexity of the sound field resulting from multipath propagation increases the resolution over free space. Therefore, TR processing can obtain a higher resolution than the conventional method in an ocean waveguide [7]. The target is usually localized in a single beamwidth and treated as a point-like scatterer with conventional transmission when its actual dimension is not too large or it localizes at a long distance. However, it is possible that the target spans several beams using TR processing, as a result of an increase in the effective aperture. In this case the target should not be treated as a point-like scatterer.

For the MIMO radar system, it is assumed that the target is large with complex bodies composed of many small elemental scatterers [5]. For our purposes in this paper, the target has a regular aspect. The main cause of target scintillations is the wave propagation effect in a shallow water waveguide. Although TR processing can enhance echo-to-reverberation ratios compared to the conventional method [3], target scintillations can possibly degrade detection performance more when TR processing is used because it has a higher resolution than the conventional method.

In this paper, we consider the use of both TR processing and spatial diversity to enhance target detectability. The rest of the paper is organized as follows: In Section 2, a signal model based on a distributed target using both TR processing and spatial diversity is introduced. Section 3 presents experimental results of a laboratory waveguide. A summary and concluding remark then follow in Section 4.

2. PROBLEM FORMULATION

In order to choose a right wave front for TR transmission, the target transfer function should be known a-priori in a shallow water waveguide. We assume that there is a probe source near the target or that information about the (approximate) locations of the targets of interest is available and the environmental parameters such as sound speed profile, water depth, and bottom characteristics are exactly known.

Therefore, the target transfer function can be obtained from the probe source or propagation model to achieve TR detection. In practice, it is difficult for us to exactly know the target transfer function. Several approaches can solve the problem resulting from an unknown target transfer function. For example, iterative time reversal and decomposition of the time reversal operator are two common practical TR methods [1,2]. In addition, a collection of TRM-to-probe source transfer functions constitutes a virtual source array, and then steers to focus at a selected location for which the transfer function is not known *a priori* [8].

The point source assumption currently dominates signal models used in sonar theory. For a large range between the target and the array, the point source assumption is an adequate approximation using the conventional method even in a shallow water waveguide. For TR processing, a transverse resolution can be nine or more times better than the free-space limit prediction, which relies on the waveguide nature of acoustic propagation and attenuation in the bottom [7]. Therefore, the point source model is not suited to depict the received signal for a TR detector, and a more detailed model must be developed.

Assuming that the source-receiver array (SRA) has N elements, the transfer matrix $\mathbf{K}(t)$ is a $N \times N$ matrix with $k_{lm}(t)$ being the signals received at the lth element after a temporal Dirac function is applied to the mth element. Consider a target composed of many small elemental scatterers that are distributed over an area denoted by $\mathbf{S}$. We can define a function $\sigma(r_i, z_i, t)$, which represents the apparent reflectivity depending on the reflectivity of the elemental scatterer and the propagation effects. r_i and z_i represent respectively the range and depth of the ith scatterer $(r_i, z_i) \in \mathbf{S}$, and $\sigma(r, z)$ is zeros outside $\mathbf{S}$. Provide that the target is composed of p elemental scatterers, the transfer matrix can be expressed as

$$\mathbf{K}(t) = \mathbf{G}(t) * \Sigma(t) * \mathbf{G}^T(t), \tag{1}$$

where $\mathbf{G}(t)$ is an $N \times p$ matrix, and each column $\mathbf{g}_i(t) = [g(r_i, z_i, z_1, t), \ \cdots \ , g(r_i, z_i, z_N, t)]^T$ is a normalized channel Green's function between the SRA and one of the elemental scatterers, and $\Sigma(t) = diag\{\sigma(r_1, z_1, t), \ \cdots \ , \sigma(r_p, z_p, t)\}$, and * denotes the convolution product in time. When there is only one scatterer, the transfer matrix can be denoted by

$$\mathbf{K}_o(t) = \mathbf{g}_o(t) * \sigma(t) * \mathbf{g}_o{}^T(t). \tag{2}$$

Denote by $\mathbf{r}(t) = [r_1(t), \ \cdots \ , r_N(t)]^T$ the collection of the received signals at the various receiving elements, and by $\mathbf{s}(t) = [s_1(t), \ \cdots \ , s_N(t)]^T$ the collection of the transmitted signals from the various transmitting elements. The received signal vector can be described by the following simple model

$$\mathbf{r}(t) = \mathbf{K}(t) * \mathbf{s}(t) + \mathbf{n}(t). \tag{3}$$

The signal model Eq. 3 is a generalized formulation and can be used to describe many different systems. The difference among the systems is the design of the transmitted signals and the target model. For example, steering a beam toward a certain direction based on plane wave propagation and a point-like target represents the conventional detector, while using a right wave front of waveguide propagation to ensonify a location based on a point-like target is TR detection. Furthermore, if we choose the probe signals transmitted both on waveguide propagation and a distributed target model, this is a joint processing of the TR method and the spatial diversity technique.

Conventional detector: Conventional detectors are the most common sonar system used array of sensors in practice. For conventional detectors, the transmitted signal is given by $\mathbf{s}(t) = \mathbf{v}(t) * s(t)$, where $\mathbf{v}(t)$ is the transmitter steering vector which depends on the spatial parameter θ. The target is modeled as a point-like scatterer. In this case, the received signal is given by

$$\mathbf{r}(t) = \mathbf{K}_o(t) * \mathbf{v}(t) * s(t) + \mathbf{n}(t). \tag{4}$$

Time reversal detector: TR detectors are proposed to make use of multipath propagation to enhance echo to reverberation ratios. The transmitter steering vector of a TR detector is the time reversal target transfer function based on waveguide propagation such as channel Green's function $\mathbf{g}_o(T-t)$ (time reversal in the time domain is equal to complex conjugation in the frequency domain). The received signal becomes

$$\mathbf{r}(t) = \mathbf{K}_o(t) * \mathbf{g}_o(T-t) * s(t) + \mathbf{n}(t). \tag{5}$$

Substituting Eq. 2 into Eq. 5 and considering that $\mathbf{g}_o(t)$ is a normalized function, the received signal now becomes

$$\mathbf{r}(t) = \sigma(t) * \mathbf{g}_o(t) * s(t) + \mathbf{n}(t). \tag{6}$$

Before any signal processing methods are adopted at the receiver side, the received signals in Eq. 6 get the highest average element signal-to-noise ratio by TR transmission matched to the waveguide. At the same time, the spatial coherent processing can be achieved as long as a good focus is obtained at the location of the target for a monostatic configuration. Therefore, TR detection in theory is an "optimal" detector which splits signal processing between the transmitter side and the receiver side, and both sides realize an "optimal" spatial filter matched to the waveguide transfer function.

Joint detector of time reversal processing and spatial diversity: Provided that the target spans several beamwidths, it can be viewed as a distributed target composed of several elemental scatterers. We can either use a probing signal design such as maximizing the power around the locations of the scatterers[6], or selectively focus on each of the scatterers and incoherent sum the ultimate outputs at the receiver side[9].

Let us assume that there are p elemental scatterers. Without loss of generality, they are assumed to be at locations $\{(r_i, z_i)\}_{i=1}^{p}$. A time-division method is proposed for selectively focusing target spatial separation. Suppose that the multiple signals are equally spaced by T, the pulse can then be expressed as

$$\mathbf{s}(t) = \sum_{i=1}^{p} \mathbf{g}_i(t-iT) * s_i(t-(i-1)T) \tag{7}$$

The received signal can be described as

$$\mathbf{r}(t) = \sum_{i=1}^{p} r_i(t-iT) = \sum_{i=1}^{P} \mathbf{K}(t-iT) * \mathbf{g}_i(t-iT) * s_i(t-(i-1)T) + \mathbf{n}(t) \tag{8}$$

If T is large enough so that there is no overlap between echoes of different pulses, a TR detection would be realized for point-like scatter over a time duration less than T

in which the echoes are expected. The detector of joint time reversal processing and spatial diversity creates p virtual TR detectors based on a point-like assumption, and then adds their outputs non-coherently.

3. WAVEGUIDE EXPERIMENT

3.1. Experimental Setup

The experiment was performed in a laboratory waveguide (see Fig.1) using a vertical SRA of 32 elements equally spaced by 0.04 *m* from 0.08 to 1.32 *m* in depth. The laboratory waveguide is 14 *m* in length, 1.2 *m* in width and 1.44 *m* in height. Three sides of the pool are covered by sound absorption material and the other side is a steel sheet located 13 *m* away from the SRA. The pool bottom is sand 22 *cm* thick. The target is an air-filled, steel cylinder shell 0.51 *m* long, 0.15 *m* in diameter and 3 *cm* in wall thickness. It is suspended 8.6 *m* in range and ranges from 0.76 to 1.27 *m* in depth.

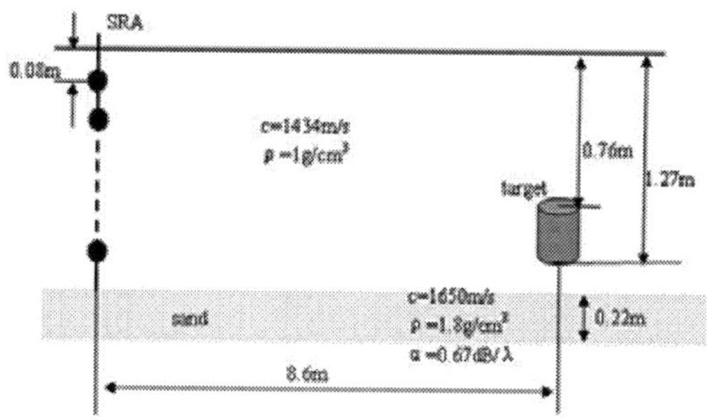

FIGURE 1. Laboratory waveguide experimental setup

3.2. Experimental Setup

The transmitted signal is a 1 *ms* linear frequency modulated (LFM) signal with a center frequency of 12 *kHz* and 4 *kHz* bandwidth. According to the geoacoustic parameter, the spatial resolution can be obtained using Eq. 20 and Eq. 21 in reference. The depth resolution for the 3 *dB* beamwidth of center frequency is about 0.11 *m* and the range resolution for the 3 *dB* beamwidth of center frequency is about 0.09 *m*. Because the size of the target is much larger than the depth resolution, it can not be modeled as a point-like scatterer. In order to fully ensonify the different parts of the target, it is modeled as a distributed target composed of four elemental scatterers located respectively at 0.80 *m*, 0.94 *m*, 1.08 *m* and 1.22 *m* in depth.

In order to investigate the waveguide propagation effect on target detection, TR detection based on a point-like target assumption for each elemental scatterer is implemented. At the transmitter side, the excitation weight vector is the target transfer function of ith scatterers obtained via forward modeling. At the receiver side, the echo received by the SRA with N sensors is coherently added using spatial matched filtering and becomes a signal with one dimension (a sequence in time domain). Then temporal matched filtering is implemented in order to enhance the range resolution. This is the procedure of data processing for TR detection.

Fig. 2 shows the outputs of TR detectors for the shallowest scatterer and the deepest one. Note that the target at range of 8.6 *m* is obvious whether we ensonify the shallowest part of the target or the deepest part of the target, while the steel sheet (about 13 *m* in range) is barely visible. The reason is that the focusing capability of TR transmission can realize spatial and temporal focusing at the place of the target, and at the same time use less energy to ensonify other places besides the target. Therefore, the TR transmission reduces the generation of the reverberation. Comparing Fig. 2(a) with (b), the output of the deepest part is obviously better than the shallowest one. The target is a cylinder with regular aspect, and the reflectivities of the four elemental scatterers have little difference in free space. However, the detection performance differs significantly for ensonifying different parts of the target. As mentioned above, the main factor lies in the propagation effects between the scatterers and the SRA.

For detection by joint TR processing and spatial diversity, the transmitted multiple signals are given in Eq. 7 via forward modeling. The outputs of the four virtual TR detectors are obtained and then non-coherently added to get the ultimate output, as shown in Fig. 3. It can be seen from Fig. 3 that the detection performance is greatly improved by making good use of the spatial diversity of the distributed target. The main advantage of this processing is that it can guarantee the stability of target detection particularly when there are target scintillations.

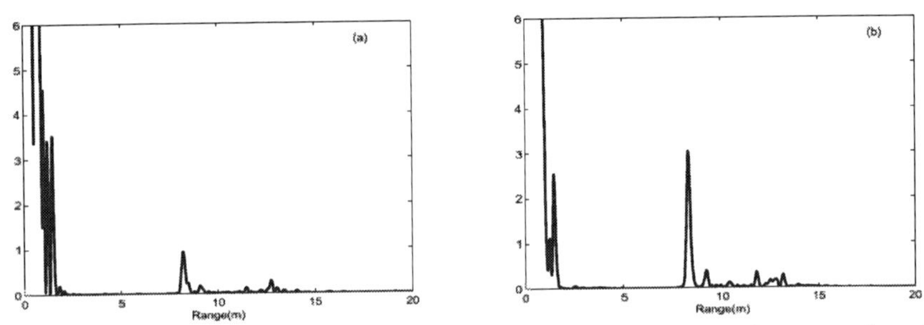

FIGURE 2. The output of the TR detector with the distributed target at 8.6 *m* in range and ranging from 0.76 to 1.27 *m* in depth. (a)the excitation weight vector for TR transmission is the target transfer function of the shallowest scatterer, (b) the excitation weight vector for TR transmission is the target transfer function of the deepest scatterer.

4. CONCLUSIONS

In this paper, we discuss distributed target detection in a shallow water waveguide. In order to take full advantage of wave propagation in a shallow water waveguide and guarantee the stability of target detection at the same time, joint processing of TR processing and spatial diversity is introduced. The waveguide experiment showed that (1) Even though the target has a regular aspect, the detection performances differ significantly when different parts of the target are ensonified due to the wave propagation effects in a shallow water waveguide; (2) The detection performance of joint TR processing and spatial diversity is significantly improved, which can guarantee the stability of target detection particularly when there are target scintillations. Note that we only investigate the detection performance of the "optimal" detectors, which can rarely be used in practical applications because we do not know exactly the transfer function in an uncertain ocean environment. The topic of practical time reversal implementation is left for future research.

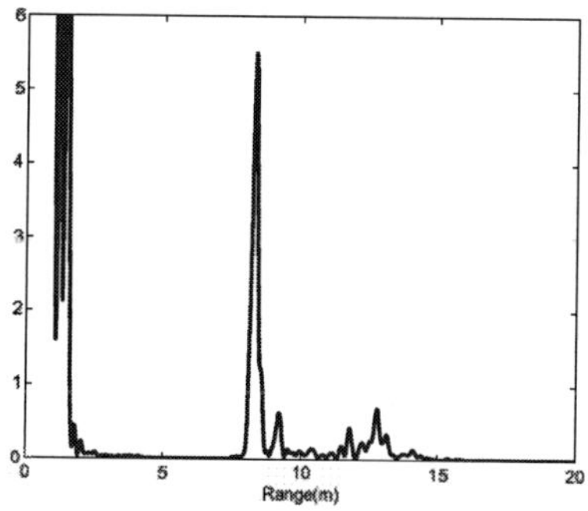

FIGURE 3. The output of the detector of joint TR processing and spatial diversity with the distributed target at 8.6 *m* in range and ranging from 0.76 to 1.27 *m* in depth.

ACKNOWLEDGMENTS

The authors would like to acknowledge the support of the waveguide experiment from Hangzhou Applied Acoustics Research Institute. This work was partly supported by the National Natural Science Foundation of China (Grant Number 60872066 and 60872067).

REFERENCES

1. K. G. Sabra, P. Roux, H. C. Song, W. S. Hodgkiss, W. A. Kuperman, T. Akal, and J. M. Stevenson, "Experimental demonstration of iterative time-reversed reverberation focusing in a rough waveguide. Application to target detection,"J. Acoust. Soc. Am., vol. 120, no. 3, 2006, pp. 1305-1314.
2. C. Prada, J. D. Rosny, D. Clorennec, J. G. Minonzio, A. Aubry, and M. Fink, "Experimental detection and focusing in shallow water by decomposition of the time reversal operator," J. Acoust. Soc. Am., vol. 112, no. 2, 2007, pp. 761-768.
3. S. Kim, W. A. Kuperman, W. S. Hodgkiss, H. C. Song, G. Edelmann, and T. Akal, "Echo-to-reverberation enhancement using a time reversal mirror," J. Acoust. Soc. Am., vol. 115, no. 4, 2004, pp. 1525-1531.
4. J. Li, P. Stoica, "MIMO radar with colocated antennas," IEEE SIGNAL PROCESSING MAGAZINE, SEPTEMBER, 2007, pp. 106-114.
5. A. M. Haimovich, R. S. Blum, L. J. Cimini, "MIMO radar with widely separated antennas," IEEE SIGNAL PROCESSING MAGAZINE, JANUARY, 2008, pp. 116-129.
6. P. Stoica, J. Li, Y. Xie, "On probing signal design for MIMO radar," IEEE TRANSACTIONS ON SIGNAL PROCESSING, vol. 55, no. 8,2007, pp. 4151-4161.
7. S. Kim, G. F. Edelmann, W. A. Kuperman, W. S. Hodgkiss, and H. C. Song, "Spatial resolution of time reversal arrays in shallow water," J. Acoust. Soc. Am., vol. 110, no. 2, 2001, pp. 820-829.

8. S. C. Walker, P. Roux, and W. A. Kuperman, "Synchronized time-reversal focusing with application to remote imaging from a distant virtual source array," J. Acoust. Soc. Am., vol. 125, no. 6, 2009, pp. 3828-3834.
9. J. C. Guey, and M. R. Bell, "Diversity waveform sets for delay-doppler imaging," IEEE TRANSACTIONS ON INFORMATION THEORY, vol. 44, no. 4, 1998, pp. 1504-1522.

Sonar detection range index estimation approach in uncertain environments

Fan Li, Sheng-ming Guo and Yao-ming Chen

Institute of Acoustics, Chinese Academy of Sciences, Beijing, China 100190

Abstract. The traditional detection range index prediction of sonar systems assumes a deterministic environment and causes overestimation of the detection range index. The realistic ocean environment consists of a quantitative measure of environmental uncertainty such as sound speed profile, sea depth and so on. An estimation approach that incorporates the effects of environmental uncertainty into the sonar detection range index is proposed in this paper. The sonar detection range index prediction has been implemented by using Monte Carlo simulation. In simulation, the sound speed gradient, sea depth and bottom geo-acoustic parameters as important uncertainty environmental parameters are generalized to stochastic variables and satisfy the normal distribution. The sonar detection range index with unknown source depth is also considered.

Keywords: detection range index; uncertainty environmental parameter; unknown source depth

PACS: 43.30.Vh, 43.30.Bp

INTRODUCTION

Detection range is an important performance index for sonar systems. The traditional detection range index prediction of sonar systems assumes an isospeed, deterministic geo-acoustic parameters and bathymetry environment, and they ignore the effects of uncertain environmental parameters, causing overestimation of the detection range index. The overestimation cannot be resolved by advanced computational routines. Ocean environmental parameters such as water depth, sound speed profile, bottom geo-acoustic parameters have strong spatial and temporal variability, whose values are not known precisely. In the actual ocean environment, the sound speed profile often becomes the negative or positive sound speed gradient. Furthermore, the sea floor is more variable in the acoustic properties even over short ranges. Source position motion also limits the accuracy of the estimate of the detection

CP1272, *Shallow-Water Acoustics, Second International Shallow-Water Acoustics Conference*
edited by J. Simmen, E. S. Livingston, J. X. Zhou, and F. H. Li

range. Detection range can be predicted based on the classic sonar equation for a certain figure of merit (FOM). Since the FOM quantitatively equals the transmission loss under the conditions specified in the sonar equation, the FOM can give an immediate indication of the range at which a passive sonar can detect its target when corresponding transmission loss curves are already available.

Here an estimation approach that incorporates the effects of environmental uncertainty into the sonar detection range index prediction is proposed. Referring to the work of the Capturing Uncertainty directed by ONR, uncertainties of environment parameters will be transferred to uncertainty in the acoustic field prediction. For the certain FOM of a specific sonar system, the different transmission loss of the uncertain environments condition corresponds to the different sonar detection range. The credible detection range index in certain probability conditions will be achieved through a range–integrated approach. Detection range evaluations have been implemented using Monte Carlo simulations. Monte Carlo data samples over environmental uncertainties are generated considering a fine-scale variety in accordance with the actual ocean environment. The uncertainty of three uncertain parameters, including sound speed profile, water depth and bottom geo-acoustic parameters, are assumed to be statistically independent a priori with Gaussian distribution. Furthermore, the paper also considers the uncertainty of target source depth influence on the detection range index prediction. Referring to the M.B. Porter paper on selecting optimum receive depth, uncertain target source depth is assumed to be statistically independent a priori with uniform distribution.

The organization of the paper is as follows. The detection range prediction methodology is presented in Sec II. In Sec III, the uncertain shallow–water experimental environment is introduced to predict the detection range index. According to the simulated calculation, the range index comparison results between the traditional method and the paper method for uncertain environmental are present. The prediction results also consider the uncertain target source depth.

DETECTION RANGE INDEX PREDICTION METHODOLGY

The sonar equations are simple algebraic expressions used to quantify various aspects of sonar performance. The passive sonar equation with noise background is as follows:

$$SL - TL(r, z, z_s) = NL - DI + DT \tag{1}$$

The combination of terms labeled Figure of Merit (FOM) combines environmental, sonar and target parameters into one convenient quantitative measure of sonar performance.

$$FOM = SL - (NL - DI) - DT \tag{2}$$

Since the FOM quantitatively equals the transmission loss under the conditions specified in the sonar equation, the FOM can give an immediate indication of the range at which the passive sonar can detect its target. FOM is associated with the source level (SL), the noise level (NL), the directivity index of the array (DI) and a detection threshold (DT). We define the signal excess[1] (SE) representing the degree to which the source stands out above the noise background as follow:

$$SE(r \mid z_r, z_s) = FOM - TL(r \mid z_r, z_s) \tag{3}$$

In the traditional method, the detection range of a sonar system is considered to arrive at the maximum value when $SE(r_{\max}) = 0$. From signal detection theory,

$$r < r_{\max}, SE > 0 \text{ corresponds to } p_D = 1 \tag{4}$$

$$r > r_{\max}, SE < 0 \text{ corresponds to } p_D = 0$$

p_D: the probability of detection.

The ocean acoustic field is the complex-valued random acoustic pressure field, and here is assumed a complex Gaussian random variable. We will associate the transmission loss with the probability of detection, and the probability of detection is defined:

$$p_D(SE \mid r, z_r, z_s) = \frac{1}{\sigma\sqrt{2\pi}} \int_{-\infty}^{SE} e^{-0.5x^2/\sigma^2} dx \tag{5}$$

σ is related to the signal detection performance at a specific range and represents transitional efficiency between able-detection region and unable-detection region. For a certain set of source depth and receive depth, we can calculate the transmission loss at the different range. When the FOM is known, the SE of a realization of the uncertain environment parameters at the different range will be obtained by Eq. (3). The probability of detection, $p_D(r)$, which corresponds to the different SE will be generated in Eq.(5). Finally, we will obtain the specific detection range values of the uncertain parameter realization in Eq. (6) according to the $p_D(r)$.

$$r_{\max}(FOM \mid z_r, z_s) = \int_0^{\infty} p_D(SE \mid r, z_r, z_s)dr \tag{6}$$

The target source position is usually unknown precisely other than the uncertain ocean environment. When the target source depth is uncertain, the probability of detection is defined in Eq. (7):

$$\tilde{p}_D(SE \mid r, z_s) = \int_0^{\infty} p_D(SE \mid r, z_r, z_s)p_z(z_s)dz_s \tag{7}$$

$p_z(z_s)$ is the probability density function of target source depth, which is assumed to follow the uniform probability density function in the depth area in this paper. The maximum detection range of the case of unknown source depth is defined in Eq. (8):

$$\tilde{r}_{\max}(FOM \mid z_r) = \int_0^{\infty} \tilde{p}_D(SE \mid r, z_r)dr \tag{8}$$

The maximum detection range varies with the different set of environmental parameters. For certain environmental parameters, the occurrence probability of each detection range is $p_r(r_{\max})$. In order to get the detection range index that satisfies the different demands, we define r_D as the credible detection range under the condition of the probability D. When the threshold D is fixed, the credible range index will be obtained. The condition D=1 corresponds to the minimum value of all possible ranges. The threshold of detection range is defined in Eq. (9):

$$D=\sum_{i=1}^{N} p_r(r_{\max,i})|_{r_{\max,i}\geq r_D} \quad (9)$$

NUMERICAL SIMULATION

Simulation scenarios extended from the shallow water acoustic tomography experiment in January 2002 were used to illustrate the estimation approach. Figure 1 reviews the certain environmental parameter model, where the source depth is 10m and the receiver depth is 30m. Table I lists the symbols shown in Fig. 1.

In Table I, the ranges of environmental uncertainties were present. We only consider one receiver depth, and the sediment density was assumed to be fixed at the standard values in the simulations, since the sediment density is insensitive to the parameters for range estimation.

Since the source depth is one of the most sensitive parameters, we consider two groups of uncertain environmental scenarios: group A is the certain source depth and other four uncertain parameters; group B is the uncertain source depth and other four uncertain parameters. Source depth is assumed with uniform probability density function[6]. Uncertain source depth varies from 5m to 15m. The other four uncertain parameters are Gaussian random variables[4]. According to the experimental analysis, the sediment is assumed to the sand-mud-clay type[2,3].

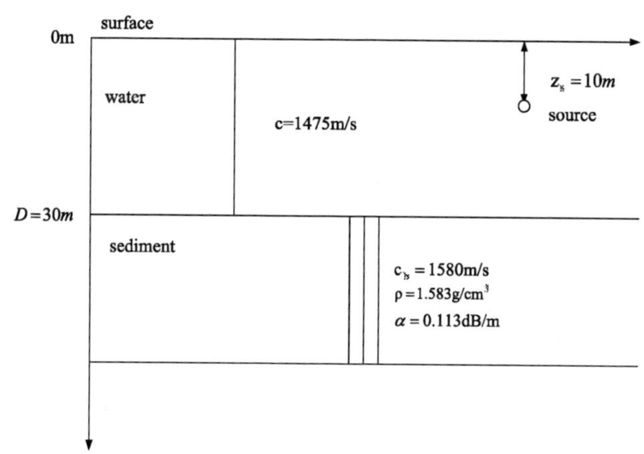

FIGURE 1. the shallow water acoustic tomography experiment in January 2002

TABLE 1. Parameters of uncertain shallow-water environments.

Environmental parameter	Standard value	Value range
Bottom depth	D=30m	D±5%D
Sound speed gradient	C=1475m/s	0.1m/s/m
Sediment sound speed	Cp=1580m/s	Cp±5%Cp
Sediment density	1.583g/cm^3	1.583g/cm^3
Sediment attenuation	α=0.113dB/m	α±20%α
Source depth	Sd=10m	Sd±5m
Receive depth	30m	30m

A. Certain source depth environmental scenarios

Group A is the certain source depth and other four uncertain parameters. Source depth is fixed at 10m, and other uncertain parameters values refer to Table 1. We consider two frequency bands, one is from 300Hz to 600Hz, the other is from 2kHz to 3kHz. The figure of merit (FOM) is 80dB. The sonar detection range probability distribution[5] with certain source depth and four uncertain parameters is shown in Figure 2. In Table 2, we choose the different the threshold D and compare the range index result between the traditional method and the paper method.

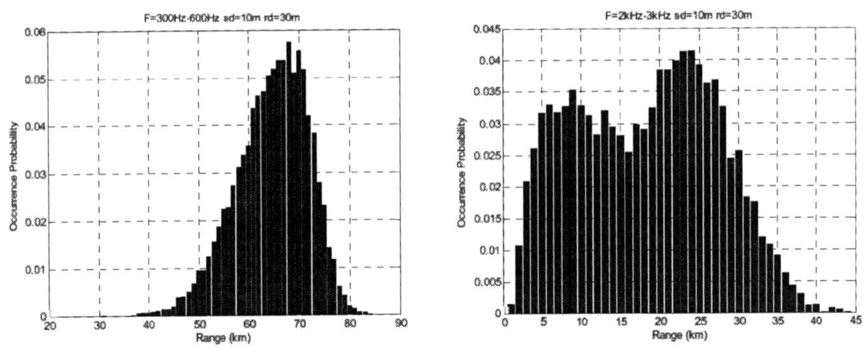

FIGURE 2. Sonar detection range probability distribution with four uncertain environmental parameters

TABLE 2. The range index comparison between two methods for fixed source depth and four uncertain environmental parameters

Frequency band	The paper method				Traditional method
	D=1	D=0.8	D=0.5	D=0.3	
300Hz~600Hz	30.9km	59.0km	66.0km	69.0km	66.66km
2kHz~3kHz	1.0km	9.0km	19.0km	24.0km	19.03km

B. Uncertain source depth environmental scenarios

Group B is uncertain source depth and other four uncertain parameters. Source depths vary from 5m to 15m and other uncertain parameters values are the same as for group A. The sonar detection range probability distribution with uncertain source depth and four uncertain parameters is shown in Figure 3. The range index comparison results for uncertain source depth and four environmental parameters are present in Table 3.

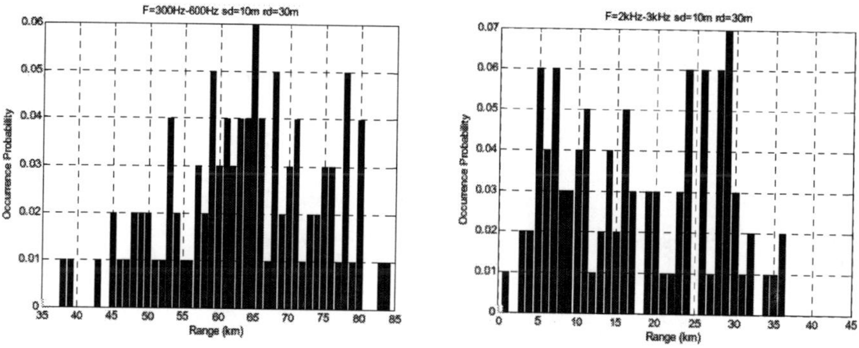

FIGURE 3. Sonar detection range probability distribution with four uncertain environmental parameters and uncertain source depth

TABLE 3. The range index comparison between two methods for uncertain source depth and four environmental parameters

Frequency band	The paper method				Traditional method
	D=1	D=0.8	D=0.5	D=0.3	
300Hz~600Hz	38.3km	55.0km	65.0km	70.0km	65.8km
2kHz~3kHz	1.1km	8.0km	17.0km	35.0km	19.04km

CONCLUSIONS

A sonar detection range index estimation approach is derived in this paper. The approach includes uncertain environmental parameters and target position parameter such as water depth, sound speed profile, bottom geo-acoustic parameters and source depth. The approach provides a detection range area for a sonar system, not a range value, in an uncertain ocean environment. This is closer to the actual problem and more reasonable than the traditional method because it allows for the effect of uncertainty. In the future, investigating and choosing a reasonable threshold D for the sonar detection range estimation approach in an uncertain environment is a key problem that is of practical interest.

REFERENCES

1. Ferla C. M and Porter M.B, *IEEE J. Oceanic Eng*, **16**, 267-278(1991).
2. WANG Dezhao and SHANG Erchang, *Underwater acoustics* , Beijing, Science Press, 1981,pp.52.
3. LIU Bosheng, LEI Jiayu, *Principle of underwater acoustics*, Harbin, Press of Harbin Engineering University, 2002, pp.69
4. Liewei Sha and Loren W.Nolte, *J.Acoust.Soc.Am.*, **117**, 1942-1953(2005)
5. F.B.Jensen, W.A.Kuperman, M.B.Porter and H.Schmidt, *Computational Ocean Acoustic, AIP series in Modern Acoustic and Signal Processing*, New York, AIP Press ,1994
6. James,K.R.,and Dowling,D.R.A, *J.Acoust.Soc.Am,* **118**,*2802-2810*(2005)

Analysis on Sound Field Recorded with Vector Sensors[1]

Fenghua Li, Mei Sun, and Renhe Zhang

State Key Laboratory of Acoustics, Institute of Acoustics, Chinese Academy of Sciences, Beijing 100190

Abstract. From data collected with vector sensors at several sites, the transmission losses of particle velocities and pressures are examined. In particular, the components of the particle velocity fields both normal and tangential to the horizontal plane are studied. Theoretical and experimental results show that both vertical and horizontal particle velocities can be regarded as the summation of normal modes. The lower modes dominate for the horizontal particle velocity, while higher modes are relatively important for the vertical particle velocity. The intensity of the vertical velocity decreases faster than that of the horizontal particle velocity. The experimental results also show that the same modes of the horizontal and vertical particle velocity have similar losses, but different amplitudes and phases. Theoretical analyses indicate that the difference of the mode amplitudes between horizontal particle velocity and vertical particle velocity depends on the eigenvalue and receiver depths, which can provide information on the estimation of bottom parameters. A geo-inversion scheme by the matched-field processing with a vector array is developed. The theoretical and experimental results indicate that the proposed inversion method can decrease the uncertainty of inversion in comparison with that by hydrophone arrays.

Keywords: Vector Sensor, Geo-acoustic Inversion, Normal mode
PACS: 43.30Bp, 43.30Re

INTRODUCTION

Vector sensors have attracted much attention in recent years. In comparison with hydrophone, vector sensors can provide measurements of three dimensional particle velocities, which have been widely used in the Direction of Arrival (DOA) estimation and improvement of Signal Noise Ratio (SNR)[1-4]. However, few papers have been published on the analyses of transmission losses of particle velocities and geo-acoustic inversion by vector sensors[5-6]. In this paper, based on normal mode theory, a concise expression of the transmission losses of vertical particle velocities have been derived. The theoretical results indicate that the vertical particle velocity has larger loss in comparison with the horizontal particle velocity, which is in agreement with the experimental results. A Matched Field Processing (MFP) with a vector sensor array has been applied to the data, which can reduce the inversion uncertainty in comparison with that of the hydrophone array.

[1] Supported by National Natural Science Foundation of China (10734100)

CP1272, *Shallow-Water Acoustics, Second International Shallow-Water Acoustics Conference*
edited by J. Simmen, E. S. Livingston, J. X. Zhou, and F. H. Li

THEORY OF VECTOR FIELDS

With normal mode theory, the pressure, horizontal and vertical particle velocities excited by a homogenous point source can be expressed as

$$\mathrm{P} = \sqrt{\frac{8\pi}{r}} e^{i\pi/4} \sum_l \Phi_l(z_s)\Phi_l(z) \frac{1}{\sqrt{\mu_l}} e^{i\mu_l r - \beta_l r} \tag{1}$$

$$\mathrm{V}_r = i\sqrt{\frac{8\pi}{r}} e^{i\pi/4} \sum_l \Phi_l(z_s)\Phi_l(z) \sqrt{\mu_l}\, e^{i\mu_l r - \beta_l r} \tag{2}$$

$$\mathrm{V}_z = \sqrt{\frac{8\pi}{r}} e^{i\pi/4} \sum_l \Phi_l(z_s)\Phi_l'(z) \frac{1}{\sqrt{\mu_l}} e^{i\mu_l r - \beta_l r} \tag{3}$$

where r is the range between the source and receiver, $\Phi_l(z)$, μ_l and β_l are the eigenfunction, eigenvalue, and eigen-attenuation, ' denotes the derivation over z. V_r and V_z are the horizontal and vertical particle velocities, respectively. In this paper, the factor $i/(\rho\omega)$ has been neglected in the expression of V_r and V_z.

By the smooth average theory, equations (1) to (3) in the Pekeris channel can be approximated as (in the three-half law region)

$$<|\mathrm{P}|^2> = \left(\frac{\pi}{QH}\right)^{1/2} r^{-3/2} \tag{4}$$

$$<|\mathrm{V}_r|^2> = k^2\left(\frac{\pi}{QH}\right)^{1/2} r^{-3/2} \tag{5}$$

$$<|\mathrm{V}_z|^2> = k^2\left(\frac{\pi}{QH}\right)^{1/2} r^{-3/2} \frac{H}{2Qr} \tag{6}$$

It can be seen from Equations (4) to (6) that in general the vertical particle velocity V_z is much smaller than the horizontal particle velocity V_r, and V_z has a larger loss than V_z. Fig. 1 shows the comparison of the losses between the measured pressure, the horizontal and vertical particle velocities. The x-axis is the range between the source and the receiver, and the y-axis is the transmission loss. The circles are the measured pressures, and the triangles are the measured vertical particle velocities, the crosses are the measured horizontal particle velocities. The lines in the figure are the numerical simulations. It can be seen in the figure that the vertical particle velocity has a larger transmission loss than that of horizontal transmission loss, because high order normal modes contribute more to vertical particle velocity due to their large grazing angle; on the other hand, the low-order normal modes contribute more to the pressure and the horizontal particle velocity. It can also be seen in the figure that the measured data are in good agreement with the simulations.

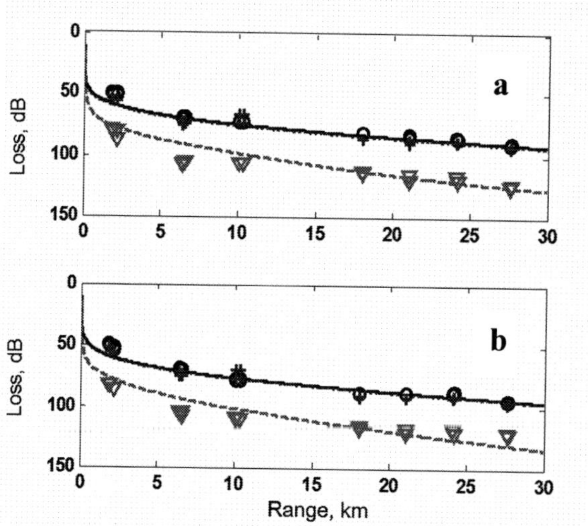

FIGURE 1. Measured and simulated pressure, particle velocities. (a): 600 Hz, (b): 800 Hz.

MODE ANALYSIS WITH VECTOR SENSOR

To further investigate the differences between the transmissions of the horizontal and vertical particle velocities, different mode amplitudes of the particle velocities are analyzed. Fig. 2 shows the sound speed profile measured in an experiment. An air-gun was used as a wide-band sound source from 5 km to 12 km. The depth of the vector sensor was 18.1 m, and the water depth was 32.5 m.

The recorded waveforms are shown in Fig. 3. The solid curves are the recorded waveforms and the dashed curves are the corresponding envelopes. The frequency range is from 100 Hz to 140 Hz. The range between the source and the receiver is 5.25 km. The upper plot is the measured horizontal particle velocity V_r, and the lower plot is the measured vertical particle velocity V_z. It can be seen clearly in the upper plot, which corresponds to V_r, that there are two normal modes in the recorded signal, and the amplitude of normal mode 1 is larger than that of normal mode 2. However in the lower plot, which corresponds to V_z, the amplitude of normal mode 1 is much smaller than that of normal mode 2. It can also be seen from the figure that the waveform of normal mode 2 of V_r has a $\pi/2$ phase shift with that of V_z.

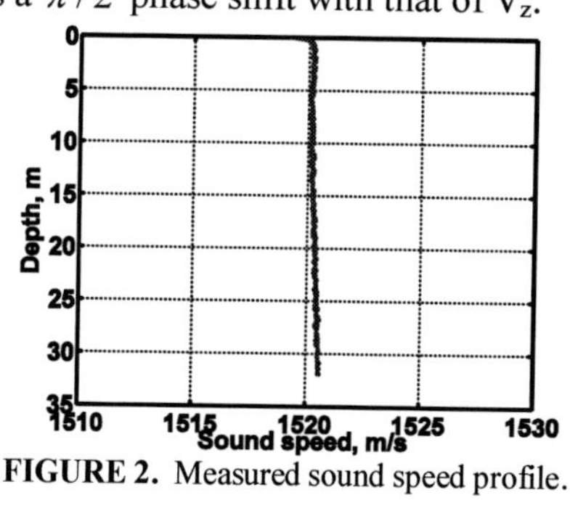

FIGURE 2. Measured sound speed profile.

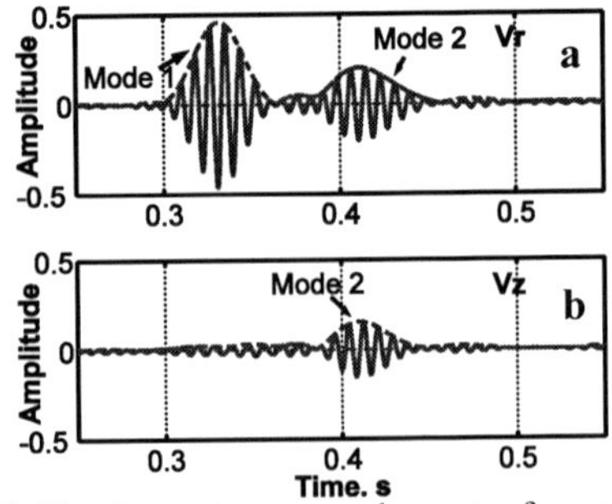

FIGURE 3. Waveform recorded by the vector sensor, the center frequency is 120 Hz. (a): Vr, (b): Vz.

The ratio between the amplitudes of the lth normal mode of V_r and V_z is defined as

$$RMHV_l = \frac{\text{Amplitude of } l\text{th mode of } V_r}{\text{Amplitude of } l\text{th mode of } V_z} = \frac{\Phi_l(z)\mu_l}{\Phi_l'(z)} \tag{7}$$

The measured ratios between the amplitudes of normal mode 2 of the horizontal and vertical particle velocity (RMHV) at 120 Hz (upper plot) and 200 Hz (lower plot) at different ranges are shown in Fig. 4. The circles are the measured RMHV. The lines are the average of all the measured RMHVs. It can be seen from Eq. 7 and Fig. 4 that the ratio $RMHV_l$ is a function of eigenfunction, eigenvalue, receiver depth, and independent of the range and source depth, which means that the same normal modes of the horizontal and vertical particle velocities have the same transmission losses. Fig. 3 also shows that high normal modes contribute more to the vertical particle velocity V_z than the horizontal particle velocity V_r. However, high normal modes have relatively larger transmission losses than the lower normal modes in general. Therefore, V_z has larger losses that V_r, which is in agreement with the measured data shown in Fig. 1.

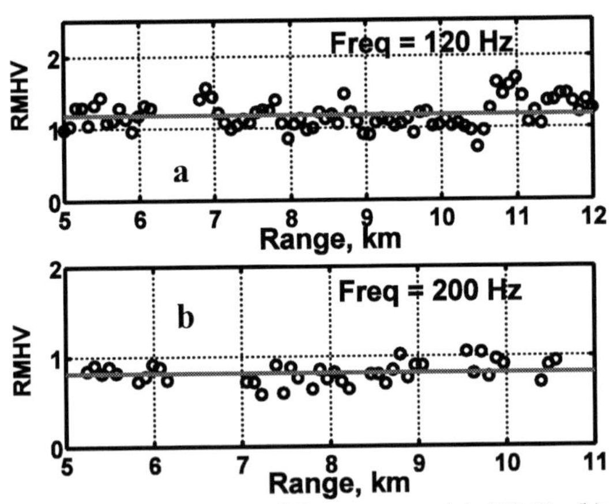

FIGURE 4. Measured RMHV at different ranges. (a): 120 Hz, (b): 200 Hz.

MATCHED-FIELD PROCESSING WITH VECTOR SENSOR ARRAY

Since the vertical and horizontal particle velocities have different combinations of the normal modes, this information can be applied to geo-acoustic inversion. In this paper, matched-field processing with a vector sensor array is applied to estimate the bottom parameters. The cost function of MFP geo-acoustic inversion with hydrophones can be defined as

$$C_p = \frac{\sum_j \left| \sum_i P(z_i, f_j) Q^*(z_i, f_j) \right|}{\sqrt{\sum_j \sum_i P(z_i, f_j) P^*(z_i, f_j) \sum_j \sum_i Q(z_i, f_j) Q^*(z_i, f_j)}} \tag{8}$$

where $P(z_i, f_j)$ and $Q(z_i, f_j)$ are the measured and the simulated sound pressures at receiver depth z_i and frequency f_j.

Similarly, the cost function of MFP geo-acoustic inversion with pressure and vertical particle velocity is defined as

$$C_{p,v_z} = \sqrt{C_p \times \frac{\sum_j \left| \sum_i V_z(z_i, f_j) v_z^*(z_i, f_j) \right|}{\sqrt{\sum_j \sum_i V_z(z_i, f_j) V_z^*(z_i, f_j) \sum_j \sum_i v_z(z_i, f_j) v_z^*(z_i, f_j)}}} \tag{9}$$

where $V_z(z_i, f_j)$ and $v_z(z_i, f_j)$ are the measured and simulated vertical particle velocity at receiver depth z_i and frequency f_j.

To verify the differences between Eqs. (8) and (9), another shallow water acoustic experiment is performed. The experimental configuration is shown in Fig. 5. A four-vector-sensor array is used in the experiment. In the geo-acoustic inversion, a half-infinite liquid bottom model is used. The density and attenuation is set as 1.78 g/cm^3 by bottom core, and the attenuation is set as 0.02 dB/m since it is not sensitive to the MFP inversion. The frequency band used in the inversion is from 200 Hz to 300 Hz. Therefore the unknown parameters are the water depth and bottom sound speed. Fig. 6 shows the cost function as a function of the water depth and bottom sound speed. The ranges of the color bars in the figure are from the maximum of the calculated cost function minus 0.05 to the maximum. The cost function shown in the upper plot is defined by Eq. (8) and the cost function of the lower plot is defined by Eq. (9). It can be seen from the figure that the uncertainty of the MFP inversion with vector array is largely reduced in comparison with that of a hydrophone array.

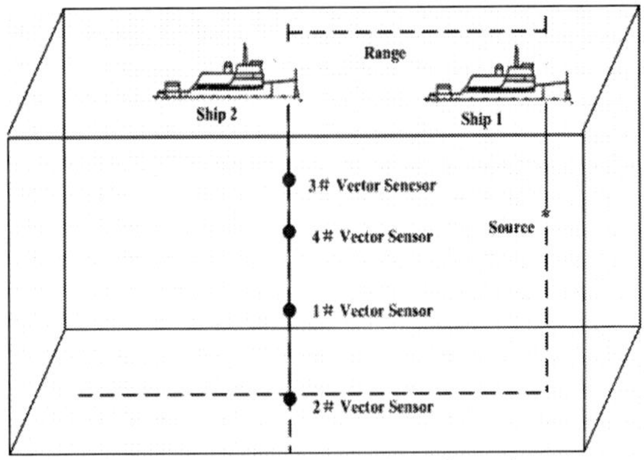

FIGURE 5. Experimental configuration.

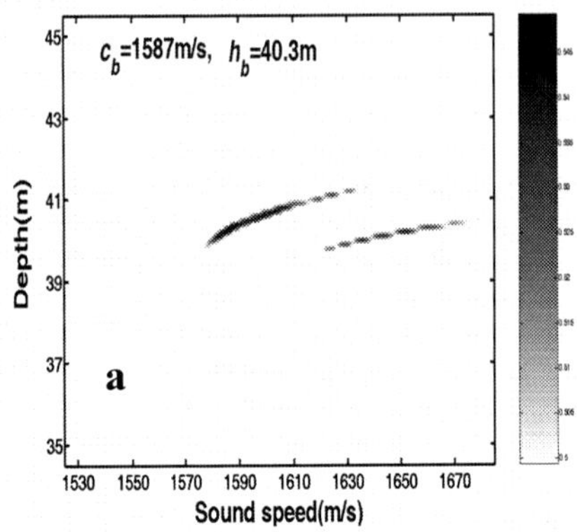

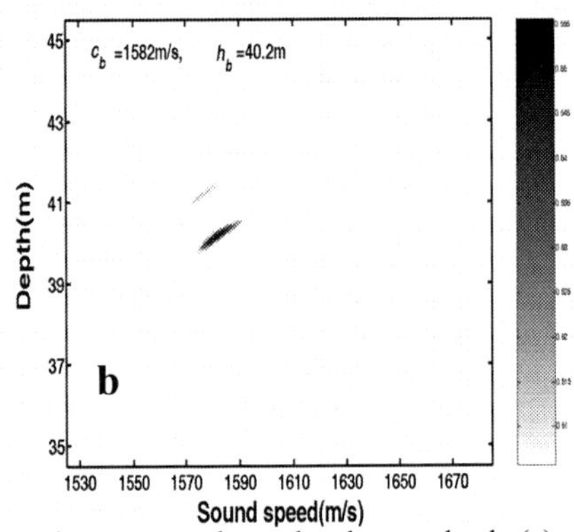

FIGURE 6. Cost function vs. bottom sound speed and water depth. (a): cost function defined as Eq.8, (b): cost function defined as Eq.9.

SUMMARY

The theoretical analyses and the experimental data show that the high order normal modes contribute more to vertical particle velocity due to their large grazing angle, and Vz has a larger loss than Vr. A MFP-based geo-acoustic inversion method has been developed in this paper. The results indicate that the MFP method with a vector sensor array can reduce the uncertainty of inversion more effectively than a hydrophone array.

ACKNOWLEDGMENTS

This work is supported by the National Natural Science Foundation of China 10734100. The authors thank the staff who attended the sea experiment to obtain the valuable data for this paper.

REFERENCES

1. M. Hawkes and A. Nehorai, *IEEE trans on Signal Processing*, **46**, 2291-2304 (1998).
2. G. Sun and D. Yang, **Acta Acoustica**, **28**, 66-72 (2003) (in Chinese).
3. A. Nehorai, *IEEE trans on Signal Processing*, **42**, 2481-2491 (1994).
4. M. Hawkes and A. Nehorai, *IEEE Journal of Oceanic Engineering*, **26**, 337-347 (2001).
5. H. S. Peng and F. H. Li, *Chin. Phys. Lett.*, **24**, 1977-1980 (2007).
6. J. Y. Hui and H. Liu, *Acta Acoustica*, **25**, 303-307 (2000) (in Chinese).

Inversion of Ocean Environmental Variations via Time Reversal Acoustics

Jianlong Li, Wen Xu*, and Liling Jin

Department of Information Science and Electronic Engineering, Zhejiang University
Hangzhou, 310027, China
**Email: wxu@zju.edu.cn*

Abstract. Time reversal (TR) processing is derived from the invariance of the wave equation for a lossless medium to change in the sign of the time variable. By retransmitting the TR version of the time-dispersed received signal propagated from a probe source (PS) to a source-receiver array, one can reacquire the transmitted pulse at the PS location (time compression and spatial focusing) when the waveguide environment is time invariant. However, if some environmental variations occur between the two transmissions, the retrofocusing signal will be defocused. This paper presents a novel method of environmental variation inversion by the focused/defocused signals measured at the PS location with environmental perturbations between the PS transmission and the SRA retransmission. Because the sound speed profile (SSP) plays a critical role in an uncertain ocean environment, inversion of the SSP represented in terms of the empirical orthogonal functions is developed and discussed in detail. Examples of the method applied to synthetic data are provided to demonstrate the method's feasibility.

Keywords: Time reversal, Inversion, Sound speed profile, Empirical orthogonal functions
PACS: 43.30.-k

INTRODUCTION

The ocean environment usually has strong temporal variability. For example, many experiments conducted in shallow water have shown significant variability in the acoustic data caused by strong internal waves [1]. The uncertainty of the acoustic environment is one of the major obstacles to adapting model-based sonar processing frameworks, such as matched-field processing (MFP), to the coastal environment [2-3].

A typical inverse problem is to find an environment in which a forward model can produce a replica with a good match to the observed data [4]. This paper presents an approach of inverting ocean environmental variations using the observed data without producing a replica (in fact, it takes an oceanic channel as a replica). This novel approach is based on time reversal (TR) acoustics, which is derived from the invariance of the wave equation for a lossless medium to change in the sign of the time variable and has been studied widely in medical imaging, radar, and underwater acoustics over the past 15 years. One may refer to an extensive overview of TR applications [5]. According to the TR principle, by retransmitting the TR version of the time-dispersed received signal propagated from a probe source (PS) to a source-receiver array (SRA), one can reacquire the transmitted pulse at the PS location (time

CP1272, *Shallow-Water Acoustics, Second International Shallow-Water Acoustics Conference*
edited by J. Simmen, E. S. Livingston, J. X. Zhou, and F. H. Li

compression and spatial focusing) when the waveguide environment is time invariant. However, if some environmental variations occur between the two transmissions, the retrofocusing signal will be defocused. In most TR applications, defocusing is undesirable, but in this paper we try to exploit the defocused data to invert the environmental perturbations between the PS transmission and the SRA retransmission.

INVERSION OF SSP VARIATION VIA TR

The SSP in the water column is one of the most important parameters in determining the sound waveguide propagation, for example, in the case of the existence of duct propagation or the fraction of energy down to the bottom. It is, however, difficult to treat this as deterministic due to both its temporal and spatial variations. We present the procedure of TR processing under random SSP perturbations using the approach in Refs. 3 and 4.

Signal Field Sampled by the SRA

The SSP in a range-dependent environment can be described by a mean profile, $c^{(0)}(z)$ (z: depth), plus some zero-mean random perturbations, $\Delta c^{(1)}(z,r)$ (r: range), i.e.,

$$c^{(1)}(z,r) = c^{(0)}(z) + \Delta c^{(1)}(z,r), \tag{1}$$

where $c^{(0)}(z)$ is assumed as known *a priori*, random perturbations $\Delta c^{(1)}(z,r) \ll c^{(0)}(z)$, the superscript $^{(0)}$ indicates the association with the mean profile, and superscript $^{(1)}$ indicates the profile when the PS transmits the signal. The SRA is deployed at $r=0$, consisting of I sensors uniformly located from z_1 to z_I with spacing Δz, D is a sufficiently great depth, at which the boundary can be treated as perfectly rigid, and the PS of frequency f is located at range r_s and depth z_s. Assume the coherent time length of the SSP variation is much larger than the wave propagation time from the PS to the SRA. Then the SSP variation can be neglected in the time interval of the acoustic wave propagation from the PS to the SRA. The waveguide transfer function from the source to the receiver is the unit point source solution to the wave equation, which is a function of the source-receiver configuration as well as environmental parameters. According to the adiabatic modal phase approximation, the Green's function can be expressed as a sum of normal modes [3]

$$p^{(1)}(z_i; r_s, z_s) = \sqrt{2\pi} \sum_{n=1}^{N} \frac{\varphi_n^{(0)}(z_s)\varphi_n^{(1)}(z_i,0)}{\sqrt{k_n^{(0)} r_s}} e^{jk_n^{(0)} r_s + j\int_0^{r_s} \Delta k_n^{(1)}(r)dr}, \tag{2}$$

where $k_n^{(0)}$, $n=1,\cdots,N$ are modal horizontal wave numbers; N is the number of the propagation modes; $\varphi_n^{(0)}(z)$, $n=1,\cdots,N$ are modal shape functions with $c^{(0)}(z)$; $\varphi_n^{(1)}(z,0)$, $n=1,2,\cdots,N$ are modal shape functions at the SRA with random perturbations; and

$$\Delta k_n^{(1)}(r) = -\frac{1}{k_n^{(0)}}\int_0^D |\varphi_n^{(0)}(z)|^2 \frac{\Delta c^{(1)}(z,r)}{c^{(0)}(z)} k^{(0)}(z)^2 dz, \tag{3}$$

which is attributed to the SSP perturbation and affects the Green's function through an added modal phase variation.

To reduce the degrees of freedom, zero-mean random perturbations $\Delta c^{(1)}(z,r)$ is often expressed in terms of a set of empirical orthogonal functions (EOFs), so that

$$\Delta c^{(1)}(z,r) = \sum_{l=1}^{L} w_l^{(1)}(r)\psi_l(z), \tag{4}$$

where $w_l^{(1)}(r)$, $l = 1,\cdots,L$ are random coefficients associated with SSP $c^{(1)}(z,r)$ and $\psi_l(z)$, $l = 1,\cdots,L$ are EOFs. Substituting Eq. (4) into Eq. (3), the perturbation in modal horizontal wave number under the current configuration becomes

$$\Delta k_n^{(1)}(r) = \sum_{l=1}^{L} w_l^{(1)}(r)\int_0^D \frac{-|\varphi_n^{(0)}(z)|^2}{c^{(0)}(z)k_n^{(0)}} k^{(0)}(z)^2 \psi_l(z)dz. \tag{5}$$

The Approach of Inversion with TR Acoustics

The SRA receives the signal, and with a time reversal operation, retransmits the signal. Assume the SSP at this time changes to

$$\begin{aligned} c^{(2)}(z,r) &= c^{(0)}(z) + \Delta c^{(2)}(z,r) \\ &= c^{(0)}(z) + \sum_{l=1}^{L} w_l^{(2)}(r)\psi_l(z), \end{aligned} \tag{6}$$

where superscript $^{(2)}$ indicates the profile when the retransmitted signal propagates from the SRA back to the PS location, and $w_l^{(2)}(r)$, $l = 1,\cdots,L$ are random coefficients associated with SSP $c^{(2)}(z,r)$. Because the TR operation is equivalent to a phase conjugation in the frequency domain, let $p^{(1)}(z_i;r_s,z_s)^*$, $i = 1,\cdots,I$ be the retransmitted signal of the SRA, where * denotes a conjugate operation, and let $p^{(2)}(z_s;r_s,z_i)$ be the Green's function from $(0,z_i)$ to (r_s,z_s). According to the acoustic wave equation, the pressure sampled at the PS location is

$$p_{tr}(r_s,z_s) = \sum_{i=1}^{I} p^{(1)}(z_i;r_s,z_s)^* p^{(2)}(z_s;r_s,z_i), \tag{7}$$

where similar to Eq. (2), $p^{(2)}(z_s;r_s,z_i)$ can be obtained via the adiabatic modal phase approximation. Substituting $p^{(1)}(z_i;r_s,z_s)$ and $p^{(2)}(z_s;r_s,z_i)$ into Eq. (7), one has

$$p_{tr}(r_s,z_s) = 2\pi\sum_{m=1}^{N}\sum_{n=1}^{N}\sum_{i=1}^{I}\frac{\varphi_n^{(0)}(z_s)\varphi_n^{(1)}(z_i,0)\varphi_m^{(0)}(z_i)\varphi_m^{(2)}(z_s,0)}{\rho(z_i)\rho(z_s)r_s\sqrt{k_n^{(0)}k_m^{(0)}}} e^{ir_s(k_m^{(0)}-k_n^{(0)})} e^{i\int_0^{r_s}[\Delta k_m^{(2)}(r)-\Delta k_n^{(1)}(r)]dr}. \tag{8}$$

For an array which adequately samples most of the modes, we may approximate the sum of sources as an integral and invoke orthonormality specified by

$$\int_0^D \frac{\varphi_m^{(1)}(z,0)\varphi_n^{(1)}(z,0)}{\rho(z)} dz = \delta_{mn}. \tag{9}$$

Note that $\varphi_n^{(1)}(z,0)$, $n=1,\cdots,N$ and $\varphi_m^{(0)}(z)$, $m=1,\cdots,N$ in Eq. (9) are not associated with the same environment. However, by the following derivation, we can find that with a small SSP perturbation the orthonormality still holds. Assume the perturbation $\Delta c^{(1)}(z,r)$ causes the perturbations $\Delta\varphi_n^{(1)}(z,0)$, $n=1,\cdots,N$ in mode functions, i.e.,

$$\varphi_n^{(1)}(z,0)=\varphi_n^{(0)}(z)+\Delta\varphi_n^{(1)}(z,0),\ n=1,\cdots,N. \tag{10}$$

From Eq. (9), one has

$$\int_0^D \frac{\varphi_m^{(1)}(z,0)\partial\varphi_n^{(1)}(z,0)/\partial c^{(1)}}{\rho(z)}dz=0, \tag{11}$$

which means the change of the mode function must be orthogonal to the mode function itself. With the help of Eqs. (10) and (11), the orthonormality with a small perturbation can be derived, i.e.,

$$\int_0^D \frac{\varphi_m^{(0)}(z)\varphi_n^{(1)}(z,0)}{\rho(z)}dz=\delta_{mn}. \tag{12}$$

According to Eq. (12), Eq. (8) becomes

$$p_{tr}(r_s,z_s)=\frac{2\pi}{\rho(z_s)r_s\Delta z}\sum_{n=1}^{N}\frac{\varphi_n^{(0)}(z_s)\varphi_n^{(2)}(z_s,0)}{k_n^{(0)}}e^{i\int_0^{r_s}[\Delta k_n^{(2)}(r)-\Delta k_n^{(1)}(r)]dr}. \tag{13}$$

Hence, the perturbation of the SSP affects the variation of the signal phase at the PS location. Note that the SSP at the PS location is assumed as known (i.e., by a measurement made at the PS), and thus $\varphi_n^{(2)}(z_s,0)$, $n=1,\cdots,N$ can be calculated by numerical methods. Alternatively, the mode function at the PS location can be directly measured. With the help of Eq. (5), Eq. (13) changes to

$$\begin{aligned}p_{tr}(r_s,z_s)&=\frac{2\pi}{\rho(z_s)r_s\Delta z}\sum_{n=1}^{N}\frac{\varphi_n^{(0)}(z_s)\varphi_n^{(2)}(z_s,0)}{k_n^{(0)}}e^{i\Delta\theta_n(r_s)}\\&\triangleq f(\overline{\mathbf{w}};r_s,z_s),\end{aligned} \tag{14}$$

where

$$\begin{aligned}\Delta\theta_n(r_s)&=\int_0^{r_s}[\Delta k_n^{(2)}(r)-\Delta k_n^{(1)}(r)]dr\\&=\sum_{l=1}^{L}\overline{w}_l(r_s)\int_0^D\frac{-|\varphi_n^{(0)}(z)|^2}{c^{(0)}(z)k_n^{(0)}}k^{(0)}(z)^2\psi_l(z)dz,\end{aligned} \tag{15}$$

$$\overline{\mathbf{w}}=\left[\overline{w}_1(r_s)\ \cdots\ \overline{w}_l(r_s)\ \cdots\ \overline{w}_L(r_s)\right]^T, \tag{16}$$

in which

$$\overline{w}_l(r_s)=\int_0^{r_s}w_l(r)dr, \tag{17}$$

$$w_l(r)=w_l^{(2)}(r)-w_l^{(1)}(r). \tag{18}$$

Assume the number of the PSs is Q ($Q\geq L$) and the position of the q th PS is r_s in range and z_{sq} in depth. Then, according to Eq. (18), $\overline{\mathbf{w}}$ can be evaluated via

$$\hat{\overline{\mathbf{w}}}=\arg\min_{\overline{\mathbf{w}}}\|\hat{\mathbf{p}}_{tr}-\mathbf{f}(\overline{\mathbf{w}})\|, \tag{19}$$

where

$$\hat{\mathbf{p}}_{tr} = \left[\hat{p}_{tr}(r_s, z_{s1}) \quad \cdots \quad \hat{p}_{tr}(r_s, z_{sq}) \quad \cdots \quad \hat{p}_{tr}(r_s, z_{sQ}) \right]^T \tag{20}$$

is the signal vector received by the PSs and

$$\mathbf{f}(\overline{\mathbf{w}}) = \left[f(\overline{\mathbf{w}}; r_s, z_{s1}) \quad \cdots \quad f(\overline{\mathbf{w}}; r_s, z_{sq}) \quad \cdots \quad f(\overline{\mathbf{w}}; r_s, z_{sQ}) \right]^T . \tag{21}$$

Finally, the variation of the SSP along the range between the PS transmission and the SRA retransmission can be inverted by

$$\int_0^{r_s} \Delta c(z,r) dr = \sum_{l=1}^{L} \hat{\overline{w}}_l \psi_l(z) . \tag{22}$$

For a range-independent environment, we have for the SSP perturbation

$$\Delta c(z) = \sum_{l=1}^{L} w_l \psi_l(z) . \tag{23}$$

The SSP coefficients are no longer a function of range and the integration in Eq. (17) becomes $w_l r_s$. Then

$$\mathbf{w} = \left[w_1 \quad \cdots \quad w_l \quad \cdots \quad w_L \right]^T = \overline{\mathbf{w}} / r_s . \tag{24}$$

EXAMPLES

To demonstrate the developed concept, we present a typical example of SSP representation. As shown in Fig. 1, the stratified waveguide model comes from the shelf break PRIMER experiment, which was conducted south of New England in the Middle Atlantic Bight during summer 1996 and is typical of shallow water propagation [6]. The water depth is 92 m. The SSP is sampled by 47 discrete depth points from the surface (0 m) to the bottom (92 m), equally spaced by 2 m. This spacing is chosen from the available SSP data measurements, and can certainly be larger for the purpose of representing either the SSP variations or the modal shape functions. The mean SSP is also plotted in Fig. 1. In the sediment, the sound speed, attenuation and density are $c_1 = 1750$ m/s, $\alpha_1 = 0.05$ dB/λ and $\rho_1 = 1.7$ g/cm^3, respectively.

The variability in the water column sound speed is modeled using EOFs. The background sound speeds required to generate the EOFs were obtained from the onsite measurements along the range-independent track from core 1 to 3 (cf. [6], Fig. 4). In terms of energy, the first three EOFs describe 98.65% of the sound speed variation. The SRA consists of 18 elements deployed from 0 m to 90 m in depth, spaced 5 m apart. The frequency of each PS at a range of 10 km is set to 100 Hz for long range propagation.

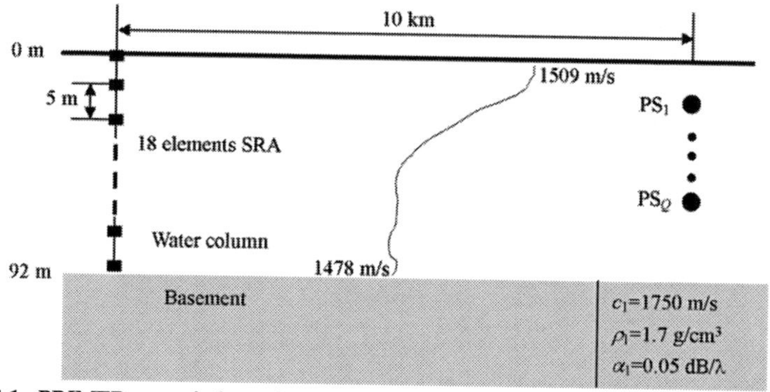

FIGURE 1. PRIMER range-independent track [6]. PSs and receiver configuration are also shown.

Choose two groups of SSPs measured at consecutive times as the real SSPs when the PSs transmitted and the SRA retransmitted, respectively. The corresponding coefficients of EOFs are $\mathbf{w}^{(1)} = [-10.9 \quad 6.7 \quad -0.6]^T$ and $\mathbf{w}^{(2)} = [-11.8 \quad 7.9 \quad 0.0]^T$, respectively. Hence, the variation of coefficients $\mathbf{w} = \mathbf{w}^{(2)} - \mathbf{w}^{(1)}$ induced by the SSP perturbation is $\mathbf{w} = [-0.9 \quad 1.2 \quad 0.6]^T$ and the number of the inverted parameters is $L = 3$. The maximum value of the SSP perturbation between the PS transmission and the SRA retransmission is 0.75 m/s. Assume there are four probing sources ($Q = 4$) equally spaced spanning the entire water column, and the signal-to-noise ratio (SNR) of the received data at each PS location is 10 dB. Figure 2 plots one dimensional (1-D) cross sections of w_1, w_2 and w_3. For each figure, the parameters are held fixed at their optimal value evaluated via Eq. (19). The SSP perturbation between the PS transmission and the SRA retransmission can be inverted by Eq. (23). Fig. 2 also shows the results of different SNRs. As expected, when the SNR decreases, the error of each inverted value increases.

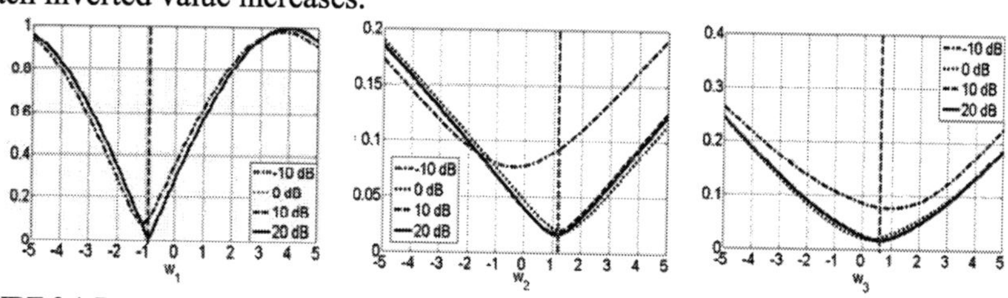

FIGURE 2 1-D cross sections of w_1, w_2 and w_3 for different SNRs. In each case, the parameters that are not varied are held at their value at the global minimum.

CONCLUSIONS

In this paper, based on TR processing, we derived a novel approach that can invert the environmental variation induced by environmental perturbations between the PS transmission and the SRA retransmission via the focused/defocused signals measured at the PS location. Because the SSP plays a critical role in an uncertainty ocean environment, inversion of the SSP represented in terms of the EOFs was developed in detail. An example simulation is given to demonstrate the validity of the approach. The advantages of the inversion method can be summarized as (i) great signal-to-noise ratios (SNRs) can be achieved at the PS location so that the estimation error can be decreased; (ii) variations of the environment at different times can be directly inverted by repeatedly retransmitting either the same received signal generated by one PS transmission or the updated received signals generated by updated PS transmissions.

ACKNOWLEDGMENTS

This work was supported by Chinese 863 high-tech program under Grant 2009AA093601 and Zhejiang Provincial Natural Science Foundation of China under Grant Y5080150.

REFERENCES

1. J. R. Apel, M. Badiey, C. S. Chiu, S. Finette, R. Headrick, J. Kemp, J. F. Lynch, A. Newhall, M. H. Orr, B. H. Pasewark, D. Tielbuerger, A. Turgut, K. Von Der Heydt, and S. Wolf, *IEEE J. Ocean. Eng.* **22**, 465-500 (1997).
2. H. Schmidt, A. B. Baggeroer, W. A. Kuperman, and E. K. Scheer, *J. Acoust. Soc. Am.* **88**, 1851-1862 (1990).
3. J. L. Krolik, *J. Acoust. Soc. Am.* **92**, 1408-1419 (1992).
4. S. D. Rajan, J. F. Lynch, and G. V. Frisk, *J. Acoust. Soc. Am.* **82**, 998-1017 (1987).
5. M. Fink, D. Cassereau, A. Derode, C. Prada, P. Roux, M. Tanter, J. L. Thomas, and F. Wu, *Rep. Prog. Phys.* **63**, 1933-1995 (2000).
6. G. Potty, J. Miller, J. Lynch, and K. Smith, *J. Acoust. Soc. Am.* **108**, 973-986 (2000).

Underwater Time Service and Synchronization Based on Time Reversal Technique

Hao Lu [1,2], Hai-bin Wang [1], Abdeldjalil Aissa-El-Bey [3], and Ramesh Pyndiah [3]

1. State Key Laboratory of Acoustics, Institute of Acoustics, Chinese Academy of Sciences, Beijing 100190, China; Email: whb@mail.ioa.ac.cn

2. Graduate University of Chinese Academy of Sciences, Beijing 100190, China

3. TELECOM Bretagne, Technopôle Brest Iroise, CS 83818, 29238 Brest cedex 3, France

Abstract. Real time service and synchronization are very important to many underwater systems. But the time service and synchronization in existence cannot work well due to the multi-path propagation and random phase fluctuation of signals in the ocean channel. The time reversal mirror technique can realize energy concentration through self-matching of the ocean channel and has very good spatial and temporal focusing properties. Based on the TRM technique, we present the Time Reversal Mirror Real Time service and synchronization (TRMRT) method which can bypass the processing of multi-path on the server side and reduce multi-path contamination on the client side. So TRMRT can improve the accuracy of time service. Furthermore, as an efficient and precise method of time service, TRMRT could be widely used in underwater exploration activities and underwater navigation and positioning systems.

Keywords: time reversal mirror, underwater time service.

PACS: 43.30

INTRODUCTION

Time base is very important for synchronization in many underwater systems. At present, there are already several available time service and synchronization systems, such as the Global Navigation Satellite System (GNSS) and the Global Positioning System (GPS) which generally use electromagnetic waves for time dissemination. However, radio signals cannot propagate deeply into water. As a result, conventional radio time service methods do not work well under water. For some underwater acoustic time service and synchronization (UWATS) systems, multi-path propagation and random phase fluctuation of acoustic signals in the ocean make it difficult to obtain precise timing information. So we have to use other methods to solve this problem.

The time reversal mirror is a kind of channel matching technique which has been widely studied during the past twenty years for its good spatial and temporal focusing

CP1272, *Shallow-Water Acoustics, Second International Shallow-Water Acoustics Conference*
edited by J. Simmen, E. S. Livingston, J. X. Zhou, and F. H. Li

properties. It is found that time reversal systems can not only be free of multi-path contamination but also obtain energy concentration. At present, the time reversal technique has been widely utilized in many acoustic applications including ultrasonic therapy, underwater communication and detection, nondestructive testing, three-dimensional imaging, and so on[1]-[4]. Because the time reversal technique can reduce multi-path contamination, we could use this property to implement real time service and synchronization which is called Time Reversal Mirror Real Time service and synchronization (TRMRT). In this method, the client sends time service signal to the server; then the received signal on the server side is time reversed and retransmitted to the client. From a detailed analysis it can be determined that the retransmitting time is just the middle of the whole time service task. TRMRT can bypass the processing of multi-path on the server side. On the other hand, multi-path contamination on the client side is greatly reduced by TRM. TRMRT can improve the accuracy of time service and greatly decrease the complexity of the system.

THEORY ANALYSES

The TRMRT system consists of two parts. One is the server which provides the real time; the other is the client which calibrates its own local clock through information interaction with the server. The server consists of an electronic subsystem and an underwater subsystem. The electronic subsystem is composed of a signal receiving module, a signal processing module, and a signal transmitting module. The underwater subsystem is a vertical transceiver array which consists of N transceiver elements. Each element is the combination of a hydrophone and a transmitting transducer. Similarly, the client includes an electronic subsystem and an underwater subsystem. The difference lies in that the client's underwater subsystem has only one transceiver element.

To explain TRMRT effectively, we set up a global viewpoint of the system and construct a unified time coordinate system for both the server and the client. When the client needs to calibrate its own clock, it sends out the requesting signal $s_{cs}(t)$ and records this time as t_{cs} which is read from its own clock.

Here, in order to keep generality, $s_{cs}(t)$ can be denoted by a wideband signal $s(t)$, $t \in [0, T_o]$. In the time and frequency domains, we have

$$s_{cs}(t) = s(t) \Leftrightarrow S_{cs}(w) = S(w). \tag{1}$$

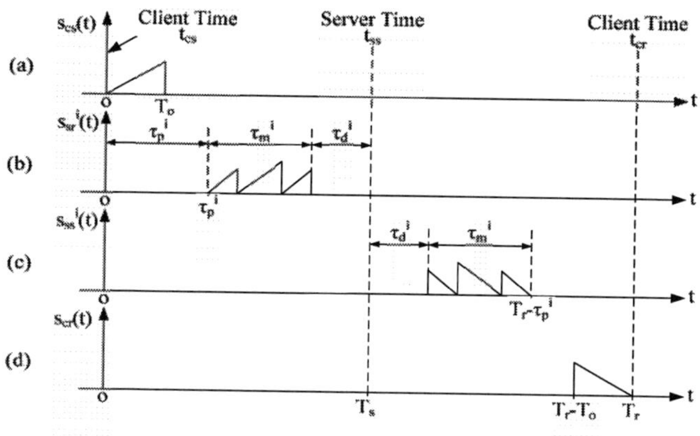

Figure 1. Four signals of TRMRT method

After $s_{cs}(t)$ propagates through the ocean channel, it is received by the ith channel of the server. This received signal $s_{sr}{}^i(t)$ can be written as

$$\begin{aligned} &s_{sr}{}^i(t) = s_{cs}(t) \otimes g_t(r_s{}^i, z_s{}^i; r_c, z_c) \\ &\Leftrightarrow S_{sr}{}^i(w) = S_{cs}(w) G_w(r_s{}^i, z_s{}^i; r_c, z_c), i = 1,2,\ldots,N \end{aligned} \quad (2)$$

Note that the Green's function $g_t(r_s{}^i, z_s{}^i; r_c, z_c)$ is the channel impulse-response from the client to the ith channel of the server. And $G_w(r_s{}^i, z_s{}^i; r_c, z_c)$ is the corresponding Green's function in the frequency-domain.

For the ith channel, suppose $\tau_p{}^i$ is the signal propagation time. And $\tau_m{}^i$ is the length of the multi-path structure. After a proper delay time $\tau_d{}^i$, the signal is retransmitted at time T_s. The corresponding absolute real time reading from the server's clock is t_{ss}, which must be recorded by the server for the future calibration process. The delay time $\tau_d{}^i$ can be adjusted to ensure that every channel is aligned to the retransmitting time T_s, then

$$T_s = \tau_p{}^i + \tau_m{}^i + \tau_d{}^i, i = 1,2,\ldots N. \quad (3)$$

As shown in Figure 1, the time reversed signal of the ith channel must be:

$$s_{ss}{}^i(t) = s_{sr}{}^i(T_r - t) \Leftrightarrow S_{ss}{}^i(w) = [S_{sr}{}^i(w)]^* e^{-jwTr}. \quad (4)$$

Here, note that $T_r = 2T_s$.

After signal $s_{ss}{}^i(t)$ back-propagates through the ocean channel, the signal received by the client can be written as

$$s_{cr}(t) = \sum_{i=1}^{N} s_{ss}{}^i(t) \otimes g_t(r_c, z_c; r_s{}^i, z_s{}^i) \Leftrightarrow S_{cr}(w) = \sum_{i=1}^{N} S_{ss}{}^i(w) G_w(r_c, z_c; r_s{}^i, z_s{}^i) \quad (5)$$

where $g_t(r_c, z_c; r_s{}^i, z_s{}^i)$ and $G_w(r_c, z_c; r_s{}^i, z_s{}^i)$ are Green's functions in the time and frequency domain respectively from the ith channel of the server to the client.

Substituting Eqs. (2) and (4) into Eq. (5), we will get

$$s_{cr}(t)=\sum_{i=1}^{N}s_{cs}(T_r-t)\otimes g_{-t}(r_s^i,z_s^i;r_c,z_c)\otimes g_t(r_c,z_c;r_s^i,z_s^i)$$
$$\Leftrightarrow S_{cr}(w)=\sum_{i=1}^{N}[S_{cs}(w)G_w(r_s^i,z_s^i;r_c,z_c)]^* e^{-jwTr}G_w(r_c,z_c;r_s^i,z_s^i) \quad (6)$$

According to the reciprocity principle of wave propagation in the ocean, $G_w(r_s^i,z_s^i;r_c,z_c)=G_w(r_c,z_c;r_s^i,z_s^i)$. Therefore,

$$Q(w)=\sum_{i=1}^{N}[G_w(r_s^i,z_s^i;r_c,z_c)]^* G_w(r_c,z_c;r_s^i,z_s^i)=\sum_{i=1}^{N}|G_w(r_s^i,z_s^i;r_c,z_c)|^2$$
$$\Leftrightarrow q(t)=\sum_{i=1}^{N}Rg_t(r_s^i,z_s^i;r_c,z_c) \quad (7)$$

Here $Rg_t(r_s^i,z_s^i;r_c,z_c)$ is the autocorrelation of the Green's function $g_t(r_s^i,z_s^i;r_c,z_c)$. It can be applied approximately with the result that [5]

$$q(t)\approx\delta(t). \quad (\delta(t) \text{ is the Dirac function}) \quad (8)$$

Following Eqs. (6) and (8), the signal received by the client must be:

$$s_{cr}(t)=s(T_r-t). \quad (9)$$

By making a convolution of the received signal $s_{cr}(t)$ with the original time service signal $s(t)$, we obtain $R(t)$ which is relatively free of multi-path contamination. Ideally, there is only one peak at time T_r, making it easy to detect the time of arrival (TOA) which is indicated by this peak. Reading from its own clock again, the client must record this time as t_{cr}.

From Figure 1, the total duration of the time service operation is T_r, so the half time is $T_r/2=T_s$. Note this is just the time when the server sends signals back. The corresponding absolute real time has been recorded by the server as t_{ss}. The server must send t_{ss} to the client through an acoustic communication system. The client can calibrate its own clock by subtracting the following time error:

$$e=(t_{cs}+t_{cr})/2-t_{ss}. \quad (10)$$

In this way, the client's can be calibrated.

NUMERICAL SIMULATION

To validate the TRMRT method, some simulation results are given in this section. In this simulation, we consider a range-independent waveguide. The sound-velocity profile (SVP) has a constant value, 1510 m/s. The depth of the water column is 100 m. The depth of the sediment layer is 8 m with a sound speed increasing from 1583 to 1687 m/s, a density of 1.86g/cm^3 and an attenuation of 0.14 dB/(kmHz). The bottom has a constant sound speed of 1750m/s, a density of 2.11g/cm^3 and an attenuation of 0.56 dB/(kmHz), as shown in Figure 2.

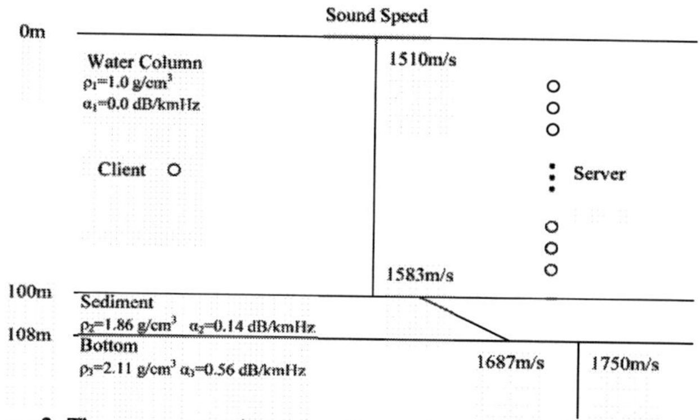

Figure 2. The ocean acoustic propagation environment for the simulation.

The client and server are launched into the water. The range between them is 30km. The server consists of 16 elements which are equally spaced over a depth from 10 to 85 m beneath the surface. The inter-space between two neighboring elements is 5m. And the client is placed at 47.5m, which is the center of the server.

The assumptions in the single channel TRMRT system are: the source level (SL) is from 150dB to 170dB; the noise level (NL) is from 50dB to 100dB; the phase fluctuation of the channel is from 0 to 180 degrees; the relative amplitude fluctuation is from 0 to 100%.

As a comparison, now we introduce another time service: the Two Way Travel Time (TWTT) method without TRM. First, the client sends a "requesting" signal at time t_{cs}. The server receives the "requesting" signal, and then calculates the arrival time t_{sr}. After some processing time, the server sends a "replying" signal back to the client at time t_{ss}. Finally, the client receives the "replying" signal at time t_{cr}. So the time error of the client can be given by

$$e = t_{cr} - \{t_{ss} + [(t_{cr} - t_{cs}) - (t_{ss} - t_{sr})]/2\}. \quad (11)$$

Figure 3 shows the relationship between the time service error vs. SL and NL. The corresponding values of this figure are given in Table 1. We can see that, under the ideal high SNR conditions, the time error could reach zero. And when the SNR goes down, the time error will appear. These simulation results prove that the theory of TRMRT method is correct.

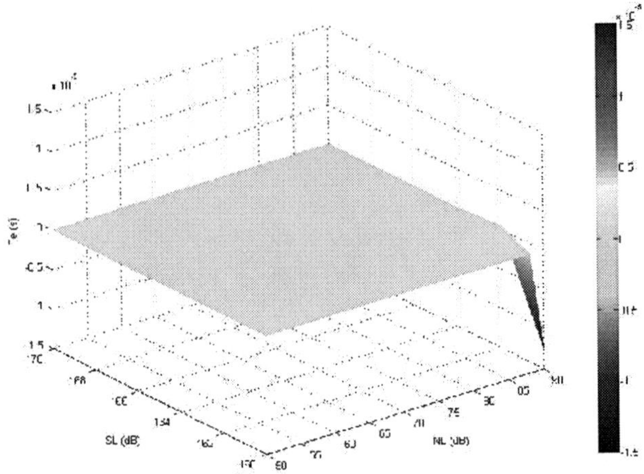

Figure 3. The relationship between time service error vs. SL and NL. (TRMRT)

Table 1. The relationship between time service error vs. SL and NL. (TRMRT)

SL(dB) / NL(dB)	160	162	164	166	168	170
50-84	0s	0s	0s	0s	0s	0s
86	0s	0s	0s	0s	0s	0s
88	0s	0s	0s	0s	0s	0s
90	-1.29e-5s	3.91e-7s	0s	0s	0s	0s

We assume that the power of all time service signals of TWTT and TRMRT are equal. The performance of TWTT is shown in Figure 4 with corresponding values of the figure given in Table 2. From this figure we can see the time service error of TWTT is higher than TRMRT.

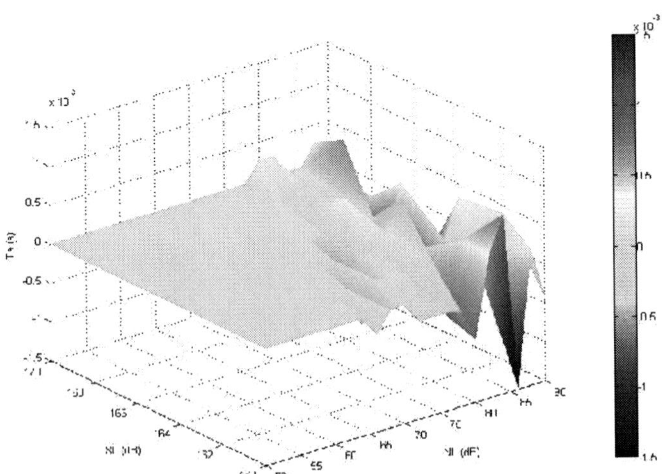

Figure 4. The relationship between time service error vs. SL and NL. (TWTT)

Table 2. The relationship between time service error vs. SL and NL. (TWTT)

SL(dB) / NL(dB)	160	162	164	166	168	170
50-64	0s	0s	0s	0s	0s	0s
66	-2.52e-4s	0s	0s	0s	0s	0s
68	0s	0s	0s	0s	0s	0s
70	2.34e-5s	-5.05e-4s	0s	0s	0s	0s
72	-2.52e-4s	2.348e-5s	0s	0s	0s	0s
74	-2.52e-4s	0s	-5.27e-4s	0s	0s	0s
76	-2.52e-4s	0s	-4.69e-5s	0s	0s	0s
78	-2.97e-4s	-2.53e-4s	-2.35e-5s	-2.53e-4s	0s	0s
80	-7.02e-4s	4.69e-5s	9.61e-5s	2.98e-4s	-2.52e-4s	2.52e-4s
82	3.32e-4s	-7.23e-4s	5.03e-4s	-2.30e-4s	-2.19e-5s	2.52e-4s
84	7.89e-4s	1.84e-4s	-1.84e-4s	-4.69e-5s	2.34e-5s	0s
86	1.49e-3s	-5.47e-5s	-7.78e-4s	-6.50e-4s	6.20e-4s	0s
88	2.40e-4s	-1.21e-4s	-1.84e-4s	2.86e-4s	-1.21e-3s	2.75e-4s
90	-4.34e-4s	3.93e-4s	3.07e-4s	-9.90e-4s	-2.16e-4s	-2.12e-4s

CONCLUSION

In traditional time service systems, server and client must deal with the ambiguity problem of TOA caused by multi-path propagation. But TRMRT can bypass the processing of multi-path on the server side. At the same time, multi-path contamination on the client side is greatly reduced. Therefore, this method can improve the accuracy of time service.

REFERENCES

1 Mathias Fink, Gabriel Montaldo, Mickael Tanter, "Ultrasonic Time Reversal Mirrors," *HIGH FREQUENCY OCEAN ACOUSTICS: High Frequency Ocean Acoustics Conference. AIP Conference Proceedings,* Volume 728, pp. 514-521, 2004.

2 H. C. Song, W. S. Hodgkiss, W. A. Kuperman, T. Akal, and M. Stevenson, "Multiuser communications using passive time reversal," *IEEE Journal of Oceanic Engineering,* Vol. 32, No. 4, pp. 915-926, 2007.

3 Estelle Kerbrat, Claire Prada, Didier Cassereau, and Mathias Fink, "Ultrasonic Nondestructive Testing of Scattering Media Using the Decomposition of the Time-Reversal Operator," *IEEE Transactions on Ultrasonics, Ferroelectrics, and Frequency Control,* Vol. 49, No. 8, pp. 1103-1113, 2002.

4 T. C. Yang, "Temporal Resolutions of Time-Reversal and Passive-Phase Conjugation for Underwater Acoustic Communications," *IEEE Journal of Oceanic Engineering,* Vol. 28, No. 2, pp.229-245, 2003.

5 Kuperman W. A., Hodgkiss W. S., Hee Chun Song, et al. Phase conjugation in the ocean: Experimental demonstration of an acoustic time-reversal mirror [J]. *J. Acoust. Soc. Am.*(S0001-4966), 1998, 103(1):25-40.

Single-mode Closed-loop Excitation in Shallow Water Using Two Vertical Hydrophone Arrays

Dayong Peng, Tianfu Gao, Juan Zeng, Haijun Liu,
Haifeng Li, Wenyao Zhao

Key Laboratory of Underwater Acoustic Environment, Chinese Academy of Sciences, Beijing, 100190, China
Institute of Acoustics, Chinese Academy of Sciences, Beijing, 100190, China

Abstract. An algorithm for single-mode close-loop excitation in shallow water using two vertical hydrophone arrays is presented. Single mode excitation is a powerful tool for studying a great variety of oceanographic processes. The desired single-mode sound field is the only variable that needs to be given in single-mode close-loop excitation. In order to achieve an exact desired single-mode sound field, two vertical hydrophone arrays are used to search out the correct vertical wave number. By analyzing the covariance of estimation value of Green's function matrix, an optimal source array weight matrix is presented to estimate Green's function matrix. An algorithm based on the weight matrix makes single-mode excitation converge at maximum speed and be steady. Advantages of the algorithm are confirmed by numerical simulations. Results of an experiment performed in a coastal environment are presented.

Keywords: single-mode, close loop, weight matrix, wave number
PACS: 43.30

INTRODUCTION

In a stratified ocean environment, the far field of acoustic pressure is well described by a sum of discrete modes[1]. Single-mode excitation, defined as exciting only a single mode in the acoustic field, can be realized by locating a number of transducers at different depths. These transducers transmit signals simultaneously and make different modes weaken or strengthen. Single-mode has broad applications for its properties of high usage of transmitted energy, low transmit loss, avoiding multi-path disturbance and restricting reverberation.

Single-mode excitations can be divided into open-loop and close-loop excitation according to their control techniques. Open-loop excitation has a relatively long history and usually utilizes the orthogonal characteristic of normal modes[2-4]. Russian researchers have done some further researches recently[5-7]. However, open-loop excitation needs *a priori* knowledge about the water column sound speed profile and sea floor acoustic characteristics, so it has significant limitations.

Single-mode close-loop excitation is a relatively new subject. The chief method of this technique is first evaluating Green's function matrix according to the feedback sound field received by a vertical hydrophone array; and obtaining the vertical source

CP1272, *Shallow-Water Acoustics, Second International Shallow-Water Acoustics Conference*
edited by J. Simmen, E. S. Livingston, J. X. Zhou, and F. H. Li

array weights to excite a defined single mode by an inverting technique. The sound field emitted according to the weight coefficient will be the defined single-mode sound field. The close-loop control algorithm doesn't need *a priori* information about the ocean environment between the source and the feedback array, so it's well adaptive. Close-loop control algorithm was first formulated by Buck[8]. He performed numerical simulations and laboratory experiments. Chen Jianyun *et al.*[9] reported on a single-mode inverse control algorithm.

In this paper, an algorithm for single-mode close-loop excitation in shallow water using two vertical hydrophone arrays is presented. Advantages of the algorithm are confirmed by numerical simulations. Results of an experiment performed in a coastal environment are presented.

ALGORITHM

Estimation of Green's function matrix

Let N be the number of transducers used in single-mode excitation, and L be the number of hydrophones used to observe. After one time signal emission, the normal mode matrix expression of the sound field is:

$$\boldsymbol{p} = \boldsymbol{Q}\boldsymbol{w} + \boldsymbol{e} \tag{1}$$

where $\boldsymbol{p}$ represents received sound field $L \times 1$ matrix, $\boldsymbol{Q}$ represents Green's function $L \times N$ matrix, $\boldsymbol{w}$ represents transducer emission weights $N \times 1$ matrix, and $\boldsymbol{e}$ represents noise $L \times 1$ matrix.

After k ($k \geq N$) times signal emission, we get:

$$\boldsymbol{P} = \boldsymbol{Q}\boldsymbol{W} + \boldsymbol{E} \tag{2}$$

where $\boldsymbol{P}$ represents k times received sound field $L \times k$ matrix; $\boldsymbol{W}$ represents k times transducer emission weights $N \times k$ matrix; $\boldsymbol{E}$ represents k times noise $L \times k$ matrix.

Then the least-squares system identification result of matrix $\boldsymbol{Q}$ is:

$$\hat{\boldsymbol{Q}} = \boldsymbol{P}\boldsymbol{W}^{+} = \boldsymbol{P}\boldsymbol{W}^{\mathrm{H}}(\boldsymbol{W}\boldsymbol{W}^{\mathrm{H}})^{-1} \tag{3}$$

Let the noise be Gaussian white noise, then the mean of $\boldsymbol{E}$ is zero. The covariance matrix of noise matrix $\boldsymbol{K}_{\boldsymbol{EE}} = \sigma_e^2 \boldsymbol{I}$, where $\boldsymbol{I}$ is a unit matrix.

Then Eq. (3) is an unbiased estimation, which means its expectation value $E(\hat{\boldsymbol{Q}}) = \boldsymbol{Q}$. The covariance matrix of $\hat{\boldsymbol{Q}}$ is:

$$\boldsymbol{K}_{\hat{\boldsymbol{Q}}\hat{\boldsymbol{Q}}} = E\{\hat{\boldsymbol{Q}}^{\mathrm{H}}\hat{\boldsymbol{Q}}\} - E\{\hat{\boldsymbol{Q}}\}^{\mathrm{H}} E\{\hat{\boldsymbol{Q}}\} = \sigma_e^2 (\boldsymbol{W}\boldsymbol{W}^{\mathrm{H}})^{-1} \tag{4}$$

Apply singular value decomposition (SVD) on Hermit matrix $\boldsymbol{W}\boldsymbol{W}^{\mathrm{H}}$:

$$\boldsymbol{W}\boldsymbol{W}^{\mathrm{H}} = \boldsymbol{U}\boldsymbol{S}\boldsymbol{U}^{\mathrm{H}} \tag{5}$$

where $\boldsymbol{U}$ is a unitary matrix ($\boldsymbol{U}^{-1} = \boldsymbol{U}^{\mathrm{H}}$), $\boldsymbol{S}$ is the singular value diagonal $N \times N$ matrix. Substituting Eq. (5) into Eq. (4), we obtain:

$$\boldsymbol{K}_{\hat{\boldsymbol{Q}}\hat{\boldsymbol{Q}}} = \sigma_e^2 (\boldsymbol{U}\boldsymbol{S}\boldsymbol{U}^{\mathrm{H}})^{-1} = \sigma_e^2 \boldsymbol{U}(\boldsymbol{S})^{-1}\boldsymbol{U}^{\mathrm{H}} \tag{6}$$

To evaluate the estimation result, trace of covariance matrix is usually used:

$$\mathrm{tr}\boldsymbol{K}_{\hat{\boldsymbol{Q}}\hat{\boldsymbol{Q}}} = \sigma_e^2 \mathrm{tr}(\boldsymbol{S}^{-1}) \tag{7}$$

Recursive control algorithm

For a continuously running system, there will always be some newly observed data. If we want to use these data to improve estimation precision, then a recursive algorithm is needed. Without a recursive algorithm, we will have to perform inverting calculation according to Eq. (3) each time. Since the dimensions of matrix P and W will continuously increase, it would be a waste of time.

Before the kth measurement, estimation $\hat{Q}(k-1)$ and $k-1$ times source weights matrix $W(k-1)$ are obtained. Let $V(k-1)=(W(k-1)W^{H}(k-1))^{-1}$, then the recursive algorithm of Eq. (3) is:

$$\hat{Q}(k)=\hat{Q}(k-1)+[p(k)-\hat{Q}(k-1)w(k)]K(k) \tag{8}$$

where $w(k)$ is the kth time emission weights matrix, which is estimated by $\hat{Q}(k-1)$ and desired single-mode eigefunction profile p_0:

$$w(k)=(\hat{Q}(k-1))^{+}p_0 \tag{9}$$

$$K(k)=\frac{w^{H}(k)V(k-1)}{1+w^{H}(k)V(k-1)w(k)} \tag{10}$$

$$V(k)=V(k-1)-V(k-1)w(k)K(k) \tag{11}$$

Optimal estimation algorithm

To minimize Eq. (7), we first design a W_m matrix to estimate Green's function. Then perform adaptive control process according to Eq.8-11.

Let

$$\alpha=\cos\left(\frac{2\pi}{N}\right)+i\sin\left(\frac{2\pi}{N}\right) \tag{12}$$

The element of xth row and yth column of matrix W_m is:

$$W_m(x,y)=A\cdot\alpha^{x(y-1)} \tag{13}$$

SIMULATION RESULTS

The environment used here is range-invariant and homogeneous in range. The parameters are as follows: water depth $h=14m$, sound speed in water $c_1=1522m/s$, density of water $\rho_1=1.0g/cm^3$, sound speed in sediment $c_2=1680m/s$, density of sediment $\rho_2=1.8g/cm^3$, attenuation in sediment $\alpha_2=1.0dB/\lambda$. The 8-element source array at $r=0m$ transmitted at a frequency of 600 Hz to the 12-element feedback array located at $r=100m$ downrange. The environment supports six trapped modes. A scheme of the environment is shown in Fig. 1.

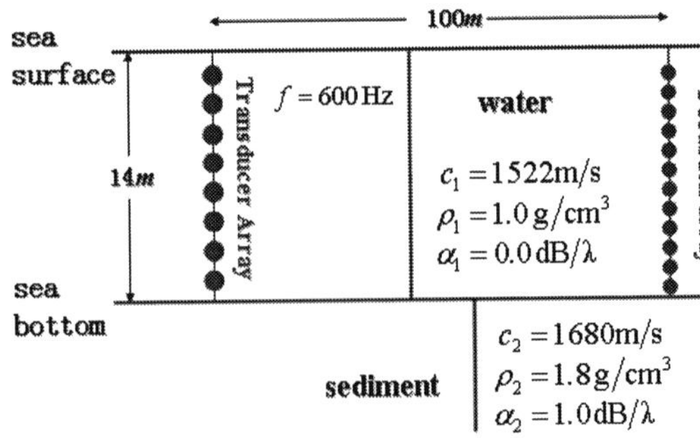

FIGURE 1. Scheme of the environment

The figure of merit used in the simulations is the signal-to-error ratio (SER), which was first used by Buck[8]:

$$\mathrm{SER} = 10\log_{10}\left(\frac{\|p_0\|^2}{\|p_0 - p(k)\|^2}\right) \tag{14}$$

Fig. 2 is the comparison of performances between pure LSM and our algorithm. It can be shown from the figure that our algorithm can speed up the converging progress and be steady.

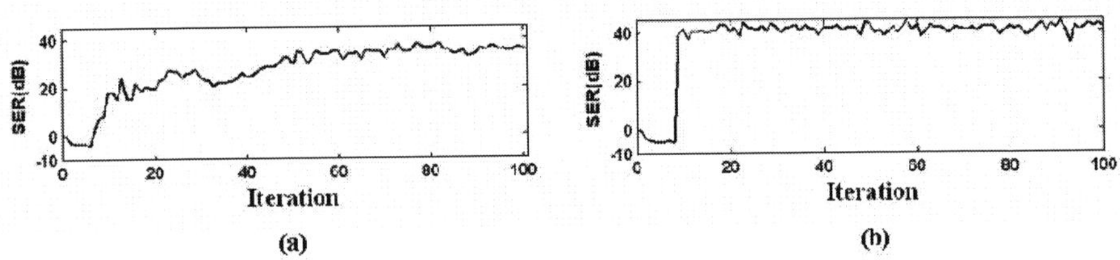

FIGURE 2. Comparison of performances between pure LSM (a) and our algorithm (b)

Table 1 lists the ratio of estimate to real value of matrix Q after 100 times iteration. The table shows that our algorithm can get a correct estimate value of Green's function matrix, and the pure LSM method can not.

TABLE 1. Ratio of estimate to real value of matrix Q

	Mean Value	Covariance	Maximum	Minimum
Pure LSM	4.6026	18.110	23.045	0.1966
Our algorithm	1.0005	0.0001	1.0353	0.9657

Fig. 3 is the comparison of vertical profiles between pure LSM and our algorithm after 100 times iteration. The upper parts of both (a) and (b) are pressure profile of received field and desired field. The lower parts of both (a) and (b) are source weight profile and eigenfunction profile.

It can be shown from Fig. 3 that both pure LSM and our algorithm can excite a single-mode sound field. However, because pure LSM can not get a correct estimate of Matrix Q, it can not obtain an optimal source weight profile. As a result, the single-mode sound field excited by pure LSM (0.016) is smaller than our algorithm (0.028).

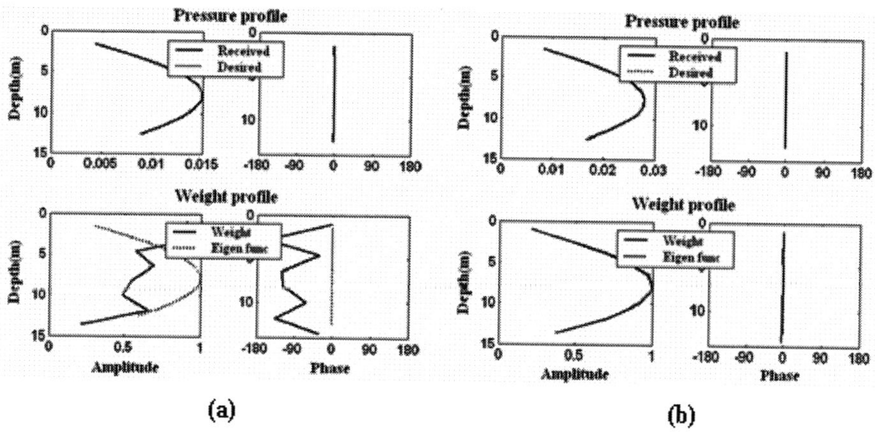

FIGURE 3. Comparison of vertical profiles between pure LSM (a) and our algorithm (b)

EXPERIMENT RESULTS

The coastal experiment was performed on Jan 16 of 2009 in Sanya Bay. The experiment site is very flat and the depth is about 14.7m. The water column is near isovelocity. Two ships are used in the experiment. The source array and feedback array are deployed fore and aft of one ship with a distance between them of about 22 meters. The observation array is deployed on the other ship. The distance between the two ships is about 200 meters. The 10-element source array transmits at a frequency of 500 Hz, depth of the first source element is 1.0 m and spacing of the source array is 1.5 m. The depth of the first element of the 14-element feedback array is 1.0m and the spacing of the feedback array is 1.0m. The depth of the first element of the 14-element observation array is 0.6 m and the spacing of the observation array is 1.0m.

During the close-loop single-mode excitation, the desired single-mode sound field profile is the only variable that needs to be given. To get the correct desired profile is very important for our experiment. The environment is horizontally stratified, so the sound field profile of real single-mode does not vary with range. Given different desired profiles, if the sound field of the feedback array and observation array are similar, then we get the real single-mode sound field.

Fig. 4 is single-mode excitation performance given different desired profiles (a), (b) and (c). Table 2 is the energy ratio of single-mode to whole sound field.

We can draw some conclusions from Fig. 4 and Table 2: (1) Energy ratio of single-mode at the feedback array is always very high, which indicate the effectiveness of the algorithm, (2) Energy ratio of the observation array varies greatly with different desired pressure profiles, which means that a correct desired pressure profile is important for the excitation. (3) High value of the energy ratio is higher than 97% for case (b), which means real single-mode is successfully excited during the experiment.

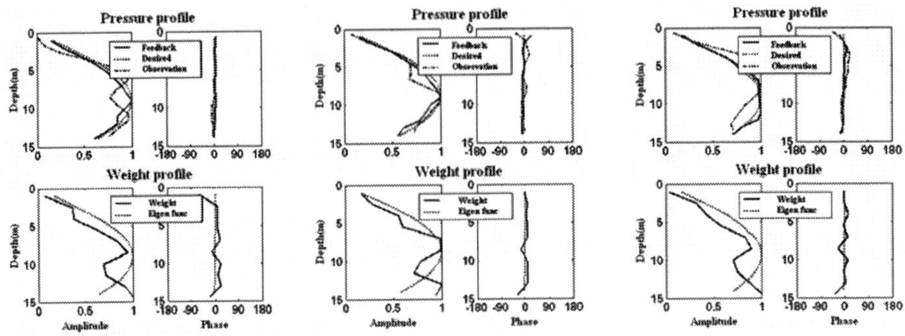

(a) (b) (c)

FIGURE 4. Single-mode excitation performance given different desired profile

TABLE 2. Energy ratio of single-mode to whole sound field

	Case (a)	Case (b)	Case(c)
Feedback array	0.99916	0.99795	0.99653
Observation array	0.96474	0.97490	0.95164

CONCLUSION

We establish an algorithm for single-mode close-loop excitation in shallow water. Advantages of the algorithm are confirmed by numerical simulations. During a coastal single-mode close-loop excitation experiment, a desired single-mode sound field is the only variable that needs to be given. In order to get an exact desired single-mode sound field, two vertical hydrophone arrays are used to search out the correct vertical wave number. Results of an experiment performed in a coastal environment are presented.

REFERENCES

1. C. L. Pekeris, *Geolog. Soc. Am. Memoirs*, **27**, 1-117 (1948).
2. C. S. Clay and K. Huang, *J. Acoust. Soc. Am.*, **67**(3), 792-794 (1980).
3. C. Gazanhes and J. L. Garnier, *J. Acoust. Soc. Am.*, **69**(4), 963-969 (1981).
4. D. F. Gingras, *J. Acoust. Soc Am.*, **103**(1), 195-204 (1998).
5. V. A. Eliseevnin, *Acoust. Phys.*, **52**(2), 227-230 (2006).
6. E. V. Golubeva and V. A. Eliseevnin, *Acoust. Phys.*, **54**(1), 52-57 (2008).
7. E. V. Golubeva and V. A. Eliseevnin, *Acoust. Phys.*, **54**(5), 664-669 (2008).
8. J. R. Buck, *J. Oceanic Engineering*, **22**(2), 281-291 (1997).
9. J. Y. Chen et al, *Acta Acoustica(in Chinese)*, **31**(4), 316-321 (2006).

Target Detection by Receiving-transmitting-united MVDR TRBF under Environmental Mismatch

Yidi Ruan[a] and Xianyi Gong[b]

[a] *China Mobile Communications Corporation Zhejiang Company Limited, Hangzhou 310006*
[b] *Department of Information Science and Electronic Engineering, Zhejiang University Hangzhou 310027*

Abstract. In this paper, the concept of Time Reversal Beamforming (TRBF) is put forward which forms a theoretical framework unifying the physical TR process and the signal processing method. Four TRBF systems are raised and their evolution illustrates that TRBF can be made more practical for target detection in real environments. The availability of these TRBF systems is shown by simulation and experiment.

Keywords: target detection, time reversal beamforming, environmental mismatch
PACS: 43.60, 43.30

INTRODUCTION

Target detection and localization in shallow water waveguides profoundly involves both acoustic propagation physics and sonar signal processing, and has been an important topic in theoretical and practical research. It is full of challenges, mainly due to the complexity and uncertainty of underwater acoustic environments. The time reversal (TR) focusing has been intensively investigated, but long after the concept of TR was introduced, it remains as a physical process[1][2]. Until recent years, for instance, Moura et al discussed a method of detection by TR in a paper in Jan. 2007 and expressed the idea of putting TR into the framework of signal processing[3][4].

We have studied the essence of TR focusing, not only as a physical process, but also as a signal processing method. We fit TR into a framework of signal processing, and develop a new concept of transmitting-receiving-united time reversal beamforming (TRBF). Then considering that the performance of TR spatio-temporal focusing is limited by environmental uncertainties, we develop several systems of TRBF and implement them for target detection in an uncertain environment.

CP1272, *Shallow-Water Acoustics, Second International Shallow-Water Acoustics Conference*
edited by J. Simmen, E. S. Livingston, J. X. Zhou, and F. H. Li

A UNIFIED FRAMEWORK FOR TRBF

TR focusing acquires channel/environment information $\mathbf{g}(\omega)$ from the forward process with a probe source (PS), and then matches the original channel/environment $\mathbf{g}(\omega)$ in the backward process, to reach the effect of focusing: $(\mathbf{g}(\omega))^{\mathrm{H}}\mathbf{g}(\omega)=|\mathbf{g}(\omega)|^2$. The whole process in essence is a *Bartlett* processing. Considering the fact that beamforming, matched filtering, matched field processing and model-based matched filtering all contain the same *Bartlett* correlation operation, we generalize the concept of beamforming, and regard all these five methods as beamforming processing under a unified theoretical framework. Therefore, TR turns into TRBF naturally. See FIGURE 1.

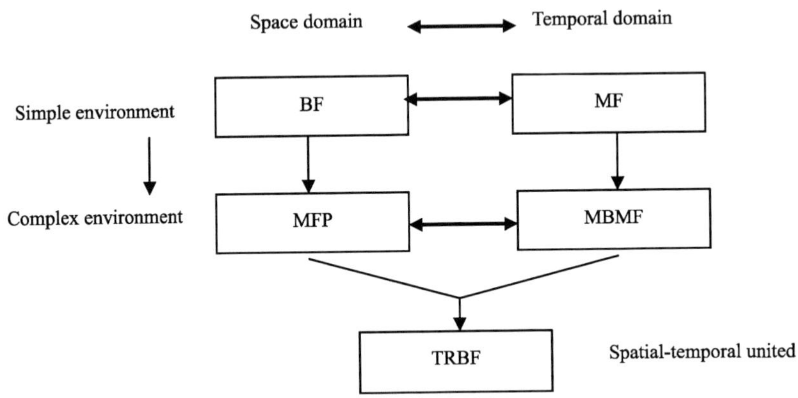

FIGURE 1. The unified framework for TRBF as a method of signal processing

However, compared with the other four types of beamforming methods, TRBF is robuster due to the invariance of both the TR solution of the wave equation and the dominant localization signature of the propagated signal[5]. Therefore there is great potential for TRBF to realize target detection in an uncertain environment.

TARGET DETECTION SYSTEM BY TRBF

Considering TR focusing in acoustic waveguides, when there is a probing source (PS) piloting the source-receiving array (SRA) to transmit the received signal time reversely, the time reversal signals will propagate backward via the forward paths to focus at PS. Here the PS actually provides information of the environment, so as to form the steering vector for the array to retransmit the received signal in a time-reversed order, which is a TRBF operation.

The approaches of TRBF target detection and localization are as follows[6]: Step I, transmitting focusing (we call it transmitting TRBF): an SRA performs a physical TR operation to form a focus at the position of the target. Step II, receiving focusing (receiving TRBF): the SRA applies an algorithmic TR operation to the echo scattered by the target and gets a focused waveform. Step III, matched filtering: use the signal transmitted by PS as a replica to match with the focused waveform from Step II, thus realizing target detection and localization. FIGURE 2 illustrates the Schematic diagram of the whole system.

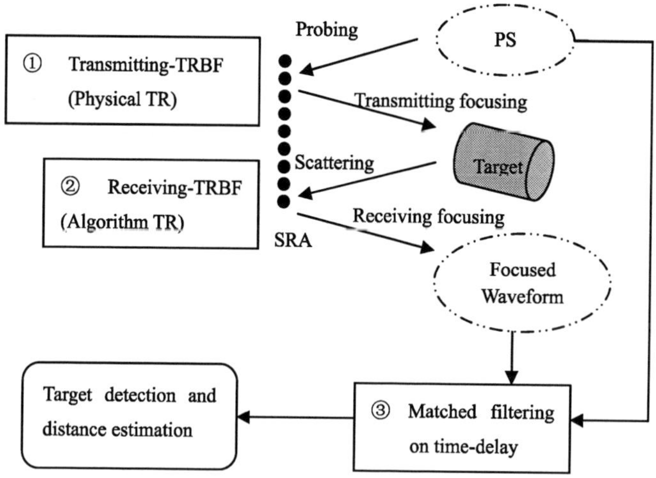

FIGURE 2. Schematic diagram of the TRBF target detection system with a probing source

But in the real environment, the assumed PS does not exist at the target place, and the environment varies with time; as a result, the steering vector of transmitting obtained by the PS may not quite match the environment in the backward process. These factors can be considered as a single problem, which is the uncertainty of the environment.

To deal with the environmental uncertainties, we first build a modeled source (MS) by *a priori* knowledge to replace the environment-probing PS, transforming the environmental uncertainties into modeling uncertainties. So the system in FIGURE 2 evolves into the one in FIGURE 3 by modeling a source at the target place.

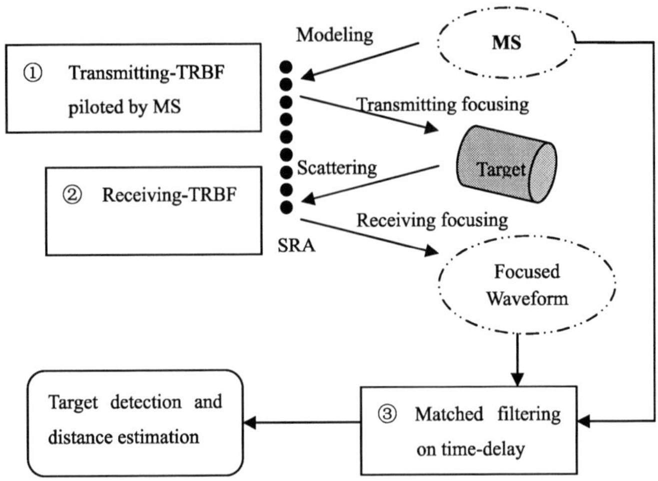

FIGURE 3. Schematic diagram of the TRBF target detection system with a modeled source

Based on good *a priori* knowledge[7], the mismatch between the modeled and the real environment can be assumed to be very small. We perform a Taylor series expansion, using the mismatch of environmental parameters as the independent variable and the mismatch of acoustic field parameters as the dependent variable, so as to model the change of signal or steering vector introduced by the mismatch as an additive interference $\mathbf{\Delta}$.

$$\mathbf{x}(D)=\mathbf{x}(D_0)+\sum_{k=1}^{\infty}\frac{1}{k!}\frac{\partial^k\mathbf{x}(D)}{\partial D^k}\bigg|_{D_0}(D-D_0)^k \tag{1}$$

$$\mathbf{x}(D)\approx\mathbf{x}(D_0)+\frac{\partial\mathbf{x}(D)}{\partial D}\bigg|_{D_0}(D-D_0)=\mathbf{x}(D_0)+\mathbf{\Delta} \tag{2}$$

where $\mathbf{x}(D)$ is the model built by one parameter D, which for example is the depth in the model, while D_0 is the true depth. In the expressions above, we assume the model depends on one parameter. However, it can be expanded to the multi-parameter case.

According to the optimal signal processing algorithm[8], we apply a robust MVDR algorithm to TRBF, in order to cope with the uncertainty of the model[9], or more precisely, the uncertainty of the modeling parameters. The corresponding system is illustrated in FIGURE 4.

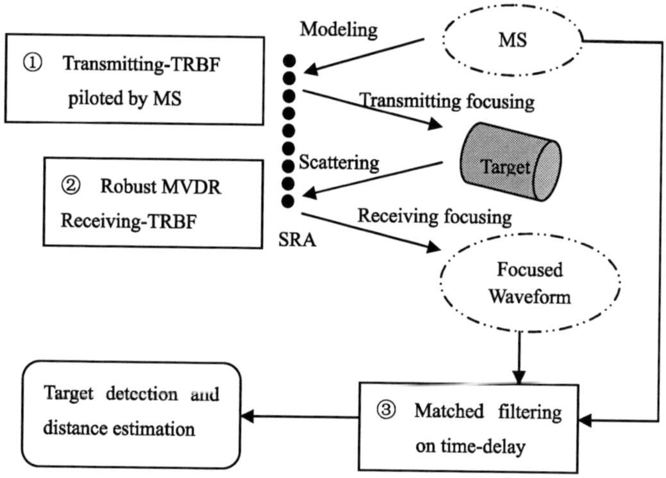

FIGURE 4. Schematic diagram of robust MVDR TRBF target detection system with a modeled source

We consider using the structure of a two-channel Generalized Sidelobe Canceller (GSC)[8] to realize the receiving MVDR TRBF, and then take the information in auxiliary channel to adjust the steering vector of transmitting TRBF. Thus a transmitting-receiving-united MVDR/GSC TRBF target detection system is established. The approach can be described as follows: Initially we perform the transmitting TRBF by an MS, and apply the GSC receiving TRBF to the receiving echo signals; then we modify the transmitting structure using the receiving steering vector, and carry out the transmitting TRBF again. These steps can be implemented iteratively. The specific flow chart is in FIGURE 5. In FIGURE 5, the signal transmitted by MS is marked as 's', and 's_1' is the scattering wave reflected by the target, 's_2' is the output of GSC receiving-TRBF. We correlate the signal 's_2' with 's', and set a threshold for the correlation to control the recursive modification of the steering vector of transmitting TRBF, thus enhancing the effect of transmitting focusing. The better the focusing effect, the better the target detection achieved.

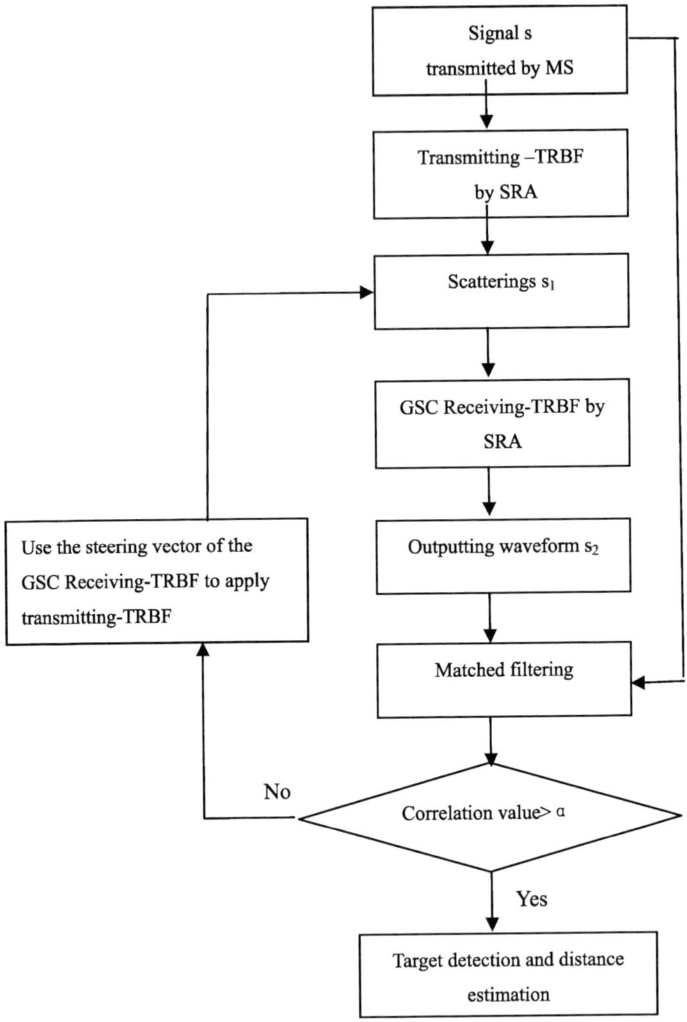

FIGURE 5. Flow chart of transmitting-receiving-united MVDR/GSC TRBF target detection system

SIMULATION AND EXPERIMENT

The evolution of the four TRBF systems illustrates the progress of TRBF in becoming more practical for target detection in real environments. We provide support for the transmitting-receiving-united MVDR/GSC TRBF target detection system by computer simulation and experimental data processing.

Simulation begins with a one-layer isospeed Pekeris waveguide with a depth of 1.41 m, but modeled with a depth of 1.39 m to create a mismatch on purpose. An MS is set at the target, and an SRA is placed at a distance to perform TRBF target detection. The focusing waveform by transmitting TRBF based on MS is shown in FIGURE 6. Then conventional receiving TRBF is applied to the echo by SRA, and the output waveform is shown in FIGURE 7. Meanwhile, GSC TRBF is conducted on the same echo, and the output is in FIGURE 8. Using the receiving steering vector to modify the transmitting structure, we carry out the transmitting-receiving TRBF again, to get the output waveform of GSC receiving TRBF applied onto the echo in the second round, which is shown in FIGURE 9.

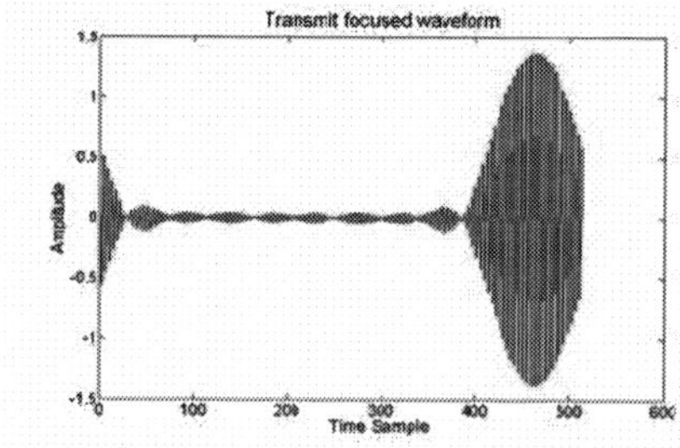

FIGURE 6. The focusing waveform by transmitting TRBF

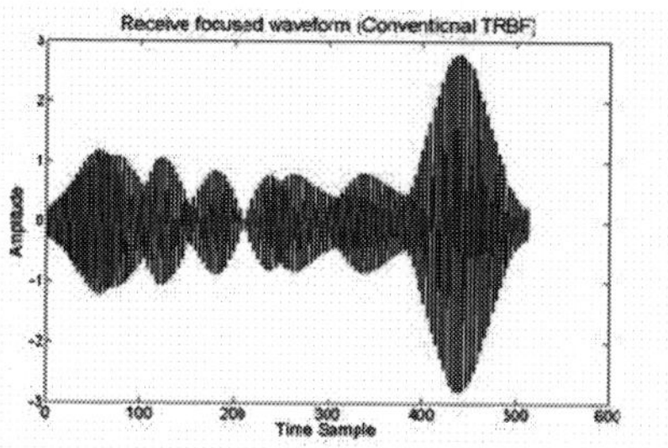

FIGURE 7. The output waveform of conventional receiving TRBF applied onto the echo

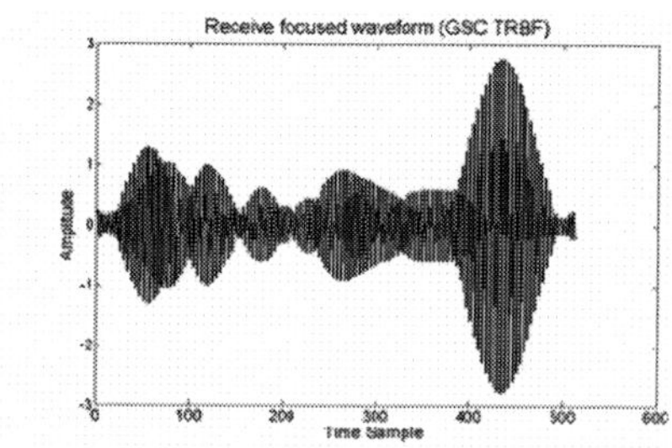

FIGURE 8. The output waveform of GSC receiving TRBF applied onto the echo

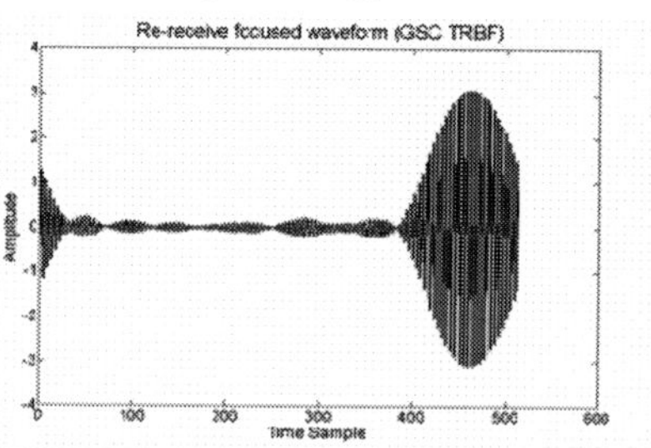

FIGURE 9. The output waveform of GSC receiving TRBF applied onto the echo in the second round

The two methods above correspond to the systems in FIGURE 3 and FIGURE 5. To compare the effects, we make a correlation with the output waveforms and the MS signal, and the result is obvious. See FIGURE 10. The second round of GSC transmitting-receiving TRBF achieves a better effect.

In the same way, a tank experiment is conducted. In the first round of transmitting and receiving, two methods are applied to give a comparison, and GSC TRBF is carried out iteratively in the second round. The results of the first round are in FIGURE 11, and those of the second round are in FIGURE 12.

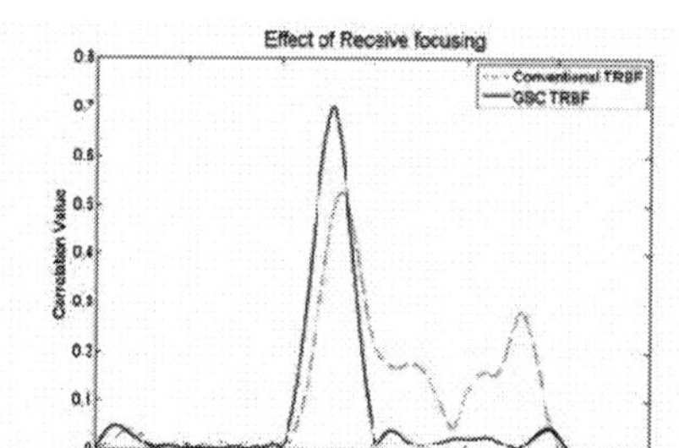

FIGURE 10. Comparison of the correlation result

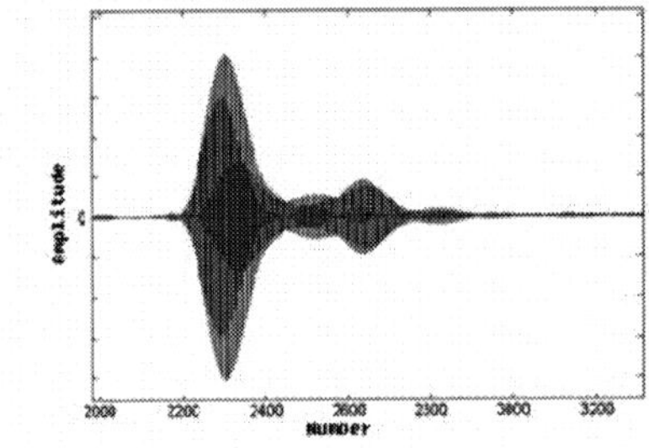

(a) The focused wave form at target

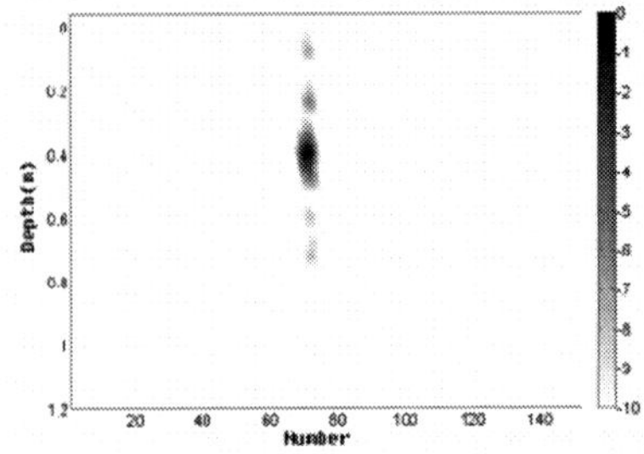

(b) Energy spectrum of signals received by a VLA at target

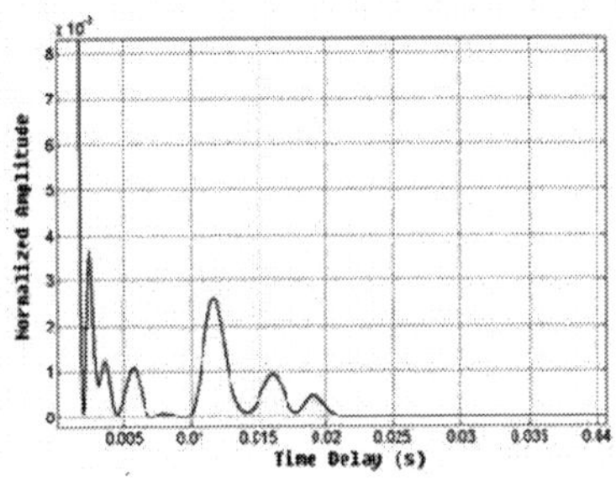

(c) The output of conventional TRBF

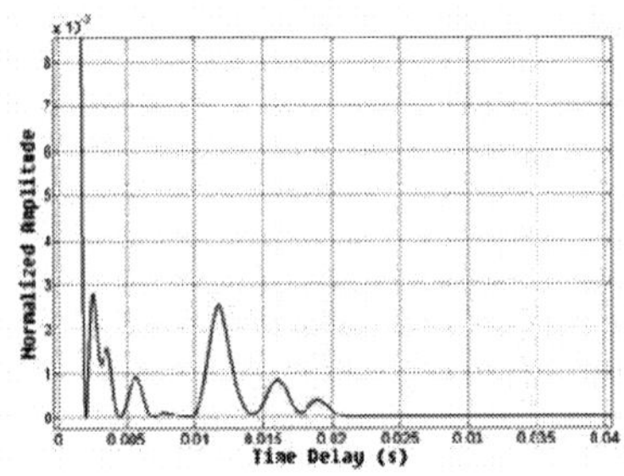

(d) The output of GSC TRBF

FIGURE 11. Results of the first round of transmitting and receiving

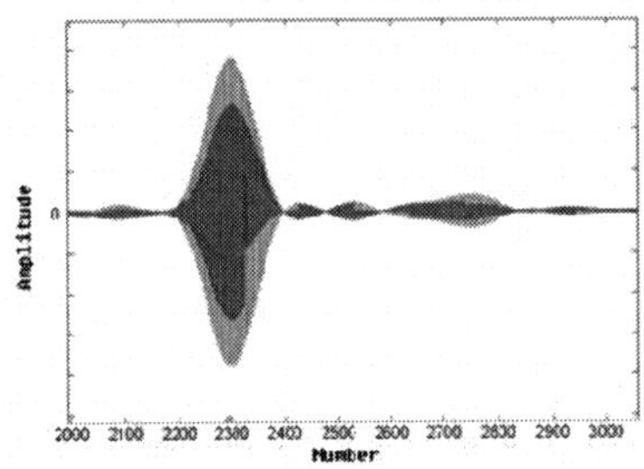

(a) The focused wave form at target

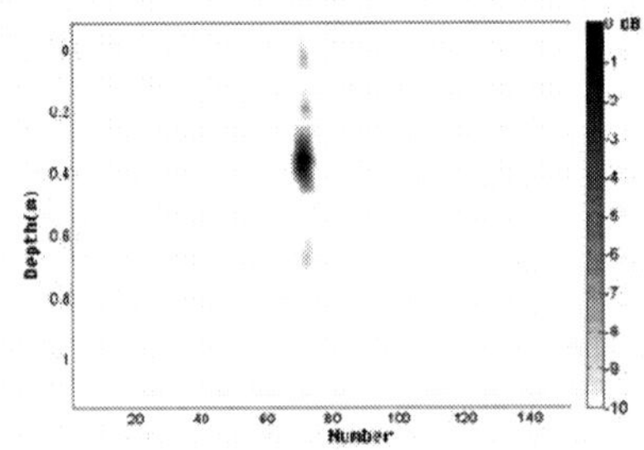

(b) Energy spectrum of signals received by a VLA at target

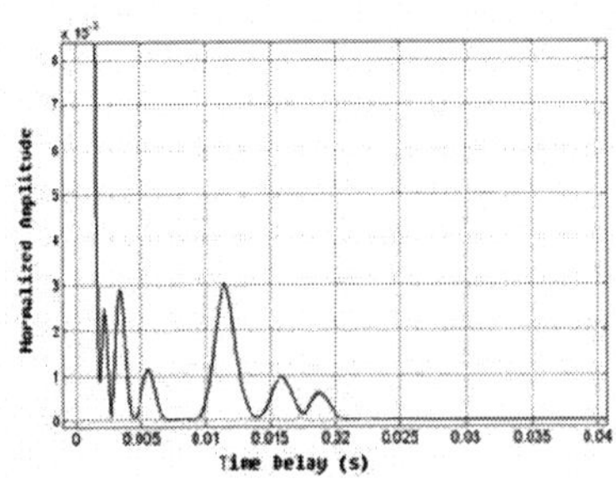

(c) The output of GSC TRBF

FIGURE 12. Results of the second round of transmitting and receiving

Comparing the results of transmitting TRBF in FIGURE 11(b) with FIGURE 12(b), it is obvious that the second round operation enhances the effect of focusing, which has a higher gain of SNR. Also the error of distance estimation in the second round is reduced. In this experiment the target is set at a distance of 8.3 m. In FIGURE 11(c), the peak at the time delay of 0s is for the direct signal from the transmitter to the receiver, and the peak at about 0.0116 s is for the target, which infers an estimation of 8.47 m (the sound speed in water is 1459.8 m/s at that time). Although the results of distance estimation in FIGURE 11(c) and (d) are almost the same, the effect by the second round of transmitting-receiving GSC TRBF turns out to be much better. The peak in FIGURE 12(c) appears at 0.0114 s, inferring an estimation of 8.32 m, which is closer to the real location of the target.

CONCLUSIONS

In this paper, TR is considered not only as a physical process, but also a signal processing method. We generalize the concept of beamforming, and develop the method of TRBF to solve the problem of target detection in an uncertainty environment. The availability of the four TRBF systems posted is shown by the simulations and experiments, and their evolution illustrates that TRBF can be developed to be more practical for target detection in real environments.

REFERENCES

1. P. Roux and M. Fink, "Time reversal in a waveguide: Study of the temporal and spatial focusing", J.Acoust.Soc.Amer., **107**(5), 2000.
2. D. R. Dowling and D. R. Jackson, "Phase-conjugation in underwater acoustics," *J. Acoust. Soc. Amer.*, **89**, p171-181,1991.
3. J. M. F. Moura, Y. W. Jin, D. Stancil, J. G. Zhu, A. Cepni, Y. Jiang and B. Henty, "Single antenna time reversal adaptive interference cancellation," *IEEE ICASSP*, 2005.
4. J. M. F. Moura and Y. W. Jin, "Time reversal imaging by adaptive interference canceling," *IEEE* Trans. Signal process., **56**(1), 2008.
5. W. A. Kuperman and D. R. Jackson, "Ocean acoustics, matched-field processing and phase conjugation," *Topics Appl.Phys*. **84**,43-96, 2002.
6. Y. D. Ruan, X. Pan, J. L. Li and X. Y. Gong, "Robust minimum-variance time-reversal beamforming for target detection and localization in uncertain environment," *ACTA ACUSTICA*, **28**(4), 2009.
7. J. J. Fuchs, "A robust matched detector", *IEEE Trans. Signal process.*, 55(11), 2007.
8. J. Li and P. Stoica, *Robust Adaptive Beamforming*, A John Wiley & Sons, Inc., Publication, 2004.
9. S. Kim, W. A. Kuperman, W. S. Hodgkiss, H. C. Song, G. F. Edelmann, T. Akal, "Robust time reversal focusing in the ocean," *J.Acoust.Soc.Amer.*, **114**(1), 2003.

An Experiment On Passive Synthetic Aperture Time Reversal Communications In Shallow Water

Bingwen Sun and Shengming Guo

Key Laboratory of Underwater Acoustic Environment, Chinese Academy of Sciences, Beijing 100190 China

Abstract. Time reversal can focus in both time and space. The temporal focusing properties have been used in underwater acoustics communications. Typical passive time reversal communication experiments use a vertical array of receivers and a sound source. The property of a horizontal time reversal array that it focuses well in the temporal domain indicates the utilization of synthetic aperture time reversal. Passive synthetic aperture processing repeats emitting same signals during the cruising and records the received signal using one or more receivers. Passive synthetic aperture time reversal communication techniques are presented and applied to results from a sea trail. The experiment consisted of a towed hydrophone and a vertical array. During the experiment, a probe signal is emitted to sample the transmit function then be used by time reversal processing. Due to the moving of the hydrophone, Doppler shift compensation and receive wave forms synchronization must be performed before the time reversal processing. The signal-to-noise ratio is increased and the temporal side lobes are decreased in the time reversal process. Additional communication algorithms can be appended after the time reversal process to increase performance. The result from the at-sea experiment confirms the ability of passive synthetic time reversal communications.

Keywords: time reversal, synthetic aperture, shallow water
PACS: 43.30, 43.60

INTRODUCTION

Time reversal is a process in which a source at one location transmits sound which is received at another location, time reversed, and retransmitted. The retransmitted sound returns to the source location focusing in both time and space. This property shows the potential use of time reversal in underwater acoustic communications. Typically a vertical array of receiver/transducer and a sound source are used. A horizontal time reversal array focuses well in the temporal domain[1]. W. J. Higley, etc performed an at-sea experiment using a two-element system and passive synthetic aperture time reversal which confirmed the feasibility of an inexpensive communication system using a synthetic aperture time reversal array at sea[2]. In synthetic aperture time reversal scheme, a moving source repeats transmitting same signal always including a probe signal and an information signal which is recorded by a one-element receiver shown in Fig. 1. This paper shows an at-sea experiment to

CP1272, *Shallow-Water Acoustics, Second International Shallow-Water Acoustics Conference*
edited by J. Simmen, E. S. Livingston, J. X. Zhou, and F. H. Li

demonstrate the ability of the two-element system and passive synthetic aperture time reversal in East China Sea.

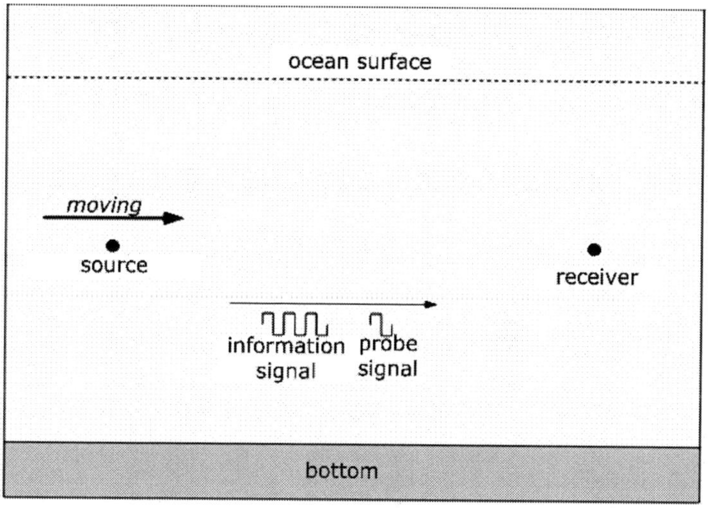

FIGURE 1. Passive synthetic aperture time reversal scheme. A moving source repeats transmitting same signal including a probe signal and an information signal. One-element receiver records the signal and does some processing.

EXPERIMENTAL DETAILS

The experiment was carried out in a shallow water region in East China Sea in May 2008. A vertical receiver array (VRA) was moored. Only one channel was used in latter processing. A source was towed by a ship at a speed of 3.4 m/s along the dotted line as shown in Fig. 2. The max distance between the VRA and the source was 10 km.

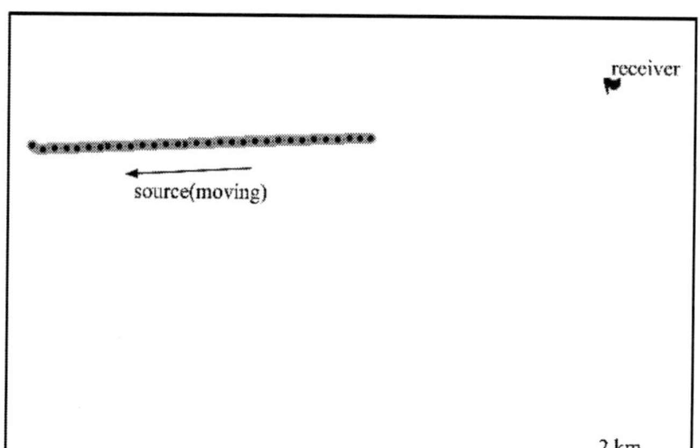

FIGURE 2. Experiment setup. A moving source was towed by a ship along the dotted line. VRA (flag) was moored.

The source transmitted same signal every 10 seconds. The signal consisted of 4 parts: sine signal, hyperbolic chirp, M-sequence signal and BPSK signal. Sine Signal's carrier frequency was 6 kHz. Frequency in hyperbolic chirp increased from 4 kHz to 8 kHz. The length of M-sequence signal was 8. The BPSK signal included 2295 symbols and was almost 1.15 seconds. One sample of the received signal is shown in Fig. 3.

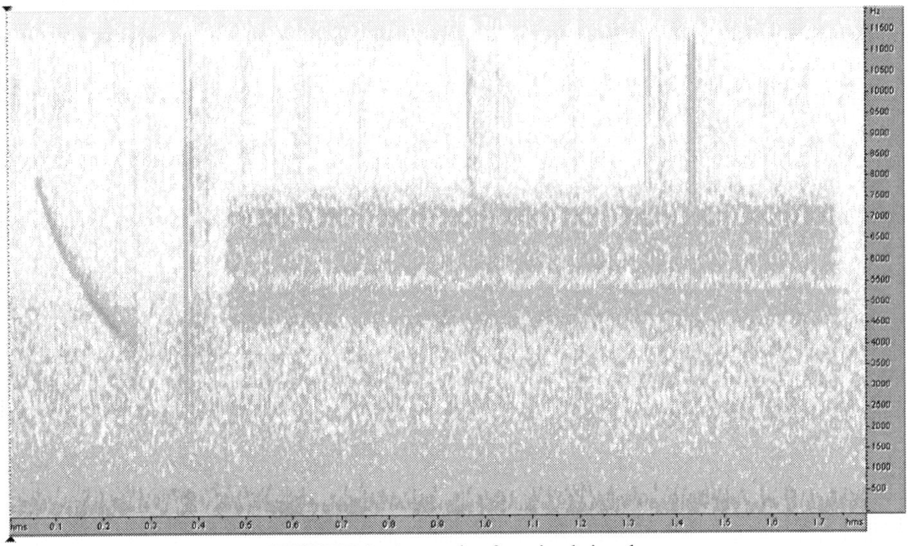

FIGURE 3. A sample of received signal.

DATA PROCESSING AND RESULT

Only one channel of VRA was used. The received signal was sampled firstly as shown in Fig 4(a). Doppler frequency shift compensation then was done to each sample. Hyperbolic chirp part was used to estimate the transmission function. The conjugation version of transmission function then was used to each information signal. All the samples then were summed up followed by a Decision Feedback Equalization (DFE) as shown in Fig 4(b).

When only one sample was used, Tab. 1 shows the result of bit error rate for 6 samples. Fig. 5(a) shows constellation diagram of one sample (ID = 1). When using 31 samples, BER reduced to 0.043573%, much lower than using one sample. Fig. 5(b) shows constellation diagram of synthetic aperture time reversal.

TABLE 1. Bit error rate (BER) of one sample.

Sample ID	BER(%)
1	0.47930
7	0.26144
13	0.21786
19	0.17429
25	0.56645
31	1.26360

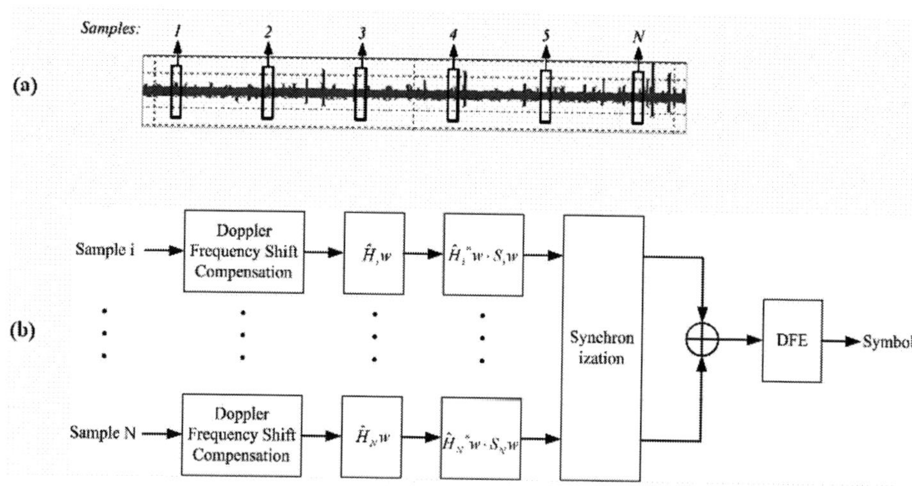

FIGURE 4. Processing of received signals. a) sampling; b) synthetic aperture time reversal

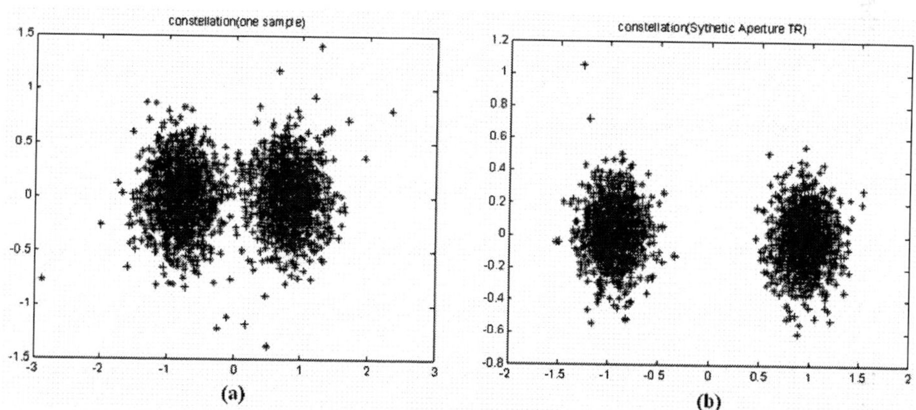

FIGURE 5. Constellation diagram. a) one sample; b) synthetic aperture time reversal

CONCLUSION

This experiment and this paper confirm the feasibility of synthetic aperture time reversal at shallow water sea. The two-element system decrease the cost compared to typical time reversal communication schemes. Repeating transmission of the communication sequence tends to lower data rate. Additional DFE can be appended for good performance.

ACKNOWLEDGMENTS

The authors are grateful for the contributions of the Institute of Oceans in carrying out the experiment.

REFERENCES

1. M. R. Dungan and D. R. Dowling, "Orientation effects on linear time reversing array retrofocusing in shallow water", J. Acoust. Soc. Am. **112**, 1842–1852(2002).
2. W. J. Higley, Philippe Roux, W. A. Kuperman, W. S. Hodgkiss, H. C. Song, T. Akal and Mark Stevenson, "Synthetic aperture time-reversal communications in shallow water: Experimental demonstration at sea", J. Acoust. Soc. Am, **118**, 2365-2372(2005).

Extraction of the waveguide invariant from a shallow-water acoustical experiment

Ling-ai Tian, Shihong Zhou, Fuchen Liu

Hangzhou Applied Acoustic Research Institute, Hangzhou, China

Abstract. A spectrogram is a visual representation of sound pressure as a function of frequency and range. The waveguide invariant describes the dispersive propagation characteristics of the field in a given waveguide. It is defined as the ratio of the striations or interference patterns. It also relates the modal group and phase velocities of a medium. Within this paper, one approach using image processing and signal processing technique for extraction of the waveguide invariant from striations in spectrogram is presented, whilst the experimental results are compared with theoretically obtained results. Waveguide invariant values at different receiver depths are extracted using shallow-water acoustical experimental data.

Keywords: interference patterns, waveguide invariant, image processing.

INTRODUCTION

In an underwater acoustic field, the interference patterns (striations) in range and frequency spacing are often distinct from the mutual constructive/destructive interference between modes. A scalar parameter called the waveguide invariant was defined by Brekhovskikh and Lysanov [1]. It is a function of the propagation characteristics of the environment and is related the slope of the striations. Assuming that the source spectra $S(\omega)$ vary with frequency smoothly, then

$$\frac{\delta r}{\delta\omega}=-\frac{r}{\omega}\frac{\partial\Delta k_{mn}(\omega)/\partial\omega}{\Delta k_{mn}(\omega)/\omega}=-\frac{r}{\omega}\frac{\partial S_g^{mn}(\omega)}{\partial S_p^{mn}(\omega)} \tag{1}$$

where $S_g^{mn}(\omega)$ and $S_p^{mn}(\omega)$ are, respectively, the differential group slowness and the differential phase slowness between mode *m* and mode *n*.

$$-\partial S_g^{mn}(\omega)/\partial S_p^{mn}(\omega)=1/\beta \tag{2}$$

where $\partial S_g^{mn}(\omega)/\partial S_p^{mn}(\omega)$ is almost invariable [2]. The waveguide invariant, herein designated as β, describes the dispersive characteristics of the field. It relates the environmental and the propagation characteristics. The β are generally negative for refracting propagation and positive for bottom-interacting propagation.

Within this paper, one approach using image processing and signal processing techniques for extracting the waveguide invariant from the striations in the spectrogram is presented.

CP1272, *Shallow-Water Acoustics, Second International Shallow-Water Acoustics Conference*
edited by J. Simmen, E. S. Livingston, J. X. Zhou, and F. H. Li

WAVEGUIDE INVARIANT

In a range-independent oceanic waveguide, a moving broadband source has a relation expression of frequency and range. It follows that

$$\frac{\delta\omega}{\omega} = \beta\frac{\delta r}{r} \tag{3}$$

In the aforementioned waveguide, the waveguide invariant is a constant, so equation (3) can be rewritten as

$$\omega(t) = \omega(t_0)\left(\frac{r(t)}{r(t_0)}\right)^{\beta} \tag{4}$$

where $r(t)$ and $\omega(t)$ are, respectively, the range and frequency of source at time t, $r(t_0)$ and $\omega(t_0)$ are the range and frequency of source at time t_0 (the reference point).

When the range from source to receiver is sufficiently large, the normalized slope can be shown to approximate the waveguide invariant. That is, the waveguide invariant can be computed by the striation slope in range and frequency spacing. The relationship can be rewritten as

$$\lim_{\frac{\Delta r}{r}\to 0} \frac{\Delta w}{w} / \frac{\Delta r}{r} = \beta \tag{5}$$

HOUGH TRANSFORM TECHNIQUE

The Hough transform [3] was first proposed in 1962; it is an effective technique in image processing. Extraction of line using the Hough transform is one method that transforms the line equation from the x-y axes to the line's coefficient axes, and converts the extraction technique to a counter technique.

The line equation in Oxy spacing can be rewritten as:

$$y=ux+v \tag{6}$$

where u and v are, respectively, the slope and intercept of line. Generally, given a line corresponding to an array (u, v), or, contrarily, given an array (u, v) corresponding to a line, y=ux+v. That is, one line y=ux+v in x-y spacing corresponds to one point in u-v spacing. Similarly, one line v=-xu+y in u-v spacing corresponds to one point (x, y) in x-y spacing. So the line can be extracted using this property.

The line equation can also be shown as a normal pattern to avoid infinite slope,

$$x\cos\theta + y\sin\theta = \rho \tag{7}$$

where ρ is the distance between the line and original point, θ is the separation angle between normal and x axes. One line in the Oxy plane corresponds to one point in the $O\theta\rho$ plane. Likewise, one point in the Oxy plane corresponds to one line in the $O\theta\rho$ plane.

EXTRACTION OF WAVEGUIDE INVARIANT FROM BROADBAND EXPLOSIVE SOUND SOURCE

The purpose of this section is to test the Hough transform for extracting the waveguide invariant. The technique was applied to the horizontal line array data of the oceanic experiment. The sound speed profile of the experimental ocean area is depicted in Figure 1. Additionally, the water-depth is 50m, and the sediment is silt.

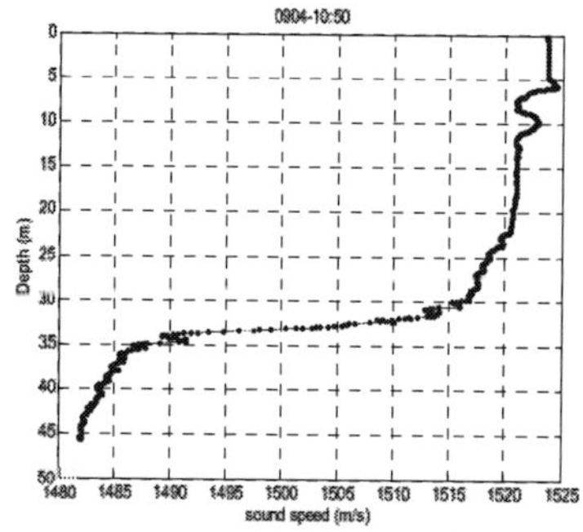

FIGURE 1 The sound speed profile

In the experiment, the broadband explosive sound source produced by a plasma of 6kJ power was at depth 12m and towed at 3 knots. The receiver array of 32 hydrophones was anchored in place. In addition, the first hydrophone was at depth 3.8m, and the space between adjacent hydrophones was 1.2m. Pulse data received at ranges from 1.6km to 2.9km was band pass filtered. The lower frequency and upper frequency are 350Hz and 750Hz. Each of receiver positions was recorded with GPS.

The received pressure spectrogram is divided into several small units in the frequency and range directions respectively. Firstly, a threshold value $th1$ is selected. The maximum of every unit spectrum is compared with $th1$. Secondly, the Hough transform is applied to values above $th1$. So we get a Hough matrix in the $O\theta\rho$ plane. Finally, the Hough matrix is also divided into several small units. The resolution of θ and ρ are, respectively, 1 degree and 1. Subsequently, another threshold $th2$ is selected. Each value is compared with $th2$. All values above $th2$ correspond to the same line in the x-y plane, that is, the line in frequency and range spacing is detected.

Figure 3 shows the interference patterns of the simulated data with the actual sound speed profile (see Figure 1). The pink lines are detected with the Hough transform. The waveguide invariant values can be calculated with the slopes of the lines. The result of the Hough transform is depicted in Figure 2. The red squares in Figure 2 are all points greater than $th2$, which correspond to the pink lines in Figure 3. The waveguide invariant value is the reciprocal value of the line slope. There are 5 lines, so the mean value of the waveguide invariant extracted from Figure 3 is 1.029.

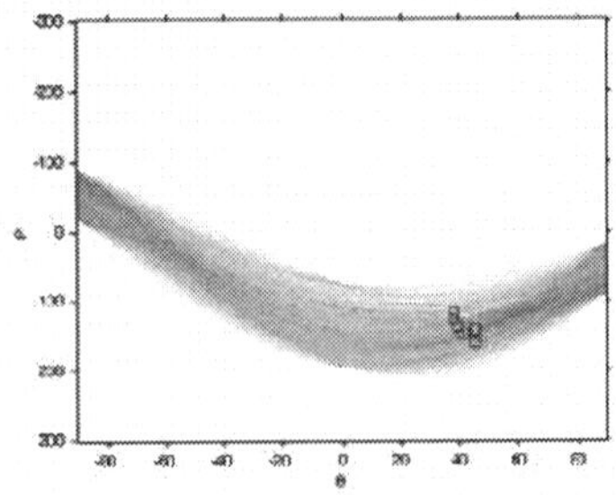

FIGURE 2 The Hough transform of simulated data

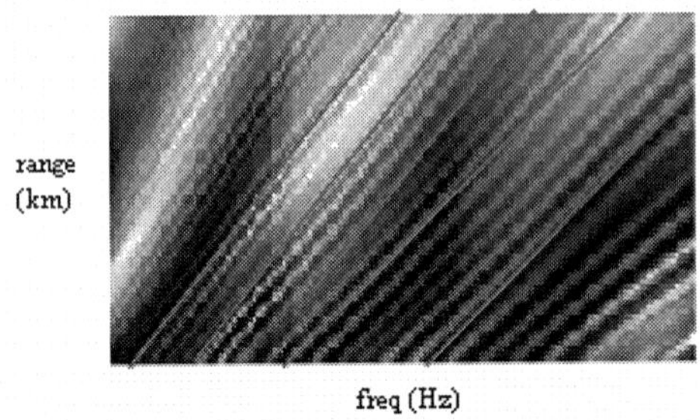

FIGURE 3 The simulated data and detected lines

Figure 4 presents a 1.3km pressure spectrogram over the 350-750Hz band, estimated from the data collected by the first hydrophone (at depth 3.8m) in the vertical line array in LoFAR'08. The map is plotted against log freq and log range. The map is divided into three maps in the frequency direction in order to detect the waveguide invariant value more exactly. The Hough matrixes of the three maps are shown in Figure 5. The red squares, values are above *th*2, correspond to the line detected in frequency and range spacing. The waveguide invariant value is the reciprocal value of the line slope. Figure 6 shows 11 lines detected over the total frequency band. So the mean value of the waveguide invariant is 0.947.

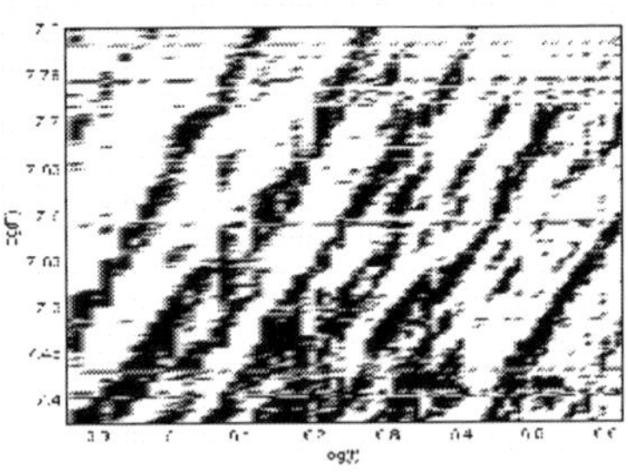

FIGURE 4 The pre-processed data of the first hydrophone

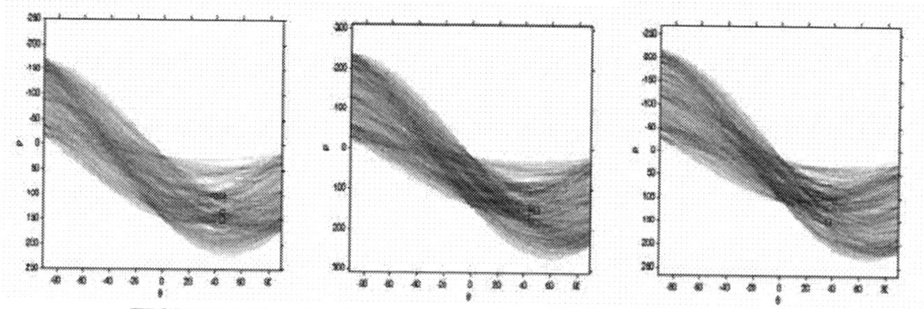

FIGURE 5 The Hough transform of data over three frequency bands

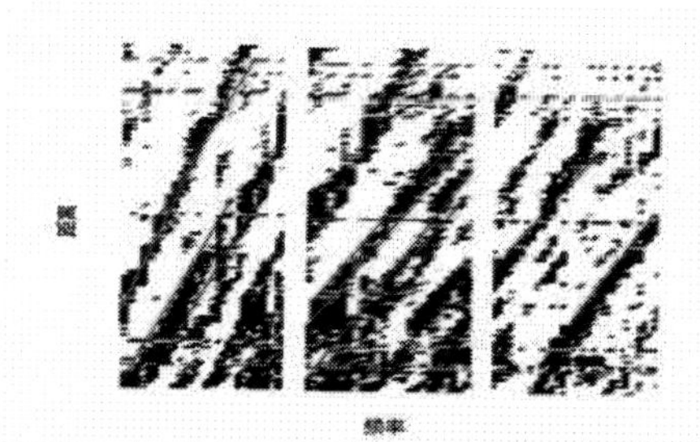

Figure 6 Extraction of waveguide invariant values of experimental data

The waveguide invariant values at different depth are obtained with the same technique. Similarly, the theoretical values at different depth are evaluated. The results are depicted in Table 1.

TABLE 1. Mean waveguide invariant value at different depth

Depth (m)	No. of line segments	Experimental β	Theoretical β
3.8	11	0.947	1.029
6.2	13	1.109	1.024
7.4	10	1.005	0.917
8.6	10	1.060	0.961
11	17	0.919	1.060
12.2	17	1.027	1.098
13.4	12	0.942	0.979
21.8	16	0.962	0.935
26.6	12	0.945	0.966
30.2	19	0.987	0.939
31.4	11	1.006	0.998
36.2	18	1.025	0.947
37.4	16	1.104	0.915
38.6	20	0.949	0.946

Table 1 shows that all experimental and theoretical values of waveguide invariant are close to 1. It also shows the experimental values almost agree with the theoretical values. That is, application of Hough transform to extract the waveguide invariant proves to be useful even though there is error between experimental and theoretical values.

CONCLUSION

The close match between the theoretical and experimental β showed that the proposed technique is effective for extraction of the waveguide invariant. The waveguide invariant can be used in geoacoustic inversion techniques to improve the low-frequency spatial correlation.

REFERENCES

1. L. M. Brekhovskikh and Y. P. Lysanov. *Ocean Acoustics in Ocean Physics*[M]. Edited by V. M. Kamenkovich and A. S. Monin (Nauka, Moscow), 1978, 2: 49-145.
2. G. L. D'Spaina and W. A. Kuperman. *Application of waveguide invariant to analysis of spectrograms from shallow water environments that vary in range and azimuth*. Journal of the Acoustical Society of America[J], 1999,106(5): 2454-2468.
3. QinXiangpei. *Image processing and interface program* [M]. Printed by Electronic Industry publisher, 2009：449-450

A Robust Separating and Tracking Method on Two Wideband Sources by Subspace Rotation with One Vector Hydrophone

De-jun Wang and Feng-hua Li

State Key Laboratory of Acoustics, Institute of Acoustics, Chinese Academy of Sciences, Beijing, 100190, China

Abstract. It has been proved theoretically that two incompletely correlated sources can be identified by linear signal processing methods. However, it is difficult in practice. A new method to separate two wideband sources with one vector sensor is presented in this paper. The method is the combination of subspace rotation and spatial matched filter. Simulations show that this method is insensitive to the initial azimuth error, independent of signal spectrum, and better than wideband focusing subspace methods at low SNR. The sea trial is performed and the experiment results show that the proposed method is effective to separate and track two wideband sources in the underwater environment.

Keywords: Vector Sensor, Wideband Sources Separating and Tracking, Subspace Rotation
PACS: 43.30.Es, 43.30.Re, 43.30.Pc, 43.30.Yj

1. INTRODUCTION

Vector hydrophones outperform traditional hydrophones in directional sensitivity and temporal-spatial gain [1]. Significant development has occurred in vector hydrophones and vector signal processing [2-3]. When a large-scale traditional hydrophone array is difficult to deploy, a single or a few vector hydrophones is an ideal substitute. When using a single vector hydrophone, platform noise or multiple interference significantly affects the performance of detection systems. If average acoustic intensity, which has high temporal-spatial gain, is used to estimate the azimuth, the compositional azimuth is obtained. And if platform noise is comparatively very strong, the compositional azimuth is pointing to the platform. If the spectra of the source signal and the interference are not overlapped, they can be separated by filtering. However, their valid spectra are usually overlapped, and multiple sources separating and tracking of this kind need to be investigated.

It has been proved that under 2nd-order statistics one vector hydrophone can separate two sources which are not fully correlated [4]. Methods such as LOFAR and histogram technique and Demon of propeller perform multiple sources separation in the frequency domain. An approach [5] based on one electromagnetic vector sensor uses the ESPRIT method to separate five incoherent narrowband sources of unknown

CP1272, *Shallow-Water Acoustics, Second International Shallow-Water Acoustics Conference*
edited by J. Simmen, E. S. Livingston, J. X. Zhou, and F. H. Li

frequencies. The above methods are invalid to separate arbitrary bandwidth and spectrum sources. An approach [6] based on one electromagnetic sensor can separate three wideband uncorrelated sources and is validated by simulations. An alternative method based on a nonlinear transcendental equation set of pressure and particle velocities' high order moments has been developed that can identify up to four sources. However, the azimuths and intersection angles are limited. Some subspace-based algorithms used in the acoustic vector array can also be applied in the case of one vector hydrophone, but more research is needed in the circumstance of low SNR and will be compared with the proposed method.

Although the above methods are effectively applicable for some underwater scenarios, the actual signals and channels are complex and unstationary. Therefore robust methods that are less dependent on *a priori* knowledge of arbitrary bandwidths are more useful. The proposed method in this paper focuses on two arbitrary bandwidth and spectrum sources, rotates in azimuth vector subspace, separates out a single source, estimates azimuths, and tracks up to two sources quickly. Based on the theory, the performance of this method is thoroughly analyzed by simulations. In addition, this method is proved to be feasible by sea trial data.

2. THEORY

Assume there are two point sources (source 1 and source 2) at θ_1 and θ_2, radiating uncorrelated or correlated wideband acoustic signals. The output pressure and x direction and y direction velocities of the two sources from one vector hydrophone are $p_1(t)$, $\upsilon_{x1}(t)$, $\upsilon_{y1}(t)$, $p_2(t)$, $\upsilon_{x2}(t)$, and $\upsilon_{y2}(t)$, respectively. The velocities are all compensated by the sound impedance. In common sense, pressure and horizontal velocity are well correlated in the far field [7] satisfying $\upsilon_{x1}(t) \approx p_1(t)\cos\theta_1$ and $\upsilon_{y1}(t) \approx p_1(t)\sin\theta_1$, taking source 1 as an example. So the signal space is composed of pressure and horizontal velocities. For source *i* (*i*=1, 2), the received signal vector $\mathbf{X}_i(t)$ can be expressed as

$$\mathbf{X}_i(t) = [p_i, \upsilon_{xi}, \upsilon_{yi}]^T = a_i(t) \cdot [1, \cos\theta_i, \sin\theta_i]^T \equiv a_i(t) \cdot \mathbf{S}(\theta_i)_{3\times1}, \quad (1)$$

where $a_i(t)$ is the pressure signal of source *i*, and

$$\mathbf{S}(\theta_i)_{3\times1} \equiv [1, \cos\theta_i, \sin\theta_i]^T, \quad (2)$$

which is defined as the azimuth vector. The output of one vector hydrophone, denoted by $\mathbf{X}(t)$, is the linear superposition of $\mathbf{X}_1(t)$, $\mathbf{X}_2(t)$ and ambient noise $\mathbf{N}(t)$

$$\mathbf{X}(t) = \mathbf{X}_1(t) + \mathbf{X}_2(t) + \mathbf{N}(t) = \sum_{i=1}^{2} a_i(t) \cdot \mathbf{S}(\theta_i) + \mathbf{N}(t). \quad (3)$$

$\mathbf{N}(t)$ is assumed to be spatial white Gaussian noise and uncorrelated to source signals, and its autocorrelation matrix is

$$E\{\mathbf{N}(t) \cdot \mathbf{N}(t)^T\} = \sigma_n^2 \mathbf{I}_{3\times3}, \quad (4)$$

where $\mathbf{I}_{3\times3}$ denotes a 3rd-order identity matrix, and σ_n^2 denotes the variance of ambient noise. Other ambient noise models, such as the isotropic model and all kinds of anisotropic models, can be adopted, and theoretical derivation is similar, but performance is a little different.

Assume the initial estimation of the azimuths are $\hat{\theta}_1$ and $\hat{\theta}_2$, and the true azimuth of source 1 is

$$\theta_1 = \hat{\theta}_1 + \delta\theta_1, \tag{5}$$

where $\delta\theta_1$ denotes the initial error. Let $\mathbf{S}'(\hat{\theta}_1)$ denote the derivation of $\mathbf{S}(\hat{\theta}_1)$ on $\hat{\theta}_1$, then we can expand $\mathbf{S}(\theta_1)$ in a 1st-order Taylor series and have

$$\mathbf{S}(\theta_1) = \mathbf{S}(\hat{\theta}_1) + \mathbf{S}'(\hat{\theta}_1)\delta\theta_1. \tag{6}$$

So the task of separating and tracking is to estimate $\delta\theta_1$.

Taking the source 1 channel as an example, weight $\mathbf{X}(t)$ linearly by $\mathbf{W}_1$, and the channel's output signal can be expressed as

$$y_1(t) = \mathbf{W}_1^T\mathbf{X}(t), \tag{7}$$

$$\text{s.t.}\ \ \mathbf{W}_1^T\mathbf{S}(\hat{\theta}_2) = 0, \quad \mathbf{W}_1^T\mathbf{S}(\hat{\theta}_1) = 1. \tag{8}$$

This is equivalent to transforming $\mathbf{X}(t)$ to $\mathbf{X}_{P1}(t)$, which is in the subspace orthogonal to historical information $\mathbf{S}(\hat{\theta}_2)$, in which we search the optimal weight $\mathbf{W}_{P1}$, then estimate $\delta\theta_1$. When iterations reach the stable state, the tracking algorithm converges to its precision. Assume the subspace transforming matrix is $\mathbf{P}_1^T$, satisfying

$$\mathbf{P}_1^T\mathbf{S}(\hat{\theta}_1) = [k_1 \ \ 0]^T, \ \ \mathbf{P}_1^T\mathbf{S}(\hat{\theta}_2) = [0 \ \ 0]^T, \ \ \mathbf{P}_1^T\mathbf{P}_1 = \mathbf{I}, \tag{9}$$

where k_1 is the normalized coefficient. $\mathbf{P}_1^T$ satisfying Eq. (9) is a 2×3 matrix, and is derived as follows. Take SVD on the two azimuth vectors:

$$\left[\mathbf{S}(\hat{\theta}_1) \ \ \mathbf{S}(\hat{\theta}_2)\right] = \mathbf{U}_{3\times3}\begin{bmatrix} \lambda_1 & 0 \\ 0 & \lambda_2 \\ 0 & 0 \end{bmatrix}\mathbf{V}_{2\times2}^T, \tag{10}$$

where both $\mathbf{U}_{3\times3}$ and $\mathbf{V}_{2\times2}$ are unitary matrices. According to Eq. (9),

$$\mathbf{P}_1^T\mathbf{U}\begin{bmatrix} \lambda_1 & 0 \\ 0 & \lambda_2 \\ 0 & 0 \end{bmatrix}\mathbf{V}^T = \begin{bmatrix} k_1 & 0 \\ 0 & 0 \end{bmatrix}, \tag{11}$$

then

$$\mathbf{P}_1^T = \begin{bmatrix} k_1\lambda_1^{-1}\mathbf{V}_{11} & k_1\lambda_2^{-1}\mathbf{V}_{12} & 0 \\ 0 & 0 & M \end{bmatrix}\mathbf{U}^T. \tag{12}$$

From $\mathbf{P}_1^T\mathbf{P}_1 = \mathbf{I}$, M=1, and

$$k_1 = \frac{1}{\sum_{i=1}^{2} \left| \mathbf{V}_{1i} / \lambda_i \right|^2}. \tag{13}$$

Do subspace transforming on the signal space by $\mathbf{P}_1^T$, and in the new subspace the signal vector $\mathbf{X}_{P1}(t)$, weight $\mathbf{W}_{P1}$, and azimuth vector $\mathbf{S}_{P1}(\hat{\theta}_1)$ are

$$\mathbf{X}_{P1}(t) = \mathbf{P}_1^T \mathbf{X}(t), \quad \mathbf{W}_1 = \mathbf{P}_1 \mathbf{W}_{P1}, \quad \mathbf{S}_{P1}(\hat{\theta}_1) = \mathbf{P}_1^T \mathbf{S}(\hat{\theta}_1). \tag{14}$$

Then deduce $\mathbf{W}_{P1}$ and $\delta\theta_1$ satisfying Eqs. (14) and (8). From Eqs. (9) and (14), the autocorrelation matrix of $\mathbf{X}_{P1}(t)$ is

$$\mathbf{R}_{P1} = E\{\mathbf{X}_{P1}\mathbf{X}_{P1}^T\} = \sigma_{s1}^2 \mathbf{S}_{P1}(\theta_1)\mathbf{S}_{P1}^T(\theta_1) + \sigma_n^2 \mathbf{I}_{2\times 2}, \tag{15}$$

where σ_{s1}^2 is the variance of $a_1(t)$. Eq. (15) indicates that $\mathbf{S}_{P1}(\theta_1)$ is the characteristic vector of the autocorrelation matrix. Let $\mathbf{W}_{P1}$ be the optimal vector of the matched filter, then we have

$$\mathbf{W}_{P1} \propto \mathbf{S}_{P1}(\theta_1). \tag{16}$$

Substitute Eqs. (9) and (6) into Eq. (14), we have

$$\begin{aligned} \mathbf{S}_{P1}(\theta_1) &= \mathbf{P}_1^T \left[\mathbf{S}(\hat{\theta}_1) + \mathbf{S}'(\hat{\theta}_1)\delta\theta_1 \right] \\ &= \begin{bmatrix} k_1 + \alpha_1 \delta\theta_1 \\ \alpha_2 \delta\theta_1 \end{bmatrix} \\ &\approx \begin{bmatrix} k_1 \\ \alpha_2 \delta\theta_1 \end{bmatrix}, \end{aligned} \tag{17}$$

where $\mathbf{P}_1^T \mathbf{S}'(\hat{\theta}_1) = [\alpha_1 \quad \alpha_2]^T$. Transform the measured data of $\mathbf{X}(t)$ by Eq. (14), calculate $\hat{\mathbf{R}}_{P1}$, perform eigenvalue decomposition, then obtain the estimated $\hat{\mathbf{W}}_{P1}$ of $\mathbf{W}_{P1}$:

$$\hat{\mathbf{W}}_{P1} = [\beta_1 \quad \beta_2]^T. \tag{18}$$

From Eqs. (16)~(18), $\mathbf{S}_{P1}(\theta_1)$ and $\hat{\mathbf{W}}_{P1}$ are normalized by their first elements respectively, and they become equivalent:

$$\mathbf{S}_{P1}(\theta_1)/k_1 = \hat{\mathbf{W}}_{P1}/\beta_1, \tag{19}$$

then

$$\delta\theta_1 = \frac{\beta_2 k_1}{\beta_1 \alpha_2}. \tag{20}$$

Similarly, $\delta\theta_2$ of source 2 is estimated in channel 2. The sources' azimuths are updated by $\delta\theta_1$ and $\delta\theta_2$, then used as the initial azimuths for the next iteration. In this way, two sources are separated and tracked in real time.

3. SIMULATIONS

First, the performance of separating two uncorrelated wideband sources is shown. Then the method is compared with wideband focusing subspace methods, especially at low SNR.

3.1 Uncorrelated Sources

Assume there are two wideband uncorrelated Gaussian noise sources, the ambient noise is spatial white Gaussian noise, bandwidth B=400Hz, integral time T=1s.

First, consider the influence of $\theta_1(t)$, $\theta_2(t)$. Assume $\theta_1(t)=40-15\cos(2\pi t/40)$ degree, $\theta_2(t)=110+15\cos(2\pi t/40)$ degree, BT=400, SNR of source 1 and 2 is 25 dB and 8 dB respectively, initial errors are -20° and 20°. Tracking results are shown in Fig. 1. The results show that it is not $\theta_1(t)$ or $\theta_2(t)$ but $|\theta_1(t)-\theta_2(t)|$ that affects the separating performance.

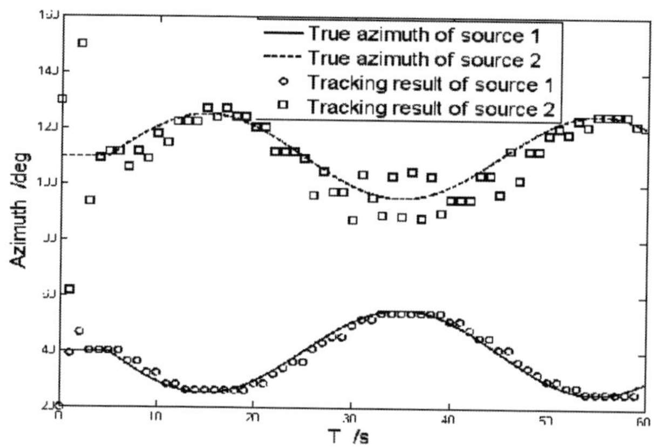

FIGURE 1. Tracking course (BT=400, source 1 SNR=25dB, source 2 SNR=8dB).

Second, consider intersection angle $|\theta_1-\theta_2|$ and minimum SNR (MSNR, the minimum SNR satisfying some tracking precision at $|\theta_1-\theta_2|$). It is an important problem in practice. Tracking precision here is defined as the summation of standard deviations of the two channels. Assume source1 and source 2 have equal strength, BT=400, initial errors are -10° and 10°. Dependence between intersection angle and MSNR is shown in Fig. 2. The curves are normalized by BT.

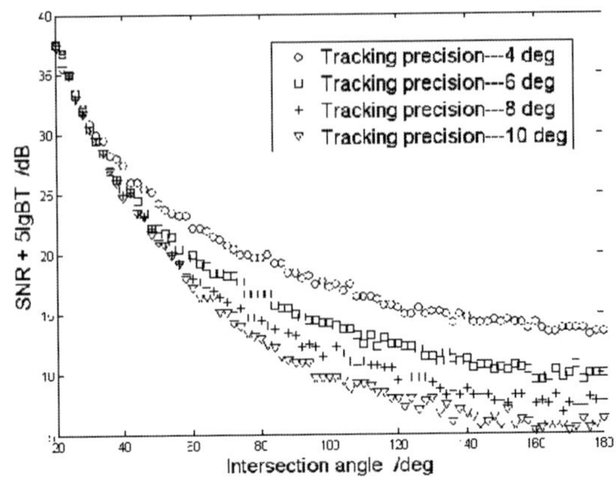

FIGURE 2. Relation between MSNR and intersection angle.

3.2 Comparison with Wideband Focusing Subspace Methods

Acoustic vector wideband MVDR and MUSIC methods have been studied recently and the spatial resampling technique is used. We apply the MUSIC method to one vector hydrophone, then compare it with the subspace rotation method.

Simulation conditions: two uncorrelated wideband sources, at 10° and 100°, normalized SNR: SNR +5lgBT variates from 8 dB to 17 dB. The result of MUSIC has a small bias below 10 dB while that of the subspace rotation method does not. Tracking precision curves are shown in Fig. 3. From this figure, if SNR is higher than 14 dB, the two methods have similar precision; when SNR is low, the performance of MUSIC degrades rapidly. For example, when SNR=9 dB, the precision of the subspace rotation method and MUSIC is 18° and 37° respectively.

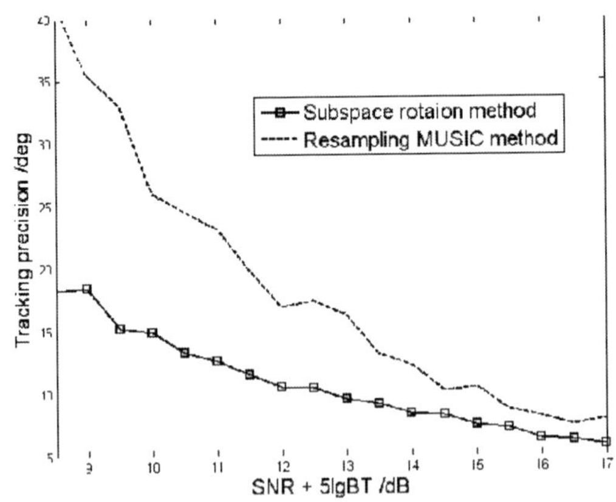

FIGURE 3. Tracking precision of uncorrelated sources.
Equal strength, azimuths: 10° and 100°.

4. SEA TRIAL

A sea trial to separate and track two ships with one vector hydrophone was carried out in the Qingdao Sea.

Given a pair of initial azimuths at 150° and 10°, the subspace rotation method and the energy flux method are used to separate and track the two ships. For the subspace rotation method, variant frequency bands are tested and we obtain similar results as shown in Fig. 4, where the frequency band is 220 Hz~400 Hz. Though the spectrum of ship 2 is submerged by ship 1, the method can separate and track the two wideband sources while the line spectrum at 148 Hz is used to track ship 2 in the energy flux method. Therefore the subspace rotation method is effective in tracking two arbitrary bandwidth and spectrum sources.

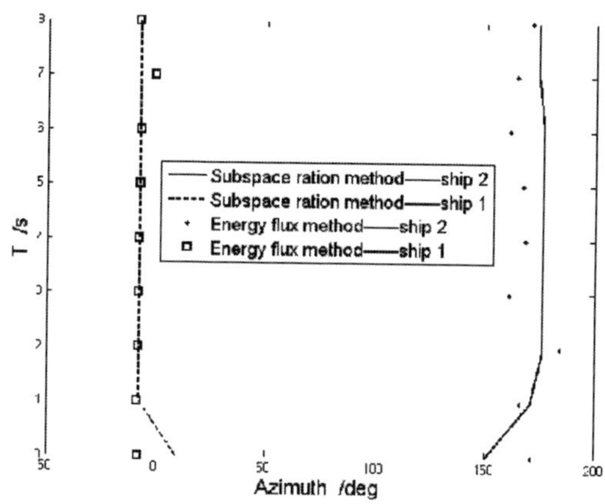

FIGURE 4. Tracking two ships with one vector hydrophone.
148Hz line spectrum is used to track ship 2 in the energy flux method.

5. CONCLUSION

In this paper, wideband sources separation with one vector hydrophone is addressed, and the method based on subspace rotation and spatial matched filter is proposed. Simulations indicate that this method is insensitive to initial azimuth errors, robust, equally efficient in all the azimuths, and has time-bandwidth-gain. In the circumstance of low SNR, this method outperforms wideband focusing subspace methods. Therefore, it is a very practical separating and tracking method on two sources in the underwater environment. The method is verified by sea trial data.

REFERENCES

1. M. Hawkes and A. Nehorai, IEEE Trans. on Signal Processing **46**, 2291-2304 (1998).
2. A. Nehorai and E. Paldi, IEEE Trans. Signal Processing **42**, 2481-2491 (1994).
3. M. Hawkes and A. Nehorai, IEEE Journal of Oceanic Engineering **26**, 337-347 (2001).

4. B. Hochwald and A. Nehorai, IEEE Trans. on Signal Processing **44**,83-95 (1996).
5. T. Wong Kainam and M.D. Zoltowski, AP-S. Digest **2**, 1368-1371 (1996).
6. C.C. Ko, J. Zhang, and A. Nehorai, IEEE Transactions on Aerospace and Electronic Systems **38**, 1109-1116 (2002).
7. O.E. Gulin and Desen Yang, CHINESE Journal of acoustics **23(1)**, 58-70 (2004).

A Design of Direct-Sequence Spread-Spectrum Acoustic Communication System

Shengjun Xiong[a], Guosong Zhang[b] and Shihong Zhou[a]

[a]*Hangzhou Applied Acoustic Research Institute, Hangzhou, China*
[b]*Department of Electronics and Telecommunications, Norwegian University of Science and Technology, Trondheim 7491, Norway*

Abstract. A direct-sequence spread-spectrum (DSSS) acoustic communication system was proposed. It used a Gold-sequence family to transmit information, which was recovered by a RAKE filter receiver. A Gold-sequence family of large size and of large main-to-side lobe ratio (MSLR) of the quasi-orthogonal function was selected, and the spectral efficiency was improved without communication reliability loss. For the transmitter, message bits were loaded in one symbol by selecting the sequence number in the Gold-sequence family and pulse position modulation (PPM) was used. For the receiver, the transmitted information was recovered using a cyclic correlation method. Field sea experiments were conducted to test this DSSS system. It was robust at low signal-to-noise ratio (SNR) and over complex multipath channels.

Keywords: Gold-sequence, DSSS, RAKE filter, underwater acoustic communication

INTRODUCTION

In recent years, underwater acoustic communication has received much attention, and research has been devoted to high rate communications using an array. For instance, current research focuses on time reversal [1] and multiple-input multiple-output (MIMO) communications [2]. Spatial diversity resulting from multipath propagation is exploited by employing a transducer array covering the water column, and robust communications of high data rates are realized. However, underwater acoustic modems with a long array are complex in realization, as such modems require complicated signal processing and consume more battery power, and so are seldom considered for underwater networks.

In this paper, we focus on the DSSS technique. Binary sequences from a Gold-sequence family were selected to increase spectral efficiency. For the receiver, a RAKE filter which improved anti-multipath interference performance was researched. Field experimental results and analysis are shown to verify its feasibility.

TRANSMISSION PRINCIPLE

A. Pseudo-noise sequence families

Many families of binary sequence of period $2^n - 1$ with good correlations have

CP1272, *Shallow-Water Acoustics, Second International Shallow-Water Acoustics Conference*
edited by J. Simmen, E. S. Livingston, J. X. Zhou, and F. H. Li

been found [3]. Sequences of three-valued cross correlation were introduced by classical families of Gold, Gold-like, small Kasami, large Kasami, and DBCH. FIGURE 1 shows the size of these families and reciprocal relationships. From a comparative study of different Pseudo-noise (PN) sequence families, a PN sequence family of large size and good correlations can be selected in the DSSS acoustic communication system. A large size can increase spectral efficiency which will be explained later.

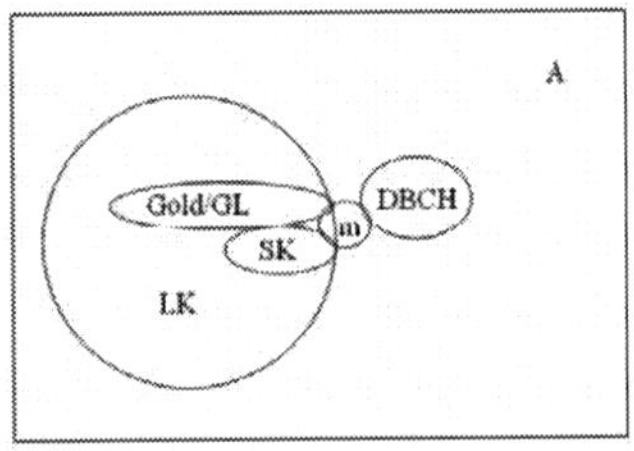

FIGURE 1 PN sequences families

$\{x\}$ is a binary (-1, 1) valued sequence of length L, and its period is L. It can be given by

$$\{x\} = \{x_0, x_1, x_2, \cdots, x_{L-1}\}. \quad (1)$$

The sequences $\{x\}, \{Tx\}, \cdots, \{T^{L-1}x\}$ are called phases of $\{x\}$, where T is an operator which shifts the vector x cyclically to the left by one chip. A family $s(x,y)$ consisting of N different linear combinations of $\{x\}$ and $\{y\}$ is obtained, i.e,

$$s(x,y) = \{\{s^0\},\{s^1\},\cdots,\{s^{N-1}\}\} = \{\{x \oplus y\},\{x \oplus Ty\},\cdots,\{x \oplus T^{N-1}y\}\}. \quad (2)$$

where $\oplus$ denotes XOR operation.

The cyclic autocorrelation function [4] $C_{x,x}(l)$ of one sequence $\{x\}$ is defined as

$$C_{x,x}(l) = \sum_{i=1}^{L}(1-2x_i)(1-2T^l x_i) \quad l = 1,2,\cdots,L. \quad (3)$$

The quasi-orthogonal function of sequence $\{s^i\}$ in family $s(x,y)$ is defined as

$$Q(l) = \sum_{j=1}^{L}[1-2(x_j \oplus T^i y_j)][1-2(x_j \oplus T^l y_j)] \quad l = 1,2,\cdots,N. \quad (4)$$

There are some properties shown by

$$T^{i+j} = T^i T^j, \quad (5)$$

$$T^i x T^i y = T^i(xy), \quad (6)$$

$$T^i xy = xT^i y, \quad (7)$$

$$T^i x + y = x + T^i y, \quad (8)$$

$$\sum_{i=1}^{L} T^i x_i = \sum_{i=1}^{L} x_i. \quad (9)$$

Substituting (5)~(9) into (4) gives

$$Q(l) = \sum_{i=1}^{L}(1-2y_i)(1-2T^{l-k}y_i) = C_{y,y}(l-i). \quad (10)$$

The quasi-orthogonal function of sequence $\{s^i\}$ is equal to the cyclic autocorrelation function $C_{y,y}(l-i)$. FIGURE 2 shows the quasi-orthogonal function of a Gold-sequence family where the length of sequence is 511, a DBCH-sequence family where the length of sequence is 255, a small Kasami-sequence family where the length of sequence is 255, and a large Kasami-sequence family where the length of sequence is 63. The PN sequence family of maximum MSLR is the Gold-sequence family. The PN sequence family of maximum size is the small Kasami-sequence family, whose MSLR is the minimal. Therefore, a Gold-sequence family could be used to improve spectral efficiency and reliability of DSSS communication.

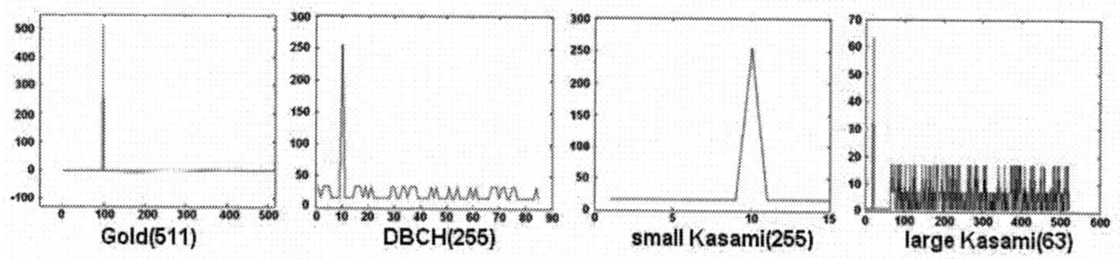

FIGURE 2 The quasi-orthogonal function of sequences.

The number-position function of sequence $\{s^i\}$ in family $s(x,y)$ is defined as

$$\begin{aligned}\Phi(l,k) &= \sum_{j=1}^{L}[1-2(x_j \oplus T^i y_j)][1-2(x_j \oplus T^l y_j)T^k] \\ &= \sum_{j=1}^{L}(1-2x_j)(1-2T^i y_j)(1-2x_j T^k)(1-2T^{l+k} y_j)\end{aligned}, \quad (11)$$

where l is the sequence number in the family and k is the number of chips cyclically shifted. The function of a Gold-sequence family is shown in FIGURE 3; the length of the Gold-sequence is 127.

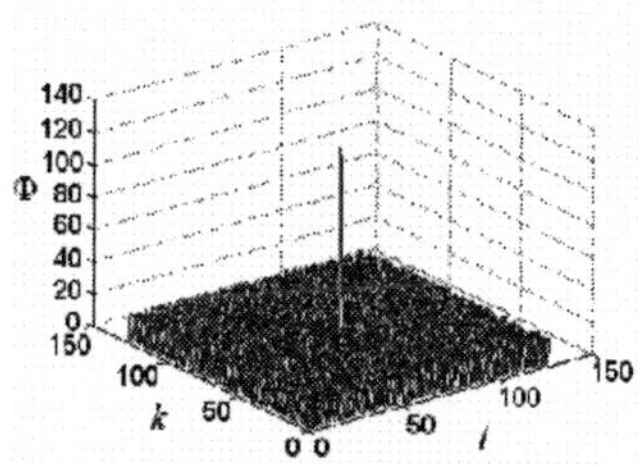

FIGURE 3 The number-position function of a Gold-sequence family

B. Modulation method

A spreading sequence S of L chips corresponds to one transmitted symbol: S denotes 1, and $-S$ denotes 0 (or -1). S can be cyclically shifted by integers from 0 to $L-1$, and one transmitted PPM symbol is generated shown in FIGURE 4.

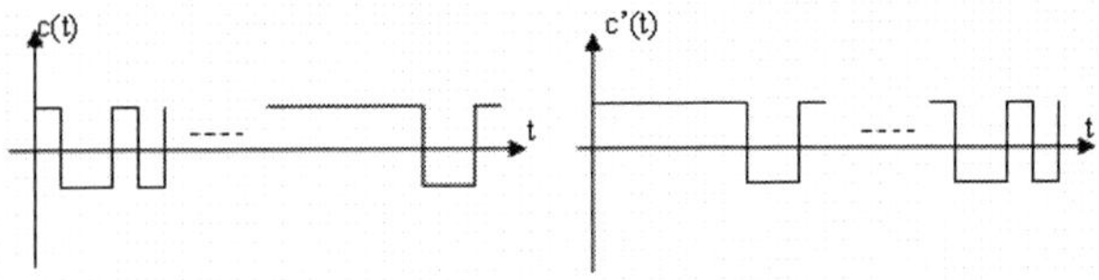

FIGURE 4 The illustration of one PPM symbol generation.

The transmitted symbol can be expressed as

$$s(t) = real(\sum_{k=0}^{L-1}(T^m s^i(k+1) + jT^n s^j(k+1))G(t - kT_{chip})e^{j2\pi f_c t}), \quad (12)$$

$$(i, j = 1,2,...,N; \quad m,n = 1,2,...,L \quad)$$

where i and j is a sequence number in the Gold-sequence family whose size is $L+2$, T_{chip} is the chip duration, and f_c is center frequency, $\{s^i\}$ is shifted cyclically to the left by m chips, and $\{s^j\}$ is shifted cyclically to the left by n chips. The $2\log_2 L + 2\log_2 N$ bits data can be transmitted within one message symbol, and the data rate is

$$R = \frac{2\log_2 L + 2\log_2 N}{LT_{chip}}. \quad (13)$$

The bandwidth efficiency is

$$\eta = \frac{2\log_2 L + 2\log_2 N}{L}. \quad (14)$$

C. A RAKE receiver

At the receiver, the real part and image part of a complex baseband signal is two Gold-sequences, the Gold-sequence is cyclically correlated with every sequence in the family, the receiver output will show a push pin in the number-position figure as in FIGURE 3. By searching the push pin position in the figure, the x axis value of the position is the transmitted message i and j in (12), the y axis value of the position is the transmitted message m and n in (12), and the message can be resumed.

The information is structured in frames, as shown by FIGURE 5. The HFM pulses are regularly inserted in order to synchronize the receiver. In every frame, channel impulse response was estimated by HFM replica correlation.

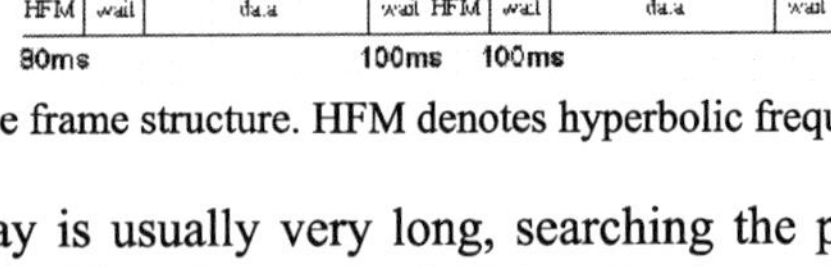

FIGURE 5 The frame structure. HFM denotes hyperbolic frequency modulated

The multipath delay is usually very long, searching the push pin position in the number-position figure will make a poor decision due to correlation peaks generated by later arrivals. Suppressing the multiple propagation arrivals, a RAKE filter can be used to combine the energy of multiple arrivals. Multipath can be estimated by correlation peaks of HFM. If a channel is static in a short period, the replica correlation peaks can be employed as taps of a RAKE filter. The correlator output of HFM is expressed as

$$R_{hfm}(t) = [h(t) \otimes s_{hfm}(t) + n(t)] \otimes s_{hfm}(-t), \tag{15}$$

where ⊗ denotes convolution, and $s_{hfm}(t)$ denotes the transmitted HFM pulse. The received data stream is matched filter with these taps, the matched filter outputs can be expressed as,

$$\begin{aligned} R_s(t) &= [h'(t) \otimes s(t) + n'(t)] \otimes R_{hfm}(-t) \\ &= h'(t) \otimes h(-t) \otimes S_{hfm}(t) \otimes S_{hfm}(-t) \otimes S(t) \\ &\quad + h'(t) \otimes s(t) \otimes S_{hfm}(t) \otimes n'(-t) \\ &\quad + h(-t) \otimes S_{hfm}(t) \otimes S_{hfm}(-t) \otimes n'(t) \\ &\quad + n'(t) \otimes n(-t) \otimes S_{hfm}(t) \end{aligned}, \tag{16}$$

$h'(t)$ is the impulse function for data stream, $n'(t)$ is the noise for data stream. Because of spreading gains in DSSS communication system, the last three items in (17) can be ignored. $S_{hfm}(t) \otimes S_{hfm}(-t)$ is the auto-correlation of HFM. If the channel is static, $h'(t) \otimes h(-t)$ is the auto-correlation of $h(t)$, whose maximum value is coherent combining of the paths. This step represents an ideal temporal correlator that has been automatically calibrated, the RAKE filter acts as the RAKE receiver [5] which combines multipath arrivals. However, the coherent RAKE receiver needs combining parameters, including arrival time and phases, which are required to be estimated. It is complicated for a receiver to obtain these parameters.

EXPERIMENTS AND ANALYSIS

Field trials were carried out in Bohai sea, P. R. China, in August 2008. The sea depth changes from 40 m to 50 m over 5 km ranges; the sea bottom is characterized by sand and mud. Two research vehicles were employed. One carried the transmitter (TX) broadcasting acoustic signals using an omni-directional transducer, and the other collected wave signals as the receiver (RX).

The communication signal was mapped by a Gold family of lengths 127, 255, 511 and 1023, the data rate was 170 bps, 305 bps, 541 bps and 945 bps, and the transmitted symbol time was 21.2 ms, 42.5 ms, 85.2 ms, and 170.5 ms, respectively.

The channel impulse response estimations over 30 s are shown in FIGURE 6a. Multipath structure is clearly shown, and it was stable over 30 seconds. The temporal coherence on a 30-seconds observation period is plotted on Fig. 6b.

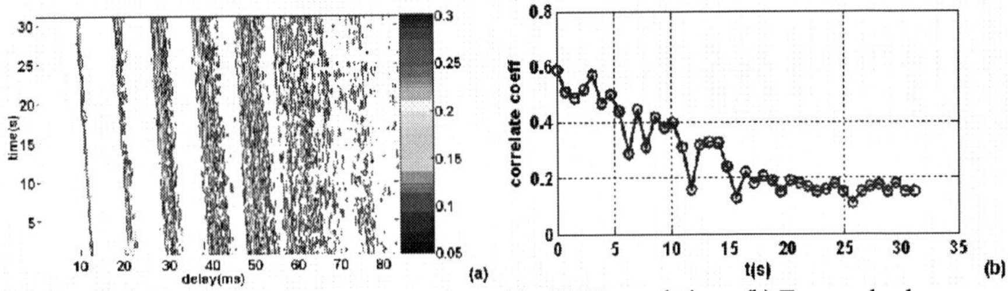

FIGURE 6 (a) Channel impulse response estimated by HFM correlations. (b) Temporal coherence.

By adjusting the transmitting amplifier, the received signal of four different SNRs, -4.07 dB, -1.37 dB, 1.63 dB and 4.63 dB, was collected, and then was processed offline in the laboratory. The performance of different data rates is shown in FIGURE 7. DSSS and RAKE filter technique was shown to be robust when the SNRs were low. DSSS communication using the shorter Gold-sequence had poorer performance when data rates were increased, due to reduced spreading gains.

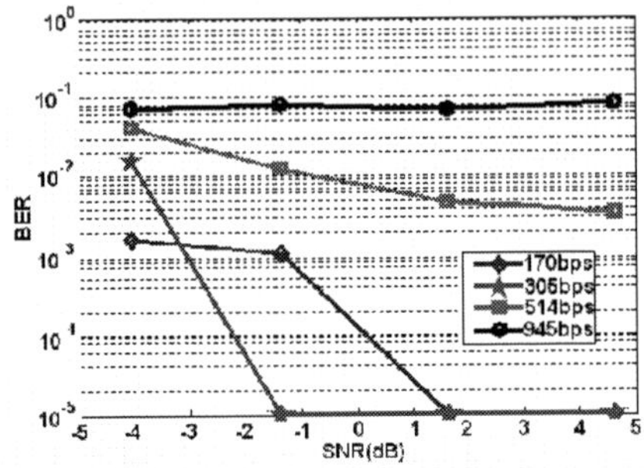

FIGURE 7 BER performance of the different data rates

A. Impact of channel variations on a RAKE filter

The channel was coherent over 0.6 s, the received signal whose frame had one HFM pulse and three message symbols was analyzed, the message symbol duration was 170.5 ms using Gold sequence of length 1023. FIGURE 8a shows that the multipath arrivals of three message symbols were similar to those of HFM pulse. The multipath interference was very serious; therefore, there is no push pin in every number-position function, as shown in the top panel of FIGURE 9. After RAKE filtering, the correlator outputs of three message symbols are shown by FIGURE 8b, the replica correlator peaks of every message symbol were combined to one peak, and the push pin appears in all number-position functions, as shown in the bottom panel of FIGURE 9.

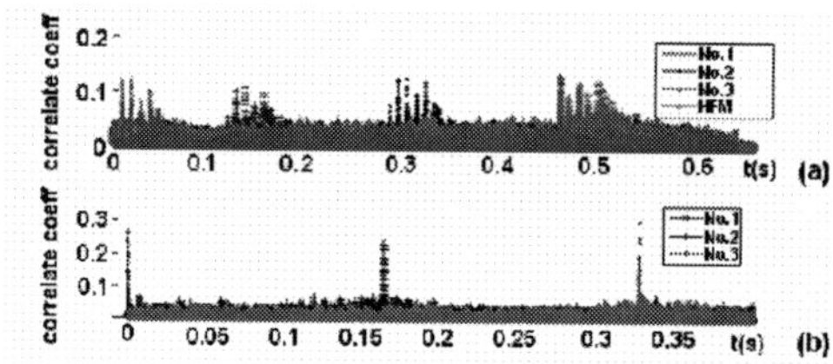

FIGURE 8 Comparison of the correlator peaks using RAKE filter (a) the correlator outputs of HFM pulse and three message symbols, (b) the correlator outputs of three message symbols after RAKE filtering

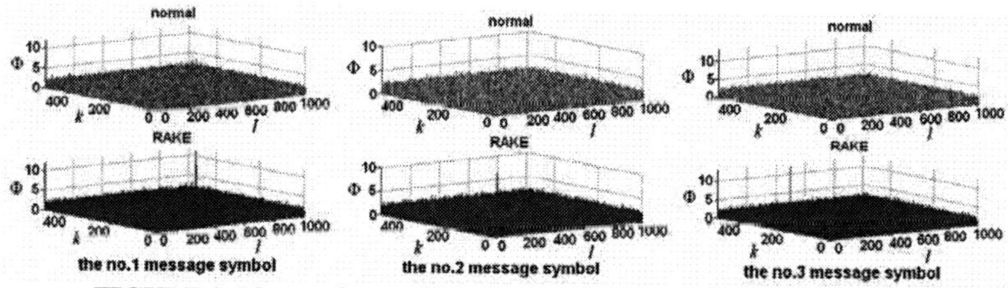

FIGURE 9 The number-position function of real part using RAKE filter

When the channel became too dynamic, the correlation peaks of HFM pulse could not act as the taps of the RAKE filter for the later data stream, and communications performances like BER will be poor [6]. This was obvious when the short sequence was used, which was proved by the following analysis.

The communication signal of 1.2 s was transmitted, the Gold sequence of length 127 was used, and there were 48 symbols. It was possible to associate errors within one frame with channel time variations. The received signal is shown in FIGURE 10a. The correlation coefficients of the message symbols after RAKE filtering are shown in FIGURE 10b. With elapsed time, the coefficient decreased due to channel variations. The errors are shown in FIGURE 10c. Therefore, the frame time must be restricted for BER performance.

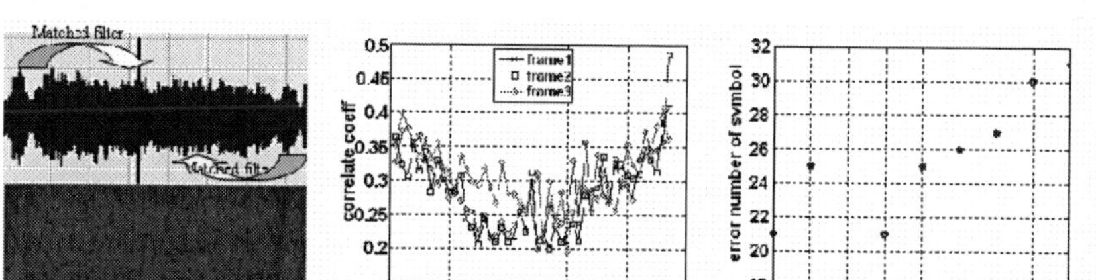

FIGURE 10 Demodulate the symbols. (a) the received signal, (b) the correlate coefficient, (c) the 9 errors in the frame

CONCLUSION

From the comparative study of the different PN sequence families, the Gold-sequence family is selected for its large size and large main-to-side lobe ratio (MSLR)

of the quasi-orthogonal function; therefore, the bandwidth efficiency is improved without effects on BER performance.

Experimental results at different SNRs with conditions of different sequence lengths have been shown and discussed. The performance indicates that the DSSS will be more robust at low SNRs. Though the data rate was high using short sequences, the performance was poor due to reduced spreading gains.

The channel contained a cluster of multipath arrivals. A RAKE filter was realized by the replica correlation outputs of HFM. Using the RAKE filter, a peak of low side lobes appears in the number-position function and multipath arrivals were combined. The underwater channel time variations affected the performances of the RAKE filter, and the frame duration is limited by channel temporal coherence.

REFERENCES

1. T. Yang, *"Temporal resolutions of time-reversed and passive-phase conjugation for underwater acoustic communications,"* IEEE J. Oceanic Enu. vol. 28, DD. .229-245, 2003.
2. D. Kilfoyle. J. Preisig. and A. Baggeroer, *"Spatial modulation over partially coherent multiple-input/multiple-output channels,"* IEEE Trans. Signal Processing, vol. 51, pp. 794-804, Mar. 2003.
3. D.**V.** Sarwate and M.**B.** Pursley, *"Crosscorrelation properties of Pseudorandom and related sequences,"* Proceedings of The IEEE, vol. 68, no. 5, pp. 593-619,1980.
4. L. Bomer and M. Antweiler, *"Periodic complementary binary sequences,"* IEEE Trans. Inform.Theory, vol. 36, pp. 1487-1494, Nov. 1990.
5. H.M.E1-Sallabi, H.L.Bertoni, and P.Vainikainen, *"Experimental evaluation of RAKE finger life distance for CDMA system,"* IEEE Antennas Wireless Propaga. Lett, vo1.1, pp50-52, 2002.
6. Guosong Zhang, Jens M. Hovem, Hefeng Dong, Shihong Zhou and Shuanping Du, *"An efficient spread spectrum pulse position modulation scheme for point-to-point underwater acoustic communication,"* Applied Acoustics 71 (2010) 11 - 16.

Active Matched Field Localization By A Horizontal Receiving Array*

Yan Zhang and Fenghua Li

State Key Laboratory of Acoustics, Institute of Acoustics, Chinese Academy of Sciences, Beijing 100190, China, Email: lfh@mail.ioa.ac.cn

Abstract. The active matched field localization by a horizontal receiving array is investigated in this paper. With explosive echo signals received by a horizontal array, a target ship can be localized by a broadband matched field processor. The signal to noise ratio (SNR) of MFP is increased in comparison with the conventional plane-wave beam-forming. Furthermore, the interference of obstructions is suppressed by MFP. The experimental results indicate that it is feasible to conduct active matched field localization with the echo signals observed by a horizontal array.

Keywords: active matched field localization, horizontal receiving array, explosive echo signals
PACS: 43.30

INTRODUCTION

Target localization is one of the basic research tasks in underwater acoustics. The conventional beam-forming method is only valid under the condition that plane waves propagate in infinite homogeneous media. However, due to the complexity of natural underwater environments, the sound field generated by a point source is very different from a plane wave field. Consequently, the performance of classical location systems significantly degrades. Matched field processing combines signal processing techniques and sound propagation theory in ocean waveguides and detects targets by matching numerically calculated sound fields to observed signals. In comparison with traditional methods, MFP can gain greater precision by fully exploiting the complexity and variety of the sound field's space distribution. MFP is currently a promising area of research in underwater acoustics.

As early as the beginning of the 1960s, Clay claimed that source directions could be estimated by matching the pulse response function of a layered waveguide with the outputs of several detectors at different depths. Since the 1970s and 1980s, the research done in MFP has been greatly promoted. However, the previous researches were usually confined to methods using vertical arrays and most of the theories and experiments concerned passive matched field localization. Although a vertical line array can be used to localize a sound source at its depth and range, it is difficult to estimate its azimuth. The findings of Booth and his collaborators indicate that the

* Supported by the National Natural Science Foundation of China (10734100)

CP1272, *Shallow-Water Acoustics, Second International Shallow-Water Acoustics Conference*
edited by J. Simmen, E. S. Livingston, J. X. Zhou, and F. H. Li

detection performance of a gradient line array is beyond that of a vertical array; these two arrays have the same length and detectors [9]. A preferable 'Γ-shaped' array is shown in [10]. Compared to vertical and partly vertical arrays, horizontal arrays are much more extended in space and the array aperture is one of the most important factors influencing their performance. It has been demonstrated that a horizontal array can also be successful in passive matched field localization [11]. As shown in [12], it is also possible to realize active matched field localization by a vertical line array.

In this paper, the feasibility of active matched field localization by a horizontal receiving array will be discussed. With explosive echo signals received by a horizontal array in a sea trial, the target ship can be localized by a broadband matched field processor. The signal to noise ratio (SNR) of MFP is higher than that of the conventional plane-wave beam-forming and the interference of obstructions is suppressed by MFP. Experimental results indicate that it is feasible to conduct matched field localization with echo signals observed by a horizontal array.

THEORY ANALYSIS

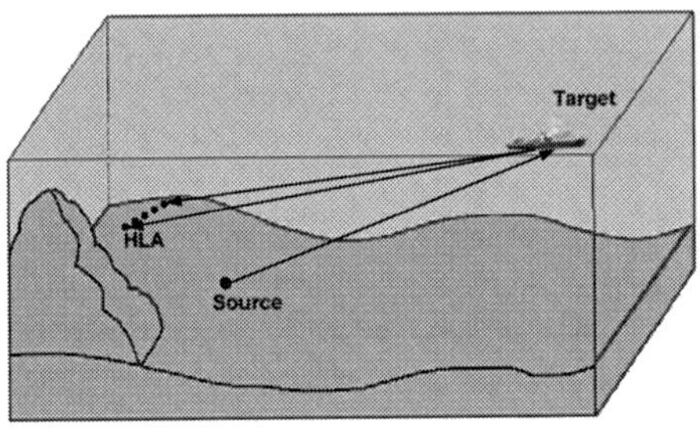

FIGURE 1. The system configuration

Figure 1 shows the system configuration of active matched field localization with a horizontal line array, which lies at the depth of z_a. The number of hydrophones composing the array is denoted by J. The signal sent out by the source at (r_s, z_s), scattered by the target whose center point is at (r_t, z_t), and received by the jth element at (r_j, z_a) is

$$P_j^{sca} = S \cdot H_j^{sca} = S \cdot H(r_t, z_t; r_s, z_s) \cdot Sca \cdot H(r_j, z_a; r_t, z_t), \tag{1}$$

where S is the source signal, and $H(\cdot)$ and Sca are the pulse response function of the water channel and the scattering function of the target, respectively.

The ambiguity function of Bartlett MFP is

$$B_{MF}(r,z) = \sum_j^J P_j^{sca} \cdot \hat{H}_j^*, \tag{2}$$

where MF denotes matched field processing, (r,z) is the field point to be scanned. The replica of sound field is denoted by $\hat{H}_j = H(r_j, z_a; r, z)$, which is the pulse response function of the channel from the field point at (r,z) to the hydrophone at (r_j, z_a). The symbol * denotes conjugation operation. Then,

$$
\begin{aligned}
P_j^{sca} \cdot \hat{H}_j^* &= \left(S \cdot H(r_t, z_t; r_s, z_s) \cdot Sca \cdot H(r_j, z_a; r_t, z_t)\right) \cdot H^*(r_j, z_a; r, z) \\
&= S \cdot Sca \cdot H(r_t, z_t; r_s, z_s) \cdot \left(H(r_j, z_a; r_t, z_t) \cdot H^*(r_j, z_a; r, z)\right)
\end{aligned} \tag{3}
$$

The intensity attenuation of the replica sound field due to range increase is not beneficial to the representation of localization results. In order to highlight the targets, the ambiguity function is usually normalized as

$$
B_{MFN}(r,z) = \frac{\sum_{j=1}^{J} P_j^{\mathrm{sca}} \hat{H}_j^*}{\left[\sum_{j=1}^{J} \left|P_j^{\mathrm{sca}}\right|^2\right]^{1/2} \left[\sum_{j=1}^{J} \left|\hat{H}_j\right|^2\right]^{1/2}}. \tag{4}
$$

According to Eq. (3) and the fact that $H \cdot H^*$ will be a positive real number when the field point lies at (r_t, z_t), $B_{MFN}(r,z)$ will have the maximum value at the target location of (r_t, z_t).

The incoherent broadband MFP in the frequency domain can be expressed as

$$
\widehat{B}_{MFN}(r,z) = \frac{\sum_{m=1}^{M} \left|\sum_{j=1}^{J} P_j^{\mathrm{sca}}(f_m) \hat{H}_j^*(f_m)\right|}{\sum_{m=1}^{M} \left[\sum_{j=1}^{J} \left|P_j^{\mathrm{sca}}(f_m)\right|^2 \sum_{j=1}^{J} \left|\hat{H}_j(f_m)\right|^2\right]^{1/2}}, \tag{5}
$$

where f_m is the mth frequency sample. The total number of frequency samples is denoted by M.

The localization effects of MFP and conventional beam-forming can be compared by contrasting their localization SNRs. The localization SNR of MFP is defined as

$$
SNR_{MF} = 10\lg\left(\frac{\widetilde{I}_{MF}(r_t, z_t, \theta_t)}{mean\left(\widetilde{I}_{MF}(r,z,\theta)\Big|_{(r,z,\theta) \neq (r_t, z_t, \theta_t)}\right)}\right), \tag{6}
$$

where $\widetilde{I}_{MF}(r,z,\theta)$ is the energy summation of the received signals of all elements and can be given by

$$
\widetilde{I}_{MF}(r,z,\theta) = \sum_{m=1}^{M} \frac{\left|\sum_{j=1}^{J} P_j^{\mathrm{sca}}(f_m) \hat{H}_j^*(f_m)\right|^2}{\sum_{j=1}^{J} \left|\hat{H}_j(f_m)\right|^2}. \tag{7}
$$

In Eq. (7), the signals are weighted and the weighting coefficients are dependent on the scanned field points. For a field point at (r,z,θ), the weighting coefficient is the conjugated value of the normalized pulse response function of the channel from the field point to each element. Notation (r_t, z_t, θ_t) is the target location and $mean(\ )$ is to average the values of all scanned points except the target point.

The localization SNR of beam-forming is defined as

$$
SNR_{BF} = 10\lg\left(\frac{\widetilde{I}_{BF}(\theta_t)}{mean\left(\widetilde{I}_{BF}(\theta)\Big|_{\theta \neq \theta_t}\right)}\right), \tag{8}
$$

where BF denotes the beam-forming and $\widetilde{I}_{BF}(\theta_t)$ is the energy summation of the signals received by all elements.

$$\widetilde{I}_{BF}(\theta) = \sum_{m=1}^{M} \frac{\left| \sum_{j=1}^{J} P_j^{\text{sca}}(f_m) e^{-i \cdot 2\pi f_m \cdot t_j} \right|^2}{J}. \tag{9}$$

In Eq. (9), the signals are also weighted and the weighting coefficients are dependent on the scanned azimuths. For an azimuth of θ, it is the normalized phase modification of each element's observed signal. Notation θ_t is the target azimuth and $mean(\ )$ is to average the values of all scanned azimuths except the target azimuth.

EXPERIMENTAL SIGNAL PROCESSING

In Oct. 2008, an underwater acoustic experiment was conducted. Using the explosive echoes observed in this experiment, broadband matched field processing is performed. Satisfactory results validate the method of active MFP localization by a horizontal receiving array.

The ocean environment in the experiment is shown in Fig. 2. The relative positions of the source, target and receiving array used in this experiment are shown in Fig. 3. The array, deployed at the bottom of the sea, contains 55 elements. A target ship floats at the sea surface.

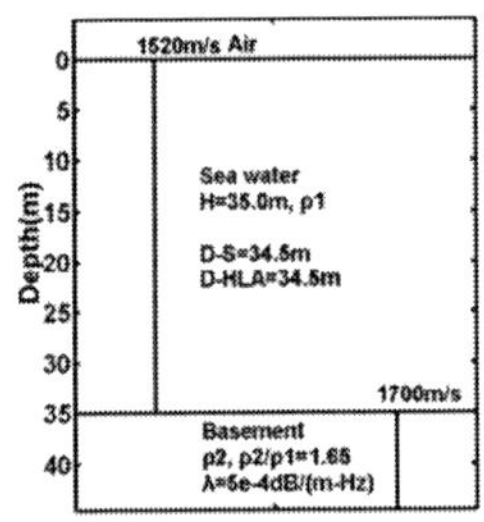

FIGURE 2. The ocean environment

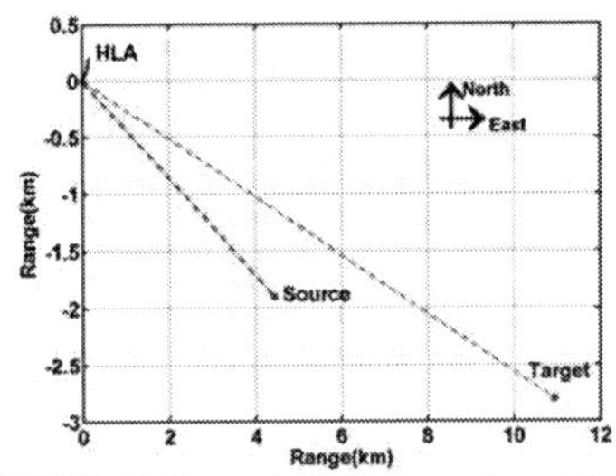

FIGURE 3. Layout of the experimental system

A bomb is exploded at the sea bottom. The observation time is long enough to ensure that the direct wave and scattered echoes are covered for each element of the receiving array. The long signals observed by the whole array are separated into successive temporal section groups. Every group of signal sections is processed with conventional beam-forming. The frequency range from 250 Hz to 450 Hz is considered here. The processing results based on beam-forming are shown in Fig. 4.

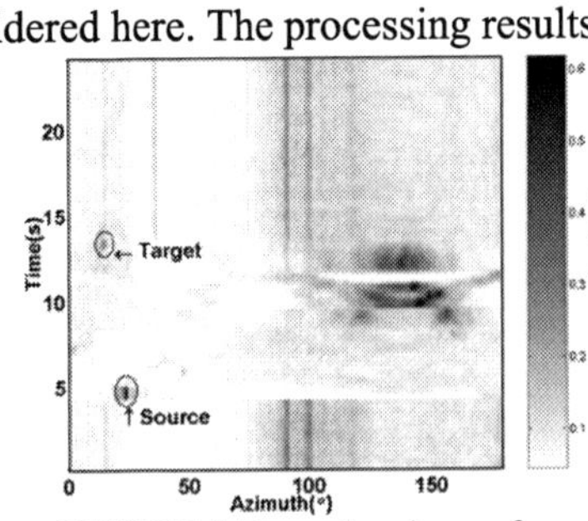

FIGURE 4. Beam-forming surface

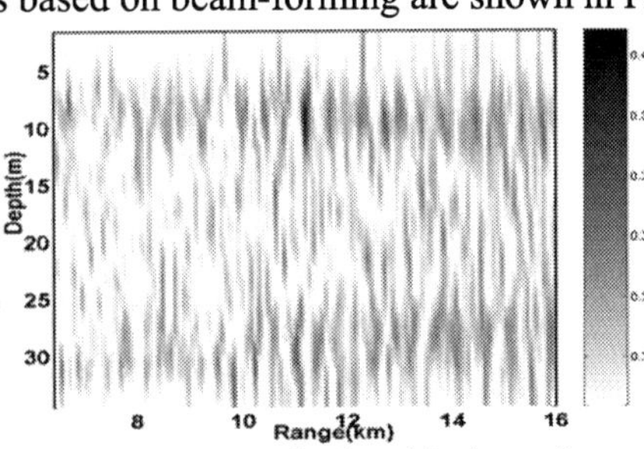

FIGURE 5. Range-depth ambiguity surface

Here, the azimuths are clockwise degrees away from the direction of east and the time coordinates correspond to the times of the signal sections. It is obvious that the maximum value appears at the time of 5 seconds and the azimuth of 24° south of east. The azimuth is totally the same as the bomb-dropping orientation. The point (24°, 5 s) reveals the source azimuth and the arriving time of the direct wave. There is also a highlight point at 15° south of east. It appears 7.9 s later than the direct wave. According to the GPS recordings, the experimental target ship is at 14.3° south of east. So the point (15°, 5+7.9 s) reveals the target azimuth and the arriving time of the echo scattered by the target. The highlight area (125° -150°, 9.5 s-13 s) reveals the azimuth bound of an obstruction in the experimental area and the arriving time of its echoes.

The echo signal of the target is extracted and MFP is conducted. Here, the replica sound field $H(r_j,z_a;r,z)$, which is the pulse response function of the channel between a field point at (r,z) and an element at (r_j,z_a), is evaluated by Kraken. The broadband matched field processor expressed by Eq. (5) is used. The locating range-depth ambiguity surface is shown in Fig. 5, where the x and y axes present the horizontal distances and depths of scanned field points, respectively. The distances cover a range of 6.4 km to 16 km. The target is indicated at the point (11.2 km, 9.5 m). According to the records of GPS, the distance between the target ship and the first element of the receiving array is 11.3 km, which is in reasonable agreement with the MFP localization result.

The localization effects of MFP and conventional beam-forming can be compared by contrasting their localization SNRs. The comparison is shown in Fig. 6. In this figure, the vertical coordinate denotes the energy summation of the received signals, which are weighted, and the left and right curves are evaluated according to $\widetilde{I}_{MF}(r_t,z_t,\theta)$ and $\widetilde{I}_{BF}(\theta)$ in Eq. (7) and Eq. (9), respectively. The SNRs of MFP and beam-forming are 3.1 dB and 2.1 dB, which are calculated by Eq. (6) and Eq. (8) respectively. The SNR of MFP is higher than that of beam-forming. Comparing these two curves we find that although the target at 13° south of east can be localized with both of the methods, the detrimental effect of the obstruction in the experimental area on target localization can be mitigated by MFP. This merit of MFP enables emphasis of the real target.

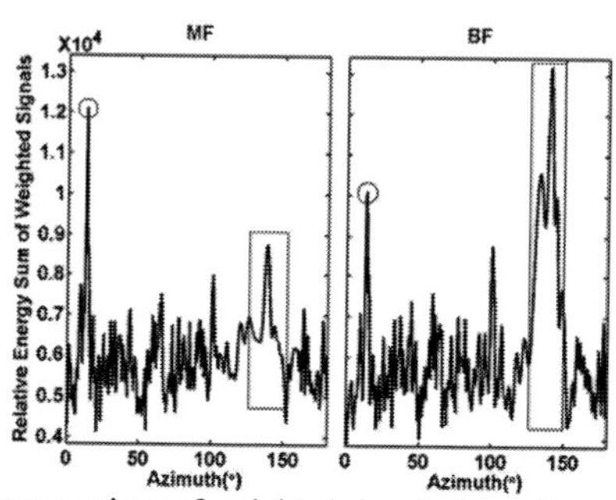

FIGURE 6. Energy summations of weighted signals depending on different azimuths

According to the MFP results, the target is 13° south of east, 11.2 km away from the first element of the receiving array and 9.5 m deep. These results basically agree with the real situation, which indicates that echo signals received by a horizontal array can be used to actively localize target ships.

SUMMARY

Target localization is one of the basic tasks in underwater acoustics. In complex environments, it is difficult for conventional beam-forming to have satisfactory localization results. Matched field localization takes full advantage of the complexity and variety of sound field distribution. It is based on the theory of matching numerically calculated sound fields to practically observed signals. Therefore, this technique can localize targets more precisely. Matched field localization by horizontal receiving arrays aims at applying MFP to active detection systems.

Experimental results indicate that:

(1) Making use of observed echo signals of a horizontal receiving array, a real ship target can be actively localized in range, depth and azimuth by a broadband matched field processor.

(2) It is feasible to conduct matched field localization by a horizontal array. The localization SNR of MFP is higher than that of conventional beam-forming and the interference of obstructions is suppressed by MFP.

ACKNOWLEDGMENTS

We would like to express our appreciation to the experimental staff, whose hard work ensures the robustness and reliability of the valuable data.

REFERENCES

1. Clay, C.S., *J. Acoust. Soc. Am* **33(7),** 865-870 (1961).
2. Baggeroer, A.B., Kuperman, W.A., and Mikhalevsky, P.N., *IEEE J.Oceanic Eng.* **18(4),** 401-424 (1993).
3. Hinich, M.J., *J. Acoust. Soc. Am.* **54(2),** 499-503 (1973).
4. Bucker, H.P., *J. Acoust. Soc. Am.* **59(2),** 368-373 (1976).
5. Fizell, R.G. and Wales, S.C., *J. Acoust. Soc. Am.* **78(Suppl.),** S57 (1985).
6. Yi He and Renhe Zhang, *Progress in Nature Science* **4(1),** 118-122 (1994).
7. Shihong Zhou, Renhe Zhang, etc, *Progress in Nature Science* **7(6),** 661-667 (1997).
8. Zhenglin Li, Renhe Zhang, etc, *Chinese Journal of Acoustics* **28(5),** 425-428 (2003).
9. N.O. Booth, A.T. Abawi, P.W. Schey and W.S. Hodgkiss, *IEEE J. Oceanic Eng.* **25,** 296-313 (2000).
10. Huiliang Ge, Ranwei Li, Xianyi Gong, *Technical Acoustics* **20(Suppl.),** 113-115 (2001).
11. Fenghua Li, Jianjun Liu, Renhe Zhang, *Technical Acoustics* **23(z1),** 29-32 (2004).

12. Renhe Zhang, Zaixiao Gong, Fenghua Li. "Possibility of Matched field Localization by Using Echo Signals". National Acoustics Conference 2002, Acoustical Society of China.

13. Hua Gao, "Hybrid Method for Scattering from an Object in a Stratified Medium", Ph.D. Thesis, Chinese Academy of Sciences, 2008.

Improvement of Longitudinal Correlation of Explosive Signals by Using Waveguide Invariance*

Yanjun Zhang[1, 2] Fenghua Li[1**] Xiaoxing Su[3]

1 State Key Laboratory of Acoustics, Institute of Acoustics, Chinese Academy of Sciences, Beijing, 100190

2 Graduate University of Chinese Academy of Sciences, Beijing 100080

3 School of Electronics and Information Engineering, Beijing Jiaotong University, Beijing, 100044

Abstract. Longitudinal correlation is a very important parameter in underwater acoustic signal processing. Normal mode interference is often the dominating effect on the decrease of the low frequency longitudinal correlation coefficient. Based on the waveguide invariance, a method was developed to enhance the correlation between the signals recorded by two horizontally separated hydrophones [1]. In this research, the method is used to deal with other experimental data. Unlike the former research, the signal is generated by an explosive source, which is more complicated than a bulb source due to the strong bubble pulse following the shock wave. The results show that the spatial correlations of low frequency signals received by a horizontal array can be improved measurably.

Keywords: Longitudinal correlation, Waveguide invariance, Frequency-shift compensation

PACS: 43.30Bp, 43.30Re

INTRODUCTION

In order to improve the array gain and obtain a high SNR, large aperture arrays with a great number of sensors at low frequency are widely used. The correlation coefficient is a very important parameter in underwater acoustics that determines the sonar performance.

Two main effects that lead to the decrease of the ocean acoustic spatial correlation are the fluctuation of the ocean channel and the normal mode interference [2, 3]. At low frequency, the normal mode interference is often the dominating effect on the decrease of the longitudinal correlation. Figure 1 shows the longitudinal correlation coefficient versus longitudinal separation in the frequency bandwidth of 200-250 Hz measured from one shallow water acoustic experiment. The circles and

---

* Work supported by the National Natural Science Foundation of China (10734100)

** To whom correspondence should be addressed. Email: lfh@mail.ioa.ac.cn

CP1272, *Shallow-Water Acoustics, Second International Shallow-Water Acoustics Conference*
edited by J. Simmen, E. S. Livingston, J. X. Zhou, and F. H. Li

the short vertical bars denote the averages and the rms errors of the correlation coefficients, respectively. The average correlation coefficients decrease to 0.5 when the separation of receivers increases to 314 m. It is well established that the decrease of the longitudinal correlation in shallow water at low frequency is mainly due to the normal mode interference.

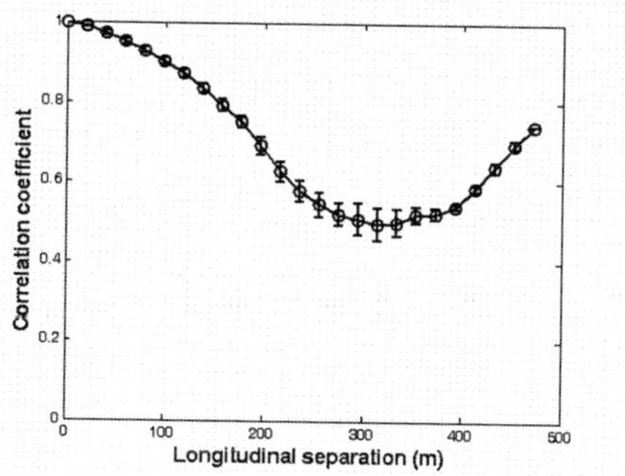

Fig. 1 Measured longitudinal correlation coefficients in the frequency band of 200-250Hz.

It was originally observed by Russian scientists that sound intensity had a steady spatial-frequency distribution in the ocean waveguide, due to the normal mode interference [4], and the concept of waveguide invariance was established [5]. Ren-he Zhang *et al.* gave a method based on the linear phase relationship of the waveguide invariance to improve the longitudinal correlation when the source spectrum varies slowly with frequency [1]. In this research, the improvement of the correlation of the signal generated by explosive sources (TNT) is studied.

First, a brief introduction of the waveguide invariance and the linear phase relationship is given. Second, the method to enhance the longitudinal correlation is introduced. Third, a method is presented to eliminate the bubble pulse in the explosive signal when the bubble pulse is as strong as the shock wave. Finally, the frequency shift compensation (**FSC**) based on the linear phase relationship is used to increase the longitudinal correlation.

I. WAVEGUIDE INVARIANCE AND FREQUENCY SHIFT COMPENSATION

Based on the waveguide invariance, for a continuous broadband signal, the spectrogram, plotted as a function of range, exhibits striated bands of intensity maxima and minima. The waveguide invariance parameter "β" describes the slope of the bands. If $H(\omega,r)$ denotes the pulse response function of the ocean channel, where r is the distance between the source and the receiver, under the condition of $\Delta r << r$, the linear phase shift relationship can be derived as follows:

$$H\left[\omega\left(1+\beta\frac{\Delta r}{r}\right),r+\Delta r\right]\approx\exp\left[j\omega\tau\left(\Delta r\right)\right]H\left(\omega,r\right),\tag{1}$$

where $\tau\left(\Delta r\right)$ is the time delay, and ω is the angular frequency.

If the source spectrum is $S\left(\omega\right)$, the sound pressure at distance r can be expressed as

$$P\left(\omega,r\right)=S\left(\omega\right)H\left(\omega,r\right).\tag{2}$$

If the source spectrum varies slowly with ω and $\Delta r << r$, one has

$$S\left[\omega\left(1+\beta\frac{\Delta r}{r}\right)\right]\approx S\left(\omega\right).\tag{3}$$

Substituting Eq. (3) and Eq. (1) into Eq. (2), we obtain the linear phase relationship between $P\left[\omega\left(1+\beta\frac{\Delta r}{r}\right),r+\Delta r\right]$ and $P\left(\omega,r\right)$ as follows:

$$P\left[\omega\left(1+\beta\frac{\Delta r}{r}\right),r+\Delta r\right]\approx\exp\left[j\omega\tau\left(\Delta r\right)\right]P\left(\omega,r\right).\tag{4}$$

The correlation coefficient between two receivers at distances of r and $r+\Delta r$ is defined as

$$\rho_0\left(\Delta r\right)=\max_{\Delta t}\left\{\frac{\mathrm{Re}\left[\int_{\omega_1}^{\omega_2}P\left(\omega,r+\Delta r\right)P^*\left(\omega,r\right)\exp\left(-j\omega\Delta t\right)d\omega\right]}{\sqrt{\int_{\omega_1}^{\omega_2}\left|P\left(\omega,r+\Delta r\right)\right|^2d\omega\int_{\omega_1}^{\omega_2}\left|P\left(\omega,r\right)\right|^2d\omega}}\right\}.\tag{5}$$

The correlation coefficient ρ_0 may be far less than 1 at large Δr due to the normal mode interference. To suppress the decorrelation by the normal mode interference, Zhang *et al.* gave a frequency-range transform (denoted as $\mathcal{F}$) as follows:

$$\mathcal{F}\left[P\left(\omega,r+\Delta r\right)\right]=P\left[\omega\left(1+\beta\frac{\Delta r}{r}\right),r+\Delta r\right].\tag{6}$$

The correlation coefficient based on the frequency-range transform is defined as

$$\rho_1\left(\Delta r\right)=\max_{\Delta t}\left\{\frac{\mathrm{Re}\left[\int_{\omega_1}^{\omega_2}P\left(\omega\left(1+\beta\frac{\Delta r}{r}\right),r+\Delta r\right)P^*\left(\omega,r\right)\exp\left(-j\omega\Delta t\right)d\omega\right]}{\sqrt{\int_{\omega_1}^{\omega_2}\left|P\left(\omega\left(1+\beta\frac{\Delta r}{r}\right),r+\Delta r\right)\right|^2d\omega\int_{\omega_1}^{\omega_2}\left|P\left(\omega,r\right)\right|^2d\omega}}\right\}.\tag{7}$$

Substituting Eq. (4) into Eq. (7), we have $\rho_1\approx 1$.

II. THE PROBLEM OF EXPLOSIVE SIGNALS AND THE SOLUTION

1. Characteristics of the Explosive Source

An explosive signal consists of a shock wave followed by a series of bubble pulses caused by the successive oscillations of the globular mass of gaseous materials. The bubble pulses have waveforms similar to the shock wave, except for the amplitude and time delay. The shock wave has a slow varying spectrum and the bubble pulses influence that spectrum. Suppose that $x(t)$ is the waveform of the shock wave, and the first bubble pulse can be expressed as $ax(t-\tau)$ approximately, where a and τ are the relative amplitude (**RABP**) and the time interval (**TIBP**) between the shock wave and the first bubble pulse, respectively. Measurements show that the peak pressure of the second bubble pulse is only about one-fifth of the peak pressure of the first [8]. So ignoring the bubble pulses after the first, the pressure waveform of the explosive source can be expressed as

$$y(t) = x(t) + ax(t-\tau). \tag{8}$$

Equation (8) can be expressed in the frequency domain

$$Y(\omega) = X(\omega)B(\omega) = X(\omega)\left(1 - ae^{-j\omega\tau}\right), \tag{9}$$

where $X(\omega)$ and $Y(\omega)$ are the spectrum of the shock wave and the whole explosive signal, respectively. $B(\omega)$ is the factor of the bubble pulse (**FBP**).

The absolute value of the explosive signal spectrum is

$$|Y(\omega)| = |X(\omega)||B(\omega)| = |X(\omega)|\sqrt{1 + a^2 - 2a\cos(\omega\tau)}. \tag{10}$$

From Eq. (10), we know that the spectrum of the explosive signal oscillates with ω due to **FBP**. The spectrum of the received signals from the explosive source at distance r can be expressed as

$$F_0(\omega,r) = Y(\omega)H(\omega,r) \approx X(\omega)B(\omega)H(\omega,r). \tag{11}$$

Figure 2 depicts the spatial-frequency distribution of $|F_0(\omega,r)|^2$, where the stripe pattern of the intensity distribution in the spatial-frequency domain deteriorates. Figure 3 shows the corresponding sound signals recorded by the horizontal line array (HLA) at 8.82 km away from the explosive source.

The **FSC** method in Eq. (7) works on the condition that the source spectrum varies slowly with ω. For this reason, the influence of the bubble pulse should be eliminated.

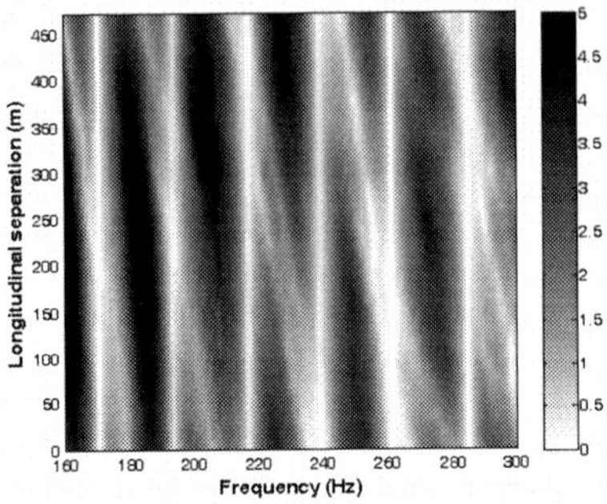

Fig. 2 Intensity distribution of the received signals from the TNT source

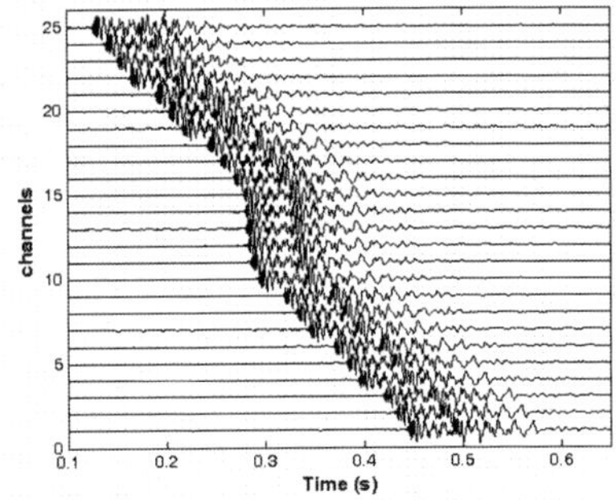

Fig. 3 Recorded sound signals at 8.82km.

2. Elimination of the Influence of the Bubble Pulse

There are many ways to eliminate the bubble pulse in an explosive signal, such as the deconvolution method, the cepstrum method [6], the adaptive method [7], and so on. The deconvolution method by **FBP** fails when **ABP** is close to 1. Both the cepstrum method and the adaptive method obtain different **FBP** from the signals of different receivers, and the linear phase relationship is broken.

In this research, a method of deconvolution is given to solve the problem when **ABP** is close to 1. First, truncate the explosive signal, and the length of the truncated signal is less than or equal to **TIBP**. The spectrum of the truncated signal at distance r can be expressed as

$$F_1(\omega,r)=\tilde{X}(\omega,r)\tilde{H}(\omega,r), \tag{12}$$

where $\tilde{X}(\omega,r)$ and $\tilde{H}(\omega,r)$ are the approximate forms of $X(\omega,r)$ and $H(\omega,r)$, respectively.

Figure 4 shows the truncated sound signals from the whole explosive signals (in Fig. 3). Fig. 5 is the intensity distribution of the truncated signals in spatial-frequency domain.

Divide the spectrum of the explosive signal by the spectrum of the truncated signal to get the approximation of **FBP** at distance r, as Eq.(13):

$$\tilde{B}(\omega,r)=\frac{F_0(\omega,r)}{F_1(\omega,r)}=\frac{X(\omega)B(\omega)H(\omega,r)}{\tilde{X}(\omega,r)\tilde{H}(\omega,r)}. \tag{13}$$

Then calculate the average of $\tilde{B}(\omega,r)$,

$$\overline{B}(\omega)\equiv\frac{1}{N}\sum_{i=1}^{N}\tilde{B}(\omega,r_i), \tag{14}$$

where N is the number of the sensors in the HLA, and $\overline{B}(\omega)$ is the approximate form of bubble pulse factor (**AFBP**).

Figure 6 shows $\left|\tilde{B}(\omega,r)\right|$ versus frequency and longitudinal separation (or sensor No. in the HLA), and the absolute value of **AFBP** $\left|\bar{B}(\omega)\right|$ is shown in Fig.7.

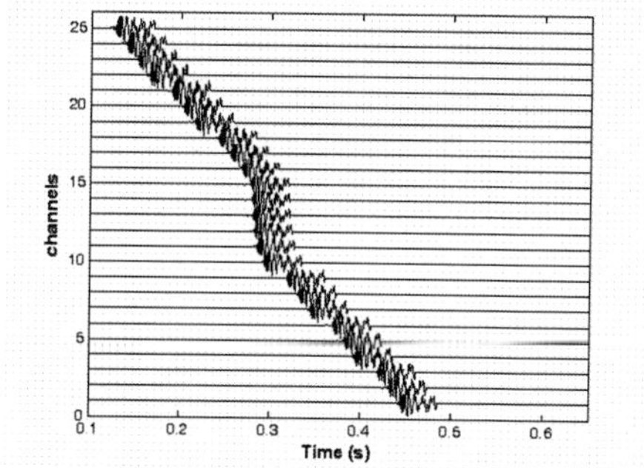

Fig. 4 Truncation signals

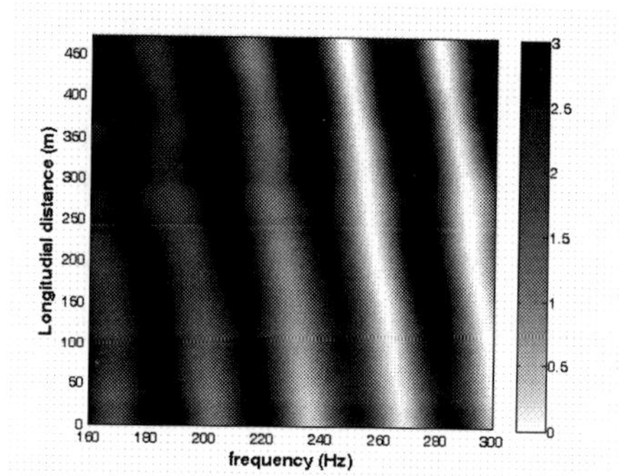

Fig. 5 Intensity distribution of the truncation signals.

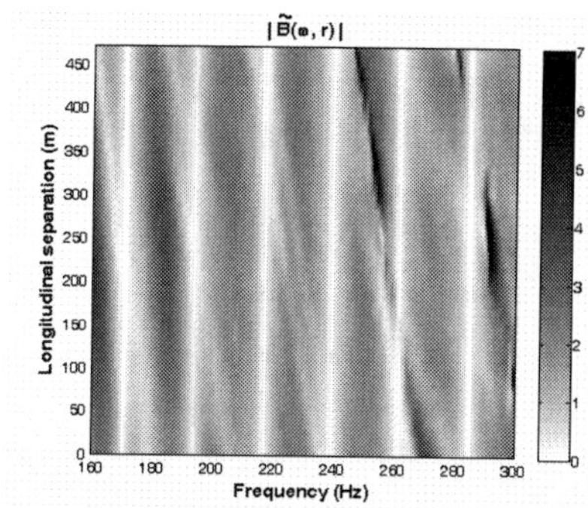

Fig. 6 $\left|\tilde{B}(\omega,r)\right|$

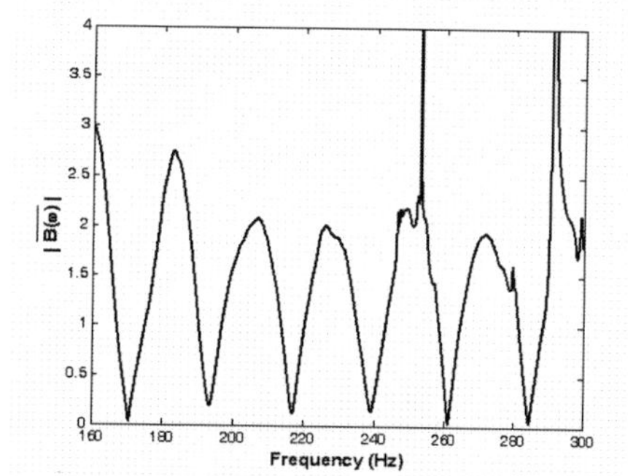

Fig. 7 $\left|\bar{B}(\omega)\right|$ versus ω

Then, dividing the explosive signal spectrum by $\bar{B}(\omega)$, one has

$$F_2(\omega,r)=\frac{F_0(\omega,r)}{\bar{B}(\omega)}=\frac{X(\omega)B(\omega)H(\omega,r)}{\bar{B}(\omega)}\approx X(\omega)H(\omega,r). \qquad (15)$$

$F_2(\omega,r)$ is approximate to the received signal spectrum after the bubble pulse has been eliminated. Figure 8 shows the spatial-frequency distribution of $\left|F_2(\omega,r)\right|^2$. Figure 9 shows the cost function, which is used to find the slope of the stripes (β) in Fig. 8.

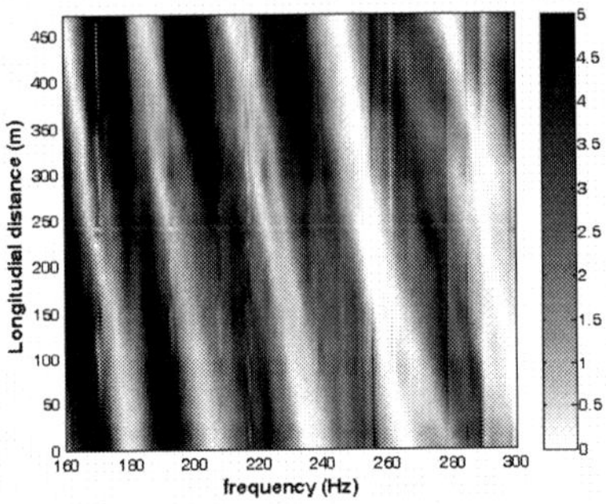

Fig. 8 The spatial-frequency distribution of the signal intensity, with the bubble pulse cleared.

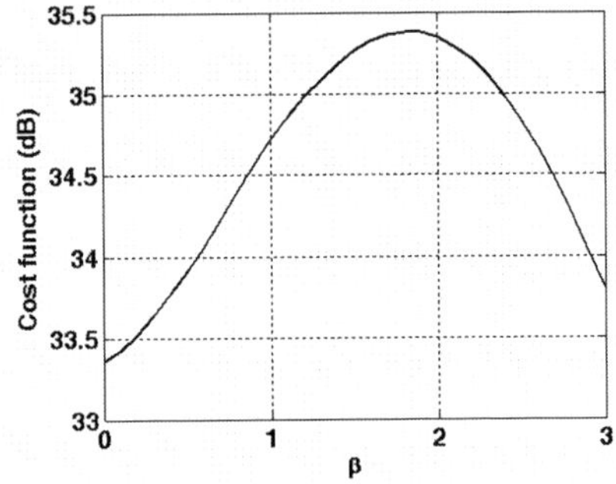

Fig. 9 The cost function to find the waveguide invariance, in the frequency band of 200-250Hz.

III. Frequency Shift Compensation

Use the FSC method to enhance the longitudinal correlation coefficients. Figure 10 shows the waveforms of p1(t), p2(t) and p3(t) in the frequency band of 200-250 Hz. Sound signals p1(t) and p2(t) are received at two longitudinal distances with the bubble pulse cleared, and p3(t) is the frequency-shift signal of p2(t). It is shown in Fig.11 that the cross-correlation coefficient between p1(t) and p3(t), which is approximately 0.95, is improved significantly compared to the cross-correlation coefficient between p1(t) and p2(t), which is approximately 0.53. Figures 12(a) and 12(b) show the comparisons of the correlation coefficients with and without frequency shift compensation versus the longitudinal separation. The frequency bands are (a) 200-250 Hz and (b) 250-300 Hz, respectively. It is also shown in Fig. 12 that the correlation coefficients are significantly improved after frequency shift compensation.

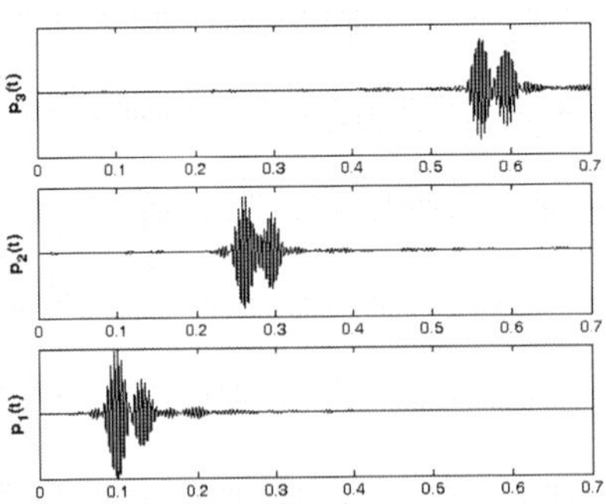

Fig. 10 Waveforms of the three signals in frequency band 200-250Hz, where $p_1(t)$ and $p_2(t)$ are the signals at two ranges, and $p_3(t)$ is the frequency-shift compensation signal of $p_2(t)$.

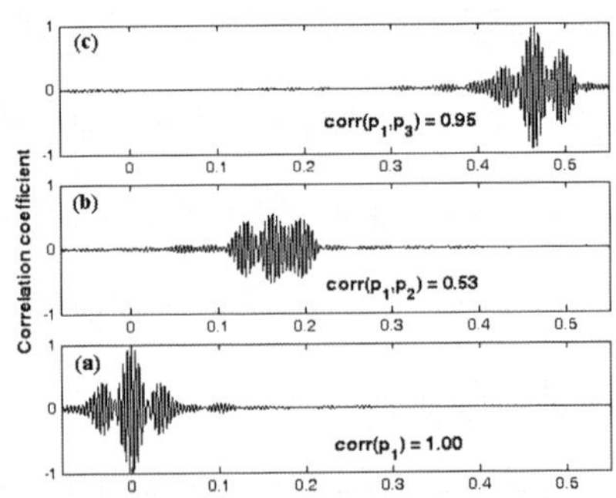

Fig. 11 (a) Auto-correlation curve of $p_1(t)$, (b) cross-correlation curve of $p_1(t)$ and $p_2(t)$, (c) cross-correlation curve of $p_1(t)$ and $p_3(t)$.

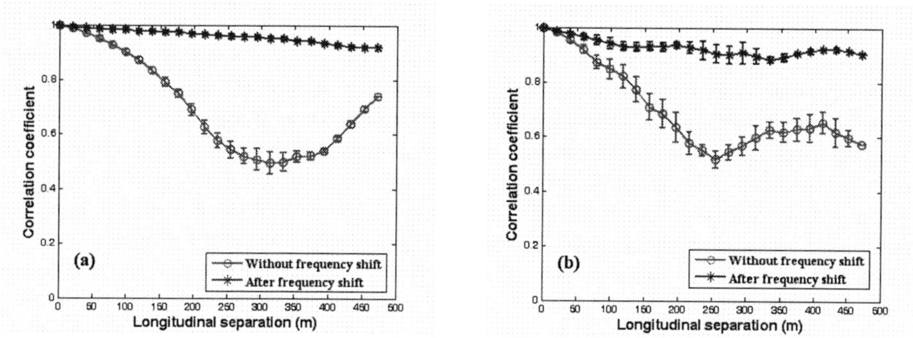

Fig.11 Longitudinal correlation coefficients of the signals with and without frequency shift. The frequency bands are (a) 200-250Hz, (b) 250-300Hz.

ACKNOWLEDGMENTS

This work is supported by the National Natural Science Foundation of China. The authors thank the staff who attended the sea experiment to obtain the valuable data for this paper.

REFERENCES

1. R. H. Zhang, X. X. Su and F. H. Li, Chin. Phys. Lett. **23** 1838-1841 (2006).
2. Q. Wang and R.H. Zhang, J. Acoust. Soc. Am. **92** 932-938 (1992).
3. X. L. Li et al, Appl. Acoust. **24** 11-14(2005) (in Chinese).
4. S. G. Gershman and Yu I. Tuzhilkin, Sov. Phys. Acoust. **1** 34-41 (1965).
5. S. D. Chuprov and N. E. Mal'tsev, Doklady Akademii Nauka SSSR **257** 475-479 (1981).
6. Y. Zhang and F. H. Li, et al, Acta Acoustica, **34**(2) 124-130 (2009).
7. F. X. Ge, Adaptive Bubble Pulse Cancellation and Its Application, (paper in the same conference).
8. R.J. Urick, Principles of Underwater Sound, New York: McGraw-Hill Book Company, 1975, pp. 80-85.

Uncertain Acoustic Field Modeling and Robust Source Localization in Shallow Water

Hangfang Zhao[a,b] Xianyi Gong[a,b] and Zibin Yu[a]

[a] *National Key Laboratory of Science and Technology on Sonar, Hangzhou Applied Acoustics Research Institute, Hangzhou City, Zhejiang Province, China*
[b] *Department of Information Science & Electronic Engineering, Zhejiang University, Hangzhou City, Zhejiang Province, China*

Abstract. Oceanic environmental uncertainty can cause significant performance degradation of the SONAR system. Understanding and modeling the uncertainty propagating from environment to acoustic field and then to steering vector is necessary for SONAR design and operation to mitigate the uncertainty effect and provide robust detection and location of targets. The statistical property of uncertainty can be described by the probability density functions or second-order moments of environmental parameters and acoustic fields. Based on the probability description, a stochastic response surface method is used to propagate the uncertainty from environment to acoustic field by polynomial chaos expansion. Then covariance matrix and associated ellipsoidal neighboring space are used to describe the uncertainty set of acoustic field and steering vector for sonar signal processing. Finally, a robust Minimum Variance (MV) matched-field processing method is derived by extending the constrained optimization of MV from single point to an uncertainty steering vector ellipsoid. We apply sea test data collected by a vertical array in shallow water to source localization.

Keywords: Uncertain acoustic field modeling, robust Minimum Variance, matched-field processing.
PACS: 43.30.Wi, 43.60.Kx

I. INTRODUCTION

Oceanic environmental uncertainty refers to incomplete or inaccurate knowledge of physics parameters, and insufficient sampling of parameters' variability in space and over time. The uncertainty of oceanic environments produces fluctuations of the acoustic field that cannot be predicted deterministically and that significantly degrades the performance of the deterministic model-based sonar signal processing known as matched-field processing (MFP). In mathematics, uncertainty is treated as a stochastic process, described by probability densities function (pdf) that reflects both knowledge and perturbation. Knowledge is denoted by the mean values of parameters or acoustic field. Perturbation around the mean values depicts the uncertainty or variability. The better the knowledge or the less the uncertainty, the more tightly the probability density groups around the mean values. The probabilistic description of acoustic field modeling allows for exploiting the uncertainty to improve the performance of sonar by incorporating it with signal processing algorithms [1, 2].

CP1272, *Shallow-Water Acoustics, Second International Shallow-Water Acoustics Conference*
edited by J. Simmen, E. S. Livingston, J. X. Zhou, and F. H. Li

Most acoustic propagation models are deterministic models, which output a deterministic acoustic field for a given set of ocean environmental parameters. These propagation models cannot propagate the uncertainty from oceanic environment to the acoustic field, but need to incorporate the uncertainty of oceanic environmental parameters to acquire the uncertainty of a predicted acoustic field. One of the means is to replace the sound velocity in the wave equation by a random variable to produce a stochastic differential equation, which is an "intrusive" approach [4]. The intrusive approach modifies the structure of the original equation, whose solution requires a new solver. Because sound velocity is the only variable in the wave equation, the intrusive approach can deal with the uncertainty of sound velocity alone. Alternatively, a "non-intrusive" stochastic response surface (SRS) model could be used [5]. The SRS method fits a general polynomial model to predict acoustic field using a least squares algorithm. Then we use a fitted response surface as a proxy for the acoustic propagating model. The non-intrusive method does not change the structure of the original equation, nor does it need a new specialized solver. Multiple sources of uncertainty can be included easily in the method. These advantages make SRS a wiser choice.

We have developed a robust model to represent the stochastic predicted acoustic field as completely as possible and with the highest possible fidelity with probabilistic representation. Given the uncertainty of acoustic field introduced through environmental parameters, uncertainty can never be removed by modeling and data processing, nor can a perfect match between measured fields and predicted fields be achieved; therefore, the most robust possible match should be pursued. Minimum variance MFP, which is a high resolution approach with single point constrained optimization, is the counterpart implementation of the optimal method in exactly known environments. MV passes a signal in a "look direction" entirely, suppressing energy in all other directions. When there is uncertainty in the steering vector, MV, due to its high resolution properties, will suppress the desired signal instead of maintaining a distortionless response to it. In this situation, MV is not an optimal method. The uncertainty of the oceanic environment translates into the uncertainty of a predicted acoustic field, and of the steering vector of sonar processing. In most cases, the "second-order" models characterized by the correlation or covariance matrices containing the main information about the signal could be a substitute for the "full-order" models by pdf. Therefore, covariance matrices and the associated ellipsoids can be used for describing the overall shape of distributions of points in multi-dimensional Euclidean space [6]. The uncertainty of steering vectors is modeling with an ellipsoid whose center is mean value, and whose shape and size denote the uncertainty. The robust match methods [7,8,9] choose a proper replica vector in this ellipsoid to acquire the best performance under the worst condition by extending the single point constrained optimization to the inequality constrained optimization in the ellipsoid, which turns out the idea of minimax principle. This idea is coincident with the philosophy of robust statistics [10] : "a robust statistical procedure should perform reasonably well at the assumed model, should be impaired only slightly by small departures from the model, and should be catastrophically impaired by somewhat larger deviations. "

Accurate knowledge of the oceanic environment including the water column and bottom contributes to modeling acoustic signals precisely, and to the effectiveness of deterministic signal processing methods. However, the uncertainty of oceanic environment requires quasi-deterministic modeling, and turns to quasi-deterministic signal processing that adapts to the modeling. The known is modeled or processed deterministically while the unknown is treated statistically [11], and this paper is concerned about the latter. The outline of this paper is as follows. In Section II, we discuss probability descriptions of uncertain oceanic environment and acoustic field, as well as uncertainty propagation from the environment to the acoustic fields and to steering vectors. The robust MV MFP method designed to localize a target by constrained optimization in the uncertain ellipsoid of replica vectors is the topic of Section III. In Section IV, a robust localization example is given to locate a source with the sound speed uncertainty. Some concluding remarks are in Section V.

II. UNCERTAINTY ACOUSTIC FIELD MODELING

A. Probabilistic descriptions of oceanic environment and acoustic field

The environmental parameters affecting acoustic signal propagation include sound speed profiles, depth of water column and geoacoustic properties of sediments. Usually the uncertainty is a small value compared to the mean value, and hence a quasi-deterministic modeling of environmental parameters can be equivalent to a neighboring space $\mathcal{B}(\mathbf{b}_0)$ with the mean value $\mathbf{b}_0$ located at center, uncertainty depicted by shape and size. Each element in the space is described by pdf $f(\mathbf{b})$, which is constructed from the measured data or a dynamics model. Likewise, the stochastic acoustic field is modeled as a neighboring space $\mathcal{P}(\mathbf{p}_0)$ of mean field $\mathbf{p}_0$, with each element quantified by pdf $f(\mathbf{p})$.

B. From uncertain environmental parameters to stochastic acoustic field

The uncertain environmental parameter vector $\mathbf{b}(r,z;\xi)$ located at the neighboring space $\mathcal{B}(\mathbf{b}_0)$ is written as

$$\mathbf{b}(r,z;\xi) = \mathbf{b}_0 + \delta\mathbf{b}(r,z;\xi), \tag{1}$$

where $\mathbf{b}_0$ is the deterministic mean field, $\delta\mathbf{b}(r,z;\xi)$ represents the stochastic perturbation with (r,z) describing the two dimension propagating in range and depth, $\xi = [\xi_1\ \xi_2 \cdots \xi_D]$ is a D-dimensional random vector that completely describes the uncertainty of SSP, depth of water column and geoacoustic description of sediments. Values of elements of ξ are drawn from probability distribution. The stochastic representation of acoustic field $p(r,z;\xi)$, which resulted from the uncertain oceanic environment, can be expanded as a mean-square convergent series with polynomial chaos as basis functionals [4,5],

$$p(r,z;\xi) = \sum_{q=0}^{\infty} \gamma_q(r,z)\Lambda_q(\xi) \tag{2}$$

where $\gamma_q(r,z)$ is a deterministic expansion coefficient, and $\Lambda_q(\xi)$ is polynomial chaos, multivariate polynomial functions of random variables ξ with a specific form that depends on the pdf $f(\xi)$. When the deterministic expansion coefficients $\gamma_q(r,z)$ are computed, Eq.(2) contains a complete probabilistic description of the stochastic process and associated uncertainty, since the polynomial chaos forms a complete orthogonal basis set in Hilbert space in the sense that the ensemble average of any two polynomials is orthogonal with respect to the pdf. By satisfying the mean square error of expansion, the infinite sum of Eq. (2) can be truncated to a finite number of terms. Assume the order of the polynomial chaos is *S*, then the number of terms in Eq. (2) is $T=(S+D)!/S!D!$, and ∞ in Eq. (2) is substituted by *T*,

$$p(r,z;\xi)=\sum_{q=0}^{T}\gamma_q(r,z)\Lambda_q(\xi). \tag{3}$$

Obviously, Eq.(3) is linear sum of polynomial basis functions weighted by the deterministic coefficients. The uncertain acoustic fields in Eq. (1) and Eq. (2) could be interpreted as a set of stochastic response surfaces with each surface representing a set of possible acoustic field values in a D-dimensional uncertainty space whose coordinate values are those of the coefficients of the random variables ξ. Once the deterministic coefficients are computed, the uncertain acoustic field is acquired using probability weighted combinations of the environmental uncertainty. The deterministic coefficients can be calculated by least-squares approach. The procedure of modeling a stochastic predicted acoustic field using SRS methods is described simply as,

(1) Determine a subset of environmental parameters $\xi^{(m)}, m=1,\cdots,M$ ($M>T$) as inputs of a propagation model.
(2) Make multiple runs of the propagating model, for example, the Kraken model, using the specific values in the subset to generate a set of predicted acoustic fields at location (r, z).
(3) Fit a general polynomial model denoted by Eq.(3) to the predicted acoustic fields using least squares method to estimate the expansion coefficients $\gamma_q(r,z)$.
(4) Run Eq. (3) using the estimated coefficients with the same or different (drawn from same pdf) subsets of environmental parameters to obtain the proxy stochastic response surface as a substitute for the propagating model.

However, since the acoustic field is a nonlinear function of environmental parameters, the pdf of $p(r,z;\xi)$ cannot be computed directly from the pdf of environmental parameters. One of the methods to estimate the pdf of $p(r,z;\xi)$ is the maximum entropy method from the stochastic response surfaces [1, 12]. Although the distribution of stochastic predicted fields may be deviated from Gaussian, the primary information does include in second order moments [2,3] the correlation or covariance matrices and the associated ellipsoids. With the SRS method, the mean and covariance matrix can be computed directly from the deterministic coefficients [4],

$$\langle p(r,z;\xi)\rangle=\gamma_0(r,z) \tag{4}$$

$$R(r_1,z_1;r_2,z_2)=\sum_{q=1}^{T}\gamma_q^*(r_1,z_1)\gamma_q(r_2,z_2) \tag{5}$$

C. From stochastic predicted acoustic fields to uncertain ellipsoid of steering vector

The uncertainty of predicted acoustic field results in the uncertainty of the steering vector of MFP. The ellipsoid model $\mathcal{A}(\mathbf{a}_0)$ of uncertain steering vectors is sufficient for MFP, which is a quadratic operation of propagating signal predicted vector with covariance matrix of array data. The ellipsoid $\mathcal{A}(\mathbf{a}_0)$ could be computed directly from mean and covariance of the stochastic predicted fields [7]. $\mathcal{A}(\mathbf{a}_0)$ is expressed as

$$\mathcal{A}(\mathbf{a}_0):\left\{\mathbf{a}=\mathbf{A}\mathbf{u}+\mathbf{a}_0 \middle| \|\mathbf{u}\|\leq 1\right\}, \tag{6}$$

where $\mathbf{a}_0$ is the assumed deterministic steering vector, and equals the mean of stochastic predicted field, $\mathbf{a}_0=\langle \mathrm{p(r,z;\xi)}\rangle=\gamma_0(\mathrm{r,z})$. $\mathbf{A}$, which represents the shape and size of the ellipsoid, is a full column rank matrix with dimension of $M\times L(L<M)$, where M is the number of elements of the array. $\mathbf{A}$ is related with the covariance matrix by $\mathbf{R}=\mathbf{A}\mathbf{A}^H$. $\mathbf{u}$ is a $L\times 1$ vector, and denotes the coordinate of the ellipsoid.

III. ROBUST MINIMUM VARIANCE MATCHED FIELD LOCALIZATION

A. Minimum Variance matched-field localization and its extension

In the known oceanic environment, the uncertainty of acoustic field vanishes, and the ellipsoidal sphere of uncertain steering vector degenerates to a point of $\mathbf{a}_0$, where the optimal localization method is Minimum Variance (MV) MFP. The basic idea of MV is to choose a weight vector to pass the signal in the "look direction," but to minimize the output power of other directions. The MV is a constrained optimization problem

$$\min_{\mathbf{w}} \mathbf{w}^H\hat{\mathbf{R}}_y\mathbf{w} \qquad \textit{subject to } \mathbf{w}^H\mathbf{a}_0(r,z)=1 . \tag{7}$$

where $\hat{\mathbf{R}}_y$ is the estimated covariance matrix from measured array data

$$\hat{\mathbf{R}}_y=\frac{1}{N}\sum_{i=k-N+1}^{k}\mathbf{y}(i)\mathbf{y}^H(i) , \tag{8}$$

moreover, the weight vector has an analytical solution, given by

$$\mathbf{w}_{MV}=\frac{\hat{\mathbf{R}}_y^{-1}\mathbf{a}_0(r,z)}{\mathbf{a}_0^H(r,z)\hat{\mathbf{R}}_y^{-1}\mathbf{a}_0(r,z)} . \tag{9}$$

The high resolution property of MV makes it sensitive to the uncertainty of steering vector $\mathbf{a}_0(r,z)$, and several techniques were developed to mitigate the sensitivity of MV to modeling uncertainty, or equivalently, to improve the robustness of MV to modeling error. For example, the uncertainty of steering vector $\mathbf{a}_0(r,z)$ is merged into the uncertainty of covariance matrix $\hat{\mathbf{R}}_y$. The modified MV is a diagonal loading MV with regularizing the covariance matrix by plus a small value μ on the diagonal of covariance matrix,

$$\min_{\mathbf{w}} \mathbf{w}^H \hat{\mathbf{R}}_y \mathbf{w} + \mu \mathbf{w}^H \mathbf{w} \qquad \textit{subject to } \mathbf{w}^H \mathbf{a}_0(r,z) = 1. \tag{10}$$

The robustness is improved through heightening the background energy,

$$\mathbf{w}_{DL} = \frac{\left(\hat{\mathbf{R}}_y + \mu \mathbf{I}\right)^{-1} \mathbf{a}_0(r,z)}{\mathbf{a}_0^H(r,z)\left(\hat{\mathbf{R}}_y + \mu \mathbf{I}\right)^{-1} \mathbf{a}_0(r,z)} \tag{11}$$

B. Robust Minimum Variance matched field processing

However, heightening the background energy is not an ultimate method to improve the robustness of MV MFP. A robust MV MFP (RMV-MFP) should be implemented in all uncertainty space by extending a single point constraint of MV to an inequality constraint in uncertain steering vector ellipsoid,

$$\begin{aligned} &\min_{\mathbf{w}} \mathbf{w}^H \hat{\mathbf{R}}_y \mathbf{w} \qquad \textit{subject to } \operatorname{Re} \mathbf{w}^H \mathbf{a}(r,z) \ge 1 \\ &\forall \mathbf{a}(r,z) \in \mathcal{A}(\mathbf{a}_0),\ \mathcal{A}(\mathbf{a}_0): \left\{\mathbf{a} = \mathbf{A}\mathbf{u} + \mathbf{a}_0 \,\middle|\, \|\mathbf{u}\| \le 1\right\} \end{aligned} \tag{12}$$

When $L = M$, the uncertain ellipsoid degenerates into a sphere, the constrained optimization problem is simplified,

$$\begin{aligned} &\min_{\mathbf{w}} \mathbf{w}^H \hat{\mathbf{R}}_y \mathbf{w} \qquad \textit{subject to } \operatorname{Re} \mathbf{w}^H \mathbf{a}(r,z) \ge 1 \\ &\left[\mathbf{a}(r,z) - \mathbf{a}_0\right]^H \mathbf{R}^{-1} \left[\mathbf{a}(r,z) - \mathbf{a}_0\right] \le 1 \end{aligned} \tag{13}$$

Then second order cone based algorithms can be used to calculate the steering vector. Alternatively, using Lagrange multiplier methodology the solution of Eq. (13) is given

$$\mathbf{a}(r,z) = \hat{\mathbf{R}}_y \left(\hat{\mathbf{R}}_y + \lambda \mathbf{A}\mathbf{A}^H\right)^{-1} \mathbf{a}_0, \tag{14}$$

where λ is Lagrange multiplier. Then the weight vector is

$$\mathbf{w} = \frac{\left(\hat{\mathbf{R}}_y + \lambda \mathbf{A}\mathbf{A}^H\right)^{-1} \mathbf{a}_0}{\mathbf{a}_0^H \left(\hat{\mathbf{R}}_y + \lambda \mathbf{A}\mathbf{A}^H\right)^{-1} \hat{\mathbf{R}}_y \left(\hat{\mathbf{R}}_y + \lambda \mathbf{A}\mathbf{A}^H\right)^{-1} \mathbf{a}_0}. \tag{15}$$

Finally, signal power estimation is

$$\hat{\sigma}_s^2 = \frac{1}{\mathbf{a}^H \hat{\mathbf{R}}_y^{-1} \mathbf{a}} \frac{\|\mathbf{a}\|^2}{\|\mathbf{a}_0\|^2}. \tag{16}$$

IV. Robust Localization example with sound speed uncertainty

A. Specification of uncertain environment

The acoustic and environmental measurements are from a sea test conducted in September 2008 in the Yellow Sea of China. The depth of the water column is 50 meters. The sound speed profiles are calculated from collections of a vertical thermometer train covering the water column for more than 30 hours. A 500 Hz acoustic source is located at a depth of 25 m and at a range of 14.5 km from the receiving array. The array is a vertical line array of 32 hydrophones uniformly spaced

1.2 m apart. The top element of the array is at 3.8 m. The sediment is 6 meters thick, with 1620.3 m/s compressive sound speed, 1.8 cm/g^3, and 0.6 dB/λ attenuation.

The measured SSPs show great variability in 30 hours. Fig.1 shows the mean value and the standard deviation of uncertain SSPs. The covariance matrix of sound speed field is calculated, then decomposed with SVD. Realizations of SSP, pictured in Fig.2, are acquired by truncating the first three eigenvalues λ_i and eigenvectors $\varphi_i(z)$ by $c(z;\xi) = c_0(z) + \delta c(z) = c_0(z) + \sum_{i=1}^{3} \sqrt{\lambda_i}\varphi_i(z)\xi_i$, where $c_0(z)$ is mean SSP, ξ is 3-dimensional random variables, with each value drawn from standard normal distribution. Assuming all other environmental parameters are known exactly, Eq.(3) is used to model the stochastic predicted acoustic field, and then the covariance matrix is computed by Eq.(5). Fig.3(a) shows the covariance matrix of predicted acoustic field, and Fig. 3(b) plots the eigenvalue of the covariance matrix, which displays a rank spreading phenomena. The uncertain steering vector ellipsoid is obtained finally from the mean values and the covariance matrix of the stochastic predicted acoustic field.

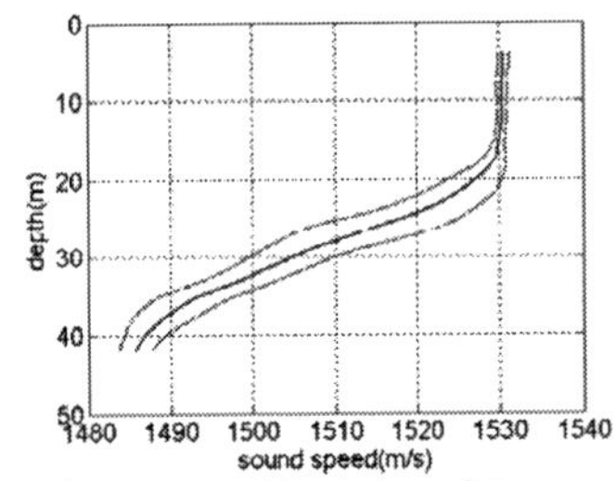

Fig.1 The mean and deviation of the uncertain SSPs

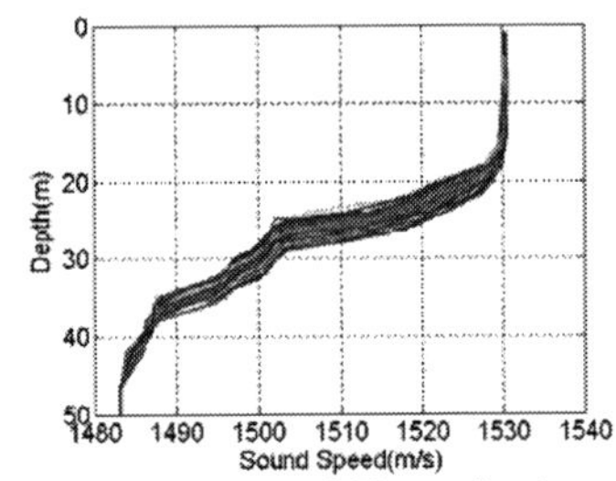

Fig.2 A set of SSP realizations

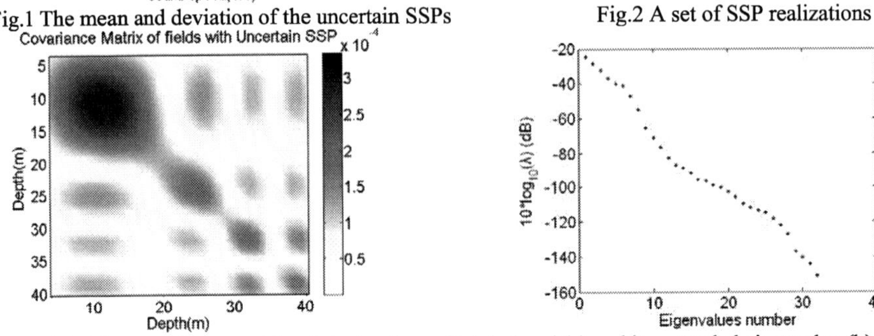

Fig.3 The covariance matrix of stochastic predicted signal (a) and its extended eigenvalue (b)

B. The measured acoustic field

A 500 Hz PCW pulse signal with 4 s duration is transmitted every 90 s. The localization is based on 13 pulse signals sampled during 20 minutes period of time. The time-frequency representation of the summation of 32 channels signal is shown in Fig. 4(a). The SNRs of the 13 pulses are shown in Table 1. Obviously, the signal has a high SNR and variability.

Table 1. The SNRs (dB) of 13 pulses

No.	1	2	3	4	5	6	7	8	9	10	11	12	13
SNR	23.61	9.82	17.89	6.88	19.24	9.83	10.92	18.29	16.05	22.49	25.08	19.69	21.65

C. Robust localization result

The ambiguity surface of RMV-MFP is shown in Fig 4(b). The estimated location is at a range of 15.1 km and a depth of 20 m. By comparison, the ambiguity surface of convenient MFP (CMFP) in which the covariance matrix of noise is treated as a scale of identity matrix, and of the diagonal loading minimum variance MFP (DL-MVMFP) are shown in Fig.4(c) and Fig.4(d), respectively. The estimated locations are at a range of 17.4 km and a depth of 21 m for CMFP and at a range of 15.1 km and a depth of 20 m for DL-MFP, respectively. It makes sense that RMV-MFP and DL-MVMFP get the same estimation of source location, because RMV-MFP could be interpreted as a diagonal loading method with the diagonal loading levels depending on the ellipsoid of uncertain steering vector to allow lower background. The uncertainty of ocean depth and the tilt of vertical array could be the main reasons for a large error in depth.

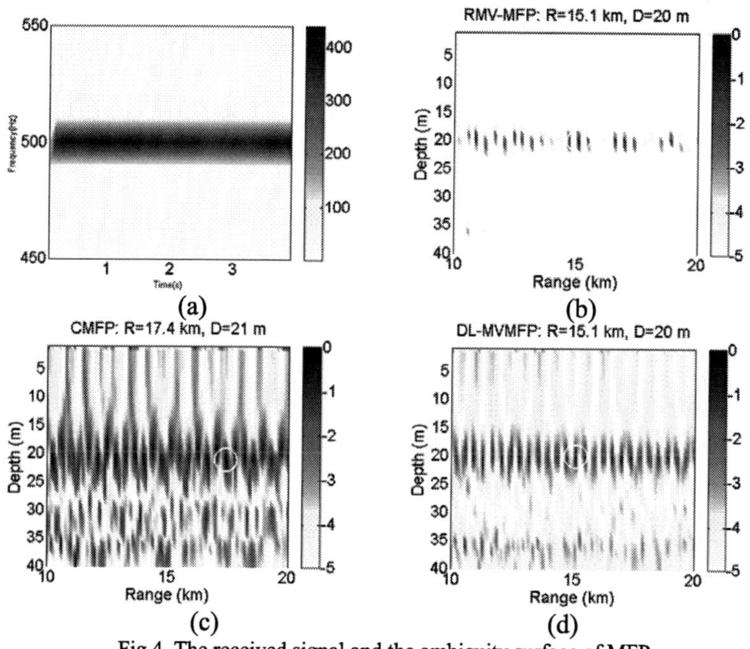

Fig.4 The received signal and the ambiguity surface of MFP
(a) the time-frequency spectrum of measured signal, (b) the ambiguity surface of RMV-MFP with the estimated location (15.1 km, 20 m), (c) the ambiguity surface of CMFP with the estimated location (17.4 km, 21 m), and (d) the ambiguity surface of DL-MVMFP with the estimated location (15.1 km, 20 m).

V SUMMARY

We have done studies in both robust modeling and robust processing to recover the localization performance in uncertain oceanic environments. For robust modeling, the model knowledge of uncertain environmental parameters is incorporated into a propagating model to get probabilistic representations of stochastic predicted fields, and ellipsoids of uncertain steering vectors. For robust processing, the minimum variance MFP is generalized by extending single point constraint to inequality constraint of uncertain space to get a robust minimum variance MFP.

REFERENCES

1. R.L.Culver and H.J.Camin, "Sonar signal processing using probabilistic signal and ocean environmental models, " JASA, 124(6), Dec., 2008.
2. A.B.Baggeroer, H.Schmidet, "Performance bounds on the detection and localization in a stochastic ocean, " in N.G.Pace and F.B.Jensen (eds.) Impact of Littoral Environmental Variability of Acoustic Predictions and Sonar Performance, 2002, P507-514.
3. A.B.Baggeroer, P.M.Daly, "Stochastic matched field array processing for detection and nulling uncertain ocean environments, " Signals, systems and computer, 2000, Conference record of the thirty-fourth Asilormar Conference on Volume 1, P662-667.
4. S.Finette, "A stochastic representation of environmental uncertainty and its coupling to acoustic wave propagation in ocean waveguide, " JASA, 120, 2006.
5. S.Finette, "A stochastic response surface formulation of acoustic propagation through an uncertain ocean waveguide environment, " JASA, 126(5), 2009.
6. P.J.Huber, "Robust covariance," in S.Gupta and D.Moore (eds.), Statistical Decision Theory and Related Topics II, 165-191, Academic Press, 1977.
7. R.G.Lorenz and S.P.Boyd, "Robust minimum variance beamforming," in J.Li and P.Stoica (eds.), *Robust adaptive beamforming*, John Wiley and Sons, 2006.
8. A.B.Gershman, Z.Q.Luo, and S.Shahbazpanahi, "Robust adaptive beamforming based on worst-case performance optimization," in J.Li and P.Stoica (eds.), *Robust adaptive beamforming*, John, Wiley and Sons, 2006.
9. J. Li, P.Stoica and Z.Wang, "Robust capon beamforming," in J.Li and P.Stoica (eds.), *Robust adaptive beamforming*, John, Wiley and Sons, 2006.
10. P.J.Huber, *Robust statistics*, John Wiley and Sons, 1981.
11. E.S.Livingston, et. al., "Guest Editorial capturing uncertainty in the tactical ocean environment," IEEE Journal of Oceanic Engineering, 31(2), 2006.
12. J.N.Kapur, and H.K.Kesavan, "*Entropy Optimization Principles with Applications*," Academic Press, 1992.

Author Index

M

N

P

Q

R

S

T

W

X

Y

Z

nal Conference
ater Acoustics
第二屆國際淺海聲
16–20 Sept 2009
Shanghai, CHINA
The conference will emphasize the following topics in shallow water acoustics:
• Acoustic propagation and reverberation models
• Effects of water column variability on acoustics
– effects due to physical structures (internal waves, fronts, turbulence...)
– scattering from biologics (fish, plankton...)
– tomography and assimilation
• Seafloor acoustic interaction
– sediment acoustics
– geo-acoustic models and inversion
• Sea-surface acoustic interaction
– effects due to bubbles and rough surfaces
• Ambient noise
• Field characteristics in the (shallow water) sound waveguide

2nd Internatio
on Shallow Wa
學會議
中國·上海
• Experimental techniques and instrumentation
Conference Co-Chairs:
Jeffrey Simmen
Applied Physics Laboratory-UW (USA)
Jing Tian
Institute of Acoustics, Chinese Academy of Sciences
Advisory Board Co-Chairs:
Ellen Livingston
Office of Naval Research (USA)
Shi-e Yang
Harbin Engineering University (China)
Organized by:
• Institute of Acoustics, Chinese Academy of Sciences
• National Laboratory of Acoustics, Chinese Academy of Sciences
• Applied Physics Laboratory, University of Washington
• Georgia Institute of Technology
Honorary Co-Chairs:
Renhe Zhang
National Laboratory of Acoustics, Chinese Academy of Sciences
Jixun Zhou
Georgia Institute of Technology (USA)
Sponsored by:
• Office of Naval Research (USA)
• Chinese Academy of Sciences
• Acoustical Society of China
• Acoustical Society of America
• National Natural Science Foundation of China
The conference language is English.
Tours and Laboratory/Institute visits in China will be arranged.
Details available at:
www.apl.washington.edu